The Evolution of Parental Care

EDITED BY

Nick J. Royle

Centre for Ecology & Conservation, University of Exeter, UK

Per T. Smiseth

Institute of Evolutionary Biology, University of Edinburgh, UK

and

Mathias Kölliker

Department of Environmental Sciences, University of Basel, Switzerland

The Evolution of Parental Care. First Edition. Edited by Nick J. Royle, Per T. Smiseth, and Mathias Kölliker.

Great Clarendon Street, Oxford, OX2 6DP,
United Kingdom

Oxford University Press is a department of the University of Oxford.
If furthers the University's objective of excellence in research, scholarship,
and education by publishing worldwide. Oxford is a registered trade mark of
Oxford University Press in the UK and in certain other countries

First Edition published in 2012

Impression: 1

British Library Cataloguing in Publication Data

Data available

Library of Congress Cataloging in Publication Data

Library of Congress Control Number: 2012936928

ISBN 978–0–19–969257–6 (hbk.)
978–0–19–969258–3 (pbk.)

Printed and bound by
CPI Group (UK) Ltd, Croydon, CR0 4YY

Contents

Foreword

Tim Clutton-Brock

In 1982, evolutionary biologists gathered in Cambridge at a commemorative meeting aimed to coincide with the centennial of Darwin's death (Bendall 1983). Richard Lewontin, doyen of population genetics, gave a perceptive but gloomy assessment of the achievements of evolutionary biologists (Lewontin 1983). John Maynard Smith responded in a characteristically humorous fashion. Perhaps, he suggested, Lewontin could agree that there had been some small advance in our understanding of the evolution of reproductive strategies and social behaviour? The suggestion was not designed to draw an enthusiastic response from Lewontin, to whom the development of sociobiology had been less than welcome. Nevertheless, there was an immediate groundswell of support from a substantial number of delegates from across a wide range of disciplines that, yes, there had been significant advances in our understanding of the evolution of animal societies and breeding systems and Lewontin was rather grudgingly persuaded to agree.

Maynard Smith's modest claim was well founded for, over the previous two decades, the integration of research in ecology, population genetics, and ethology (in which he had played a leading role) had initiated a new research field that focussed on a functional explanation of reproductive strategies and behaviour. At that stage, research focussed on six main areas: foraging strategies, the evolution of sociality, signalling systems, sexual selection and cooperative breeding, and the evolution of life histories (Krebs and Davies 1978). Parental care was being studied extensively by behavioural scientists working on developmental processes (Hinde 1975) but was not a major focus of interest in behavioural ecology and did not warrant a chapter until the third edition of *Behavioural Ecology* (Krebs and Davies 1991). Nevertheless, it was already clear that contrasts in the extent of care and the involvement of different categories of individuals had profound implications for the evolution of differences in fecundity, juvenile survival and longevity, mating systems, and the relative intensity of competition between the sexes (Triver 1972; Emlen and Oring 1977). In addition, in many groups of animals, parental care generated long lasting bonds between parents and offspring which were the precursors of the evolution of stable family groups, eusocial societies, and extended social networks (Emlen 1991). Theoretical treatments of parental care (especially (Trivers 1974)) also played an important role in promoting recognition of the extent and significance of conflicts of interest between individuals and their importance in understanding social relationships (Parker and MacNair 1978; Parker and MacNair 1979). Finally, an increasing body of empirical research was exploring the evolution of interspecific differences in parental behaviour (Orians 1980; Gross and Shine 1981) as well as adaptive aspects of intraspecific variation in parental strategies within species (Drent and Daan 1980; Stamps 1980; Clutton-Brock, Albon et al. 1981; Curio 1983; Mock 1984a; Mock 1984b).

Over the rest of the decade, research on the evolution of parental care expanded rapidly. Theoretical research continued to investigate the evolution of parent/offspring conflict (Parker 1985; Parker and Mock 1987) and extended this approach to consider the causes and consequences of conflicts of interest between siblings (Mock 1984a; Mock 1984b) as well as between parents or other care-givers (Houston and Davies 1985). Empirical studies, too, proliferated, exploring the adaptive significance of variation in parental strategies in males as well as in females (Gross and Sargent 1985), the extent to which parents adjust investment in relation to the quality and needs of offspring (Stamps et al. 1985; Drummond et al. 1986) and the effects of differential investment in sons and daughters on the sex ratio of their offspring (Clutton-Brock et al. 1984; Clutton-Brock et al. 1985).

By the late eighties, a substantial body of research had investigated the evolution of parental care but no attempt had been made to assess the general conclusions that were emerging. The *Evolution of Parental Care* (Clutton-Brock 1991) was an attempt to fill this gap. It tried to review the distribution of parental care across species and the emerging generalisations about parental strategies and tactics within the framework of evolutionary theory; though there were several important aspects of care that it did not cover, including research on the genetics, physiology and development of care. As the new *Evolution of Parental Care* shows, research on parental investment has continued to expand exponentially since 1991 and a new synthesis is overdue. Novel areas of theory have explored the possibility of cooperation between siblings (Chapter 8), the effects of kinship on the distribution of resources (Chapter 11), and the co-adaptation of traits affecting parental care and its consequences in parents and offspring (Chapter 16). Existing theory has been re-examined, revised, extended, or rejected (Chapters 6, 7, and 9) and a wide array of empirical work has tested theoretical predictions and assumptions, and extended our knowledge both of parental strategies and of the evolutionary responses of offspring (Chapters 3, 4, 5, 10, 12, and 13). Rapid advances in genetic techniques over the last twenty years have also had a significant impact on the field. At a relatively early stage, DNA analysis made it possible to check paternity and assess the proximity of kinship between individuals but, more recently, the development of gene-based pedigrees has generated powerful new techniques of measuring heritability (Kruuk 2004). Advances in quantitative genetic theory and statistical techniques have demonstrated the extent of maternal effects and their potential both to accelerate and to retard rates of evolution (Chapters 15 and 16). Genomic studies have begun to explore the mechanisms controlling the development of care in both sexes while research on epigenetic mechanisms raises the intriguing possibility that parents adjust the pre-natal development of their descendents to adapt them to the ecological or social circumstances that they are likely to encounter (Chapter 17).

This book represents an important milestone in the development of our understanding of parental care and its consequences although, there are still many important questions to be answered (Chapter 18). Both the development of novel questions, techniques and levels of analysis, and the increasing sophistication of research look set to continue. In particular, the integration of genetic and genomic research with observational and experimental studies of parental behaviour is opening up a wide range of new possibilities (Chapter 18) and may foster the use of systems where it is feasible to manipulate the mechanisms underlying developments. However, as recent research of the quantitative genetics of maternal effects illustrates, long-term field studies that can maintain recognition of large samples of individuals and explore the consequences of variation in parental care in subsequent generations under approximately natural conditions also have an important role to play (Maestripieri and Mateo 2009) and maintaining them needs to be recognized as a priority (Clutton-Brock and Sheldon 2010b).

References

Bendall, D. S., Ed. (1983). *Evolution From Molecules To Men*. Cambridge, University Press.

Clutton-Brock, T. and B. C. Sheldon (2010b). Individuals and populations: the role of long-term, individual-based studies in ecology and evolutionary biology. *Trends In Ecology & Evolution* 25: 562–73.

Clutton-Brock, T. H. (1991). *The Evolution of Parental Care*. Princeton, NJ, Princeton University Press.

Clutton-Brock, T. H., S. D. Albon, et al. (1981). Parental investment in male and female offspring in polygynous mammals. *Nature* 289: 487–9.

Clutton-Brock, T. H., S. D. Albon, et al. (1984). Maternal dominance, breeding success and birth sex ratios in red deer. *Nature* 308: 358–60.

Clutton-Brock, T. H., S. D. Albon, et al. (1985). Parental investment and sex differences in juvenile mortality in birds and mammals. *Nature* 313: 131–3.

Curio, E. (1983). Why do young birds reproduce less well? *Ibis* 125: 400–4.

Drent, R. H. and S. Daan (1980). The prudent parent: energetic adjustments in avian breeding. *Ardea* 68: 225–52.

Drummond, H., E. Gonzalez, et al. (1986). Parent-offspring cooperation in the blue-footed booby, *Sula rebouxii*: social roles in infanticidal brood reduction. *Behavioural Ecology and Sociobiology* 19: 365–72.

Emlen, S. T. (1991). Evolution of cooperative breeding in birds and mammals. *Behavioural Ecology: an evolutionary approach*. J. R. Krebs and N. B. Davies. Oxford, Blackwell Scientific Publications: 301–37.

Emlen, S. T. and L. W. Oring (1977). Ecology, sexual selection, and the evolution of mating systems. *Science* 197: 215–23.

Gross, M. R. and R. C. Sargent (1985). The evolution of male and female parental care in fishes. *Am.Zool.* 25: 807–22.

Gross, M. R. and R. Shine (1981). Parental care and mode of fertilization in ectothermic vertebrates. *Evolution* 35: 775–93.

Hinde, R. A. (1975). The concept of function. *Function and Evolution in Behaviour*. G. Baerends, C. Beer and A. Manning. Oxford, Clarendon Press: 1–15.

Houston, A. and N. B. Davies (1985). The evolution of cooperation and life-history in the dunnock *Prunella modularis*. *Behavioural Ecology: Ecological Consequences of Adaptive Behaviour*. R. M. Sibley and R. H. Smith. Oxford, Blackwell Scientific Publications: 471–87.

Krebs, J. R. and N. B. Davies (1978). *Behavioural Ecology: an evolutionary approach*. Oxford, Blackwell Scientific Publications.

Krebs, J. R. and N. B. Davies (1991). *Behavioural Ecology: an evolutionary approach*. Oxford, Blackwell.

Kruuk, L. E. B. (2004). Estimating genetic parameters in wild populations using the *'animal model'*. *Philosophical Transactions of the Royal Society of London B* 359: 873–90.

Lewontin, R. C. (1983). Gene, organism and environment. *Evolution from Molecules to Men*. D. S. Bendall. Cambridge, University Press: 273–85.

Maestripieri, D. and J. M. Mateo, Eds. (2009). *Maternal Effects in Mammals*. Chicago, University of Chicago Press.

Mock, D. W. (1984a). Infanticide, siblicide and avian nesting mortality. *Infanticide: Comparative and Evolutionary Perspectives*. G. Hausfater and S. B. Hrdy. New York, Aldine: 3–30.

Mock, D. W. (1984b). Siblicidal aggression and resource monopolization in birds. *Science* 225: 731–3.

Orians, G. H. (1980). *Some Adaptations of Marsh-nesting Blackbirds*. Princeton, NJ, Princeton University Press.

Parker, G. A. and M. R. MacNair (1978). Models of parent-offspring conflict. I. Monogamy. *Animal Behaviour* 26: 97–110.

Parker, G. A. and M. R. MacNair (1979). Models of parent-offspring conflict. IV Suppression: evolutionary retaliation of the parent. *Animal Behaviour* 27: 1210–35.

Parker, G. A. (1985). Models of parent-offpsring conflict. V Effects of the behaviour of two parents. *Animal Behaviour* 33: 519–33.

Parker, G. A. and Mock, D. W. (1987) Parent-offspring conflict over clutch size. *Evolutionary Ecology* 1: 161–74.

Stamps, J. A. (1980). Parent-offspring conflict. *Sociobiology: Beyond Nature/Nurture*. G. W. Barlow and J. Silverberg. Boulder, Colorado, AAAS Selected Symposium 35: 589–618.

Stamps, J. A., A. Clark, et al. (1985). Parent-offspring conflict in budgerigars. *Behaviour* 94: 1–39.

Trivers, R. L. (1972). Parental investment and sexual selection. *Sexual selection and the descent of man, 1871–1971*. B. Campbell. Chicago, Aldine-Atherton: 136–79.

Trivers, R. L. (1974). Parent-offspring conflict. *American Zoologist* 14: 249–64.

Preface

Parental care is a trait that shows tremendous diversity both within and across different animal taxa, and is an important topic in evolutionary biology and behavioural ecology. Parental care forms an integral part of an organism's reproduction, development, and life-history, and because caring for offspring means that parents have less time, resources, or energy available to search for or attract mates, the evolution of parental care is closely linked with sexual selection. In addition, the evolution of parental care represents an important step in the evolution of sociality as it leads to the formation of family groups, which provides a bridge to more complex forms of social structures. But because parents and offspring share only some of their genes, conflicts emerge in sexually reproducing organisms that shape the evolution of parental care and offspring strategies to demand care. As a consequence, the family also constitutes a model to understand the evolutionary tension between cooperation and conflict.

The importance of parental care in evolutionary biology has only been recognized relatively recently. Darwin did not consider parental care in great detail, except when speculating on the development of a moral sense and the role of selection operating on families in the evolution of more complex sociality: 'With respect to the origin of the parental and filial affections, which apparently lie at the base of the social instincts, we know not the steps by which they have been gained; but we may infer that it has been to a large extent through natural selection' (Page 105; *The Descent of Man*, Darwin 1871). The development of the robust conceptual framework that we have today for understanding the evolution of parental care was dependent on numerous innovations in the wider field of ecology and evolution. In particular: (i) the incorporation of kin selection to evolutionary thinking (Hamilton 1964), (ii) an appreciation of the relationship between parental care and the ecology and life-history of organisms (Lack 1968), (iii) the recognition that specific ecological conditions can drive the evolution of parental care and sociality (Wilson 1971), (iv) the insight of Trivers (1972) in making the connection between parental care, parental investment, and sexual selection, (v) his application of kin selection logic to derive genetic conflicts between parents and offspring over the amount and duration of parental care (Trivers 1974), and (vi) the introduction of evolutionary game theory to study the evolution of parental care and family conflicts by Maynard-Smith (1977). In particular the concept of parental investment (Trivers 1972) was vital in triggering a large amount of research on sexual selection and mating system evolution. As a result, Tim Clutton-Brock noted in his book *The Evolution of Parental Care*, published in 1991, that 'few areas of evolutionary biology [...] progressed as rapidly over the past two decades as our understanding of animal breeding systems' (p. 3). The underlying rationale for writing a book on the evolution of parental care by Clutton-Brock therefore centred on the importance of parental care in determining the strength of sexual selection. Since then the study of the evolution of parental care has progressed and diversified substantially.

The idea for the present book arose from discussions among the three of us about parental care research, and the realization that more than 20 years had passed since Clutton-Brock's (1991) classic book on this topic. In the intervening years since this book was published there has been growing recognition of the central importance of parental care research in behavioural and evolutionary biology, with an increasing number of papers being published and new fields of study steadily emerging. For example, there has been notable progress in the study of some aspects of parental care that were newly emerging at the time of the publication of Clutton-Brock's book, such as the physiology of maternal effects or the effect of parentage on parental care. However, there has also been notable progress in other topics that, although they had emerged at the time, had progressed rather slowly, such as within-family conflicts, cooperative breeding, and brood parasite–host co-evolution. More recently new areas of research altogether have emerged, such as the evolutionary and molecular genetics of parental care. It therefore seemed high time to have a go at synthesizing these exciting developments in the study of the evolution of parental care.

The aim of this book is to provide a comprehensive, fresh overview of research on the evolution of parental care in animals. The book integrates the major advances in the field over the last two decades since Clutton-Brock's (1991) book, focusing on establishing key concepts and on drawing general principles whilst emphasizing a broad taxonomic approach throughout. There are three main sections that represent major themes in the evolution of parental care and 18 chapters. The chapters and sections are arranged in a logical order to encourage reading of the book from front to back. Nevertheless, each chapter has been written so that it can easily be read in isolation, too.

Chapter 1: The introductory chapter by Smiseth, Kölliker, and Royle sets the stage for the book by reviewing the diversity of parental care across taxa, providing definitions of key terms and discussing some of the central concepts in the evolution of parental care.

Section I is on the **Origin and evolution of parental care**. This section deals broadly with the factors that promoted the early evolution of parental care. In **Chapter 2**, Klug, Alonzo, and Bonsall review theory and describe under which conditions parental care can evolutionarily originate, and how life-history and ecology interact to determine the favourable conditions for its spread. In **Chapter 3**, Alonso-Alvarez and Velando review the causes of variation in the evolutionary benefits and costs of parental care, especially with respect to the physiological mechanisms mediating these fitness consequences. Although we have advocated a question-driven approach throughout, any treatment of a biological phenomenon such as parental care also requires appropriate coverage of its natural history and diversity. **Chapter 4** by Balshine covers the diversity and distribution of forms of care among vertebrates, and **Chapter 5** by Trumbo the very diverse and sometimes striking and peculiar forms of care that have evolved among invertebrates. Finally, **Chapter 6** by Kokko and Jennions explains the multiple and complex evolutionary relationships between sexual selection and the sex roles in parental care.

Section II is concerned with **Cooperation and conflict in parental care** and covers the tension between conflict and cooperation that emerges in the context of parental care due to sexual reproduction and the resulting asymmetries in fitness consequences of care and/or genetic relatedness. Kilner and Hinde (**Chapter 7**) discuss how the resolution of parent–offspring conflict affects parent and offspring strategies and review the experimental evidence. **Chapter 8** by Roulin and Dreiss explores conflicts and cooperation between siblings and how these are resolved via mechanisms such as aggression, signalling, and negotiation. Lessells (**Chapter 9**) explores the theoretical underpinning and empirical evidence for conflicts between parents over the provision of care. In **Chapter 10**, Komdeur presents the ecological and social factors that alter the fitness returns of investment in sons versus daughters and the evidence for adaptive sex allocation, and **Chapter 11** by Alonzo and Klug reviews the central importance of parentage in the evolution of parental care. The two extremes of cooperation and conflict in animal families are covered in **Chapter 12** by Cant, on the evolution of cooperative breeding, and **Chapter 13** by Spottiswoode, Kilner, and Davies on the manipulation of parents and exploitation of parental care by brood parasites.

Section III covers the **Evolutionary genetics of parental care**. This section takes a change of perspective from the majority of chapters in sections I and II, which focus on how selection acts on phenotypes, towards establishing how phenotypic variation in parent and offspring traits are generated and maintained by genetic and non-genetic factors, how this variation is exposed to natural selection, and what the molecular and quantitative genetic trait architecture of parental care are. Uller (**Chapter 14**) describes how parental care can enhance adaptive evolution by generating environmentally induced variation in offspring phenotypes through processes such as phenotypic and genetic accommodation. In **Chapter 15**, Hadfield discusses the mathematical framework of quantitative genetic parental effect models to study the co-evolutionary dynamics of parental effects and offspring phenotypes. **Chapter 16** by Kölliker, Royle, and Smiseth outlines theory and experimental research exploring how the co-evolution of parents and offspring lead to co-adaptated strategies individuals use as offspring and as parents. While quantitative genetic results are discussed across the previous chapters, **Chapter 17** by Champagne and Curley presents a comprehensive overview of the current knowledge of the molecular genetics and epigenetics of parental care.

Finally, in the summarizing **Chapter 18** by Royle, Smiseth, and Kölliker we offer a summary of the previous 17 chapters, draw conclusions and discuss promising avenues for future research.

Our main target audiences are new and established researchers and students in behaviour, ecology, genetics, development, and evolution; although we hope the book may also appeal to academics and students in other, related disciplines, such as psychology and sociology.

This project would not have been possible without the contributions of the authors. The editing of this book has been a highly enjoyable process as it has allowed us to interact with some of the most talented researchers working on parental care. We would like to thank all the authors for their hard work, efforts to meet deadlines, and willingness to buy into the project with such enthusiasm. We would also like to extend our gratitude to the many reviewers, some of whom read and commented on several chapters: Kate Arnold, Matt Bell, Jon Blount, Kate Buchanan, Tim Clutton-Brock, James Curley, Sasha Dall, Jeremy Field, Scott Forbes, Simon Griffith, Uri Grodzinski, Reinmar Hager, David Haig, Ian Hardy, Ben Hatchwell, Megan Head, Fabrice Helfenstein, Camilla Hinde, Andy Horn, Clarissa House, Rufus Johnstone, Charlotta Kvarnemo, Tim Linksvayer, Clauco Machado, Joah Madden, Dario Maestripieri, Joel McGlothlin, John McNamara, Allen Moore, Geoff Parker, Tom Pike, Sarah Pryke, Eivin Røskraft, Andy Russell, Spencer Sealy, Ben Sheldon, Emilie Snell-Rood, Bård Stokke, Tamas Székely, Fritz Trillmich, Tobias Uller, Mary Jane West-Eberhard, and Jon Wright. Finally, we wish to thank the publishers at OUP, especially Helen Eaton and Ian Sherman, who provided invaluable help and advice during the whole process.

Nick Royle: On a professional level I would especially like to thank Ian Hartley and Geoff Parker, my supervisors and mentors during my first post-doc. I never realized research on conflicts could be so much fun, as well as being so educational. Thanks also to Jan Lindström, Neil Metcalfe, Craig Walling, Jason Wolf, Scott Forbes, Maggie Hall, Jon Blount, Josie Orledge, Tom Pike, Sasha Dall, Wiebke Schütt, Allen Moore, Megan Head, Paul Hopwood, Heinz Richner, Philipp Heeb, and Mathias Kölliker, amongst others, for making subsequent collaborations similarly intellectually stimulating and fun. On a personal level I need to thank my parents, John and Sheila, and my brother, Phil, for stimulating and encouraging my interest in parental care and family dynamics throughout my life. Last, but certainly not least, I thank my wife, Marieke, and my two boys, Lachlan and Lucas, for their love and support and for providing daily reminders of the joys of being a parent.

Per Smiseth: I wish to thank Svein-Håkon Lorentsen, Trond Amundsen, and Allen Moore, who acted as supervisors during various stages of my career and who introduced me to various aspects of parental care in organisms as different as grey seals, bluethroats, and burying beetles. I also wish to thank my many collaborators, whose expertise and insights have helped me develop my ideas and thinking about the evolution of parental care, including Andy Gardner, Loeske Kruuk, Mathias Kölliker, Allen Moore, Danny Rozen, Michelle Scott, and Jon Wright. I also wish to thank the members of the burying beetle group at The University of Edinburgh, Clare Andrews, Sarah Mattey, and Roni Mooney, and members of the Institute of Evolutionary Biology, The University of Edinburgh, who helped clear up many of my confusions during coffee; Jarrod Hadfield, Tom Little, and Alastair

Wilson. I also wish to thank the honours students attending the Evolution of Parental Care course at The University of Edinburgh for their contributions to developing my ideas on this subject through discussions and questions. Finally, I wish to thank my partner, Sarah, for a huge amount of patience and understanding during the 18 months of this project.

Mathias Kölliker: I would like to thank particularly Heinz Richner and Butch Brodie for their support as supervisors and mentors, and for introducing me to different perspectives in evolutionary thinking that continue to stimulate my research. I would also like to thank Sabrina Gaba, Ken Haynes, Philipp Heeb, Rufus Johnstone, Allen Moore, Ben Ridenhour, Alexandre Roulin, Nick Royle, Per Smiseth, Jean-Claude Walser, and Jon Wright for the pleasant and rewarding collaborations. Thanks also to former and current members of the earwig lab at the University of Basel for their enthusiasm and work: especially Stefan Boos, Ralph Dobler, Flore Mas, Joël Meunier, Lilian Röllin, Dimitri Stucki, and Janine Wong. On a personal level, I am grateful to my parents Léonie and Eduard Bühlmann for their parental care and for motivating me early to pursue an academic pathway. My very special thanks are to my wife Geneviève and my children Joachim and Michelle—for their love and support, for travelling with me, for having made me parental, and for having so pleasantly and importantly enriched my social environment that forms our family.

Nick Royle, Per Smiseth, and Mathias Kölliker

List of Contributors

Carlos Alonso-Alvarez, Ecology Unit, Instituto de Investigación en Recursos Cinegéticos, IREC (CSIC-UCLM-JCCM), Ronda de Toledo s/n, 13005 Ciudad Real, Spain
carlos.alonso@uclm.es

Suzanne H. Alonzo, Department of Ecology & Evolutionary Biology, Yale University, PO Box 208106, 165 Prospect St., New Haven, CT 06520, USA
suzanne.alonzo@yale.edu

Sigal Balshine, Department of Psychology, Neuroscience and Behaviour, McMaster University, 1280 Main St. West, Hamilton, Ontario, L8S 4K1 Canada
sigal@mcmaster.ca

Michael B. Bonsall, Mathematical Ecology Research Group, Department of Zoology, University of Oxford, Oxford, OX1 3PS, UK
michael.bonsall@zoo.ox.ac.uk

Michael A. Cant, Centre for Ecology & Conservation, Biosciences, College of Life & Environmental Sciences, University of Exeter, Cornwall Campus, Penryn, Cornwall, TR10 9EZ, UK
m.a.cant@exeter.ac.uk

Frances A. Champagne, Department of Psychology, Columbia University, 1190 Amsterdam Avenue, New York, NY 10027, USA
fac2105@columbia.edu

Tim H. Clutton-Brock, Department of Zoology, Downing Street, Cambridge, CB2 3EJ, UK
thcb@cam.ac.uk

James P. Curley, Department of Psychology, Columbia University, 1190 Amsterdam Avenue, New York, NY 10027, USA
jc3181@columbia.edu

Nicholas B. Davies, Department of Zoology, Downing Street, Cambridge, CB2 3EJ, UK
n.b.davies@zoo.cam.ac.uk

Amélie N. Dreiss, Department of Ecology and Evolution, Biophore, University of Lausanne, CH 1015 Lausanne, Switzerland
amelie.dreiss@unil.ch

Jarrod Hadfield, Edward Grey Institute, Department of Zoology, University of Oxford, Oxford, OX1 3PS, UK
jarrod.hadfield@zoo.ox.ac.uk

Camilla A. Hinde, Department of Animal Sciences, Wageningen University, De Elst 1, Building 122, BHE Group, 6708 WD Wageningen, The Netherlands
camillaoxford@gmail.com

Michael D. Jennions, Evolution, Ecology & Genetics, Research School of Biology, The Australian National University, Canberra, ACT 0200, Australia
michael.jennions@anu.edu.au

Rebecca M. Kilner, Department of Zoology, University of Cambridge, Downing Street, Cambridge, CB2 3EJ, UK
rmk1002@cam.ac.uk

Hope Klug, Department of Ecology & Evolutionary Biology, Yale University, PO Box 208106, 165 Prospect St., New Haven, CT 06520, USA and Department of Biological & Environmental Sciences, University of Tennessee at Chattanooga, 215 Holt Hall, Chattanooga, TN 37403, USA
hope-klug@utc.edu

Hanna Kokko, Evolution, Ecology & Genetics, Research School of Biology, The Australian National University, Canberra, ACT 0200, Australia
hanna.kokko@anu.edu.au

Mathias Kölliker, Department of Environmental Sciences, Zoology and Evolution, University of Basel, Vesalgasse 1, 4051 Basel, Switzerland
mathias.koelliker@unibas.ch

Jan Komdeur, Centre for Ecological and Evolutionary Studies, University of Groningen, Centre for Life Sciences, PO Box 11103, 9700 CC Groningen, The Netherlands
j.komdeur@rug.nl

C. M. Lessells, Netherlands Institute of Ecology (NIOO-KNAW), PO box 50, 6700AB Wageningen, The Netherlands
k.lessells@nioo.knaw.nl

Alexandre Roulin, Department of Ecology and Evolution, Biophore, University of Lausanne, CH 1015 Lausanne, Switzerland
alexandre.roulin@unil.ch

Nick J. Royle, Centre for Ecology & Conservation, Biosciences, College of Life and Environmental Sciences, University of Exeter, Cornwall Campus, Penryn, Cornwall, TR10 9EZ, UK
n.j.royle@exeter.ac.uk

Per T. Smiseth, Institute of Evolutionary Biology, School of Biological Sciences, University of Edinburgh, West Mains Road, Edinburgh, EH9 3JT, UK
per.t.smiseth@ed.ac.uk

Claire N. Spottiswoode, Department of Zoology, Downing Street, Cambridge, CB2 3EJ, UK
cns26@cam.ac.uk

Stephen T. Trumbo, Department of Ecology and Evolutionary Biology, University of Connecticut, 99 East Main St., Waterbury, CT 06702, USA
trumbo@uconn.edu

Tobias Uller, Edward Grey Institute, Department of Zoology, University of Oxford, OX1 3PS, Oxford, UK
tobias.uller@zoo.ox.ac.uk

Alberto Velando, Departamento de Ecoloxía e Bioloxía Animal, Universidade de Vigo, 36310 Vigo, Spain
avelando@uvigo.es

CHAPTER 1

What is parental care?

Per T. Smiseth, Mathias Kölliker, and Nick J. Royle

1.1 Introduction

Parents of most animals, including the vast majority of invertebrates, provide no care for their offspring beyond supplying them with a small package of yolk that serves as an initial source of nutrition until the offspring are fully capable of fending for themselves. Yet parents of some animals go to great lengths to increase their offspring's survival prospects by protecting them from predators, food shortages, desiccation, and a range of other environmental hazards. Familiar examples include mammals and birds, in which one or both parents provide elaborate forms of care that include nourishment of the developing embryo via a placenta or in the form of yolk, protection of offspring against predators and parasites, and provisioning of milk, arthropods, or some other source of food after birth or hatching. Less familiar examples of parental care are found among reptiles, amphibians, fishes, arthropods, molluscs, annelids, and other invertebrate groups. Some of these examples include elaborate forms of care comparable to those found in mammals and birds. For example, some reptiles, fishes, insects, arachnids, molluscs, brachiopods, and bryozoans nourish developing embryos via a placenta-like structure, and a small number of amphibians, fishes, insects, arachnids, crustaceans, and leeches provide food for their offspring after hatching or birth (Section 1.2). Other examples include much simpler forms of care, such as attendance of eggs until hatching, which is a common form of care in amphibians, fishes, and invertebrates (Section 1.2). The diversity in parental care among amphibians, fishes, and invertebrates make these latter groups particularly valuable as study systems for the evolution of parental care (Chapters 4 and 5).

Parental care occurs whenever parents increase the survival and growth of their offspring, often at a cost to their own survival and reproduction (for a formal definition, see Section 1.3). The study of the evolution of parental care is an important topic in its own right. Major aims are to understand the evolutionary causes of the observed diversity in the form, level, and duration of parental care, as well as the extent to which it is provided by the male, the female, or both parents (Clutton-Brock 1991). The evolution of this diversity is thought to reflect variation in the benefits and costs of different forms of parental care to males and females which, in turn, depends upon factors such as offspring dependency on care, environmental hazards, life-history traits, mating opportunities, and paternity uncertainty (e.g. Clutton-Brock 1991; Westneat and Sherman 1993). The study of parental care is also important because its evolution is closely linked with that of other key traits in evolutionary biology. For example, early studies in this field were often motivated by the suggestion that the predominance of female-only parental care leads to more intense sexual selection in males than in females (Trivers 1972; Clutton-Brock 1991). Although the traditional view that sex-differences in the intensity of sexual selection is caused by sex-differences in parental investment is now largely abandoned, the co-evolution of parental care, sexual selection, and mating systems remains an important topic in evolutionary ecology (Chapter 6). In addition, the evolution of parental care has important implications for our understanding of life-history evolution (Martin 2004; Chapters 2 and 3), sex allocation (Chapter 10), sociality (Wilson 1975), cooperation and conflict within families (Mock and Parker 1997; Chapters 7, 8, 9, and 12), phenotypic plasticity (Chapter 14), and the genetic and epigenetic inheritance of traits expressed in

The Evolution of Parental Care. First Edition. Edited by Nick J. Royle, Per T. Smiseth, and Mathias Kölliker.

social interactions among close relatives (Cheverud and Moore 1994; Chapters 15, 16, and 17).

In this chapter, we begin by providing a brief overview in Section 1.2 of the diversity among species and higher taxa in the forms of care that parents provide to their offspring. In Section 1.3, we outline the key terms that are used in the study of the evolution of parental care. Some of these terms have a very precise definition in the theoretical literature, and understanding the nuances between them is particularly important when translating between theoretical and empirical work, such as when theoretical predictions are used to guide empirical studies and when empirical findings are used to inform theoretical modelling. In Section 1.4, we discuss how to assign fitness to parents and offspring. This issue is important because mistakes in the assignment of fitness among parents and offspring may lead to erroneous conclusions about the evolution of parental care. In Section 1.5, we briefly discuss the environmental conditions that are thought to favour the origin and subsequent modifications of parental care.

1.2 Forms of parental care

In order to understand the evolution of the diversity in parental care, it is necessary to categorize parental traits into specific forms of care. The terminology used to describe diversity in parental care can sometimes be confusing because alternative schemes are used for different taxa, and the same form of care may go under different names in different taxonomic groups (Blumer 1979; Crump 1995). Here we provide a general description of the basic forms of care observed across animals arranged in chronological order throughout offspring development. Further discussions of the diversity of parental care in vertebrates and invertebrates are provided in Chapters 4 and 5, respectively.

1.2.1 Provisioning of gametes

Provisioning of energy and nutrients, such as proteins and yolk lipids, into eggs by the female is a basal form of parental care. Deposition of energy and nutrients into eggs beyond the minimum required for successful fertilization may enhance offspring fitness by increasing the offspring's size, nutrient reserves, and/or developmental stage at the time of hatching. Studies on birds and arthropods show that larger eggs often produce offspring with greater nutrient reserves at hatching, and that egg size has a greater effect on juvenile growth and survival when food is limited or when predation risk is high (Williams 1994; Fox and Czesak 2000). Although egg size has a positive impact on offspring survival and growth after hatching in many arthropods (Fox and Czesak 2000), some studies on altricial birds report no such effects, presumably because post-hatching parental care in these species has a stronger effect on offspring growth and survival that masks any initial effect due to egg size (Williams 1994).

Females deposit other substances in addition to energy and nutrients that may also enhance offspring fitness, including antibodies (Boulinier and Staszewski 2008), hormones (Groothuis and Schwabl 2008), and antioxidants (Royle et al. 1999). In great tits (*Parus major*), exposure to fleas during egg laying induces host responses, most likely mediated through maternal androgens deposited into eggs, that enhance nestling growth rate and recruitment (Heeb et al. 1998). In many insects, females coat their eggs with defensive structures or chemicals (Hilker and Meiners 2002). For example, females of the beetle *Cryptocephalus hypochaeridis* allocate a substantial amount of time and energy to coating their eggs with a combination of specialized secretions and faecal material, which forms a hard defensive structure protecting the eggs against predators (Ang et al. 2008).

Although females are responsible for the deposition of energy and nutrients into gametes in all anisogamic organisms, males may also contribute to gamete provisioning by defending resources used by females to produce eggs, providing females with nuptial gifts, transmitting nutrients or defensive chemicals via the ejaculate, or even by being eaten by females (Simmons and Parker 1989; Hilker and Meiners 2002). It may often be problematic to determine whether male contributions to female gamete provisioning should be considered a form of parental care or part of the males' mating effort (Simmons and Parker 1989). To this end, it is essential to determine whether male contributions affect the size or quality of individual gametes or the number of gametes that are produced. We suggest that male contributions to female gamete provisioning should be considered a form of parental care only if they enhance the size or quality of individual gametes, thereby increasing offspring fitness.

1.2.2 Oviposition-site selection

Oviposition-site selection is non-random choice of egg-laying sites in any oviparous animal. It includes selection of nest sites in animals that build nests in which eggs are laid, such as many birds, and selection of spawning sites in animals with external fertilization, such as many fishes and amphibians (Refsnider and Janzen 2010). Oviposition site selection may increase offspring fitness by minimizing the risk of detection of eggs by predators, parasitoids, and brood parasites, and ensuring that eggs develop in a suitable microclimate or that offspring have access to a suitable habitat after hatching in which they can hide from predators and obtain nutrients for growth and development. Studies on several species show that oviposition or nest-site selection increases offspring fitness. For example, females of the mosquito *Culiseta longiareolata* avoid ovipositing in pools that contain larval predators (Spencer et al. 2002). Female dusky warblers (*Phylloscopus fuscatus*), a small passerine breeding in mosaics of shrub and tundra habitat, show plasticity in nest-site selection, preferring safer nest-sites to those nearer to food when the risk of nest predation from Siberian chipmunks (*Tamias sibiricus*) is high (Forstmeier and Weiss 2004). However, females of some species, such as the grass miner moth *Chromatomyia nigra*, prefer oviposition sites that are safer to themselves rather than to their offspring (Scheirs et al. 2000), suggesting that oviposition site selection may sometimes increase the female parent's own fitness. Consequently, although oviposition site selection may often be considered a form of parental care, this is not the case for all species.

1.2.3 Nest building and burrowing

Nest building and burrowing is a common form of care in vertebrates as well as invertebrates. The simplest form of nest building occurs in some terrestrial snails where eggs are simply buried beneath the substrate surface (Baur 1994), and in many salmonids where eggs are covered with substrate after spawning (Blumer 1982). More elaborate forms of nest building involve the use of materials found in the environment, such as mud used by the mud-dauber wasp *Trypoxylon politum* (Brockman 1980) and swallows in the genus *Hirundo* (Winkler and Sheldon 1993), and plant materials used by weaver birds (Hansell 2000) and three-spined sticklebacks (*Gasterosteus aculeatus*) (Wootton 1976). In other species, nest building involves the use of processed plant materials, such as paper used by wasps of the subfamilies Polistinae, Vespinae, and Stenograstrinae (Hansell 1987), or materials produced by the parents themselves, such as silk used by webspinners of the order Embiidina (Edgerly 1997) and mucus used to build bubble nests by the frog *Chiasmocleis leucostict* (Haddad and Hödl 1997). Finally, some species construct nesting burrows, including the cricket *Anurogryllus muticus* (West and Alexander 1963), and martins of the genus *Riparia* (Winkler and Sheldon 1993). Nest building and burrowing may increase offspring fitness by concealing eggs and juveniles from predators, parasitoids, and brood parasites, or by buffering eggs and juveniles against environmental hazards, such as extreme temperatures, flooding or desiccation. In some birds, nest architecture appears to have evolved to serve multiple functions: the outer nest layer is constructed to conceal the nest from predators and protect eggs and nestlings from rain, while the nest lining provides insulation against cold and heat (Hansell 2000). In many species, nest building is also associated with nest sanitation and antimicrobial properties, such as the use of aromatic plant material as protection against pathogens by blue tits (*Cyanistes caeruleus*) (Mennerat et al. 2009).

1.2.4 Egg attendance

Egg attendance occurs in species where parents remain with the eggs at a fixed location, usually the oviposition site, after egg laying (Crump 1995). Egg attendance is the most common and phylogenetically widespread form of post-fertilization parental care among amphibians, fishes, and invertebrates

(Blumer 1982; Crump 1995; Costa 2006). Egg attendance may increase offspring fitness by providing protection against environmental hazards such as egg predators, oophagic conspecifics, parasitoids and pathogens, desiccation, flooding, and hypoxia. Parental removal experiments provide a simple method for establishing the adaptive value of egg attendance. For example, in the bug *Elasmucha grisea*, the experimental removal of the female led to a ten-fold increase in egg losses due to predation (Melber and Schmidt 1975; Fig. 1.1). In the harvestman *Iporangaia pustulosa*, the experimental removal of guarding males led to the complete loss of one-third of clutches to predators, and all remaining broods suffered substantial egg losses. In contrast, when guarding males were allowed to remain with the eggs, most clutches suffered no egg losses at all (Requena et al. 2009). Meanwhile, in the terrestrially breeding mountain dusky salamander (*Desmognathus ochrophaenus*), eggs attended by females suffer less water loss than unattended eggs (Forester 1984).

Figure 1.1 Egg attendance in the parent bug, *Elasmucha grisea*. This species breeds on birch trees, and female parents protect the eggs and young nymphs from parasitoids and predators. (Photo: Per Smiseth.)

Egg attendance is often associated with a range of parental behaviours directed toward particular biological or environmental hazards. In many species, parents actively defend their eggs against predators or oophagic conspecifics, a behaviour that is often termed egg guarding (Crump 1995). For example, female membracid bugs remain with their eggs when approached by a predator, and may even approach and attack the predator. In contrast, females not attending a clutch of eggs will typically attempt to escape when approached by the same predator (Hinton 1977). In birds, egg attendance is associated with incubation, which increases offspring fitness by providing the developing embryo with a source of heat that supports offspring development (Deeming 2001). Avian incubation is often accompanied by the development of a brood patch, which enhances the transfer of heat from parents to eggs (Deeming 2001). Incubation is also found in some ectotherms, such as ball pythons (*Python regius*), where incubation prevents desiccation of eggs (Aubret et al. 2005). Other behaviours associated with egg attendance include egg fanning, which increases oxygen access to eggs in fishes (Green and McCormick 2005), and active removal of microbes and fungi, which is reported from several species including the millipede *Brachycybe nodulosa* (Kudo et al. 2011).

1.2.5 Egg brooding

Egg brooding is a non-behavioural form of parental care where parents carry the eggs after laying. Some species carry eggs externally, including the giant water bug *Adebus herberti* where males carry the eggs on their backs (Smith 1976) and the deep-water squid *Gonatus onyx* where females hold the egg mass in their tentacles (Seibel et al. 2005). Other species carry eggs internally, including marsupial frogs of the genera *Amphignathodon*, *Flectonotus*, and *Gastrotheca* where parents brood the eggs within specialized pouches (Duellman and Maness 1980), and mouth-brooding fishes, including many cichlids where parents brood the eggs within their mouths (Oppenheimer 1970). Egg brooding may increase offspring fitness by providing protection against egg predators, oophagic conspecifics, parasitoids and pathogens, desiccation, flooding, and hypoxia. Egg brooding might offer some advantages over egg attendance to parents breeding in variable environments because brooding allows parents to move more freely while caring for the eggs. For example, brooding may allow parents to move their clutches away from approaching predators and/or track suitable conditions that change over time. Furthermore, costs of care may be lower for brooding than for attending parents because brooding parents are better able to forage while caring for the eggs.

1.2.6 Viviparity

Viviparity is a non-behavioural form of parental care characterized by the retention of fertilized eggs within the female reproductive tract (Clutton-Brock 1991). Viviparity is derived from oviparity, where eggs are deposited with intact shells and membranes. Viviparity is ubiquitous in marsupial and eutherian mammals, but has also evolved repeatedly from oviparity across a wide of taxa, including squamate reptiles (Blackburn 2006), fishes, (Blackburn 2005), insects (Meier et al. 1999), onychophorans, molluscs, tunicates, echinoderms, arachnids, and bryozoans (Adiyodi and Adiyodi 1989). Viviparity may enhance offspring fitness by providing effective protection against predators and harsh environmental conditions. Viviparous species show diverse forms of embryonic provisioning modes, ranging from strict lecithotrophy, where the developing embryos are provisioned solely by yolk, to extreme matrotrophy, where the embryo is primarily nourished by sources other than yolk. Lecithotrophic viviparity is often termed ovoviviparity, though this term is now largely abandoned due to confusion over its definition. There are four main forms of matrotrophy: (i) oophagy, where embryos feed on trophic (often unfertilized) eggs, such as in many lamnoid sharks (Gilmore 1993); (ii) adelophophagy, where embryos feed on sibling embryos, such as in the sand tiger shark *Carcharias taurus* (Gilmore 1993); (iii) trophodermy, where embryos absorb maternal nutrients via their skin or gut epithelia, such as in some clinid fishes (Gunn and Tresher 1991); and (iv) placentotrophy, where nutrients are transferred from the mother to embryos via a placenta, such as in marsupial and eutherian mammals (Clutton-Brock 1991). The evolution of matrotrophy is thought to depend on patterns of resource availability because matrotrophic species can spread their investment of resources into offspring over time, while lecithotrophic species must invest all resources prior to fertilization, leading to a high peak in resource requirements (Trexler and DeAngelis 2003).

1.2.7 Offspring attendance

Offspring attendance occurs in species where parents remain with their offspring after hatching either at a fixed location or by escorting the offspring as they move around, and may increase offspring fitness in a similar way as egg attendance (see above). Offspring attendance is often associated with specific parental behaviours directed towards particular environmental hazards, such as predators, pathogens, and desiccation. For example, parental removal experiments on the lace bug *Gargaphia solani* show that the vast majority of offspring survive to maturity regardless of whether the female parent is removed or not when there are no predators present. In contrast, when predators are present, the presence of a guarding female improves offspring survivorship sevenfold compared to when the female is removed (Tallamy and Denno 1981). Furthermore, in African bullfrogs (*Pyxicephalus adspersus*), attending males dig channels to adjacent ponds to prevent the pool with their tadpoles from drying out (Kok et al. 1989), and in the burying beetle *Nicrophorus vespilloides*, parents produce antimicrobial secretions that limit the growth of microbial competitors on the carrion used for breeding (Rozen et al. 2008).

1.2.8 Offspring brooding

Offspring brooding is a non-behavioural form of parental care where parents carry their offspring after hatching or birth. Some species carry their offspring externally, including scorpions (Shaffer and Formanowicz 1996) and some mammals (Altmann and Samuels 1992). Other species carry offspring internally in specialized brood pouches as in marsupial mammals (Low 1978), in their mouths as in mouth-brooding cichlids (Oppenheimer 1970), or even in their stomachs as in the now extinct gastric-brooding frogs (Tyler et al. 1983). Offspring brooding is also recorded from a wide range of marine invertebrates, including cnidarians, sipunculans, molluscs, annelids, brachiopods, bryozoans, crustaceans, and echinoderms (Adiyodi and Adiyodi 1989). Offspring brooding may increase offspring fitness in much the same way as egg brooding (see above). Some terrestrially-breeding amphibians, including most poison-arrow frogs (Dendrobatidae), have a particular form of offspring brooding where tadpoles or froglets are transported from a terrestrial oviposition site that is protected from aquatic predators to an aquatic nursing site where the tadpoles can feed and complete their development (Crump 1995).

1.2.9 Food provisioning

Food provisioning is found in species where parents provide their offspring with a source of food after hatching or birth. The simplest form of food provisioning occurs in some wading birds, such as crowned plovers (*Vanellus coronatus*) in which parents use a distinctive posture and call to attract chicks when they discover a source of food (Walters 1984). More elaborate forms of food provisioning include mass provisioning, where parents provision food for their offspring before hatching, such as in many solitary wasps and bees (Field 2005), and progressive provisioning, where parents repeatedly feed their offspring after hatching or birth, such as in mammals and many birds, as well as in some amphibians (Weygoldt 1980), insects (Field 2005), crustaceans (Diesel 1989), and leeches (Kutchera and Wirtz 1987). Progressive provisioning may be based on food that is obtained directly from the environment, as in many passerine birds which feed their offspring a diet mainly consisting of arthropods, or pre-digested food as in many seabirds which feed their nestlings regurgitated fish or squid (Clutton-Brock 1991) and burying beetles which feed their larvae regurgitated carrion (Smiseth et al. 2003; Fig. 1.2). Alternatively, it may be based on specialized food sources such as milk produced by female mammals (Clutton-Brock 1991), unfertilized trophic eggs as in the poison-arrow frog *Dendrobates pumilio* (Weygoldt 1980), and modified skin produced by females of the caecilian *Boulengerula taitanus* (Kupfer et al. 2006). The most extreme form of food provisioning is matriphagy, where the hatched offspring consume their mother, as in the spider *Diaea ergandros* (Evans et al. 1995) and the hump earwig *Anechura harmandi* (Suzuki et al. 2005).

Figure 1.2 Food provisioning in the burying beetle *Nicrophorus vespilloides*. This species breeds on carcasses of small vertebrates, and the female parent, sometimes with help from the male, provisions the larvae with pre-digested carrion. (Photo: Per Smiseth.)

1.2.10 Care after nutritional independence

Care for offspring after they have reached the age of nutritional independence is an unusual form of parental care that has mainly been reported from longer-lived vertebrates (Clutton-Brock 1991). For example, in winter flocks of Bewick's swans (*Cygnus columbianus*), parents assist their offspring in competition with other families over access to food. Cygnets that are separated from their parents spend less time feeding than cygnets that remain close to their parents (Scott 1980). In American red squirrels (*Tamiasciurus hudsonicus*), females acquire and defend a second food cache several months prior to conception, and subsequently pass this cache on to one of their offspring at independence ten months after birth (Boutin et al. 2000).

Care for offspring after nutritional independence may be more common in invertebrates, including insects, than traditionally thought. For example, in the burying beetle *Nicrophorus vespilloides*, larvae become nutritionally independent at the age of 72 hours, but female parents remain with the larvae and defend them from conspecific intruders and predators for a further 48h (Smiseth et al. 2003).

1.2.11 Care of mature offspring

Care to mature offspring is an extremely rare form of parental care that is restricted to some social vertebrates (Clutton-Brock 1991). This form of care is known from bonobos (*Pan paniscus*), where the presence of the female parent helps mature sons during competitive interactions with other males, thereby enhancing the son's social status and their mating success (Surbeck et al. 2011). For further details on this form of care, see Chapter 4.

1.3 Definition of terms

In order to understand the evolutionary causes of the diversity in forms and patterns of parental care described above, it is important to establish a clear terminology shared by theoreticians and empiricists. Clutton-Brock (1991) noted that the terminology used in the study of parental care was 'diffuse and misleading', a situation that continues to this day. Clutton-Brock (1991) identified four key terms used mainly by behavioural ecologists—parental care, parental expenditure, parental investment, and parental effort—and noted that the main difference between them is the currency used to measure the benefits and costs of parental care. Here, we provide an update on Clutton-Brock's discussion of these four key terms, identify current sources of confusion over the use of terminology, and suggest how the terminology could be used more consistently in the future (Table 1.1). In this discussion, we include a fifth term, parental effect (also known as maternal or paternal effect), which is used by evolutionary geneticists studying the evolution of parental care (Cheverud and Moore 1994; Chapter 14). Although we argue for a more consistent use of terminology in this field, we recognize that these terms are used differently in different contexts and by different authors. Thus, we encourage readers to pay particular attention to the way in which these terms are defined when reading the literature in this field.

Clutton-Brock (1991) defined *parental care* as 'any form of parental behaviour that appears likely to increase the fitness of a parent's offspring'. Parental care is a purely descriptive term that is used to describe variation in its form, level, or duration regardless of any costs to the parents. Clutton-Brock pointed out that the term parental care could be used in either a narrow sense that focuses strictly on behavioural traits, or a broad sense that includes non-behavioural traits, such as gamete provisioning, gestation, viviparity, and nests (Clutton-Brock 1991, p. 13). To distinguish between these two uses of the term, we propose that *parental behaviour* is used for strictly behavioural forms of care, while 'parental care' is used when also including non-behavioural traits. Although we broadly agree with Clutton-Brock's (1991) definition, one potential issue is that it includes parental traits that

Table 1.1 Definitions of key terms used in the study of the evolution of parental care.

Term	Definition
Parental care	Any parental trait that enhances the fitness of a parent's offspring, and that is likely to have originated and/or to be currently maintained for this function
Parental expenditure	Any expenditure of parental resources (including time and energy) on parental care of one or more offspring
Parental investment	Any investment by the parent in an individual offspring that increases the offspring's survival and reproductive success at the cost of the parent's ability to invest in other current or future offspring
Parental effort	The combined fitness costs that the parent incurs due to the production and care of all offspring in a given biologically relevant period, such as a breeding attempt
Parental effect	The causal effects that the parent's phenotype have on the offspring's phenotype, including its growth and survival, over and above direct effects due to genes inherited from parents

evolved (i.e. that originated and are currently maintained) to serve a function other than enhancing offspring fitness. For example, males of many species defend breeding territories to attract a female partner. Males with the best territories are often more successful at attracting a female, but may also produce offspring with increased growth and survival if these territories have more resources and/or are safer from predators. In this case, male territoriality should be classed as a form of parental care only if there is evidence that it originated or is currently maintained due to its beneficial effects on offspring fitness (Simmons and Parker 1989). However, if the beneficial effects of male territoriality to offspring are incidental, territoriality should be considered as part of the male's mating behaviour. To address this issue, we propose a slight modification in the definition of parental care that includes non-behavioural traits and excludes parental traits that incidentally increase offspring fitness (Table 1.1).

Other authors have proposed alternative definitions of parental care, restricting the term to specific periods of the offspring's development, such as after fertilization (Blumer 1979) or after hatching or birth (Crump 1995). These definitions exclude some forms of care on the basis that they fall outside the restricted period, while including other, often very similar, forms of care that happen to fall inside the period. For example, restricting parental care to the period after fertilization would exclude nest building because it takes place before fertilization (Hansell 2000), but include nest sanitation and nest repair that occur after fertilization. These restrictions limit the general usefulness of the term to evolutionary biologists because the limitations are based on the temporal characteristics of parental care rather than its adaptive value. For clarity, we suggest that the terms *post-fertilization* or *post-natal* are added when focusing on these specific time periods. In this chapter, we use a broad definition of parental care when describing the diverse forms of parental care in animals, but some of the following chapters use narrow definitions tailored to specific topics of interest (e.g. Chapter 2).

The second term discussed by Clutton-Brock (1991) is *parental expenditure*, defined as 'the expenditure of parental resources (including time and energy) on parental care of one or more offspring' (Table 1.1). Part of the confusion over the terminology in this field is due to many authors using the term parental effort (see below) as a synonym for parental expenditure. To promote a more consistent use of terminology, we follow Clutton-Brock's suggestion that the term parental expenditure is used to describe amount of time, energy, and resources that parents devote to care. Measures of parental expenditure, such as parental provisioning rates in birds and milk production in mammals, are sometimes used to quantify benefits of care to offspring. Indeed, Winkler (1987) argued that parental expenditure (in his terminology: parental effort) is the most relevant term when the evolution of parental care is viewed from the offspring's perspective. The reason for this is that the offspring's benefits depend on the actual amount of energy and resources received from the parent, and not on the fitness costs that parents incur from providing care. Parental expenditure is sometimes also used to describe energy and time costs of parental care in terms of increased metabolic rate, increased energy or protein intake, increased time spent at parental activities, or decreased parental body mass (Clutton-Brock 1991). For example, recent measures of energy costs of parental care as multiples of basal metabolic rate suggest that the costs of food provisioning, egg formation, and incubation are comparable (Nager 2006). Parental expenditure is sometimes calculated as a proportion of the parent's total resource budget that is allocated to care for one or more offspring. Such measures are referred to as relative parental expenditure (Clutton-Brock 1991), and should not be confused with measures of fitness costs of parental care (see below).

The third term discussed by Clutton-Brock (1991) is *parental investment*, defined by Trivers (1972) as 'any investment by the parent in an individual offspring that increases the offspring's chance of survival (and hence reproductive success) at the cost to the parent's ability to invest in other offspring' (Table 1.1). The appropriate currency of parental investment is the fitness costs that parents incur from providing care, which include reduced fecundity in the current breeding event (Maynard Smith 1977), reduced survival until future breeding attempts, reduced future mating success, and reduced reproductive success in future breeding attempts (Clutton-Brock 1991). Parental investment is perhaps the most important term in this field as it describes the direct fitness costs that a caring parent incurs from increasing its offspring's fitness, and as such underpins almost all theoretical work on the evolution of parental care (see Chapter 3). Furthermore, when combined with kin selection theory, the term provides the foundation for all theories on parent–offspring conflict over the allocation of parental resources (Trivers 1974; Mock and Parker 1997; Chapter 7). Unfortunately, it is notoriously difficult to estimate parental investment empirically as such estimates require the demonstration of both a cost of care to the parents and a benefit to their offspring (Mock and Parker 1997, p. 254). Consequently, there are still only a small number of studies providing empirical evidence that care for individual offspring incurs fitness costs to parents (e.g. Royle et al. 2002).

The final term discussed by Clutton-Brock (1991) is *parental effort*, defined as the combined fitness costs that the parent incurs due to the production and care of all offspring in a given biologically relevant period, such as a breeding attempt (Low 1978; Table 1.1). The term parental effort was originally introduced to partition Williams's (1966) term 'reproductive effort' into two components: parental effort, as defined above, and mating effort, defined as the effort allocated to the attraction of mates and/or exclusion of sexual competitors (Low 1978). An animal's total resource budget is divided between its parental effort (PE), mating effort (ME), and somatic effort (SE), the effort allocated to self-maintenance that increases future survival and reproduction, such that PE + ME + SE = 1. Thus, parental effort can only be increased at a cost to the parent's mating effort, somatic effort, or both. As pointed out by Low (1978), the terms parental effort and parental investment are closely linked, as parental effort equals the sum of parental investment across all offspring in a given brood (or during any other biologically relevant period). Like parental investment, parental effort is a key evolutionary term used in many models for the evolution of parental care, including models on how the evolution of male parental care is influenced by paternity and additional mating opportunities (Westneat and Sherman 1993; Magrath and Komdeur 2003; Chapters 6 and 11).

There are considerable inconsistencies in the use of the term parental effort, as the term is sometimes used as a synonym of parental expenditure (Winkler 1987), and parental investment is sometimes used as a synonym for parental effort (Daan and Tinbergen 1997). Parental expenditure describes the energy and time costs of care, while parental investment and parental effort describe the fitness costs associated with increasing the fitness of individual offspring and with the production and care for all offspring in a given brood, respectively. It is important to recognize that energy and time costs are not equivalent to fitness costs when parents vary with respect to condition and/or breed in variable environments. Parents that are in good condition or that breed in a favourable environment may spend more time and energy on care, yet incur lower fitness costs, than parents that are in poor condition or that breed in a harsh environment. Furthermore, the concepts of parental investment and parental effort differ in that the former refers to the fitness costs that a caring parent incurs as a consequence of investing in one of its offspring, while parental effort refers to the combined fitness costs that a parent incurs from the number of offspring produced in that breeding attempt and the average investment in each offspring. Thus, the term parental effort includes costs due to actions that increase the parent's own fitness, such as production of additional gametes, as well as actions that increase the offspring's fitness, while the term parental investment only includes costs due to the latter actions. To illustrate the distinction between the terms, consider brood size manipulation experiments that have been conducted in many birds. Such experiments often find that parents respond to an enlarged brood size by increasing their provisioning rate, that each offspring in enlarged broods receives as much food (or sometimes less) than each offspring in control broods, and that parents incur a fitness cost due to their increased workload (Parejo and Danchin 2006). In this example, the increased provisioning rate corresponds to an increase in parental expenditure, and the fitness cost to an increase in parental effort. However, there is no increase in parental investment because the parents' fitness costs are not associated with an increase in the fitness of individual offspring.

In addition to the terms parental expenditure, parental investment, and parental effort, the literature on parental care increasingly uses the term *parental effect* (also known as *maternal effect* or *paternal effect* when focusing on female or male parents, respectively). This term is used by evolutionary geneticists to describe the causal effects that the parent's phenotype has on the offspring's phenotype, including its growth and survival, over and above direct effects due to genes inherited from parents (Cheverud and Moore 1994; Mousseau and Fox 1998; Table 1.1). The term parental effect may include various mechanisms by which the parent's phenotype affects or influences the offspring's phenotype. Such mechanisms include parental care (Cheverud and Moore 1994), maternal hormones (Groothuis and Schwabl 2008), and maternal condition (Schluter and Gustafsson 1993). Although parental effects are mediated through the parent's phenotype, and as such constitute an environmental source of variation for the offspring's phenotype, they may have a partially genetic basis, in which case they might evolve in response to selection (Cheverud and Moore 1994). Parental effects that have evolved in response to selection are sometimes termed adaptive parental effects, the evolution of which has recently attracted much interest from evolutionary geneticists and evolutionary ecologists (Mousseau and Fox 1998). Traditionally, parental effects have been considered adaptive when increasing the offspring's fitness, for example by buffering offspring from harmful environmental effects (Marshall and Uller 2007). However, maternal effects may also have been selected to enhance maternal fitness, in which case they may decrease the offspring's fitness, such a when parents are under selection to produce larger clutches with smaller eggs, which effectively increases the parents' own fitness at the expense of that of their offspring (Marshall and Uller 2007). Other parental effects may be non-adaptive in the sense that they are due to incidental effects of the parents' phenotype on the offspring's fitness, such as when a parent's parasite load influences its offspring's fitness (Sorci and Clobert 1995). Although the terms parental effect and parental care are both descriptive, they differ from one another in that parental effect is a more general term and parental care represents that subset of adaptive parental effects that increases the offspring's fitness.

1.4 Assigning fitness to parents and offspring

In Section 1.3, parental care was defined as any parental trait that increases the offspring's fitness often at a cost to the parents' own fitness. An important question in any theoretical and empirical study of the evolution of parental care is how to assign fitness to parents and their offspring (Arnold 1985; Clutton-Brock 1988; Cheverud and Moore 1994; Wolf and Wade 2001). This issue deserves careful attention, as it is a common source of confusion in the study of parental care that in part stems from the different practices used by behavioural ecologists and evolutionary geneticists. Behavioural ecologists often view the evolution of parental care in light of kin selection theory, where a proportion of those fitness effects that are directly attributed to parental care are assigned from the recipient of care (the offspring) to the individual providing care (the parent) (Hamilton 1964). In contrast, evolutionary geneticists assign fitness strictly to the individual whose survival, growth, and reproduction has been affected by care (Arnold 1985). To discuss this issue in a way that considers the practices of both disciplines, we first focus on the assignment of offspring survival and reproduction to the offspring's and parents' personal fitness, and then discuss the different ways of assigning fitness benefits of parental care to parents and offspring.

1.4.1 Assigning offspring survival to offspring and parental fitness

One source of confusion over how to assign offspring survival and reproduction to the personal (or direct) fitness of offspring or their parents is due to the use of different practices in different situations. In some situations, such as in the definition of parental care as any parental trait that increases *offspring* fitness (Section 1.3), offspring survival and reproduction are clearly assigned to the offspring's fitness. Meanwhile, in other situations, parental fitness is estimated as the number of offspring that survive until recruitment and the subsequent reproductive performance of these offspring (Clutton-Brock 1988). This latter practice of assigning offspring survival and reproductive performance to the parents' fitness is justified on the grounds that parental care has a strong causal effect on offspring fitness (Clutton-Brock 1988; Grafen 1988). This practice should not be confused with the use of kin selection theory in behavioural ecology (Section 1.4.2). According to kin selection theory, it is the increase in offspring fitness which is directly attributed to parental care, once multiplied by the coefficient of relatedness between parents and offspring, that should be assigned to the parent's inclusive fitness. Evolutionary geneticists assign offspring survival and reproduction to the offspring's own fitness on the grounds that this practice is consistent with the proposal that an individual's fitness ideally should be measured as the number of zygotes it produces over its lifetime from the time of fertilization until the time of death (Arnold 1985).

Although there might be situations where it is practical to measure a parent's fitness in terms of offspring recruitment and reproductive performance (i.e. when data on zygote production are not readily available, as in mammals and other viviparous taxa), this practice is problematic in the context of the evolution of parental care. Firstly, as stated above, it is at odds with the definition of parental care as any parental trait that increases the *offspring's* fitness. Secondly, the practice risks introducing double counting of fitness if an offspring's survival is counted towards both its own and its parent's fitness. Such double counting can be avoided by identifying a specific point in an animal's life cycle before which fitness is always ascribed to parents and after which it is ascribed to offspring. For example, parental and offspring fitness can be separated at the time of fertilization (Arnold 1985), offspring independence (Grafen 1988), or offspring recruitment into the breeding population (Clutton-Brock 1988). We favour the first of these options because it is fully consistent with the definition of parental care, though practical problems associated with obtaining suitable data may necessitate other options for measuring the parents' and offspring's fitness. Regardless of which option is used, it is important that any rule for assigning parental and offspring fitness is explicitly defined and that these definitions are followed consistently.

1.4.2 Assigning costs and benefits of care to offspring and parents

The suggestion that parental and offspring fitness should be separated at the time of fertilization creates an apparent paradox when the evolution of parental care is considered from a strictly personal (or direct) fitness perspective. Given that parental care increases the offspring's fitness at a cost to parents' own fitness, a purely personal fitness perspective suggests that parents should be under selection not to provide care for their offspring. This conclusion is clearly at odds with the observation that costly parental care is widespread among animals. Behavioural ecologists and evolutionary geneticist provide somewhat different solutions to this apparent paradox, reflecting the different ways in which these two disciplines assign fitness costs and benefits of parental care to parents and offspring.

Behavioural ecologists often examine the evolution of parental care using cost–benefit analyses combined with kin selection theory. Kin selection theory provides a powerful approach for understanding the evolution of altruistic traits expressed among close relatives, such as worker sterility in eusocial hymenopterans (Hamilton 1964). Although parental care is often not considered an altruistic trait, this position is based on the practice of assigning offspring survival and growth as part of their parent's personal fitness which, as discussed above, is problematic. Hamilton (1964) himself used parental care as an example of an altruistic trait in his seminal work on kin selection. Kin selection theory applied to parental care distinguishes between direct and indirect benefits of care. Direct benefits refer to the personal fitness benefits that offspring accrue from receiving care, such as enhanced survival, while indirect benefits refer to the fitness benefits that parents accrue from providing care due to the increased survival and fecundity of their offspring. For example, if we define the parent's fitness as the number of zygotes it produces over its lifetime (Arnold 1985), the cost of care to the parent amounts to the reduction in the parents' zygote production due to the amount of parental care it provides. The direct benefit of care to offspring amounts to the increase in the offspring's zygote production that is directly caused by the amount of parental care its parent provides. The parents' indirect benefit is obtained by employing Hamilton's rule; that is, multiplying the offspring's benefit by the coefficient of relatedness between parents and offspring (Hamilton 1964). In this case, Hamilton's rule suggests that parental care can evolve in situations where the parent's indirect benefit in terms of an increase in offspring gamete production due to care, scaled by the coefficient of relatedness, outweighs the cost to parents of reduced gamete production.

Evolutionary geneticists study the evolution of parental care in terms of both selection and inheritance. They prefer separating parental and offspring fitness at the time of fertilization because it provides a clear framework for distinguishing the processes of selection and inheritance (Arnold 1985). The reason is simply that this practice ensures that effects on parental and offspring fitness are separated at the same time point in an animal's life cycle as when the genes are passed from parents to their offspring (Chapter 15). Indeed, evolutionary geneticists often reject the inclusive fitness approach used by behavioural ecologists for the reason that the assignment of indirect fitness benefits across generations confounds effects due to selection (e.g. increased offspring survival due to parental care) with effects due to inheritance (e.g. inheritance of offspring traits that increase offspring survival until recruitment).

Evolutionary geneticists study selection on parental care as the association between a parental care trait and fitness. Given that parental care is costly to parents but beneficial to offspring, parental care will be under antagonistic selection in the parental and offspring life-stages (Kölliker et al. 2010). When individuals are offspring, there is selection for receiving care because care increases offspring fitness. However, when individuals become parents, there is selection against providing care because the associated costs reduce the parents' fitness. The close relatedness between parents and offspring means that selection on offspring generates a correlated response in parental care. For example, if parental care increases offspring survival until recruitment, selection on offspring causes a correlated response in parental care because the surviving offspring have inherited genes for parental care from their parents. Thus, correlated responses in parental care due to selection on offspring play a similar role in evolutionary genetics as that played by the inclusive fitness concept in behavioural ecology: both allow parental care to evolve despite lowering the parents' fitness due to its beneficial effects on the offspring's fitness. Evolutionary geneticists focus not only on the process of selection but also on that of inheritance (Arnold 1985; Cheverud and Moore 1994), and a major benefit to separating parental and offspring fitness at the time of fertilization is that it allows for detailed studies on trait inheritance and genetic architecture that are not confounded by effects due to selection. Information on trait inheritance and genetic architecture are important in the study of parental care because they determine how parental care evolves in response to selection (Wolf and Wade 2001; Chapter 16).

1.5 Origin and evolution of parental care

Section 1.2 provides many examples of how various forms of parental care increase offspring fitness by neutralizing specific hazards that might threaten offspring survival or growth, including predators, cannibalistic conspecifics, parasites and pathogens, desiccation, flooding, hypoxia, and food limitation. A striking example is the bromeliad crab *Metopaulias depressus*, in which females deposit snail shells into bromeliad pools used for breeding, thereby effectively neutralizing the very low pH levels and boosting the low level of calcium carbonate in such pools (Diesel 1989). Although empirical studies on parental care provide good evidence for the current benefits of parental care, much less is known about its evolutionary origin (Chapters 2 and 5). For example, food provisioning, which is the focal form of care in many empirical studies, provides obvious current benefit by preventing offspring mortality due to starvation (Clutton-Brock 1991). However, this current benefit reflects that offspring are completely dependent on their parents for food, a condition that cannot explain its origin because offspring dependency on parents would only have evolved following the origin of parental food provisioning (Smiseth et al. 2003). Thus, studies on current adaptive value provide little insight into how parental food provisioning increased offspring fitness in the ancestral state where offspring still had the ability to forage independently of their parents.

Wilson (1975) made the first major attempt to understand the conditions favouring the evolutionary origin of parental care. Wilson proposed four prime environmental movers that have promoted the evolution of parental care: stable and structured habitats, harsh environmental conditions, specialized food sources, and predation risk. Tallamy and Wood (1986) later argued that Wilson's prime movers essentially boil down to variation in the distribution, persistence, abundance, richness, and physical properties of different food resources. Tallamy and Wood argued that the nature of the food resource utilized by insects determines their spatial and temporal patterns of feeding and reproduction, which in turn influence how competitors, predators, and parasites impact the offspring's survival and growth. They identified foliage, wood, detritus, dung, carrion, and living animals as resources associated with the evolution of parental care. For example, insects feeding on foliage tend to be exposed to predators and parasitoids, and many such species have evolved egg and offspring attendance. Insects feeding on wood are less exposed to predation as wood provides shelter, but have often evolved parental care to help with inoculation of offspring with gut symbionts or the wood with fungi. Finally, insects feeding on dung and carrion utilize a rich and ephemeral resource that attracts a wide range of competitors. Many such species have evolved parental care as a means to secure or protect the resource from various competitors.

Tallamy (1984) noted that parental care represents only one among several alternative solutions for how to overcome problems associated with environmental hazards that reduce offspring survival. For example, the herring (*Clupea harengus*) has evolved a long reproductive lifespan combined with high fecundity (instead of parental care for eggs and fry) to deal with highly variable rates of offspring mortality resulting from starvation, predation, and other harsh environmental conditions (Armstrong and Shelton 1990). Tallamy (1984) proposed that the likelihood that parental care evolves as a solution to these hazards depends on pre-existing traits that can be shaped by natural selection into parental traits that enhance offspring fitness. For example, the evolution of attendance and guarding of offspring against predators may have evolved from ancestral defensive or aggressive behaviours found in ancestral non-caring species (Tallamy 1984). Such ancestral forms of care are likely to be relatively simple traits (e.g. attendance and guarding of offspring), which subsequently were modified into the highly elaborate and complex forms of parental care observed today in many animal taxa (Clutton-Brock 1991).

Currently, little is known about the conditions favouring the evolution of more complex and elaborate forms of parental care. In birds and many other taxa, parental care comprises multiple parental behaviours such as nest building, incubation and protection of eggs and offspring, and provisioning of food after hatching or birth (Clutton-Brock 1991). The evolution of such elaborate forms of care has largely been ignored because most theoretical studies treat parental care as a unitary trait rather than a composite of several functionally integrated traits. A recent model suggests that the evolution of elaborate forms of care may be driven by mutual reinforcement between different components of parental care and offspring behaviours (Gardner and Smiseth 2011). For example, the origin of food provisioning may allow parents to choose safer nest sites and promote sibling competition for food provided by parents, which in turn may further drive the evolution of parental food provisioning (Gardner and Smiseth 2011). Thus, the evolution of elaborate forms of care may be driven by mutual reinforcement between different components of parental care and offspring traits, leading to a unidirectional trend from simple ancestral forms of care towards increasingly complex derived forms of care with very few reversals.

1.6 Conclusion

In this chapter, we have provided a brief overview of the tremendous diversity among species and higher taxa in the forms of care that parents provide to their offspring. We have discussed the terminology used in the study of the evolution of parental care, identified sources of confusion over the use of terminology, and suggested how the terms should be used in the future to improve translation between theory and empirical work in this field. We have also addressed how to assign fitness, and fitness benefits and costs, to parents and offspring. Finally, we have briefly discussed the environmental conditions that are thought to favour the origin and subsequent modifications of parental care.

Acknowledgements

We thank Matt Bell, Tim Clutton-Brock and Alastair Wilson for valuable comments on earlier drafts of the chapter.

References

Adiyodi, K. G. and Adiyodi, R. G. (1989). *Reproductive Biology of Invertebrates. Volume IV: Fertilization, Development, and Parental Care*, parts A and B. Wiley, New York.

Altmann, J. and Samuels, A. (1992). Costs of maternal care: infant-carrying in baboons. *Behavioral Ecology and Sociobiology* 29, 391–8.

Ang, T. Z., O'Luanaigh, C., Rands, S. A., Balmford, A., and Manica, A. (2008). Quantifying the costs and benefits of protective egg coating in a Chrysomelid beetle. *Ecological Entomology* 33, 484–7.

Armstrong, M. J. and Shelton, P. A. (1990). Clupeoid life-history styles in variable environments. *Environmental Biology of Fishes* 28, 77–85.

Arnold, S. J. (1985). Quantitative genetic models of sexual selection. *Experientia* 41, 1297–310.

Aubret, F., Bonnet, X., Shine, R., and Maumelat, S. (2005). Why do female ball pythons (*Python regius*) coil so tightly around their eggs? *Evolutionary Ecology Research* 7, 743–58.

Baur, B. (1994). Parental care in terrestrial gastropods. *Experientia* 50, 5–14.

Blackburn, D. G. (2005). Evolutionary origins of viviparity in fishes. In H. J. Grier and M. C. Uribe, eds. *Viviparous fishes*, pp. 287–301. New Life Publications, Homestead, Florida.

Blackburn, D. G. (2006). Squamate reptiles as model organisms for the evolution of viviparity. *Herpetological Monographs* 20, 131–46.

Blumer, L. S. (1979). Male parental care in the bony fishes. *Quarterly Review of Biology* 54, 149–61.

Blumer, L. S. (1982). A bibliography and categorization of bony fishes exhibiting parental care. *Zoological Journal of the Linnean Society* 76, 1–22.

Boulinier, T. and Staszewski, V. (2008). Maternal transfer of antibodies: raising immuno-ecology issues. *Trends in Ecology and Evolution* 23, 282–8.

Boutin, S., Larsen, K. W., and Berteaux, D. (2000). Anticipatory parental care: acquiring resources for offspring prior to conception. *Proceedings of the Royal Society of London, Series B* 267, 2081–5.

Brockman, H. J. (1980). Diversity in the nesting behavior of mud-daubers (*Trypoxylon politum* Say; Sphecidae). *Florida Entomologist* 63, 53–64.

Cheverud, J. and Moore, A. J. (1994). Quantitative genetics and the role of the environment provided by relatives in behavioral evolution. In C. Boake, ed. *Quantitative Genetic Studies of Behavioral Evolution*, pp. 67–100. University of Chicago Press, Chicago.

Clutton-Brock, T. H. (1988). *Reproductive Success*. University of Chicago Press, Chicago.

Clutton-Brock, T. H. (1991). *The Evolution of Parental Care*. Princeton University Press, Princeton.

Costa, J. T. (2006). *The Other Insect Societies*. Belknap, Harvard University Press, Harvard.

Crump, M. L. (1995). Parental care. In H. Heatwole and B. K. Sullivan, eds. *Amphibian Biology Volume 2: Social Behavior*, pp. 518–67. Surrey Beatty and Sons, Chipping Norton, Australia.

Daan, S. and Tinbergen, J. M. (1997). Adaptation of life histories. In J. R. Krebs and N. B. Davies, eds. *Behavioural Ecology: An Evolutionary Approach*, 4th edition, pp. 311–33.

Deeming, D. C. (2001). *Avian Incubation: Behaviour, Environment and Evolution*. Oxford University Press, Oxford.

Diesel, R. (1989). Parental care in an unusual environment: *Metopaufias depressus* (Decapoda: Grapsidae), a crab that lives in epiphytic bromeliads. *Animal Behaviour* 38, 561–75.

Duellman, W. E. and Maness, S. J. (1980). The reproductive behavior of some hylid marsupial frogs. *Journal of Herpetology* 14, 213–22.

Edgerly, J. S. (1997). Life beneath silk walls: a review of the primitively social Embiidina. In J. C. Choe and B. J. Crespi, eds. *Social Behavior in Insects and*

Arachnids, pp. 14–25. Cambridge University Press, Cambridge.

Evans, T. A, Wallis, E. J., and Elgar, M. A. (1995). Making a meal of mother. *Nature* 376, 229.

Field, J. (2005). The evolution of progressive provisioning. *Behavioral Ecology* 16, 770–8.

Forester, D. C. (1984). Brooding behavior by the mountain dusky salamander: can the female's presence reduce clutch desiccation? *Herpetologica* 40, 105–9.

Forstmeier, W. and Weiss, I. (2004). Adaptive plasticity in nest-site selection in response to changing predation risk. *Oikos* 104, 487–99.

Fox, C. W. and Czesak, M. E. (2000). Evolutionary ecology of progeny size in arthropods. *Annual Review of Entomology* 45, 341–69.

Gardner, A. and Smiseth, P. T. (2011). Evolution of parental care driven by mutual reinforcement between of parental food provisioning and sibling competition. *Proceedings of the Royal Society of London, Series B* 278, 196–203.

Gilmore, R. G. (1993). Reproductive biology of lamnoid sharks. *Environmental Biology of Fishes* 38, 95–114.

Grafen, A. (1988). On the uses of data on lifetime reproductive success. In T. H. Clutton-Brock, ed. *Reproductive Success*, pp. 454–71, University of Chicago Press, Chicago.

Green, B. S. and McCormick, M. I. (2005). O_2 replenishment to fish nests: males adjust brood care to ambient conditions and brood development. *Behavioral Ecology* 16, 389–97.

Groothuis, T. and Schwabl, H. (2008). Hormone-mediated maternal effects in birds: mechanisms matter but what do we know of them? *Philosophical Transactions of the Royal Society of London B* 363, 1647–61.

Gunn, J. S. and Tresher, R. E. (1991). Viviparity and the reproductive ecology of clinid fishes (Clinidae) from temperate Australian waters. *Environmental Biology of Fishes* 31, 323–44.

Haddad, C. F. B. and Hödl, W. (1997). New reproductive mode in anurans: bubble nest in *Chiasmocleis leucosticta* (Microhylidae). *Copeia* 1997, 585–8.

Hamilton, W. D. (1964) The genetical evolution of social behaviour. *Journal of Theoretical Biology* 7, 1–52.

Hansell, M. (1987). Nest building as a facilitating and limiting factor in the evolution of eusociality in the Hymenoptera. *Oxford Surveys in Evolutionary Biology* 4, 155–81.

Hansell, M. (2000). *Bird Nests and Construction Behaviour*. Cambridge University Press, Cambridge.

Heeb, P., Werner, I., Kölliker, M., and Richner, H. (1998). Benefits of induced host responses against an ectoparasite. *Proceedings of the Royal Society B* 265, 51–6.

Hilker, M. and Meiners, T. (2002). *Chemoecology of Insect Eggs and Egg Deposition*. Blackwell, Oxford.

Hinton, H. E. (1977). Subsocial behaviour and biology of some Mexican membracid bugs. *Ecological Entomology* 2, 61–79.

Kok, D., du Preez, L. H., and Channing, A. (1989). Channel construction by the African bullfrog: another anuran parental care strategy. *Journal of Herpetology* 23, 435–7.

Kölliker, M., Ridenhour, B. J., and Gaba, S. (2010). Antagonstic parent-offspring co-adaptation. *PLoS ONE* 5, e8606.

Kudo, S.-I., Akagi, Y., Hiraoka, S., Tanabe, T., and Morimoto, G. (2011). Exclusive male egg care and determinants of brooding success in a millipede. *Ethology* 117, 19–27.

Kupfer, A., Müller, H., Antoniazzi, M. M. Jared, C., Greven, H., Nussbaum, R. A., and Wilkinson, M. (2006). Parental investment by skin feeding in a caecilian amphibian. *Nature* 440, 926–9.

Kutchera, U. and Wirtz, P. (1987). A leech that feeds its young. *Animal Behaviour* 34, 941–2.

Low, B. S. (1978). Environmental uncertainty and the parental strategies of marsupials and placentals. *American Naturalist* 112, 197–213.

Magrath, M. J. L. and Komdeur, J. (2003). Is male care compromised by additional mating opportunity? *Trends in Ecology and Evolution* 18, 424–30.

Marshall, D. J. and Uller, T. (2007). When is a maternal effect adaptive? *Oikos* 116, 1957–63.

Martin, T. (2004). Avian life-history evolution has an eminent past: does it have a bright future? *Auk* 121, 289–301.

Maynard Smith, J. (1977). Parental investment: a prospective analysis. *Animal Behaviour* 25, 1–9.

Meier, R., Kotrba, M., and Ferrar, P. (1999). Ovoviviparity and viviparity in the Diptera. *Biological Reviews* 74, 199–258.

Melber A. and Schmidt, G. H. (1975). Ecological effects of the social behavior in two *Elasmucha* species (Heteroptera: Insecta). *Oecologia* 18, 121–8.

Mennerat, A., Mirleau, P., Blondel, J., Perret, P., Lambrechts, M. M., and Heeb, P. (2009). Aromatic plants in nests of the blue tit *Cyanistes caeruleus* protect chicks from bacteria. *Oecologia* 161, 849–55.

Mock, D. W. and Parker, G. A. (1997). *The Evolution of Sibling Rivalry*. Oxford University Press, Oxford.

Mousseau, T. A. and Fox, C. W. (1998). *Maternal Effects as Adaptations*. Oxford University Press, Oxford.

Nager, R. G. (2006). The challenges of making eggs. *Ardea* 94, 323–46.

Oppenheimer, J. R. (1970). Mouthbreeding in fishes. *Animal Behaviour* 18, 493–503.

Parejo, D. and Danchin, E. (2006). Brood size manipulation affects frequency of second clutches in the blue tit. *Behavioral Ecology and Sociobiology* 60, 184–94.

Refsnider, J. M. and Janzen, F. J. (2010). Putting eggs in one basket: ecological and evolutionary hypotheses for variation in oviposition-site choice. *Annual Reviews in Ecology, Evolution and Systematics* 41, 39–57.

Requena, G. S., Buzatto, B. A., Munguía-Steyer, R., and Machado, G. (2009). Efficiency of uniparental male and female care against egg predators in two closely related syntopic harvestmen. *Animal Behaviour* 78, 1169–76.

Royle, N. J., Hartley, I. R., and Parker, G. A. (2002). Sexual conflict reduces offspring fitness in zebra finches. *Nature* 416, 733–6.

Royle, N. J., Surai, P. F., McCartney, R. J., and Speake, B. R. (1999). Parental investment and egg yolk lipid composition in gulls. *Functional Ecology* 13, 298–306.

Rozen, D. E., Engelmoer, D. J. P., and Smiseth, P. T. (2008). Antimicrobial strategies in burying beetles breeding on carrion. *Proceedings of the National Academy of Sciences of the United States of America* 105, 17890–5.

Scheirs, J., De Bruyn, L., and Verhagen, R. (2000). Optimization of adult performance determines host choice in a grass miner. *Proceedings of the Royal Society B* 267, 2065–9.

Schluter, D. and Gustafsson, L. (1993). Maternal inheritance of condition and clutch size in the collared flycatcher. *Evolution* 47, 658–67.

Scott, D. K. (1980). Functional aspects of prolonged parental care in Bewick's swans. *Animal Behaviour* 28, 938–52.

Seibel, B. A., Robison, B. H., and Haddock, S. H. D. (2005). Post-spawning egg care by a squid. *Nature* 438, 929.

Shaffer, L. R. and Formanowicz, D. R., Jr (1996). A cost of viviparity and parental care in scorpions: reduced sprint speed and behavioural compensation. *Animal Behaviour* 51, 1017–24.

Simmons, L. W. and Parker, G. A. (1989). Nuptial feeding in insects: mating effort versus paternal investment. *Ethology* 81, 332–43.

Smiseth, P. T., Darwell, C. T., and Moore, A. J. (2003). Partial begging: an empirical model for the early evolution of offspring signalling. *Proceedings of the Royal Society of London, Series B* 270, 1773–7.

Smith, R. L. (1976). Male brooding behavior of the water bug *Abedus herberti* (Hemiptera: Belostomatidae). *Annals of the Entomological Society of America* 69, 740–7.

Sorci, G. and Clobert, J. (1995) Effects of maternal parasite load on offspring life-history traits in the common lizard (*Lacerta vivpara*). *Journal of Evolutionary Biology* 8, 711–23.

Spencer, M., Blaustein, L., and Cohen, J. E. (2002). Oviposition habitat selection by mosquitoes (*Culiseta longiareolata*) and consequences for population size. *Ecology* 83, 669–79.

Surbeck, M., Mundry, R., and Hohmann, G. (2011). Mothers matter! Maternal support, dominance status and mating success in male bonobos (*Pan paniscus*). *Proceedings of the Royal Society of London B* 278, 590–8.

Suzuki, S., Kitamura, M., and Matsubayashi, K. (2005). Matriphagy in the hump earwig, *Anechura harmandi* (Dermaptera: Forficulidae). increases the survival rate of offspring. *Journal of Ethology* 23, 211–13.

Tallamy, D. W. (1984). Insect parental care. *BioScience* 34, 20–4.

Tallamy, D. W. and Denno, R. F. (1981). Maternal care in *Gargaphia solani* (Hemiptera: Tingidae). *Animal Behaviour* 29, 771–8.

Tallamy, D. W. and Wood, T. K. (1986). Convergence patterns in sobsocial insects. *Annual Review of Entomology* 31, 369–90.

Trexler, J. C. and DeAngelis, D. L. (2003). Resource allocation in offspring provisioning: an evaluation of the conditions favoring the evolution of matrotrophy. *American Naturalist* 162, 574–85.

Trivers, R. L. (1972). Parental investment and sexual selection. In B. Campbell, ed. *Sexual Selection and the Descent of Man, 1871–1971*, pp. 136–79. Aldine, Chicago, Illinois.

Trivers, R. L. (1974). Parent-offspring conflict. *American Zoologist* 14, 249–64.

Tyler, M. J., Shearman, D. J. C., Franco, R., O'Brian, P., Seamark, R. F., and Kelly, R. (1983). Inhibition of gastric acid secretion in the gastric brooding frog, *Rheobatrachus silus*. *Science* 220, 609–10.

Walters, J. R. (1984). The evolution of parental behavior and clutch size in shorebirds. In J. Burger and B. L. Olla, eds. *Shorebirds: Breeding behavior and populations*, pp. 243–87, Plenum Press, New York.

West, M. J. and Alexander, R. D. (1963). Sub-social behavior in a burrowing cricket *anurogryllus muticus* (De Geer). *Ohio Journal of Science* 63, 19–24.

Westneat, D. F. and Sherman, P. W. (1993). Parentage and the evolution of parental behavior. *Behavioral Ecology* 4, 66–77.

Weygoldt, P. (1980). Complex brood care and reproductive behavior in captive poison-arrow frogs, *Dendrobates pumilio* O. Schmidt. *Behavioral Ecology and Sociobiology* 7, 329–32.

Williams, G. C. (1966). *Adaptation and Natural Selection*. Princeton University Press, Princeton, New Jersey.

Williams, T. D. (1994). Intraspecific variation in egg size and egg composition in birds: consequences on offspring fitness. *Biological Reviews* 68, 35–59.

Wilson, E. O. (1975). *Sociobiology*. Belknap, Harvard University Press, Harvard, Massachusetts.

Winkler, D. W. (1987). A general model for parental care. *American Naturalist* 130, 526–43.

Winkler, D. W. and Sheldon, F. H. (1993). Evolution of nest construction in swallows (Hirundinidae): a molecular phylogenetic perspective. *Proceedings of the National Academy of Sciences of the United States of America* 90, 5705–7.

Wolf, J. B. and Wade, M. J. (2001). On the assignment of fitness to parents and offspring: whose fitness is it and when does it matter? *Journal of Evolutionary Biology* 14, 347–56.

Wootton, R. J. (1976). *The Biology of the Sticklebacks*. Academic Press, London.

SECTION I

Origin and Evolution of Parental Care

CHAPTER 2

Theoretical foundations of parental care

Hope Klug, Suzanne H. Alonzo, and Michael B. Bonsall

2.1 Introduction

In 1871, Darwin noted that parental care is likely the foundation of social behaviour, yet remarkably little was known about how and why care evolved (Darwin 1871). Since this time, a large body of work has enhanced our understanding of the evolution of care. Studies have examined why parental care is present in some species but not others, factors that drive variation in which sex provides care, and selective pressures that give rise to particular forms of care. Collectively, and as Darwin alluded to, such work has revealed that the evolution of parental care is intimately linked with other forms of social behaviour, including mate attraction and competition, mating, and group living. Likewise, species-specific life-history and ecological and environmental conditions also influence patterns of care.

In this chapter we provide an overview of the ultimate factors promoting the origin and maintenance of parental care. We focus on theoretical treatments of parental care and discuss how verbal arguments, mathematical models, and the link between theory and data contribute to our conceptual understanding of parental care. To do this, we concentrate on connecting theoretical predictions with empirical patterns, rather than providing details of any particular theoretical approach.

2.1.1 Defining parental care

Parental care has been most broadly defined as 'any parental trait that appears likely to increase the fitness of a parent's offspring, and that is likely to have originated and/or is currently maintained for this function' (Chapter 1) or 'any form of parental behaviour that appears likely to increase the fitness of a parent's offspring' (Clutton-Brock 1991, p. 8). Under these broad definitions, parental care includes allocating resources to eggs prior to mating, offspring provisioning after birth or hatching, waste removal, nest tending and guarding (Clutton-Brock 1991; Chapter 1). While such a broad definition is useful in the sense that it is all-encompassing, it can also be problematic because it becomes difficult to distinguish parental care from other behaviours. Under these broad definitions, one could argue that any species that produces eggs exhibits parental care. If this is the case, it becomes meaningless to discuss the origin of care from an ancestral state of no care. Likewise, the process of selecting a high quality mate could be considered parental care if it increases offspring fitness under the definition proposed by Clutton-Brock (1991). For these reasons, we utilize a more focused definition of parental care. Throughout this chapter, we consider parental care to be parental behaviour that 1) occurs post-fertilization (or after the production of daughter cells if reproduction is asexual), 2) is directed at offspring, and 3) appears likely to increase offspring lifetime reproductive success. This definition allows for the discussion of the origin of care and the presence or absence of care across species or sexes. We further define parental effort as investment that is primarily related to increasing offspring survival or reproduction, with no assumption regarding costs to parents (Stiver and Alonzo 2009). Parental investment is any parental expenditure (time, energy, or other resources) that benefits the fitness of offspring but reduces the ability of a parent to invest in other components of fitness (Trivers 1972). Thus, parental

The Evolution of Parental Care. First Edition. Edited by Nick J. Royle, Per T. Smiseth, and Mathias Kölliker.

care is a form of parental investment if and when care increases offspring fitness and is costly to the parent providing it.

2.1.2 The role of modelling in parental care theory

Mathematical modelling has contributed substantially to our understanding of parental care evolution. Models of parental care have two general goals. First, some generate *a priori* predictions about patterns of care that we would expect given particular assumptions. For example, theoretical work has aimed to understand the general life-history conditions that give rise to the origin of care (Klug and Bonsall 2010). The predictions of such work can then be compared to empirical data. Other models explain existing empirical patterns. For instance, classic theory predicts that maternal care will be the norm (Trivers 1972). However, in many species males also care and in some species, care is solely paternal (e.g. paternal care is most common in fishes; Clutton-Brock 1991; Chapters 4 and 5). This empirical observation has in turn spurred a large body of theoretical work that attempts to explain existing patterns by identifying conditions that favour care by each or both sexes (Maynard Smith 1977; Sargent and Gross 1985; Queller 1997; Webb et al. 2002; Kokko and Jennions 2008).

With both predictive and explanatory models of evolution, it is important to distinguish between the origins and maintenance of behaviour. This is nicely illustrated when considering the evolution of parental care. Understanding the origins of care involves determining demographic, ecological, and environmental factors that allow care to evolve from a state of no care. In the early evolution of care, life-history traits that lead to positive population growth when rare (and hence density-dependent processes are weak) will promote the evolution of care. As traits increase in density and spread through a population, selection is affected by a different set of genetic, physiological, and environmental factors than those that influence the origin of the trait. Thus, understanding the maintenance of care often necessitates consideration of different processes from those that drive the origin of care.

2.1.3 General theoretical questions

There are four general questions that theoretical work on parental care evolution addresses:

1. When should care be provided?
2. Which sex should provide care?
3. How much care should be provided and to whom?
4. Why do we see a specific form of care behaviour in a given population or species?

In the remainder of the chapter, we discuss each of these questions.

2.2 When should care be provided?

Why some animals provide parental care, whereas others provide none, is a central question in evolutionary ecology. There is huge variation in the presence/absence of parental care within and between taxonomic groups, and parental care has independently evolved from a state of no care numerous times. For example, in ray-finned fishes care has emerged at least 33 times (Mank et al. 2005). A large amount of work has focused on identifying the conditions that favour the origin of parental care (Sargent et al. 1987; Clutton-Brock 1991; Winemiller and Rose 1992; Webb et al. 2002; Mank et al. 2005; Klug and Bonsall 2007, 2010). Costs and benefits of care, life-history characteristics, and ecological, environmental, and evolutionary dynamics are each hypothesized to affect the origin of parental care (Fig. 2.1).

2.2.1 Costs and benefits of parental care

Parental care will be favoured only when the fitness benefits to the caring parent(s) outweigh the costs associated with care (reduced parental survival and/or future reproduction). Parental care is beneficial to parents if it increases offspring survival, growth and/or quality, and ultimately offspring lifetime reproductive success. There are two broad, non-mutually exclusive ways in which parental care can increase offspring reproductive success.

First, parents can increase offspring survival. In several species, increased survival occurs during the life-history stage(s) in which parents and offspring are physically associated. Offspring guard-

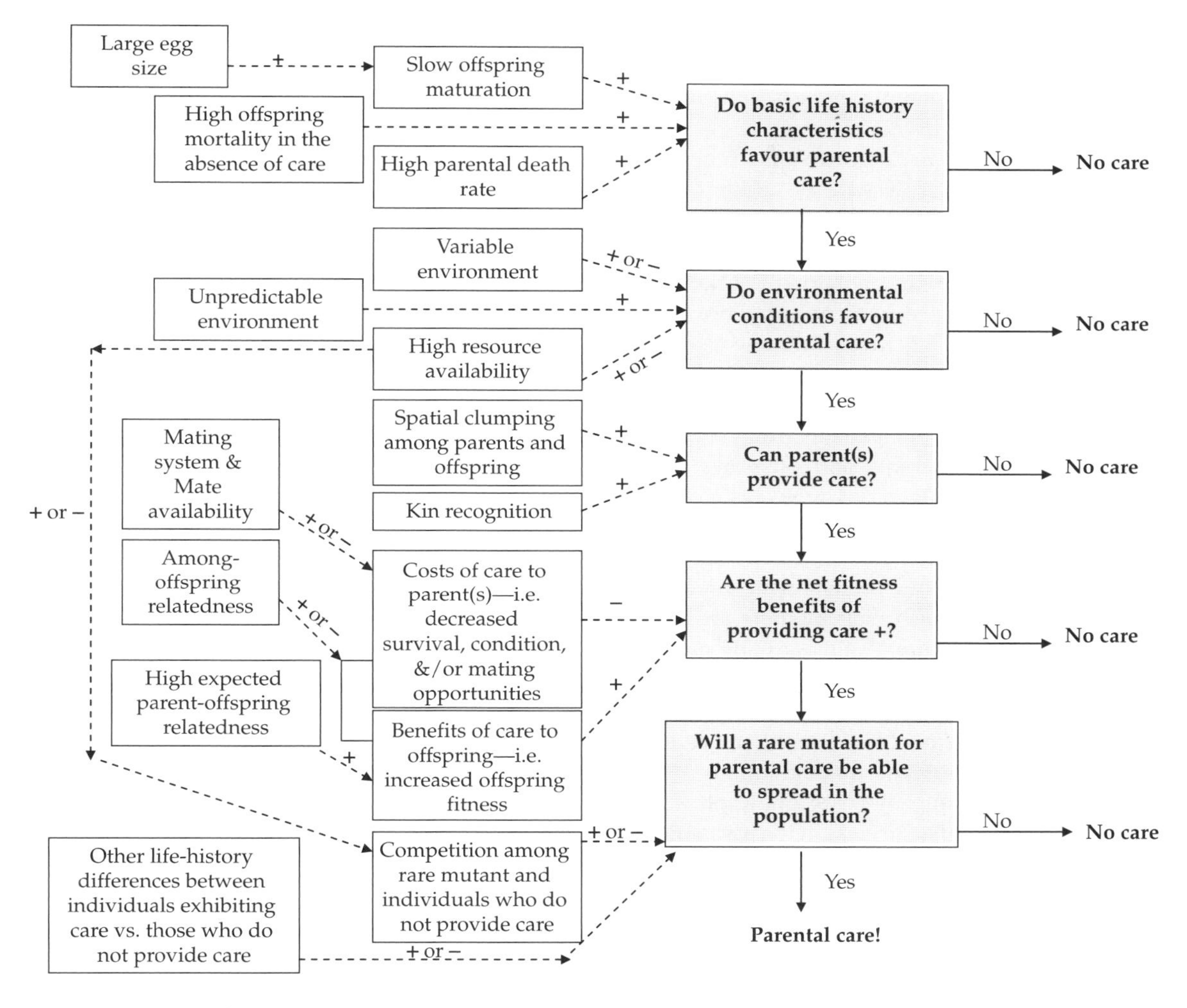

Figure 2.1 The origin of parental care. For parental care to evolve from an ancestral state of no care several conditions must be met (shaded boxes): basic life-history conditions (i.e. stage specific mortality and maturation rates) and environmental conditions favour care, parents are physically able to care for offspring, and the benefits of parental care outweigh the costs to the parent(s) providing care. Finally, parental care, which is initially rare in the population, must be able to spread. A number of factors will determine whether these conditions are met (un-shaded boxes). Here, we outline how various factors are currently predicted to affect the conditions favouring care (dashed lines). Differences between males and females can also affect the origin of care, and explicit discussion of this is provided elsewhere. (+ indicates a positive relationship between two variables,–indicates a negative relationship)

ing (Zink 2003; Klug et al. 2005), provisioning (Barba et al. 1996; Eggert et al. 1998; Kölliker 2007), waste removal (Hurd et al. 1991), and protection from parasites and disease (Croshaw and Scott 2005) increases offspring survival during one or more life-history stages in a number of animals. Parents can also improve offspring quality, which increases subsequent offspring survival when parents and offspring are no longer in close proximity. For example, provisioning increases offspring weight at independence in burying beetles (*Nicrophorus orbicollis*, Eggert et al. 1998) and some spiders (*Stegodyphus lineatus*, Salomon et al. 2005), which presumably increases subsequent offspring survival. Alternatively, parents might increase lifetime survival of their offspring by directly manipulating the relative amount of time offspring spend in various life-history stages in response to expected offspring mortality. Regardless of the mechanism, increased offspring survival will be beneficial to parents only if it increases the lifetime reproductive success of their young.

Second, parents can increase offspring reproductive success in the absence of effects on survival. In some cases care improves offspring quality, and this leads to increased offspring reproduction when parents and offspring are no longer associated. For instance, paternal presence at the nest increases offspring mass, immune response, and likelihood of breeding the following year in great tits (*Parus major*, Tinne et al. 2005). In other cases, parents directly increase offspring reproductive success when parents and offspring are in close proximity. Female vervet monkeys (*Chlorocebus pygerythrus*) are subjected to less aggression, have a higher pregnancy rate, and more surviving infants if their mothers are present in the same troop versus the case in which their mothers are absent (Fairbanks and McGuire 1986). As with survival, increased offspring quality benefits parents only if it increases offspring lifetime reproductive success.

Parental care is thought to be costly, and care can decrease parental survival, mating, and/or reproductive success (Chapter 3). For example, providing care decreases parental survival in the assassin bug *Rhinocoris carmelita* (Gilbert et al. 2010) and future maternal reproductive success in the treehopper *Publilia concava* (Zink 2003).

In some cases, parental care also decreases mating success (Stiver and Alonzo 2009). Many fishes, birds, and mammals enter a 'care-only' phase after mating in which they will not or cannot accept new mates. However, recent work suggests that the trade-off between current and future mating success might not be as ubiquitous as previously assumed (reviewed in Stiver and Alonzo 2009). For example, in the sand goby (*Pomatoschistus minutus*), females prefer males that provide parental care and Lindström et al. (2006) hypothesized that the evolution of paternal care might be favoured evolutionarily through mate choice. Theoretical work supports this hypothesis (Hoelzer 1989; Alonzo 2012): If care is favoured through sexual selection, costs are likely substantially reduced.

In models of parental care, increased offspring survival during one or more life-history stages is typically the only benefit of parental care considered (Maynard Smith 1977; Webb et al. 1999, 2002; McNamara et al. 2000; Klug and Bonsall 2007, 2010; Kokko and Jennions 2008; Bonsall and Klug 2011a,b). Likewise, many models of care assume a trade-off between current and future mating success. The specific costs and benefits of care that one assumes will directly influence predictions regarding patterns of care, yet relatively little theoretical work has examined how different costs and benefits of care affect predicted patterns of care. Thus, there is a need for better understanding of how different costs and benefits affect the evolution of care (Chapter 3). Likewise, the particular functional relationship between the level of care provided and the benefits to offspring or costs to parents can affect predicted care patterns (see related work by Smith and Fretwell 1974).

Importantly, understanding the fitness benefits of parental care—which directly determines the strength of selection on individual(s) providing care—is only one step in understanding the origin of parental care. As we consider in the next section, for care to evolve from an ancestral state of no care, both ecological and life-history conditions must also favour care (Stearns 1976; Klug and Bonsall 2010).

2.2.2 Life-history, ecology, and the origin of care

Parental care is linked closely with life history and ecology (Stearns 1976; Clutton-Brock 1991; Klug and Bonsall 2010; Bonsall and Klug 2011a,b). Imagine a population in which a rare mutation for parental care arises. For care to persist in the population, the few individuals exhibiting parental care must have the resources and life-history characteristics that make care 1) possible and 2) able to spread in the population given any age- or stage-structured dynamics and resource competition between the rare individuals who care and the more common individuals who do not care.

For parental care to be possible, parents and offspring must be associated in close proximity. Care is thus more likely to evolve when parents recognize or regularly encounter their offspring (Lion and van Baalen 2007). Theory suggests that altruistic and/or cooperative behaviours such as parental care will be more prevalent in populations that are spatially-structured and 'viscous' (i.e. have low dispersal; Hamilton 1964; Lion and van Baalen 2007) if the benefits of care are not outweighed by costs associated with increased competition among closely related individuals (West et al. 2002).

There are numerous examples of kin recognition in animals (e.g. Evans 1998, Fellowes 1998). Further, offspring often remain in close proximity to their parents. This is obviously the case in viviparous animals, who provision young before birth. Likewise, in many species, eggs are spawned in breeding territories, creating a close physical association between parents and offspring. Such breeding territories are hypothesized to have preceded the evolution of care in fishes and birds (Williams 1975; Baylis 1981; Gross and Shine 1981; Wesolowski 1994).

Basic life-history characteristics (i.e. stage specific mortality and maturation rates) also heavily influence the origin of parental care. When individuals pass through multiple life-history stages, theoretical work suggests that parental care that increases offspring survival or quality during one or more life-history stages will be most strongly favoured when offspring need care the most (Wilson 1975; Clutton-Brock 1991; McNamara et al. 2000; Kokko and Jennions 2008; Klug and Bonsall 2010). When egg mortality in the absence of care is high, egg care will be favoured, and when juvenile mortality in the absence of care is high, juvenile care will also be selected for (Webb et al. 2002; Klug and Bonsall 2010). This is true even when the magnitude of the benefit in terms of increased offspring survival is equivalent at high vs. low values of offspring survival in the absence of care. Empirical work supports this prediction. For example, Clutton-Brock (1991) noted that across invertebrate species, parental care tends to occur in species in which offspring cannot survive in the absence of care.

The duration of various life-history stages is also expected to affect the evolution of parental care. If egg mortality in the absence of care is high, selection for egg care will be stronger when the duration of the egg stage is relatively long (Shine 1978, 1989; Nussbaum 1985, 1987; Kolm and Ahnesjö 2005; Klug and Bonsall 2010). This is because the longer offspring remain in a stage, the greater their chances of dying. Shine (1978) noted that there is a positive correlation between propagule size and the presence of parental care in animals. As an explanation for this pattern, Shine (1978, 1989) and Sargent et al. (1987) suggested that parents can 1) make the egg stage relatively safe for offspring by providing parental care, 2) increase the amount of time offspring spend in the egg stage by producing large eggs, and in doing so, 3) decrease the proportion of time offspring spend in the relatively 'high risk' juvenile stage. Alternatively, Nussbaum (1985, 1987) argued that the evolution of larger eggs might instead precede the evolution of care. Larger eggs take longer to develop and thus have greater overall mortality, which will in turn subsequently select for parental care. In salamanders and frogs, it appears that the evolution of larger eggs precedes the evolution of parental care, which is consistent with the prediction that longer egg stages will select for care (Nussbaum 1985, 1987; Summers et al. 2006). Similarly, Winemiller and Rose (1992) found that highly developed parental care in fishes is associated with slow adult and offspring growth and a lack of parental care is correlated with faster adult and offspring growth.

Evolutionary theory predicts that individuals with reduced expected future reproductive success should invest more in their current offspring (Williams 1966a; Sargent and Gross 1985; Coleman and Gross 1991). Theoretical work has shown that parental care will be more likely to evolve from an ancestral state of no care when adult death rate is high (Klug and Bonsall 2010; Bonsall and Klug 2011a). There is empirical support for this prediction in fishes. Winemiller and Rose (1992) examined the relationships among various life-history traits for 216 North American fish species and found that highly developed parental care is correlated with short lifespan, whereas the lack of parental care is correlated with long lifespan.

Environmental conditions are also hypothesized to affect the origin of parental care. Verbal arguments have suggested a role for environmental variation, and in particular resource availability, in the evolution of care. Wilson (1975) and Clutton-Brock (1991) suggested that care is most likely to evolve when environmental conditions are harsh and competition for resources is intense, as these are the conditions under which the benefits of care are likely to be large. Following r- and K-selection theory, Stearns (1976) predicted that parental care will be associated with constant environments, whereas the fitness benefits of parental care will be decreased in variable environments. When the environment varies, Stearns (1976) suggested that parental care is more likely to be needed when the variability in the environment is unpredictable rather than predictable.

More recent theoretical work has found that environmental variability, the life-history traits affected by such variability, and the specific costs of care, interact to determine whether care will be favoured. Bonsall and Klug (2011a) showed using a mathematical model that resource variability *per se* does not affect the likelihood of some form of care invading from an ancestral state of no care. If, however, offspring or parental mortality is affected by environmental variability, there can be either an increase or decrease in the fitness associated with providing care relative to that in a constant environment, depending on the specific costs of care (Bonsall and Klug 2011a).

Interactions among offspring also affect selection on parental care. For example, among-offspring relatedness (the level of relatedness between offspring in a given brood) affects competition between siblings, which can in turn affect the reproductive value of offspring. As a result, interactions among offspring are expected to influence the origin of parental care (reviewed in Mock and Parker 1997 and Parker et al. 2002; Royle et al. 1999; Gardner and Smiseth 2010; Bonsall and Klug 2011b).

2.3 Which sex should provide care?

Another key question is who should provide care, if any exists. Empirically, this question is motivated by the observation that female care is more common than male care and that maternal investment is greater than paternal investment in many species with biparental care (Clutton-Brock 1991; Kokko and Jennions 2003; McDowall 2003; Cockburn 2006; Chapters 4, 5, and 6). Yet, species across a range of taxa exist where either both sexes invest in parental care or even male-biased or male-only parental care occurs (Woodroffe and Vincent 1994; Tallamy 2000; McDowall 2003; Ah-King et al. 2005; Mank et al. 2005). Extensive research has focused on explaining variation within and among species in the degree to which males and females invest in parental effort.

As discussed above, parental effort typically decreases the future reproductive success of parents. Costs of parental effort can be related to current and/or future parental survival, mate attraction, and/or reproduction. When the relative fitness advantages versus disadvantages of parental care differ between the sexes, we expect sex differences in parental care to arise (Box 2.1; Chapter 6).

Males and females differ in many ways, with one of the most fundamental differences being the size of gametes they produce (termed anisogamy). Bateman (1948) argued that anisogamy causes males and females to differ in the degree to which their reproductive success is limited by access to mates. In a study in fruit flies (*Drosophila melanogaster*), he found that while the number of offspring sired by a male increased with the number of mates, female reproductive success did not continue to increase with the number of mates. Based on this pattern, Trivers (1972) argued that sex differences in parental care could be understood by considering the differences in reproductive investment between the sexes, as females produce larger more energetically expensive gametes than males. While these arguments are intuitively appealing, we now know that past investment alone cannot explain future differences in investment (Dawkins and Carlisle 1976; Kokko and Jennions 2008). Thus, we do not expect males to provide less care than females simply because they invest less energy per sperm than females invest per egg. However, the general argument remains that the sexes may differ in many ways that influence selection on parental investment and this could explain sex differences in parental care, especially if males pay a higher cost of lost mating success when providing care (Box 2.1).

Using a game theoretical model, Maynard Smith (1977) considered how differences between the sexes in the cost of providing parental care might affect the conditions under which males and females care. In his now classic 'Model II', Maynard Smith (1977) considered that both males and females either care or desert offspring. He assumed that female expected fecundity was less when

Box 2.1 Understanding sex differences in parental care

There is striking variation across higher taxonomic groups in which sex provides most of the parental care. Below, we outline key predictions explaining sex-specific care patterns.

Predictions:

1. **Anisogamy does not directly determine which sex invests more heavily in parental care.**
   Earlier work suggested that females are more likely to care because they invest more in gametes than males. We now know that past investment will not directly drive differences between the sexes in future investment.
   References: Dawkins and Carlisle 1976; Kokko and Jennions 2008.
2. **Males and females differ in many ways. Such differences can influence selection on parental investment by males and females.**
   Past investment does not directly drive future care. However, sex differences (some of which might be related to past investment) can affect care. Some examples include:
   - **Differences between males and females in the cost of providing care**: One sex might experience greater costs of care.
   - **Differences in the benefits offspring receive when care is provided by males vs. females**: One sex might be more effective at providing care.
   - **Constraints in which sex can care**: One sex may be unable to provide essential types of care (e.g. male mammals cannot lactate) or be physically disassociated from offspring (e.g. due to internal fertilization).
   - **Certainty of parentage**: Differences between males and females in expected parentage can make one sex more likely to care on an evolutionary time scale.
   - **Strong sexual selection**: Strong sexual selection on one sex, such that some individuals of that sex can remate faster, can favour reduced care by that sex.
   - **Mating preference for care by one sex**: If parental care is preferred in mate choice, sexual selection can favour increased care by the non-choosy sex
   - **Differences in the relative abundance of males and females that are available to mate (i.e. a biased operational sex ratio, OSR)**: A biased OSR can lead to frequency-dependent selection in which the sex that faces stronger competition for mates will be selected to increase parental investment.

   References: Hoelzer 1989; Queller 1997; Houston and McNamara 2002; Houston et al. 2005; Fromhage et al. 2007; Kokko and Jennions 2008; Alonzo 2010.
3. **Differences in the risks associated with caring for offspring versus competing for mates affect the level of care that is provided by males and females.**
   Such differences can affect the evolution of sex-specific parental care in two ways:
   - **Competing for mates is associated with higher mortality than providing parental care:** In this scenario, the earlier deserting sex will become rare in the population because individuals of that sex spend more time risking death. The fact that each offspring has one mother and one father favours increased desertion by the sex that is relatively rare. Such a process could lead to maternal care if females are initially selected to care for other reasons or paternal care if males are initially more likely to care.
   - **Providing parental care is associated with higher mortality than competing for mates:** If parental care is dangerous, individuals of the deserting sex will experience less mortality, become more common in the population, and have difficulty finding a mate. This will select for increased parental investment by the deserting sex. As a result, when care is the riskier activity, selection favours increased care by both sexes and males and females are less likely to differ in the amount of parental care provided.

   Reference: Kokko and Jennions 2008.

caring than when deserting offspring. In contrast, males were assumed to have some additional probability of remating with a second female if they desert offspring from their first mate. In general, this model predicts that males will care when the probability of remating is low and the benefits (in terms of increased offspring survival) are high. In contrast, female care is predicted when fecundity costs are low but the benefits are high. However, there is a fundamental problem with this model. In Maynard Smith (1977), the probability that a male will remate is fixed and independent of female care patterns (McNamara et al. 2000; Houston and McNamara 2002, 2005; Houston et al. 2005). When females provide care they are often unavailable to remate. It is now recognized that it is important to consider 'model consistency' and how key parameters, such as paternity or mating rates, arise from the biological interactions of the model (Chapter 11).

Queller (1997) revisited the question of why females care, developing a self-consistent model. Using the basic fact that male and female fitness must be equal at the population level, he demonstrated that whenever males are not fully related to the offspring in their care (because females sire offspring with multiple males) the fitness gain to males of providing care is reduced, while higher variance in male mating success compared to females increases the costs of providing care. More recent theory shows that while many of the basic conclusions remain the same (Houston and McNamara 2002, 2005; Houston et al. 2005; Cotar et al. 2008), important differences emerge from self-consistent analyses, such as the prediction that females may sometime prefer low quality males (Cotar et al. 2008) and that paternity affects selection on male care over evolutionary time even if it does not vary between reproductive bouts (Queller 1997; Houston et al. 2005; Box 2.1).

An additional issue is that some arguments relating sexual selection (or sex differences in mating) and parental effort are circular. For example, if we assume sex differences in the distribution of mating success then it follows that these sex differences may lead to differences between males and females in parental investment. Ideally, these differences would arise out of models as emergent properties rather than be assumed *a priori*. Kokko and Jennions (2008) developed a model that considered the evolution of male and female parental effort. The results of this model echo the conclusions made by Queller (1997) arguing that sexual selection on males and mixed paternity due to female multiple mating disfavour the evolution of male care, while also adding important insights regarding the complex nature of the interactions between competition, adult sex ratio, and the evolution of care (Box 2.1; Chapter 6).

While this explains the prevalence of maternal care (assuming sexual selection is typically stronger for males than females), this does not explain the evolution of paternal care in systems where sexual selection on males occurs (Alonzo 2010; Alonzo and Heckman 2010). The common explanation for male care is that selection favouring parental care due to increased offspring survival outweighs the inherent selection against paternal care due to variation in mating success among males. However, it is important to note that this theory does not consider sexual selection due to female choice or the co-evolution of indicator traits of parental quality and the evolution of parental effort per se. While there is some theory examining how female preferences for indicators of male parental quality may co-evolve (Houston et al. 1997; Wolf et al. 1997; Kokko 1998; Soltis and McElreath 2001; Seki et al. 2007; Cotar et al. 2008; Kelly and Alonzo 2009) further theory is needed to examine thoroughly how sexual selection arising from the evolution of traits related to mate choice, mate attraction, and intrasexual competition interact with the co-evolution of male and female care patterns (Alonzo 2010). Ideally, tests of this theory using comparative analyses and carefully-controlled experiments are needed to understand the complex interaction between sexual selection and parental care evolution (Chapter 11).

The mode of fertilization is also thought to influence which sex provides care (Williams 1975; Baylis 1981; Gross and Shine 1981; Chapter 11). Some have argued that internal fertilization in the mother's body will make paternal care unlikely because it creates uncertainty of paternity or because males are not physically associated with offspring. There are strong relationships between which sex provides care and fertilization mode in teleost fishes (Gross and Shine 1981; Mank et al. 2005) and amphibians (Gross and Shine 1981). In part, the relationship between paternal care and external fertilization in these groups appears to be related to male territoriality and the physical association between males and offspring that results from external fertilization. Effects of fertilization mode on certainty of parentage can also affect parental care. The effect of parentage on parental effort is complex, and this topic is discussed in detail in Chapter 11.

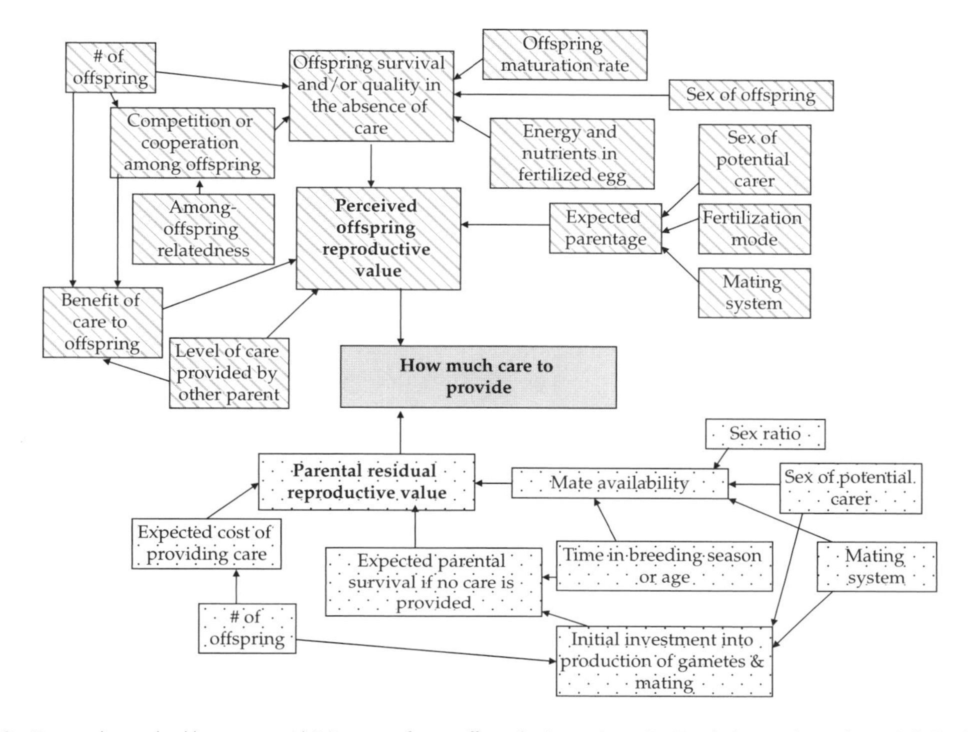

Figure 2.2 How much care should parents provide? Numerous factors affect selection on the optimal level of parental care that an individual provides for a given brood. Such variables can be categorized as affecting offspring reproductive value (diagonal grey bars) or parental residual reproductive value (black dots). This is not an exhaustive list. For example, resource availability, predation, parasitism, and disease likely also affect offspring survival and competition and adult survival and future reproductive opportunities.

2.4 How much care should be provided?

When one or both parents provide care, there is a dilemma regarding how much care to provide to current offspring and how much parental resource (time, energy) to save for future reproductive opportunities (Trivers 1972; Gross and Sargent 1985; Winkler 1987; Dale et al. 1996). Numerous factors are hypothesized to affect the amount of care provided (Fig. 2.2). Most of these variables can be categorized as affecting offspring reproductive value or parental residual reproductive value.

2.4.1 Offspring reproductive value

Offspring reproductive value is a measure of how much current offspring are expected to contribute to a parent's overall fitness. Offspring reproductive value (sometimes referred to as brood reproductive value, that is the sum of reproductive values of all offspring in a given brood) is affected by offspring survival in the absence of care, the number of offspring present, expected future survival or quality of offspring, and offspring age (reviewed in Dale et al. 1996; Fig. 2.2). Offspring reproductive value is argued to affect parental care in two ways.

First, some authors have suggested that the fitness benefits of caring for broods of greater reproductive value will be higher than those of lower reproductive value. If this is the case, parental care is expected to increase as offspring reproductive value increases (Andersson et al. 1980; Winkler 1987; Montgomerie and Weatherhead 1988; Clutton-Brock 1991; Hale 2006). This means that parents are expected to provide more care to higher quality offspring (e.g. larger offspring), older offspring, larger broods, and offspring born earlier in the breeding season that have greater survival prospects (Winkler 1987; Hale 2006). In some fishes and birds, care increases as brood size (Townshend and Wooton 1984; Windt and Curio 1986) or offspring age (Sargent and Gross 1986; Montgomerie and Weatherhead 1988) increases. If parentage is uncertain, parents should provide greater care for offspring that are more likely to be their own if there is variation in expected parentage within a brood or across reproductive episodes (Maynard Smith 1978; Baylis 1981; Winkler 1987; Westneat and Sherman 1993; Sheldon 2002, Chapter 11). There is, however, large variation in whether expected paternity affects the level of care provided across empirical studies (Alonzo 2010).

In contrast, others have noted that caring for offspring of lower reproductive value will result in greater fitness benefits to parents in some cases (Dale et al. 1996; Hale 2006). Dale et al. (1996) proposed the 'harm to offspring hypothesis', which suggests that the marginal benefit of investment in care will be highest when offspring are expected to have relatively low survival in the absence of care. They argued that older offspring or offspring in relatively good condition will have high survival in the absence of care. If this is the case, parents should provide more care to poor quality and younger offspring because they are the ones that will benefit the most from care (Dale et al. 1996). This hypothesis is consistent with theoretical work that predicts an increase in care as egg mortality in the absence of care increases (Webb et al. 2002; Klug and Bonsall 2010), as well as several empirical studies. For example, pied flycatchers (*Ficedula hypoleuca*) return to the nest sooner to feed poorer condition nestlings in comparison to offspring in better condition (Listøen et al. 2000). Likewise, provisioning by parents is greater when chicks are relatively small in Antarctic petrels (*Thalassoica antarctica*; Tveraa et al. 1998).

In general, authors agree that the effect of offspring reproductive value on the level of parental care is complex and this complexity is expected to affect patterns of care both within and across species (Clutton-Brock 1991; Dale et al. 1996; Hale 2006). Offspring reproductive value can have a positive, negative, or no effect on the fitness benefit of providing care (Clutton-Brock 1991), and this in turn is expected to affect both the origin and maintenance of parental care. Hale (2006) thus argued that understanding the relationship between offspring characteristics and parental care requires explicit examination of how the increase in the reproductive value of young once care is provided is influenced by brood characteristics, rather than focusing only on reproductive value either before or after receiving care.

2.4.2 How much care the other parent provides

In species with biparental care, the level of care provided by one parent affects offspring reproductive value and is predicted to affect the amount of care provided by the other parent. Chase (1980) and Houston and Davies (1985) developed models of biparental care in which each parent makes a decision about how much care to provide. In these models, each parent provides a fixed level of effort that maximizes their own fitness, given the effort of its mate. Over time, an evolutionarily stable state is reached in which each parent provides a specific level of care. If one parent changes their effort, the other parent is expected to compensate partially. The models by Chase (1980) and Houston and Davies (1985), which are played-out over evolutionary rather than behavioural time, provide some insight into the level of care that each parent should provide. However, costs of providing care are expected to lead to conflict among parents, and more recent work illustrates the importance of negotiation between parents (Chapter 9).

McNamara et al. (2003) demonstrated that negotiation affects the amount of care each parent provides and the overall level of care that offspring receive. If parents can negotiate they are expected to compensate to a lesser extent for a change in the other parent's effort in comparison to the case in which negotiation is impossible. When parents negotiate, the overall level of care offspring receive is expected to be less than that in the absence of negotiation (McNamara et al. 2003). Further, when parents are highly responsive to each other's effort, offspring will in some cases be better off when only a single parent provides care (McNamara et al. 2003).

There is mixed empirical evidence regarding the prediction that parents should adjust their own level of effort in response to that of their partner

(Hinde 2006; Harrison et al. 2009). Across studies, parents have been found to compensate fully for a change in their partner's care, partially compensate, match or mirror the change made by their partner, or show no response at all, and in some studies males and females responded differently to a change in their partner's effort (Johnstone and Hinde 2006; Harrison et al. 2009). To understand such variation, Johnstone and Hinde (2006) extended earlier models to incorporate uncertainty in brood value or need into parental negotiation.

Johnstone and Hinde (2006) argued that changes in effort by one partner will affect the care provided by the other parent in two ways. First, and as previously recognized, a change in effort by one parent impacts the benefits of additional investment by its mate (care by one parent changes offspring reproductive value). Second, a change in effort by one parent can convey information about the value and/or need of the brood. When parents have partial information about brood value and need, greater investment by one parent can serve as a signal to the other parent. Johnstone and Hinde (2006) demonstrated that using a partner's care as a signal of brood need favours a matching response in which the individual responds with a change of effort in the same direction as the focal parent. In contrast, compensation is favoured if a change in effort impacts the marginal value of investment by the other parent. A recent meta-analysis revealed that while there is a large amount of variation in how parents respond to their partner's effort, partial compensation appears to be the average response in birds (Harrison et al. 2009). As Harrison et al. (2009) note, it will be important to assess how brood size and age, together with time in the breeding season, affect the response of a parent to its mate's effort.

2.4.3 Parental residual reproductive value

Parental residual reproductive value is a measure of how much future reproduction is expected to contribute to a parent's fitness. Numerous factors determine a parent's expected future reproductive success, and these factors are expected to affect the level of care that a parent provides to current offspring (Fig. 2.2). All else being equal, parents should decrease care as it becomes more costly. However, it is rare that all else is equal, and in particular, the costs and benefits of care typically covary. Parents are expected to provide a level of care that maximizes the net benefit of their investment. Thus, the relationship between the costs and benefits of care directly influences the level of care that parents are expected to provide (Williams 1966b; Winkler 1987).

Parental characteristics unrelated to costs or benefits of care also affect future reproductive opportunities. Parents that are older, near the end of a final breeding season, or have relatively low survival even when they do not provide care, will have relatively low residual reproductive value and are predicted to invest more into current offspring (Williams 1966a; Stearns 1976; Sargent and Gross 1985; Winkler 1987; Clutton-Brock 1991; Klug and Bonsall 2010). This hypothesis is consistent with some empirical patterns. For example, older burying beetle females produce larger clutches and provide more food to offspring than younger females (Creighton et al. 2009). Importantly, theory does not necessarily predict that parents will provide more resources or care to each individual offspring. Smith and Fretwell (1974) suggested there is an optimal amount of resources that should be invested into each offspring. In this case, selection will favour the production of more offspring when possible, which will in turn lead to a greater overall level of care provided to the entire brood, rather than greater investment in each offspring. In contrast, other theoretical work has found that clutch size can decline with maternal age if egg size is constrained (Begon and Parker 1986) and that optimal parental investment can be affected by the environment, parental phenotype, and competition among offspring (Parker and Begon 1986). For instance, sibling competition can in some cases favour constant clutch sizes, with parents investing more heavily in each individual offspring (Parker and Begon 1986). Additionally, Dall and Boyd (2002) found that as the risk of parental starvation increases, parents are expected to forage for themselves more. As a result, provisioning is expected to decrease when the risk of starvation is great under some conditions.

A parent's sex, mating system dynamics, and population adult sex ratio also affect future reproduction, and hence the optimal level of care (Queller 1997; Houston and McNamara 2002; Cotar et al. 2008; Kokko and Jennions 2008; Box 2.1; Chapter 6). All else being equal, the more common sex in the population will be selected to invest more in care (Chapter 6). If parental care is costlier than mate competition, then the sex that provides more care will become rarer in the population. This will make it more difficult for the sex that provides less care to find a mate. Selection is then expected to favour relatively high levels of care by both parents (Kokko and Jennions 2008; Chapter 6). If, in contrast, mate competition is more costly than caring, the sex that competes more heavily will become rare

Box 2.2 When to care for, abandon, or eat offspring

Sometimes a parent is faced with a dilemma: they have offspring that they have begun to care for, the offspring need care, but continuing to invest in those offspring is not the optimal strategy for the parent—that is the parent's fitness will be maximized by abandoning some or all current offspring in order to increase future reproduction. Parents can either terminate care for all current offspring or for some fraction of young. Whole-clutch and partial-clutch termination have distinct biological significance. Any benefits of whole-clutch termination can only be seen in future reproductive success, whereas the benefits of partial-clutch care termination can be seen in current and future reproductive success. To terminate care parents either abandon (spatially disassociate from), kill, and/or or consume offspring.

Below, we provide an overview of factors that give rise to offspring abandonment, abortion, infanticide, and filial cannibalism. These hypotheses are not mutually exclusive and many factors are expected to affect the termination of care.

Offspring abandonment, abortion, infanticide, and/or filial cannibalism can be driven by:

1. **Density-dependent offspring survival.** When offspring survival is density-dependent, parents can increase the survival of remaining offspring by removing or killing some number of offspring. Such behaviour will be favoured if it increases overall parental reproductive success. Density-dependent offspring survival is potentially related to the environment or increased benefits of parental care to remaining offspring. (References: Lack 1954; Payne et al. 2004; Klug et al. 2006; Klug and Bonsall 2007.)
2. **Variation in offspring phenotype**. If offspring vary, parents might improve their fitness by killing, consuming, or abandoning young with traits that are associated with relatively low fitness. If parents terminate care for some number of lower quality offspring they can potentially allocate more resources to future reproduction or better care for remaining offspring. Alternatively, parents might kill or consume offspring that are taking a relatively long time to mature if the presence of those offspring inhibits future reproduction. (References: Forbes and Mock 1998; Klug and Bonsall 2007; Klug and Lindström 2008.)
3. **Energetic or nutritional need of the parent**. If a parent lacks the energetic or nutritional reserves to effectively care for offspring, parental fitness might be improved by abandoning, killing, or consuming offspring. Abandoning offspring will be favoured if offspring have some chance of surviving in the absence of care. Consuming and/or killing offspring will be favoured if the energetic gain from consuming young increases current or future reproductive success, or if there is some cost to the parent of allowing the offspring to live—for example if the offspring would damage a nest site that the parent plans to reuse. (References: Rohwer 1978; Sargent 1992; Manica 2002; Klug and Bonsall 2007.)
4. **Current vs. expected future fitness gains of parental care.** Parents should balance investment in current versus future reproduction to maximize fitness. If a parent has 1) low expected fitness gains from current investment and 2) relatively high expected future reproductive opportunities, it can be beneficial for parents to terminate care for some or all offspring and invest more heavily in future reproduction. Parents are expected to abandon, kill, or consume offspring more frequently earlier in the breeding season (when the parent(s) have greater chance of reproducing again), when parents are relatively young, and when offspring are relatively young (because younger offspring require greater relative parental investment to become independent). (References: Williams 1966a,b; Rohwer 1978; Sargent and Gross 1985; Coleman and Gross 1991; Sargent 1992; Manica 2002.)

in the population. This will select for decreased care by that sex (typically males) (Kokko and Jennions 2008; Box 2.1; Chapter 6).

Other theory suggests that if females are the choosier sex, they should preferentially mate with males that provide parental care, in which case sexual selection will lead to increased male care (Hoelzer 1989; Alonzo 2012). Empirical work in the sand goby found that males increase the level of care they provide in the presence of females relative to the case in which females are absent (Pampoulie et al. 2004) and females exhibit a preference for males that provide more care (Lindström et al. 2006).

In summary, parents are expected to weigh the benefits of current investment in parental care against expected future reproduction. In some cases, parents are faced with the situation in which they are providing care, offspring need care, but continuing to provide care is no longer the strategy that will maximize lifetime fitness of the individual. When this occurs, there are a number of ways in which parents terminate care, which we discuss in Box 2.2.

2.5 Why do we see a specific form of care behaviour in a given population or species?

2.5.1 What type of care to provide?

Some models aim to explain the evolution of particular types of care behaviour in one or more species. For example, food provisioning is prevalent in birds and mammals. Gardner and Smiseth (2010) found that provisioning will be favoured evolutionarily when parental provisioning is more efficient than offspring self-feeding, more effective in comparison to other care behaviour, and when parents provide high quality food that would otherwise be too costly for offspring to obtain.

Dall and Boyd (2004) used a dynamic optimization model to explore the transition from provisioning young with food to provisioning young from body reserves. They found that selection favours lactation when food supplies are uncertain and future reproductive opportunities are low. They argue that the unreliable lifestyle of small mammal-like reptiles likely favoured the evolution of lactation, and this prompted the evolution of true mammals (Dall and Boyd 2004). They also note that mass constraints related to flight might limit the body reserves that female birds can provision with, and this might explain major differences in provisioning between birds and mammals.

Together, these studies illustrate how the costs and benefits of provisioning have likely influenced key life-history differences among higher taxonomic groups, such as birds and mammals.

2.5.2 Explaining complex patterns of care in a given system: three examples

Other models of care attempt to explain complex or variable care in a given species. Such models explicitly link empirical data with model predictions. Below, we describe three examples.

Case study 1: Flexible care in a fish. In St. Peter's fish (*Sarotherodon galilaeus*) males, females, or both parents care. To understand why patterns of care are so plastic, Balshine-Earn and Earn (1997) developed a game-theoretical model to explore the relationship between the pay-offs of care for each sex and whether paternal, maternal, or biparental care is the evolutionarily stable state (ESS). Their model suggests that paternal care is the only ESS and that the observed empirical patterns of care were unstable. However, the model also revealed that operational sex ratio (OSR) strongly affects whether an ESS existed, and if so, whether it was paternal, maternal, or biparental care. Balshine-Earn and Earn (1997) noted that the OSR varies spatially and temporally in the population of St. Peter's fish, and the specific conditions confronting a given mating pair might explain variation in parental care behaviour in this system.

Case study 2: Brood parasitism in dung beetles. Female dung beetles (*Onthophagus taurus*) lay their eggs in tunnels beneath a dung pat. Once a female finds a pat, she digs a tunnel, and either prepares a brood ball in which to lay an egg, or steals the brood ball of a prior female, destroys the existing egg, and lays her egg in the existing dung ball. After laying an egg, the female seals the tunnel, and in some cases adults guard the tunnel against brood parasitism. Crowe et al. (2009) developed a game-theoretic model to predict when it is beneficial for a mother to steal a brood ball and how long a female should guard her eggs. The results suggest that the decision to steal a brood ball is heavily influenced by the time it takes to steal versus construct a new ball. Previous empirical work found that stealing a brood ball is faster than constructing a new one (discussed in Crowe et al. 2009). Thus Crowe et al. (2009) hypothesize that high levels of brood parasitism will be favoured in dung beetles. Additionally, their model suggests that females should either guard the entire time that eggs are susceptible to brood parasitism, or not guard at all. Crowe et al. (2009) further suggested that there is a critical population size below which no guarding occurs and above which all eggs are guarded, and they outline ways in which this prediction could be tested empirically.

Case study 3: Communal nesting. Many animals nest communally (Robertson et al. 1998). For example, burying beetles conceal and defend carcasses that they use as a food resource for offspring. Often, a mating pair defends the carcass, but in some cases two unrelated females communally defend and share the carcass (Eggert and Müller 1997). To understand the ecological conditions that favour the evolution of mutual tolerance among unrelated nesting females, Robertson et al. (1998) developed a genetic-algorithm model employing game theory. In their model, females can fight for sole use of the nest, tolerate the other female and breed communally but fight if attacked, or leave in search of a new site. Nests varied in the number of offspring that could be supported and the probability of nest failure. The model predicts that females will nest communally when additional nests are rare, females have limited clutch sizes, and dominant females can skew reproduction in their favour. Robertson et al. (1998) compared these predictions to previously published findings in the burying beetle. As predicted, reproductive skew was present on relatively small carcasses, but contrary to model predictions, reproductive skew was absent on larger carcasses (Eggert and Müller 1997). It is possible that dominant females cannot effectively limit reproduction of subordinates on large carcasses or that large carcasses can support all offspring present (Robertson et al. 1998).

The above examples illustrate how mathematical models can be used to understand perplexing patterns of care. Interestingly, the observed behaviour in St. Peter's fish and the burying beetle varied from model predictions: flexible care is not predicted to be an ESS in St. Peter's fish and equitable reproduction was not predicted in the burying beetle. The authors suggest alternative explanations not considered in the modelling frameworks utilized. These examples nicely highlight how models can be used to evaluate our current understanding of behaviour and identify additional critical factors, and they also illustrate that models are rarely perfect or complete. In general, it is the feedback between theoretical predictions and empirical patterns that makes modelling such a valuable tool in understanding animal behaviour.

2.6 Future directions and challenges

Existing theoretical work provides a solid foundation for our understanding of the evolutionary dynamics of parental care. Mathematical and verbal models have led to numerous predictions, which have been tested empirically, and such empirical work has in turn spurred new theoretical development. There are, however, many aspects of the evolution of parental care that we do not yet fully understand. Below, we discuss three avenues of future research.

2.6.1 Linking theory and data

Evaluating models of parental care is critical, but this is not an easy task. Traits are expected to co-evolve, particularly during the early evolution of care. It is thus particularly difficult to test predictions regarding the origin of care. For example, if eggs begin receiving care, the egg stage becomes relatively safe and selection might favour increased investment in the production of eggs (Shine 1978, 1989). As such it then becomes difficult to make inferences on the relationship between egg size and the origin or care. Likewise, models predict that care will be most likely to evolve from an ancestral state of no care when offspring survival in the absence of care is low (Klug and Bonsall 2010). To evaluate such a prediction, one must have knowledge about offspring survival prior to the evolution of care. Fortunately, phylogenetic methods provide insight into the past. Mank et al. (2005) used phylogenetic methods to examine evolutionary pathways of parental care in ray-finned fishes. Their analyses suggest that maternal, paternal, and biparental care has arisen multiple times, and the most common evolutionary transitions in fishes are 1) external fertilization to paternal care and 2) external fertilization to internal fertilization to maternal care. Likewise, Summers et al. (2006) used comparative analyses to examine the relationship between egg size and the evolution of parental care in frogs. Their results suggest that large egg size precedes the evolution of parental care, which is consistent with theoretical predictions of Nussabaum and Shultz (1989) but contrary to predictions stemming from previous theoretical work (e.g. Shine 1978).

Given the increased availability of phylogenetic data and methods, comparative analyses will play a significant role in enhancing the link between theoretical predictions and empirical patterns of care. In particular, phylogenetic methods will prove useful in examining the evolution of male care, which has historically been difficult to explain. Given that many factors interact to influence the evolution of care, it will be important to examine the relationships between life history (e.g. mortality, maturation rate, fertilization mode, stage structure), ecology (e.g. predation, resource availability), multiple mating, mating system, and the emergence of paternal, maternal versus biparental care.

2.6.2 Origin versus maintenance of care

As discussed above, co-evolution among traits is expected to occur early in the evolution of care. As a result, the conditions that give rise to the origin of care are not necessarily similar to the conditions that maintain care. Thus, it is important and interesting to distinguish between the origin and maintenance of parental care both in the devel-

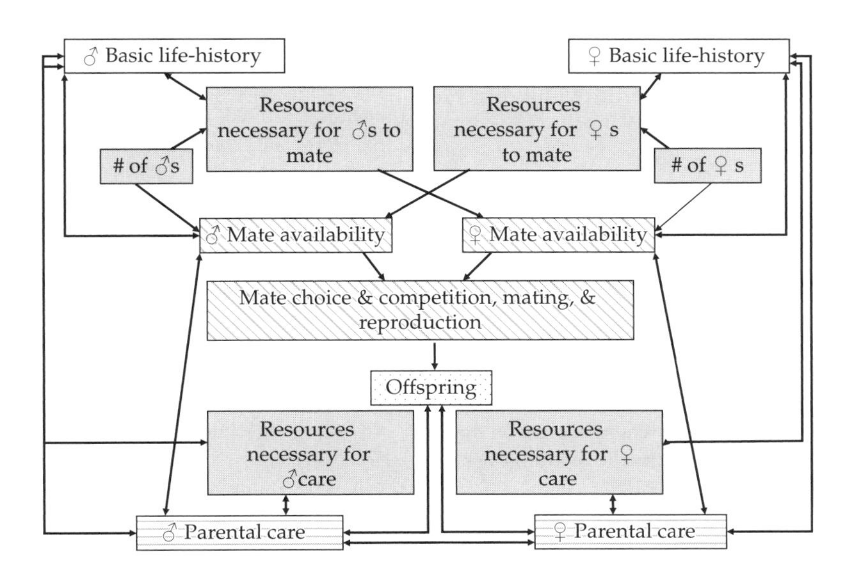

Figure 2.3 Feedback between 1) basic life-history (i.e. stage-specific mortality, maturation rates—white shading), 2) ecology (grey shading), 3) sexual selection, mating, and reproduction (diagonal lines), and parental care (horizontal lines). Basic life-history characteristics affect the types of resources required for mating and caring, competition associated with attaining those resources, mate availability, and patterns of male and female care. Ecological factors, including resource availability and male and female density, affect mate availability, which in turn affects mate choice, competition, and reproduction. Likewise, the type and abundance of resources that are necessary for care, as well as offspring number and traits (black dots), directly affect the patterns of parental care provided by each sex. In addition to these direct effects, there will be ecological and evolutionary feedback among the various factors, which is expected to lead to co-evolution among traits. Understanding the feedback between basic life-history, ecological factors such as density and resource use, sexual selection and mating behaviour, and offspring traits is essential for understanding the evolution of parental care.

opment of models and empirical tests of theory. Many models of parental care do not make this distinction explicit, and this is often appropriate for the hypotheses addressed. However, to understand parental care more fully, it will be important to develop models that explicitly consider and identify the dynamics associated with the origin, early evolution, and maintenance of care. Specifically, life-history traits (e.g. mortality, maturation rates, fertilization mode, stage structure), the costs and benefits of care, ecological factors (resource need and competition among individuals), and coevolution among traits will likely vary throughout the evolution of care.

2.6.3 Ecological and evolutionary feedback

Finally, recent work has begun to directly incorporate the feedback between ecology and evolution in models of parental care and sex roles. Such work illustrates how sexual selection, certainty of parentage, costs and benefits of care for each sex, and life-history can interact and be used to explain variation in parental care across species. There is still much work to be done in this area, and it will be critical to explore additional feedback both on evolutionary and ecological time scales (Fig. 2.3).

Acknowledgements

We are grateful to Geoff Parker and Sasha Dall for their insightful comments.

References

Ah-King, M., Kvarnemo, C., and Tullberg, B. S. (2005). The influence of territoriality and mating system on the evolution of male care: A phylogenetic study on fish. *Journal of Evolutionary Biology* 18, 371–82.

Alonzo, S. H. (2010). Social and coevolutionary feedbacks between mating and parental investment. *Trends in Ecology and Evolution* 25, 99–108.

Alonzo, S. H. (2012). Sexual selection favours male parental care, when females can choose. *Proceedings of the Royal Society of London Series B-Biological Sciences* 279, (1734) 1784–90.

Alonzo, S. H. and Heckman, K. (2010). The unexpected but understandable dynamics of mating, paternity and paternal care in the ocellated wrasse. *Proceedings for the Royal Society of London series B* 277, 115–22.

Andersson, M., Wiklund, C. G., and Rundgren, H. (1980). Parental defence of offspring: a model and an example. *Animal Behaviour* 28, 536–42.

Balshine-Earn, S. and Earn, D. J. D. (1997). An evolutionary model of parental care in St. Peter's fish. *Journal of Theoretical Biology* 184, 423–31.

Barba, E., Lopez, J. A., and Gil-Delgado, J. A (1996). Prey preparation by adult Great Tits *Parus major* feeding nestlings. *Ibis* 138, 532–53.

Bateman, A. J. (1948). Intra-sexual selection in *Drosophila*. *Heredity* 2, 349–68.

Baylis, J. R. (1981). The evolution of parental care in fishes, with reference to Darwin's rule of male sexual selection. *Environmental Biology of Fishes* 6, 223–51.

Begon, M. and Parker, G. A. (1986). Should egg size and clutch size decrease with age? *Oikos* 47, 293–302.

Bonsall, M. B. and Klug, H. (2011a). The evolution of parental care in stochastic environments. *Journal of Evolutionary Biology* 24, 645–55.

Bonsall, M. B. and Klug, H. (2011b). Effects of among-offspring relatedness on the evolution of parental care and filial cannibalism. *Journal of Evolutionary Biology* 24, 1335–50.

Chase, I. D. (1980). Cooperative and noncooperative behaviour in animals. *American Naturalis*, 115, 827–57.

Clutton-Brock, T. H. (1991). *The Evolution of Parental Care: Monographs in Behavior and Ecology*. Princeton University Press, Princeton.

Cockburn, A. (2006). Prevalence of different modes of parental care in birds. *Proceedings of the Royal Society B-Biological Sciences* 273, 1375–83.

Coleman, R. M. and Gross, M. R. (1991). Parental investment theory: the role of past investment. *Trends In Ecology and Evolution* 6, 404–6.

Cotar, C., McNamara, J. M., Collins, E. J., and Houston, A. I. (2008). Should females prefer to mate with low-quality males? *Journal of Theoretical Biology* 254, 561–7.

Creighton, J. C., Heflin, N. D., and Belk, M. C. (2009). Cost of reproduction, resource quality, and terminal investment in a burying beetle. *American Naturalist* 174, 673–84.

Croshaw, D. A. and Scott, D. E. (2005). Experimental evidence that nest attendance benefits female marbled salamanders (*Ambystoma opacum*) by reducing egg mortality. *American Midland Naturalist* 154, 398–411.

Crowe, M., Fitzgerald, M., Remington, D. L., Ruxton, G. D., and Rychtar, J. (2009). Game theoretic model of brood parasitism in a dung beetle *Onthophagus taurus*. *Evolutionary Ecology* 23, 765–76.

Dale, S., Gustavsen, R., and Slagsvold, T. (1996). Risk taking during parental care: a test of three hypotheses applied to the pied flycatcher. *Behavioral Ecology and Sociobiology* 39, 31–42.

Dall, S. R. X. and Boyd, I. L. (2002). Provisioning under risk of starvation. *Evolutionary Ecology Research* 4, 883–96.

Dall, S. R. X. and Boyd, I. L. (2004). Evolution of mammals: lactation helps mothers to cope with unreliable food supplies. *Proceedings of the Royal Society of London Series B-Biological Sciences* 271, 2049–57.

Darwin, C. (1871). *The Descent of Man and Selection in Relation to Sex*, chapter 4. D. Appleton and Company, New York.

Dawkins, R. and Carlisle, T. R. (1976). Parental investment, mate desertion and a fallacy. *Nature* 262, 131–3.

Eggert, A.-K. and Müller, J. K. (1997). Biparental care and social evolution in burying beetles: Lessons from the larder. In Choe, J. C., and Crespi, B. J. eds. *The Evolution of Social Behavior in Insects and Arachnids.* Cambridge University Press, Cambridge, pp. 216–36.

Eggert, A. K., Reinking, M., and Müller, J. K. (1998). Parental care improves offspring survival and growth in burying beetles. *Animal Behaviour* 55, 97–107.

Evans, T. A. (1998). Offspring recognition by mother crab spiders with extreme maternal care. *Proceedings of the Royal Society of London Series B-Biological Sciences* 265, 129–34.

Fairbanks, L. A. and McGuire, M. T. (1986). Age, reproductive value, and dominance-related behaviour in vervet monkey females: cross-generational influences on social relationships and reproduction. *Animal Behaviour* 34, 1710–21.

Fellowes, M. D. E. (1998). Do non-social insects get the (kin) recognition they deserve? *Ecological Entomology* 23, 223–7.

Forbes, S. L. and Mock, D. W. (1998). Parental optimism and progeny choice: when is screening for offspring quality affordable? *Journal of Theoretical Biology* 192, 3–14.

Fromhage, L., McNamara, J. M., and Houston, A. I. (2007). Stability and value of male care for offspring: is it worth only half the trouble? *Biology Letters* 3, 234–6.

Gardner, A. and Smiseth, P. T. (2010). Evolution of parental care driven by mutual reinforcement of parental food provisioing and sibling competition. *Proceedings of the Royal Society of London Series B-Biological Sciences* 278, 196–203.

Gilbert, J. D. J., Thomas, L. K., and Manica, A. (2010). Quantifying the benefits and costs of parental care in assassin bugs. *Ecological Entomology* 35, 639–51.

Gross, M. R. and Sargent, R. C. (1985). The evolution of male and female care in fishes. *American Zoologist* 25, 807–22.

Gross, M. R. and Shine, R. (1981). Parental care and mode of fertilization in ectothermic vertebrates. *Evolution* 35, 775–93

Hale, R. E. (2006). 'A re-examination of the influence of offspring reproductive value on parental effort', in *Context-dependent natural and sexual selection on male nesting activity in flagfish (Jordanella floridae)*. Doctoral Dissertation, University of Florida, Gainesville, FL.

Hamilton, W. D. (1964). The genetical evolution of social behavior II. *Journal of Theoretical Biology* 7, 17–52.

Harrison, F. Barta, Z., Cuthill, I., and Székely, T. (2009). How is sexual conflict over parental care resolved. A meta-analysis. *Journal of Evolutionary Biology* 22, 1800–12.

Hinde, C. A. (2006). Negotiation over offspring care? A positive response to partner-provisioning rate in great tits. *Behavioral Ecology* 17, 6–12.

Hoelzer, G. A. (1989). The good parent process of sexual selection. *Animal Behaviour* 38, 1067–78.

Houston, A. I. and Davies, N. B. (1985). The evolution of cooperation and life history in the dunnock Prunella modularis. In R. M. Sibly and R. H. Smith RH, eds. *Behavioural Ecology: Ecological Consequences of Adaptive Behaviour*, pp. 471–87. Blackwell Scientific Publications, Oxford.

Houston, A. I., Gasson, C. E., and McNamara, J. M. (1997). Female choice of matings to maximize parental care. *Proceedings of the Royal Society of London Series B-Biological Sciences* 264, 173–9.

Houston, A. I. and McNamara, J. M. (2002). A self-consistent approach to paternity and parental effort. *Philosophical Transactions of the Royal Society of London—Series B: Biological Sciences* 357, 351–62.

Houston, A. I. and McNamara, J. M. (2005). John Maynard Smith and the importance of consistency in evolutionary game theory. *Biology & Philosophy* 20, 933–50.

Houston, A. I., Szekely, T., and McNamara, J. M. (2005). Conflict between parents over care. *Trends in Ecology & Evolution* 20, 33–8.

Hurd, P. L., Weatherhead, P. J., and McRae, S. B. (1991). Parental consumption of nestling feces: good food or sound economics? *Behavioral Ecology* 2, 69–76.

Johnstone, R. A. and Hinde, C. A. (2006). Negotiation over offspring care—how should parents respond to each others effort. *Behavioral Ecology* 818–27.

Kelly, N. B. and Alonzo, S. H. (2009). Will male advertisement be a reliable indicator of parental care, if offspring survival depends on male care? *Proceedings in the Royal Society of London Series B* 276, 3175–83.

Klug, H. and Bonsall, M. B. (2007). When to care for, abandon, or eat your offspring: The evolution of parental care and filial cannibalism. *American Naturalist* 170, 886–901.

Klug, H. and Bonsall, M. B. (2010). Life history and the evolution of parental care. *Evolution* 64, 823–35.

Klug, H., Chin, A., and St. Mary, C. M. (2005). The net effects of guarding on egg survivorship in the flagfish, *Jordanella floridae*. *Animal Behaviour* 69, 661–8.

Klug, H. and Lindström, K. (2008). Hurry-up and hatch: selective filial cannibalism of slower developing eggs. *Biology Letters* 4, 160–2.

Klug, H., Lindström, K., and St. Mary, C. M. (2006). Parents benefit from eating offspring: density-dependent egg survivorship compensates for filial cannibalsim. *Evolution* 60, 2087–95.

Kokko, H. (1998). Should advertising parental care be honest? *Proceedings of the Royal Society of London Series B-Biological Sciences* 265, 1871–8.

Kokko, H. and Jennions, M. D. (2008). Parental investment, sexual selection and sex ratios. *Journal of Evolutionary Biology* 21, 919–48.

Kokko, H. and Jennions, M. J. (2003). It takes two to tango. *Trends in Ecology and Evolution* 18, 103–4.

Kölliker, M. (2007). Benefits and costs of earwig (*Forficula auricularia*) family life. *Behavioral Ecology and Sociobiology* 9, 1489–97.

Kolm, N. and Ahnesjö, I. (2005). Do egg size and parental care coevolve in fishes? *Journal of Fish Biology* 66, 1499–515.

Lack, D. (1954). *The Natural Regulation of Animal Numbers*, Oxford University Press, Oxford.

Lindström, K., St. Mary, C. M., and Pampoulie, C. (2006). Sexual selection for male parental care in the sand goby, *Pomatoschistus minutus*. *Behavioral Ecology and Scociobiology* 60, 46–51.

Lion, S. and van Baalen, M. (2007). From infanticide to parental care: why spatial structure can help adults be good parents. *American Naturalist* 170, E26–46.

Listøen, C., Karlsen, R. F., and Slagsvold, T. (2000). Risk taking during parental care: a test of the harm-to- offspring hypothesis. *Behavioral Ecology* 11, 40–3.

Manica, A. (2002). Filial cannibalism in teleost fish. *Biological Reviews* 77, 261–77.

Mank, J. E., Promislow, D. E. L., and Avise, J. C. (2005). Phylogenetic perspectives in the evolution of parental care in ray-finned fishes. *Evolution* 59, 1570–8.

Maynard Smith, J. (1977). Parental investment: A prospective analysis. *Animal Behaviour* 25, 1–9.

Maynard Smith, J. (1978). *The Evolution of Sex*. Cambridge University Press, Cambridge.

McDowall, R. M. (2003). In defence of the caring male. *Trends in Ecology and Evolution* 18, 610–11.

McNamara, J. M., Houston, A. I., Barta, Z., and Osorno, J-L. (2003). Should young ever be better off with one parent than with two? *Behavioral Ecology* 14, 301–10.

McNamara, J. M., Székely, T., Webb, J. N., and Houston, A. I. (2000). A dynamic game-theoretic model of parental care. *Journal of Theoretical Biology* 205, 605–23.

Mock, D. and Parker, G. (1997). *The Evolution of Sibling Rivalry*. Oxford University Press, Oxford.

Montgomerie R. and Weatherhead, P. J. (1988). Risks and rewards of nest defence by parent birds. *Quarterly Review of Biology* 63, 167–87.

Nussbaum, R. (1985). The evolution of parental care in salamanders. *Miscellaneous Publication of the Museum of Zoology—University of Michigan* 169, 1–50.

Nussbaum, R. (1987). The evolution of parental care in salamanders: an examination of the safe harbor hypothesis. *Researches on Population Ecology (Kyoto)* 29, 27–44.

Pampoulie C., Lindström K., and St. Mary C. M. (2004). Have your cake and eat it too: male sand gobies show more parental care in the presence of female partners. *Behavioural Ecology* 15, 199–204.

Parker, G. A. and Begon, M. (1986). Optimal egg and clutch size: effects of environment and maternal phenotype. *American Naturalist* 128, 573–92.

Parker, G. A., Royle, N. J., and Hartley, I. R. (2002). Intrafamilial conflict and parental investment: a synthesis. *Philosophical Transactions of the Royal Society of London—Series B: Biological Sciences* 357, 295–307.

Payne, A. G., Smith, C., and Campbell, A. C. (2004). A model of oxygen-mediated filial cannibalism in fishes. *Ecological Modelling* 174, 253–66.

Queller, D. C. (1997). Why do females care more than males? *Proceedings of the Royal Society of London Series B-Biological Sciences* 264, 1555–7.

Robertson, I. C., Robertson, W. G., and Roitberg, B. D. (1998). A model of mututal tolerance and the origin of communal association between unrelated females. *Journal of Insect Behavior* 11, 265–86.

Rohwer, S. (1978). Parent cannibalism of offspring and egg raiding as a courtship strategy. *American Naturalist* 112, 429–40.

Royle, N. J., Hartley, I. R., Owens, I. P. F., and Parker, G. A. (1999). Sibling competition and the evolution of growth rates in birds. *Proceedings of the Royal Society of London Series B-Biological Sciences* 266, 923–32.

Salomon, M., Schneider, J., and Lubin, Y. (2005). Maternal investment in a spider with suicidal maternal care, *Stegodyphus lineatus* (Araneae, Eresidae). *Oikos* 109, 614–22.

Sargent, R. C. (1992). Ecology of filial cannibalism in fish: theoretical perspective, In M. A. Elgar and B. J. Crespi eds. *Cannibalism: Ecology and Evolution Among Diverse Taxa*, pp. 38–62. Oxford University Press, Oxford.

Sargent, R. C. and Gross, M. R. (1985). Parental investment decision rules and the Concorde fallacy. *Behavioral Ecology and Sociobiology* 17, 43–5.

Sargent, R. C. and Gross, M. R. (1986). Williams' principle: an explanation of parental care in telest fishes. In T.J Pitcher, ed. *The Behaviour of Teleost Fishes*, pp. 275–93. Croom Helm, London.

Sargent, R. C., Taylor, R. D., and Gross, M. R. (1987). Parental care and the evolution of egg size in fishes. *American Naturalist* 129, 32–46.

Seki, M., Wakano, J. Y., and Ihara, Y. (2007). A theoretical study on the evolution of male parental care and female multiple mating: Effects of female mate choice and male care bias. *Journal of Theoretical Biology* 247, 281–96.

Sheldon, B. C. (2002). Relating paternity to paternal care. *Philosophical Transactions of the Royal Society of London—Series B: Biological Sciences* 357, 341–50.

Shine, R. (1978). Propagule size and parental care- safe harbor hypothesis. *Journal of Theretical Biology* 75, 417–24.

Shine, R. (1989). Alternative models for the evolution of offspring size. *American Naturalist* 134, 311–17.

Smith, D. E. and Fretwell, S. D. (1974). Optimal balance between size and number of offspring. *American Naturalist* 108, 499–506.

Soltis, J. and McElreath, R. (2001). Can females gain extra paternal investment by mating with multiple males? A game theoretic approach. *American Naturalist* 158, 519–29.

Stearns, S. C. (1976). Life-history tactics - Review of ideas. *Quarterly Review of Biology* 51, 3–47.

Stiver, K. and Alonzo, S. H. (2009). Parental and mating effort: Is there necessarily a trade-off? *Ethology* 115, 1101–26.

Summers, K., McKeon, C. S., and Heying, H. (2006). The evolution of parental care and egg size: a comparative analysis in frogs. *Proceedings of the Royal Society of London Series B -Biological Sciences* 273, 687–92.

Tallamy, D. W. (2000). Sexual selection and the evolution of exclusive paternal care in arthropods. *Animal Behaviour* 60, 559–67.

Tinne, S., Rianne, P., and Marcel, E. (2005). Experimental removal of the male parent negatively affects growth and immunocompetence in nestling great tits. *Oecologia* 145, 165–73.

Townshend, D. S. and Wooton, R. J. (1984). Effects of food supply on the reproduction of the convict cichlid, *Cichlasoma nigrofasciatum*. *Animal Behaviour* 33, 494–501.

Trivers, R. L. (1972). Parental investment and sexual selection. In B. Campbell, ed. *Sexual Selection and the Descent of Man*, pp. 136–79. Aldine-Atherton, Chicago.

Tveraa, T., Sæther, B. E., Aanes, R., and Erikstad, K. E. (1998). Regulation of food provisioning in the Antarctic petrel; the importance of parental body condition and chick body mass. *Journal of Animal Ecology* 67, 699–704.

Webb, J. N., Houston, A. I., McNamara, J. M., and Székely, T. (1999). Multiple patterns of parental care. *Animal Behaviour* S8, 983–93.

Webb, J. N., Székely, T., Houston, A. I., and McNamara, J. M. (2002). A theoretical analysis of the energetic costs and consequences of parental care decisions. *Philosophical Transactions of the Royal Society of London—Series B: Biological Sciences* 357, 331–40.

Wesolowski, T. (1994). On the origin of parental care and the early evolution of male and female parental roles in birds. *American Naturalist* 143, 39–58.

West, S., Pen, I., and Griffin, A. S. (2002). Cooperation and competition between relatives. *Science* 296, 72–5.

Westneat, D. F. and Sherman, P. W. (1993). Parentage and the evolution of parental behavior. *Behavioral Ecology* 4, 66–77.

Williams, G. C. (1966a). *Adaptation and Natural Selection: A Critique of Some Current Thought*. Princeton University Press, Princeton.

Williams, G. C. (1966b). Natural selection, the costs of reproduction, and a refinement of Lack's principle. *American Naturalist* 100, 687–90.

Williams, G. C. (1975). *Sex and Evolution*. Princeton University Press, Princeton.

Wilson, E. O. (1975). *Sociobiology: The New Synthesis*. Harvard University Press, Cambridge.

Windt, W. and Curio, E. (1986). Clutch defense in great tit (*Parus major*) pairs and the Concorde fallacy. *Ethology* 72, 236–42.

Winemiller, K. O. and Rose, K. A. (1992). Patterns of life-history diversification in North American fishes-Implications for population regulation. *Canadian Journal of Fisheries and Aquatic Sciences* 49, 2196–218.

Winkler, D. W. (1987). A general model for parental care. *American Naturalist* 130, 526–43.

Wolf, J. B., Moore, A. J., and Brodie, E. D. (1997). The evolution of indicator traits for parental quality: The role of maternal and paternal effects. *American Naturalist* 150, 639–49.

Woodroffe, R., and Vincent, A. (1994). Mothers little helpers: Patterns of male care in mammals. *Trends in Ecology and Evolution* 9, 294–7.

Zink, A. G. (2003). Quantifying the costs and benefits of parental care in female treehoppers. *Behavioral Ecology* 14, 687–93.

CHAPTER 3

Benefits and costs of parental care

Carlos Alonso-Alvarez and Alberto Velando

3.1 Introduction

In order to explain the huge variation in parental behaviour, evolutionary biologists have traditionally used a cost–benefit approach, which enables them to analyse behavioural traits in terms of the positive and negative effects on the transmission of parental genes to the next generation. Empirical evidence supports the presence of a number of different trade-offs between the costs and benefits associated with parental care (Stearns 1992; Harshman and Zera 2007), although the mechanisms they are governed by are still the object of debate. In fact, Clutton-Brock's (1991) seminal book did not address the mechanistic bases of parental care and most work in this field has been conducted over the last 20 years. Research on mechanisms has revealed that to understand parental care behaviour we need to move away from traditional models based exclusively on currencies of energy or time. Despite repeated claims, the integration of proximate mechanisms into ultimate explanations is currently far from successful (e.g. Barnes and Partridge 2003; McNamara and Houston 2009). In this chapter we aim to describe the most relevant advances in this field.

In this chapter, we employ Clutton-Brock's (1991) definitions of the costs and benefits of parental care. Costs imply a reduction in the number of offspring other than those that are currently receiving care (i.e. parental investment, Trivers 1972), whereas benefits are increased fitness in the offspring currently being cared for (see also Chapter 1). Benefits may be derived directly from resources allocated to the offspring (e.g. food, temperature), indirectly from protection against predators, or from the modification of the environment in which the offspring are developing (Chapter 1). We begin this chapter by reviewing the traditional idea of resource allocation trade-offs, and also explore how trade-offs need not be based on resources, and the relevance of cost-free resources. We then analyse in more detail studies of the benefits and costs of parental behaviour and, above all, work that combines mechanistic and functional explanations. Finally, we address the control systems that translate cues perceived by the organism about costs and benefits allowing individuals to take decisions.

3.2 Trade-offs and the nature of the parental resources

The idea of evolutionary trade-offs in the expression of different traits is intrinsically associated with the cost–benefit approach. The more a parent spends on caring for an individual offspring, the less it will spend in caring for other offspring in current or future reproduction attempts. The fitness cost–benefit differential can thus be measured in terms of the number of offspring and allows for comparisons between individuals in the same currency. Selection pressures on parental care may act on both resource acquisition and allocation (Fig. 3.1). We describe below the traditional views of resource-based allocation trade-offs in parental care and also provide some alternative or complementary perspectives.

3.2.1 Limiting resources

The allocation of resources required for parental duties may be constrained directly or indirectly by the negative effects they have on the allocation to other fitness-related traits (Stearns 1992; Roff 2002). Many resources used in parental care are subject to the principle of allocation and are consid-

The Evolution of Parental Care. First Edition. Edited by Nick J. Royle, Per T. Smiseth, and Mathias Kölliker.

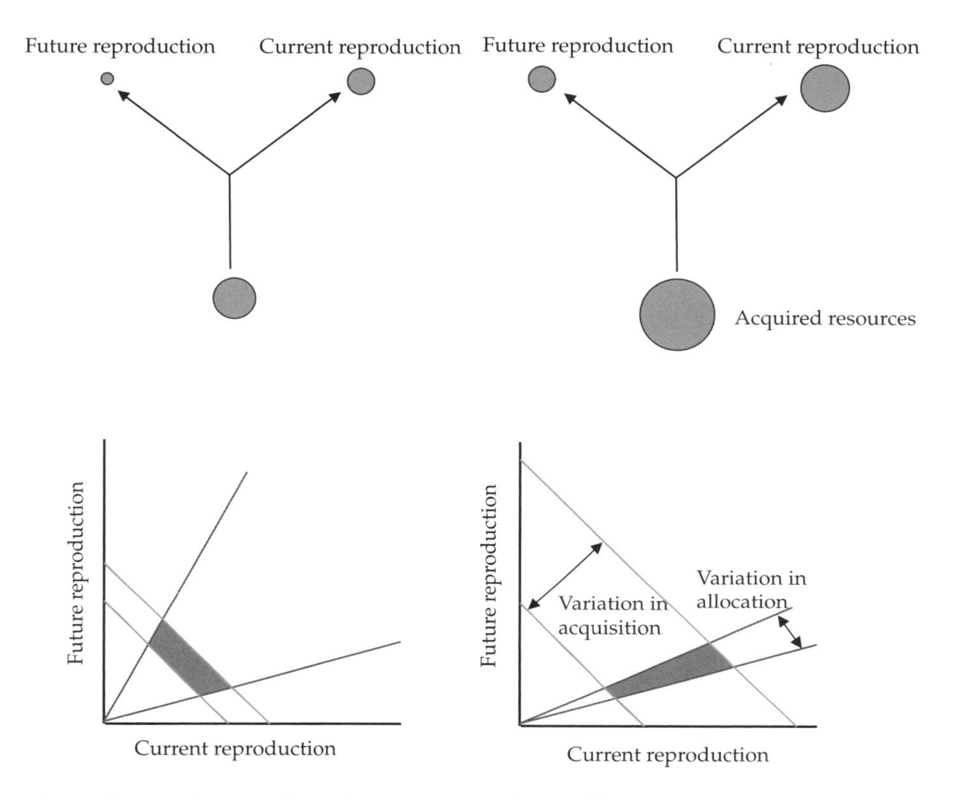

Figure 3.1 A schematic illustration of resource allocation between current reproduction and future reproduction in relation to resource acquisition (bottom panels, from van Noordwijk and de Jong 1986, used with permission from University of Chicago Press). Within individuals, the resources should be invested between competing functions (negative correlation). Among individuals, it will depend on variation in resource acquisition (top panels), and despite negative genetic correlation in resource allocation, the phenotypic correlation between competitive functions may be positive (right bottom panel).

ered limited resources that can only be spent once (Fig. 3.1; van Noordwijk and de Jong 1986). To maximize their fitness, parents should distribute these resources optimally, with two major trade-offs: between current and future offspring, and between the quantity and quality of offspring (Stearns 1992).

Energy and time are the resources most commonly used in theoretical models to exemplify the currency to be traded off between different priorities. Energy and time can also be easily combined into 'energy per unit of time' (Parker and Maynard-Smith 1990; Clutton-Brock 1991). Allocation of more energy/time to parental duties is thought to reduce the energy/time available for self-maintenance and hence for future offspring (Section 3.4). Energy acquisition and allocation are, however, complex traits affected by different factors and are therefore difficult to measure. Animals may provision current offspring using previously accumulated energy stores (capital breeders) or using energy gained contemporaneously (income breeders); nevertheless, weighing up the contribution of both processes in the same individual is difficult (Drent and Daan 1980; Stearns 1992).

Recent evidence suggests that in addition to molecules used to provide energy (i.e. macronutrients: carbohydrates, lipids, and proteins), small amounts of certain non-energetic substances such as essential aminoacids, carotenoids, flavonoids, vitamins, and minerals (i.e. micronutrients) that are not synthesized by the organism, may also need to be traded-off between competing functions. Many of these micronutrients benefit offspring growth and development, as well as parental survival. For instance, in fish and bird species, carotenoids increase fecundity and parental care (e.g. Pike et al.

2007; Tyndale et al. 2008; Morales et al. 2009), but are also required for parental immune or antioxidant defences (e.g. Pérez-Rodríguez 2009; see also Section 3.4.2.2).

3.2.2 Non-linear relationships between resource allocation and fitness

Optimal parental care is dependent on the shape of the functions described by fitness (costs or benefits) plotted against the resources allocated to parental care (Stearns 1992; Roff 2002). The shape of these relationships is commonly taken for granted, despite the fact that empirical evidence is often weaker than is acknowledged. Although simple monotonic relationships between resources allocated to parental care and fitness have been reported, the most common cases probably involve sigmoid-saturating relationships (Clutton-Brock 1991). For instance, the benefits accrued from non-energetic micronutrients show concave trends, with diminishing fitness returns as allocation to care increases. In the Chinook salmon (*Oncorhynchus tshawytscha*), hatching success is positively correlated with the amount of carotenoids deposited by the female in the egg yolk, although survival benefits decrease asymptotically (Tyndale et al. 2008). Similarly, in the diet of the Argentine ant (*Linepithema humile*) queens the size of the pupae (a fitness proxy) positively correlate with the availability of macronutrients (protein), although this effect reaches a plateau when their availability is experimentally increased (Aron et al. 2001).

Research on physiological mechanisms has also revealed the presence of thresholds that, when exceeded, lead to a switch in physiological pathways and, ultimately, control of allocation strategies (see Section 3.5). For example, a minimum level of food needs to be currently available to stimulate reproduction in income breeders (e.g. Schradin et al. 2009), while a critical level of fat stores is necessary to initiate egg-laying in capital breeders (e.g. Alisauskas and Ankney 1994).

3.2.3 Limitations of the resource allocation trade-off perspective

Current reproduction may divert resources away from maintenance (resource allocation, Fig. 3.1), but increasing evidence suggests that reproduction directly alters physiological homeostasis, which in turn causes somatic damages (Fig. 3.2). Furthermore, links between resource acquisition (diet) and metabolism may also explain the trade-off between current and future reproduction (Fig. 3.2). A first problem of resource allocation models is that the resources required for offspring may differ from the resources needed for the somatic maintenance

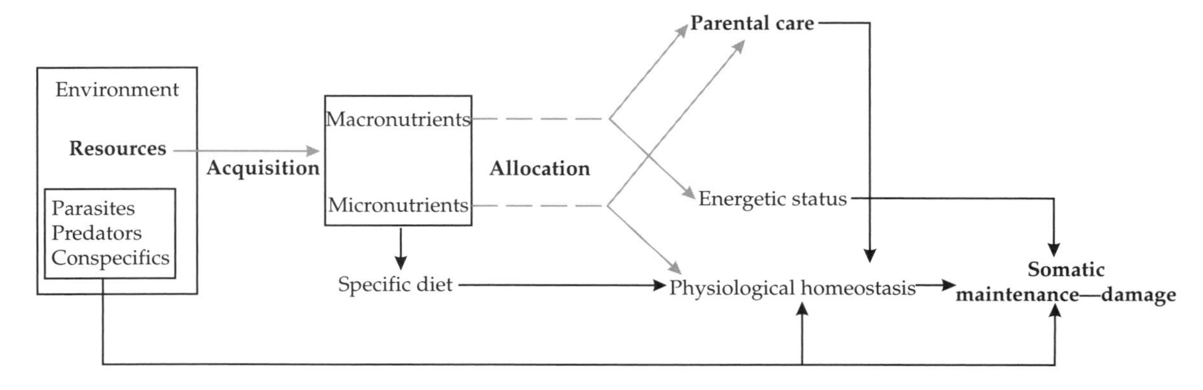

Figure 3.2 Hypothetical mechanisms linking parental care and somatic maintenance (i.e. survival). The trade-off between these life-history traits may rise from a variety of separate mechanisms. Environmental factors, including food availability, physical conditions, risk of predation or parasitism, and competitor abundance, may impose acquisition costs, limiting macro and micronutrients needed for both functions. Traditionally, the models have been centred on energy limitation, that is the allocation of macronutrients between current and future reproduction (Fig. 3.1), but reproduction can also be limited by specific micronutrients. Importantly, parental care may cause direct somatic damage via physiological imbalance (e.g. due to some signalling pathways or stress). Environmental factors governing resource acquisition may also affect physiological homeostasis (e.g. stress). Moreover, the diet composition that maximizes parental care may differ from that maximizing somatic maintenance due to the different damage effects of different subproducts of nutrient metabolism.

of parents. Indeed, parents may provide offspring with a different kind of food to that they use for their own maintenance (e.g. Cherel et al. 2005). Differences in the currency of resources that are involved in many trade-offs during parental care is known as the 'common currency problem' (Houston and McNamara 1999). Theoretical biologists approached this problem by modelling the effect of predators or parasites on foraging: animals should weigh up the benefits of these simultaneous goals, that is, energy collected vs. foraging time and mortality risk (e.g. McNamara and Houston 1986). This problem can be tackled by introducing state variables that characterize the current physiological state (e.g. hunger, size, damage, territory size, etc; Clark and Mangel 2000). The state variable may be, for example, the level of damage, which must not exceed a certain threshold (see above), while the variable to be maximized is fitness.

Aside from limiting resources, a number of mechanisms underlying parental care have been discovered when exploring the physiological complexities of organisms. This is the case for oxidative stress, which is an imbalance between the production of reactive oxygen species (ROS) during cell metabolism and the state of antioxidant and repair machineries, and leads to oxidative damage (Kirkwood and Austad 2000). Parental care may increase cell metabolism and hence ROS production and oxidative damage over time, thereby accelerating senescence (Kirkwood and Austad 2000; Metcalfe and Alonso-Alvarez 2010). Limiting substances such as antioxidants or energy required for repair mechanisms may be subject to the principle of allocation constraining parental care (Edward and Chapman 2011). Nevertheless, ROS are highly reactive (Finkel and Holbrook 2000) and above a certain threshold antioxidant/repair systems are inefficient and soma damage may be unavoidable (Fig. 3.2). Oxidative damage (as an internal state leading to somatic deterioration) may explain the link between uncoupled life-history traits, that is, between activities separated in time and therefore not subject to a direct trade-off. Costs and constraints of nutrient-sensing signalling systems may also be independent of resources, although current evidence is inconclusive (see Section 3.5).

Environmental challenges may also imply trade-offs that are independent of limiting resources (Fig. 3.2). Examples include trade-offs derived from risky, damaging, or stressful conditions during care (Clutton-Brock 1991; Harshman and Zera 2007), some of which are obviously 'all-or-nothing' trade-offs (e.g. predation risk). Mechanistic approaches have also revealed that resource acquisition has intrinsic trade-offs in diet components. For instance, a recent study by Lee et al. (2008) on fruit flies shows that the protein:carbohydrate ratio that maximizes egg production differs from the ratio that maximizes lifespan. Authors suggested that high ratios favour reproduction but impair survival since the organism suffers damage caused by subproducts of protein metabolism such as reactive oxygen species and nitrogenous breakdown substances (Lee et al. 2008). Since no diet maximizes both functions, a trade-off between reproduction and maintenance may be the inevitable outcome of resource acquisition, rather than the effect of energy allocation as is proposed by traditional models.

3.2.4 Cost-free resources and resources not involved in care

The distinction between costly and cost-free resources is critical to the understanding of the evolution of parental care since costly resources reduce parents' ability to produce other offspring (Trivers 1972). In contrast, the production of a particular form of cost-free care probably depends on its context-dependent benefits for offspring development. For example, female birds deposit hormones in the egg yolk, but the cost for mothers is unknown and perhaps negligible (Gil 2008). Nevertheless, hormone deposition may have environmental or sex-specific effects on offspring fitness, which may explain differences in hormone levels among the eggs in a single clutch (Groothuis et al. 2005; Gil 2008).

It should be noted that the allocation of substances to offspring may influence offspring fitness, although this act should not be regarded as parental care if it is a byproduct of the parental environment (Chapter 1). For example, the mothers of many species passively transfer pollutants into eggs, which may in fact be beneficial since they prepare the offspring phenotype for a polluted environment (Ho and Burggren 2010). In this case, selection has probably acted on the developmental pathways of offspring rather than on parents' behaviour.

3.3 Benefits of parental care: mechanistic basis

The evolutionary benefit of parental resource expenditure is to increase offspring survival during development (short-term benefits) or to improve offspring survival and fecundity in the long term (delayed benefits). Here we address both cases and also examine how parents seem to be able to actively prepare offspring phenotypes for future environmental challenges.

3.3.1 Short-term benefits of parental care

In a variety of species, parents improve offspring short-term survival by actively protecting descendants from harsh environments (predators, conspecifics, infections) or by allocating (or regulating) limiting resources that favour their development (Chapters 1 and 2). Many short-term benefits may also lead to longer-term benefits. Here we briefly describe some representative examples of those benefits clearly obtained in the short term.

In terms of protection, parents may prepare and maintain suitable nesting sites or directly defend offspring from predators, brood parasites, or conspecifics. Orange-crowned warblers (*Vermivora celata*) construct their nests close to the ground concealed in the vegetation, but build them higher in schrubs when the perceived risk of predation is high (Peluc et al. 2008). In some fish, offspring are guarded and protected in one of the parent's mouths (i.e. 'mouth brooding'; e.g. Balshine-Earn and Earn 1998). 'Gastric brooding' has also been described in some frogs (Tyler et al. 1983) and an asteroid (Komatsu et al. 2006). Aggressive offspring protection is found in many taxa (Chapter 1), while examples of birds being able to discriminate parasitic eggs by visual cues and reject them, thereby preventing offspring mortality, are also well documented (Chapter 13).

In terms of limiting resources, parents may improve offspring viability by regulating the availability of thermal energy, water, oxygen, and energetic and non-energetic nutrients. Social insects regulate nest temperature by metabolic heat production, fanning, and water evaporation (reviewed in Jones and Oldroyd 2007); likewise, clutch thermoregulation by parents has been commonly reported in vertebrates such as reptiles and birds (e.g. Deeming 2004). In tree frogs, the capacity of males to find enough large water pools for egg deposition favours tadpole development (e.g. Brown et al. 2010), whereas the parents of some crab and fish species enhance survival by oxygenating eggs via fanning (e.g. Baeza and Fernandez 2002; Green and McCormick 2005). Nutrients are supplied in multiple forms (see Chapter 1). For example, mothers of social insects and spiders, as well as tree frogs and certain fish species produce non-developing eggs or egg-like structures that are used to feed offspring (i.e. 'trophic eggs'; reviewed in Perry and Roitberg 2006). The nourishment of offspring by the maternal body ('matriphagy') has been described in arachnids and some insects (e.g. Suzuki et al. 2005; Salomon et al. 2011), while foetuses of viviparous caecilian amphibians are known to scrape lipid-rich secretions and cellular materials from their hypertrophied maternal oviducts (e.g. Wake and Dickie 1998). Finally, parents also provide offspring with non-energetic compounds. A good example is the transfer of carotenoids and vitamins to eggs in many vertebrates, which protect embryos from oxidative stress induced by their high anabolic activity (e.g. Surai 2002; Tyndale et al. 2008). Males of some fish species have specialized anal glands or integuments that secrete bactericide (lysozyme-like) compounds to the eggs (Knouft et al. 2003; Giacomello et al. 2006). Males may also transfer substances via their sperm. The males of the Australian field cricket (*Teleogryllus oceanicus*) produce sperm with certain proteins that can be absorbed by eggs and ultimately improve the embryo's development and the offspring chances of survival, probably via a form of signalling activity (Simmons 2011).

3.3.2 Long-term benefits of parental care

Parental care may have a strong influence throughout an offspring's lifespan, but some benefits may be delayed and become evident only after care has ceased. There are many examples whereby individuals born in good conditions accrue fitness advantages later in life as a consequence ('silver-spoon' effect; Grafen 1988). Malnutrition may permanently alter morphology, physiology, and/or metabolism during adulthood and cause long-term effects on fitness (reviewed in Monaghan 2008). For example, in zebra finches (*Taeniopygia guttata*) maternal micronutrients in the egg (carotenoids) influence sexual ornamentation displayed by offspring during adulthood (McGraw et al. 2005), whereas a lack of macro- and micronutrients (proteins and antioxidants, respectively) as a nestling reduces reproductive capacities in adulthood (Blount et al. 2006).

Parents may influence offspring fitness by affecting their brain development, thereby positively helping perceptual, cognitive, and learning capabilities in adulthood (e.g. Meaney 2001; Law et al. 2009). In many passerine species, parents provide spiders to chicks despite their relatively low energy content. Spiders contain high amounts of taurine, a free sulphur amino acid that is required for brain development (Arnold et al. 2007); blue tit (*Cyanistes caeruleus*) nestlings that were experimentally supplied with taurine later exhibited greater abilities in spatial learning than control birds (Arnold et al. 2007). In species with prolonged parental care, offspring may devote more time to learning how to forage and practising social skills, and to being taught by their parents (Hoppitt et al. 2008). Early learning can lead to more effective foraging, antipredator behaviour, defence against brood parasites, and mate choice during adulthood (Curio 1993; Brown and Laland 2001; Davies and Welbergen 2009) so can increase fitness (Mateo and Holmes 1997).

3.3.3 Parental care and offspring phenotypic adjustment

Genotypes can produce different phenotypes (reaction norms) in response to distinct environmental conditions (i.e. 'phenotypic plasticity'; Pigliucci 2001). A growing literature exists on the effects of the parental phenotype on the phenotype of the offspring (known as parental effects; see Mousseau and Fox 1998; Chapter 14). Here, we concentrate on those parental effects (usually maternal) that have a positive causal influence on the offspring via phenotypic adjustment to the environment they are likely to encounter.

In fluctuating environments with short-term predictability, parents can program offspring development to cope with particular situations (Uller 2008). Parents may produce different offspring phenotypes by affecting developmental pathways or by providing morph-specific resources (reviewed in Badyaev 2009). Parental influence can have long-lasting consequences due to phenotypic reorganization or epigenetic changes leading to differential gene expression (West-Eberhard 2003; Ho and Burggren 2010). Early programming is the consequence of both parental behaviour and plasticity in development pathways. Development pathways, especially in adverse environments, may explain how early conditions can affect offspring phenotypes without active parental effects (Monaghan 2008). We here address the effects on parents that may be subject to selection.

The phenotypic adjustment of progeny by parents is based on two important assumptions: 1) that environmental cues experienced by parents predict the environmental conditions that their offspring will encounter and 2) that phenotypic plasticity in offspring development is sensitive to signals produced by parents (Mousseau and Fox 1998). Exposure to signals during embryonic development may thus be particularly likely to cause adaptive adjustments in the offspring phenotype (i.e. without genetic changes; 'phenotypic accommodation') because a disproportionately large part of phenotypic organization occurs during such a relatively brief stage in the offspring's life history (West-Eberhard 2003; Chapter 14). Parental effects may also prepare offspring for the level of care they will receive. In a cross-fostering experiment, Hinde et al. (2010) found that foster canary (*Serinus canaria*) chicks grow better if they beg at a level similar to that of original chicks. These results suggest that mothers increase offspring fitness by matching offspring demands to parental capacity. Below, we summarize some relevant studies on how parents may enable offspring phenotypes to deal with pathogens, predators, and other adverse conditions.

3.3.3.1 *Pathogens*

Mothers can transfer information about the pathogens that offspring will encounter ('transgenerational immune priming'; Grindstaff et al. 2003): for example, mammals transfer antibodies to descendants via placenta, colostrum or breast milk, whereas birds use the egg yolk (Hanson et al. 1998; Coste et al. 2000; Boulinier and Staszewski 2008). Recently it has been shown that parents of some invertebrates (mostly insects) may also transfer some specific immune factors to their offspring (Freitak et al. 2009 and references therein). A novel study also challenges the long-held idea that fathers do not transmit immune information to their offspring: in the red flour beetle (*Tribolium castaneum*) offspring sired by males exposed to heat-killed bacteria were more resistant to a pathogen infection than offspring from non-exposed males (Roth et al. 2010). Seminal substances, probably proteins, genomic imprinting, and/or micro RNAs in the sperm could explain these findings (Roth et al. 2010).

3.3.3.2 *Predators*

Many animals learn antipredatory behaviour from conspecifics (e.g. Curio 1993; Mateo and Holmes 1997), although it is still a subject of controversy whether or not parents actually teach their offspring how to cope with predators (see Hoppitt et al. 2008). Parents may also transfer such information via their eggs (Storm and Lima 2010). In three-spined sticklebacks (*Gasterosteus aculeatus*), maternal exposure to a dummy or a natural predator prior to egg-laying has an important influence on offspring antipredator behaviour such that the offspring of predator-exposed mothers exhibit closer shoaling behaviour (Giesing et al. 2011). These effects would seem to be mediated by the maternal transfer of high levels of hormones with organizational effects such as glucocorticoids (Giesing et al. 2011).

3.3.3.3 *Other adverse environmental conditions*

In many invertebrates, females favour diapause in their offspring (e.g. via manipulation of lipid content of eggs) as a response to a short photoperiod, low temperatures, or a scarcity of potential hosts, thereby increasing offspring survival probability (Mousseau and Fox 1998; Huestis and Marshall 2006; Scharf et al. 2010). In the bryozoan *Bugula neritina*, females living in crowded or polluted environments produce larvae with a higher dispersal potential (Marshall 2008). In crowded environments parents of some avian species may produce competitive and/or aggressive offspring by depositing testosterone in their eggs (Groothuis et al. 2005; Gil 2008). Parents may also prepare offspring for future harsh environmental conditions by altering their epigenome. In rats (*Rattus norvegicus*) maternal care (pup licking and grooming) influences the stress tolerance of their pups by increasing gene expression in the promoter region of the glucocortocoid-receptor gene (Weaver et al. 2004; also Section 3.5.2 and Chapter 17). These epigenomic changes persist into adulthood and across generations. Offspring unattended by mothers are more likely to keep still and maintain a low profile, responding quickly to stress, which may be advantageous when food is scarce and danger is high, but is less beneficial when food is abundant.

3.4 The costs of parental care

Explanations for the evolution of parental care are usually based on variation in the cost of behaviour (Clutton-Brock 1991). There is some confusion in the literature regarding the use of the term cost. For example, it is commonly stated that parents transfer the cost of parental care to current offspring; yet cost can be only measured in terms of the offspring sacrificed due to the current care. When parents desert, cannibalize, or decrease provisioning to the current brood, the current offspring pay a cost, while parents only lose the potential future benefits (Section 3.3). In some cases parental care imposes a cost in terms of reduced numbers of brood-mates; this cost rises as clutch size increases and is known as 'depreciable care' (Clutton-Brock 1991). An example is egg/young provisioning, which constrains clutch size in a variety of species (Stearns 1992). On the other hand, other forms of parental care—for example, antipredator behaviour—can benefit all offspring in a brood ('non-depreciable care', Clutton-Brock 1991) and the costs only depreciate future reproduction ('residual reproductive value').

From an empirical point of view, four approaches can be used to assess the cost of parental care: phenotypic correlations between traits (Fig. 3.1), phenotypic manipulations, genetic correlations, and selection experiments (Clutton-Brock 1991; Reznick 1992). Over last two decades a huge body of literature has been produced on the first two methods. By contrast, work on the latter two has been restricted to analyses of life-history traits such as the negative genetic correlation between growth and fecundity (e.g. Roff 2000), and very few studies have been conducted on the mechanisms underpinning these traits (but see Kim et al. 2010). Consequently, in the following sections we review the costs of parental care addressed using phenotypic correlations and manipulations as approaches.

Due to their diversity, it is difficult to classify the costs of parental care. Here we broadly divide these costs into non-physiological and physiological costs. The former are mostly related to resource acquisition in the environment, which arise from exposure to predators, rivals, conspecific or interspecific parasites, and from a reduced amount of time for future mating or reproduction. On the other hand, physiological costs are mostly linked to resource allocation and arise from trade-offs between parental effort and somatic or mating effort, whether or not they are based on limiting resources (Fig. 3.2). In terms of fitness components, parental care ultimately entails reduced survival, fewer mating opportunities, and poorer capacity to invest in future offspring for parents. All the above-mentioned mechanistic costs are closely interrelated. Reductions in body energy stores or key micronutrients impair immune-capacity, favour stress, and may lead to a greater susceptibility to infection (e.g. Nordling et al. 1998), which in turn may reduce the capacity to escape from predators or to avoid reproductive (brood) parasites, thereby increasing the risk of body injuries. This implies that selection may act simultaneously, whether directly or indirectly, across a variety of different mechanistic costs (Moore and Hopkins 2009).

3.4.1 Non-physiological costs

Non-physiological costs due to parental duties may include high infection and predation risk, injuries, reproductive parasitism and consumption of time potentially devoted to other activities.

In the first case, experimental evidence suggests a positive correlation between infection risk and parental effort (e.g. Knowles et al. 2009), although the causal relationship between infection intensity due to parental effort and future reproduction or mortality has still to be conclusively demonstrated. The best evidence probably comes from birds suffering haemoparasitic infections (parasitaemia): wild female collared flycatchers (*Ficedula albicollis*) rearing enlarged broods had higher levels of blood parasites than control birds; these levels were in turn correlated with overwinter survival (Nordling et al. 1998). Nevertheless, the fitness of the experimental females was not studied (i.e. data on survival was obtained from other sample of birds). In wild great tits (*Parus major*), females with experimentally enlarged broods had increased parasitaemia and poorer overwinter survival rates, although the parasitaemia and survival were not directly correlated (Stjernman et al. 2004). However, these correlations do not necessarily imply causation. High infection may be due to increase exposure to pathogens *per se* or instead indirectly due to a reduced physiological state (e.g. loss of energetic reserves and/or immunocompetence). Thus, for instance, in a study of common eiders (*Somateria mollisima*) female survival was negatively associated with clutch size, but only during an avian cholera epizootic outbreak, thereby suggesting that parental effort reduced resistance to infection and consequently negatively affected fitness (Descamps et al. 2009). In this line, we did not find any study clearly demonstrating that increased infection due to parental care was fully independent from an indirect impairment of parental physiological state.

Parental care may increase the risk of predation, and predation obviously reduces fitness. Examples of an increase of predation risk due to parental activities are particularly common among invertebrates, where egg-brooding species suffer higher predation than non-carrying individuals (e.g. Reguera and Gomendio 1999; Li and Jackson 2003), probably due to their conspicuousness, lower escape capacity (e.g. Shaffer and Formanowicz 1996), and/or higher energetic value for predators. In the pipefish (*Nerophis ophidion*), males carrying their brood in a pouch suffer higher predation rates than females, a finding that seems to be related to their greater conspicuousness (Svensson 1988). Clutch or litter burdens also impair escape capacity, which has been well demonstrated in vertebrates (e.g. fish: Ghalambor et al. 2004; reptiles: Cox and

Calsbeek 2010; birds: Veasey et al. 2001; mammals: Schradin and Anzenberger 2001). In lizards and birds this effect seems to be mediated by muscle condition (e.g. Veasey et al. 2001; Olsson et al. 2000). In birds, fat reserves required for egg production may impair take-off and flight capacity, increasing predation risk (Witter and Cuthill 1993), although to the best of our knowledge the link between escape capacity and mortality has only been demonstrated to date in reptiles (Miles et al. 2000; Cox and Calsbeek 2010).

Parents may also suffer injuries while defending their reproductive investment from hetero- or conspecifics. For instance, burying beetles (*Nicrophorus pustulatus*) suffered more injuries when protecting their young without help from their mate (Trumbo 2007). Parental care activity may lead to wear and tear of integuments too: for example, collared flycatchers rearing experimentally enlarged broods suffered greater wear on their primary feathers and the intensity of this feather damage was positively correlated to post-breeding mortality (Mërila and Hemborg 2000).

Reproductive conspecific or interspecific parasites may also impair parents' survival or future reproduction (Chapter 13). In the former case, examples can again be found in birds (reviewed in Lyon and Eadie 2008). However, experiments have so far found little evidence of any long-term cost of conspecific parasitism, a finding that is not particularly surprising since all of these studies used precocial species in which the cost of rearing additional offspring tends to be lower (Lyon and Eadie 2008). In the latter case (interspecific brood parasitism) Hoover and Reetz (2006) reported reduced returning rates in prothonotary warblers (*Protonotaria citrea*) parasitized by brown-headed cowbirds (*Molothrus ater*). However, in certain species—including many insects and fishes that do not expend energy feeding their offspring—hosts may not necessarily suffer a cost when receiving eggs from conspecifics or heterospecifics (reviewed in Tallamy 2005; see also Chapter 13).

Finally, parental care consumes time that could be devoted to remating, conducting new reproductive events, and/or self-maintenance. The trade-off between parental care and new mating opportunities has generated a prolific literature focused on the evolution of sexual conflict and biparental care (see Chapters, 6, 9, and 11). In the case of the time dedicated to produce more offspring, it has been experimentally demonstrated in captive lace bugs (*Gargaphia solani*) that the time invested in protecting eggs is traded against fecundity in subsequent clutches (Tallamy and Denno 1982). In the case of time invested in self-maintenance, water striders (*Aquarius remigis*) that bred only once a year (univoltine life cycle) had time to recover lipid stores and survived the winter better than breeders that had two reproductive attempts per year (bivoltine cycle); the latter even had lower lifetime fecundity and longevity (Blanckenhorn 1994). Blue tits that produced a second clutch when the first was experimentally removed delayed their moult and produced a plumage with poor insulation capacity, and subsequently had lower overwinter mortality and less reproductive success the next season (Nilsson and Svensson 1996). We should however note that in the last two cases the evidence is merely correlational and could be confounded by energetic constraints.

3.4.2 Physiological costs

3.4.2.1 Energetic costs

Physiological costs have been primarily studied in terms of a loss of limiting resources such as energy or nutrients. Using a variety of different techniques the allocation of resources can be estimated by measuring energy expenditure (oxygen consumption, metabolic rates, doubly labelled water, etc; Speakman 2001). The increase of energy expenditure during parental care is particularly relevant in income breeders. However, most organisms also stockpile energy in their bodies and changes in total body mass or growth rates (in indeterminate growers) may also be used as a way of estimating energy loss (Speakman 2001). A third option is to assess the state of body energy stores that accumulate macronutrients (usually fat and muscles).

To our knowledge, in vertebrates an increase in energy expenditure associated with an increased intensity of particular parental care behaviours has only ever been experimentally demonstrated in birds and mammals. In mammals, studies have been conducted above all on small female rodents in captivity or in semi-captive conditions during gestation and lactation (Gittleman and Thompson

1988; Speakman 2008). However, despite the variety of studies, a link between the energy expenditure in current care and parents' survival and/or future reproductive success is only supported by two experiments on birds (Table 3.1). By contrast, experiments reporting body mass loss or growth delay as a cost of parental care have been performed for fish, reptiles, and birds, and have succeeded in linking such costs to fitness (particularly in birds; Table 3.1). Finally, some reptile and bird studies also have experimentally demonstrated changes in specific body energy stores, although only two have ever reported a link with fitness (Table 3.1), probably due to technical limitations in the assessment of body composition, which usually requires sacrifice of the individuals involved (Speakman 2001).

3.4.2.2 *Micronutrients*

Here we only describe those micronutients—calcium, carotenoids, and methionine—for which a link with parental behaviour has been established. In the case of calcium, allocation to the eggshell in oviparous species or milk and foetal bones in mammals has been particularly well studied. Calcium levels drop during gestation and lactation in mammals (Speakman 2008 and references therein), although we have found no report of a decline in calcium levels due to parental care in other taxa. Carotenoids are used in physiological functions (e.g. as antioxidants, detoxificants and immunoenhancers), as well as pigments of integuments (e.g. Perez-Rodriguez 2009). The egg yolk of fish, reptiles, and birds contains large amounts of carotenoids that protect the embryo from the effects of oxidative stress that result from growth (e.g. Surai 2002). Nevertheless, evidence that increased parental effort depletes maternal carotenoid levels has only been suggested from a negative correlation between clutch size and circulating carotenoid values at the end of laying in red-legged partridges (*Alectoris rufa*; Bortolotti et al. 2003). Finally, methionine stimulates fecundity (egg production) in female fruit flies, but only in a specific ratio with other essential amino acids (Grandison et al. 2009). When such a proportion is not met, methionine can become pro-oxidant, reducing parental survival and reproductive success (Grandison et al. 2009). This exemplifies the concept of nutritional geometry, whereby certain nutrients must be present in particular proportions to favour reproduction (see also Section 3.2.4). The lack of such adjustment therefore implies a fitness cost to females (Table 3.1).

3.4.2.3 *Physiological stress*

Parental care may lead to an exhaustion of energy stores, which in turn leads to physiological stress (Section 3.5.2). Physiological stress may also be triggered by other environmental stressors (Wingfield and Sapolsky 2003). Physiological stress ultimately provokes damage in the parents. This damage has been estimated by assessing levels of 'heat shock proteins' (HSPs), molecules that repair protein damage induced by a variety of stressors (Sorensen et al. 2003). High HSP values have been related to decreased fecundity in fruit flies (reviewed in Sorensen et al. 2003). As far as we know, only one study has related parental care to HSPs: in blue tits, parents whose brood was experimentally enlarged had increased blood HSP levels (Merino et al. 2006). In vertebrates, glucocorticoids levels in the blood (acute or baseline; Section 3.5.2) are the most analysed proxy of physiological stress, high values revealing high stress levels. Experiments in birds and fish support the idea that an increase in glucocorticoid levels is a consequence of parental effort (e.g. Magee et al. 2006; Golet et al. 2004). Nonetheless, recent reviews have questioned the link between this effect and fitness (e.g. Breuner et al. 2008; Bonier et al. 2009) and in fact we have only found one study that supports this assertion (Table 3.1).

3.4.2.4 *Oxidative stress*

The cost of parental care in terms of oxidative stress is supported by correlative evidence, mostly in mammals (e.g. Upreti and Misro 2002; Myatt and Cui 2004), but also by some experiments on birds. In the latter case, zebra finches whose parental effort was increased by brood enlargement had less resistance to ROS (Section 3.2.3) following reproduction than controls (Alonso-Alvarez et al. 2004; Wiersma et al. 2004). However, although medical studies have suggested that oxidative stress generated during gestation in mammals compromises the life of the mother during birth (e.g. Myatt and Cui 2004), a link between

Table 3.1 Support for a link between parental care, physiological costs, and fitness.

Mechanistic cost	Order	Species	Parental care	Exp/Corr	Capt/Wild	PC→PHC	PC→FC	PC→FC	PHC→FC	PHC→FC	References
							Reprod	Surv	Reprod	Surv	
Energy expenditure	*Aves*	*Falco tinnunculus*	Brood care and feeding	Exp	Wild	m (f: UT)	n.s.(b)	*	UT	UT	Dijkstra et al. 1990
		Falco tinnunculus	Brood care and feeding	Exp/Corr	Wild	b	UT	n.s.(b)	UT	* (b)	Deerenberg et al. 1995
Body mass loss	*Actinopterygii*	*Ambloplites rupestris*	Brood guarding	Exp/Corr	Wild	m (f: NA)	UT	*	UT	*	Sabat 1994
	Reptilia	*Vipera verus*	Egg production and gestation	Corr	Wild	f (m: NA)	NA	*	UT	*	Madsen and Shine 1993
		Urosaurus ornatus	Egg production	Exp/Corr	Wild	f (m: NA)	*	*	*	UT	Landwer 1994‡
		Anolis sagrei	Egg production	Exp	Wild	f (m: NA)	UT	*	UT	UT	Cox and Calsbeek 2010‡
			Egg production	Exp	Wild	f (m: NA)	UT	*	UT	UT	Cox et al. 2010‡
	Aves	*Stercorarius skua*	Egg production	Exp	Wild	f (m: UT)	*	n.s.	UT	UT	Kalmbach et al. 2004
		Somateria mollisima	Incubation	Exp	Wild	f (m: NA)	*	n.s.	UT	UT	Hanssen et al. 2005
		Rissa tridactyla	Incubation, brood care, and feeding	Exp	Wild	b	n.s.(b)	*	UT	UT	Golet et al. 1998
		Rissa tridactyla	Incubation, brood care, and feeding	Exp/Corr	Wild	b	*	*	*	n.s.(b)	Golet et al. 2004
		Rissa tridactyla	Brood care and feeding	Exp	Wild	f (m: n.s.)	UT	*(f, m: n.s.)	UT	UT	Jacobsen et al. 1995
		Branta c. canadensis	Brood care and feeding	Corr	Wild	f (m: n.s.)	*	n.s.(b)	UT	UT	Lessells 1986
		Larus glaucescens	Brood care and feeding	Exp/Corr	Wild	b	n.s.(b)	*	UT	n.s.(b)	Reid 1987
		Falco tinnunculus	Brood care and feeding	Exp	Wild	f (m: n.s.)	n.s.	*	UT	UT	Dijkstra et al. 1990
		Streptopelia risoria	Brood care and feeding	Exp	Capt	b	*	UT	UT	UT	ten Cate et al. 1993
		Ficedula albicollis	Brood care and feeding	Exp	Wild	b	* (f, mUT)	n.s.(f, mUT)	UT	UT	Török et al. 2004
		Cyanistes caeruleus	Brood care and feeding	Exp	Wild	f (m: n.s.)	*	*(f, m n.s.)	UT	*(f, m n.s.)	Nur 1984, Nur 1988†

		Cyanistes caeruleus	Brood care and feeding	Exp	Wild	f (m: UT)	UT	*	UT	n.s.	Stjernman et al. 2004
		Parus major	Brood care and feeding	Exp/Corr	Wild	b	*	UT	UT	*(f, mUT)	Tinbergen and Verhulst 2000
Loss of energy body stores	*Reptilia*	*Eulamprus tympanum*	Overall reproduction	Corr	Capt/Wild	f (m: NA)	UT	n.s.	*	UT	Doughty and Shine 1998
	Aves	*Stercorarius skua*	Egg production	Exp	Wild	f (m: NA)	*	n.s.	UT	UT	Kalmbach et al. 2004
Micronutrient adjustment	*Insecta*	*Drosophila melanogaster*	Egg production	Exp	Capt	f (m: NA)	UT	UT	UT	*	Grandison et al. 2009
Physiological stress	*Aves*	*Rissa tridactyla*	Incubation, brood care, and feeding	Exp	Wild	b	*	*	UT	n.s.(b)	Golet et al. 2004††
Oxidative stress	*Insecta*	*Drosophila melanogaster*	Egg production	Exp	Capt	f (m: NA)	UT	*	NA	*	Salmon et al. 2001?
		Drosophila melanogaster	Egg production	Exp	Capt	f (m: NA)	UT	*	NA	*	Wang et al. 2001?
	Aves	*Taniopygia guttata*	Overall reproduction	Corr	Capt	b	UT	UT	UT	*	Alonso-Alvarez et al. 2006#?
Immunosuppression	*Reptilia*	*Anolis sagrei*	Egg production	Exp	Wild	f (m: NA)	UT	*	UT	UT	Cox et al. 2010
	Aves	*Somateria mollissima*	Incubation	Exp	Wild	f (m: NA)	*	n.s.	UT	UT	Hanssen et al. 2005
		Tachycineta bicolor	Brood care and feeding	Exp/Corr	Wild	f (m: UT)	UT	UT	UT	*	Ardia et al. 2003

To create this table a systematic simultaneous search (Web of Science, Thompson Reuters) of the term 'fitness' plus the truncated term 'cost' plus parental care (in any of its potential terms), and plus each potential mechanistic cost (e.g. energetic and non-energetic costs, predation, etc.) was performed, using different truncated combinations. This search produced about 500 references.

Abbreviations and notes:

PC→PHC: Parental care inducing the physiological cost
PC→FC: Parental care inducing fitness costs (reproductive cost or reduced survival)
PHC→FC: link between physiological and fitness costs
m: male; f: female; b: both sexes.
NA: not applies
UT: untested
‡ Body growth as measure of body mass variability
†Both studies report findings on the same dataset
††The survival cost of the manipulation was not detected in the subsample where stress was tested
Resistance to oxidative stress was negatively associated with the number of previous breeding events, and predicted subsequent short-term longevity
? Design limitations (see Section 3.3.2)
$Individuals whose parental effort was manipulated were not the same that those tested for fitness effects

reproductive oxidative stress and fitness is only supported by a limited number of experiments and correlations (Table 3.1). When exposed to a pro-oxidant agent (paraquat), the female fruit flies that were experimentally stimulated to produce eggs died faster than non-breeders (Table 3.1). However, the authors of this experiment did not test whether oxidative stress was experimentally increased. Nonetheless, it has been shown that antioxidants inhibit paraquat-induced mortality in *Drosophila* (Bonilla et al. 2006). In zebra finches a negative correlation between the number of breeding events and resistance to oxidative stress has been reported by a study (Table 3.1) that also found that resistance to oxidative stress was negatively related to short-term mortality. However, in the latter study parental effort could also have included mating effort. Finally, it has been found that males of two reef-fish species (*Apogon fragilis* and *Apogon leptacanthus*) that protect their broods in their mouths suffer from hypoxia (Östlund-Nilsson and Nilsson 2004). Hypoxia could be an alternative cost of parental care but is probably also associated with oxidative stress (Metcalfe and Alonso-Alvarez 2010).

3.4.2.5 *Immunosuppression*

Parental care may also lead to immunosuppression, mostly as an indirect consequence of other physiological costs. In vertebrates, immunity is reduced in response to high energy expenditure, loss of body energy stores, micronutrient depletion, glucocorticoid stress response, and oxidative stress (e.g. French et al. 2007; Bourgeon et al. 2009; Perez-Rodriguez 2009). Examples from other taxa are scarce (e.g. insects: Fedorka et al 2004). The impact of parental effort on immunocompetence is well supported by experiments on birds, in which incubation and brood-rearing efforts were manipulated and the capacity to establish innate or acquired immune responses were impaired accordingly (Table 3.1). Immunosuppression is also associated with implantation and gestation in mammals (e.g. Medina et al. 1993). Immunosuppression protects the embryo from maternal immune defences, although the consequences for maternal fitness are still unclear (Speakman 2008). In fact, the link between immunosuppression and fitness is only supported by a handful of studies (Table 3.1).

3.4.2.6 *Regulatory systems*

Endogenous (e.g. neuroendocrine) control systems involved in parental decisions (Section 3.5) may *per se* create constraints and costs (Lessells 2008). For example, signal molecules or signal-producing machinery may be expensive to produce or maintain. However, selection may favour simple costless parental rules that, albeit not optimal in all situations, perform well on average (McNamara and Houston 2009). Some molecular signals involved in reproductive activities may have a negative effect on soma maintenance (Leroi 2001; Edward and Chapman 2011). A number of studies on fruit flies and the nematode *Caenorhabitis elegans* suggest that the negative effect of reproduction on longevity arises from a signalling pathway (involving a steroid or insulin-like hormone) rather than from direct resource competition (reviewed in Edward and Chapman 2011). Nevertheless, molecular signals may also activate the physiological mechanisms needed for reproduction that in turn generate damage (Barnes and Partridge 2003). Hence, it is still to be established whether the neuroendocrine control system mediates or creates costs in parental care (Lessells 2008).

3.5 Costs and benefits in the balance

A given level of care reflects the balance between its costs and benefits in a given environment. To reach this balance, organisms have evolved control systems that translate the environmental cues perceived by the sense organs into molecular (neuroendocrine) signals that influence physiology, gene expression, and behaviour (Harshman and Zera 2007; Hau and Wingfield 2011).

Parents must balance costs and benefits by taking decisions that maximize fitness, although control mechanisms that integrate environmental cues may produce sub-optimal reaction norms in some circumstances (McNamara and Houston 2009). Indeed, reaction norms are subject to constraints and costs of control regulating systems (Section 3.4.2). This may explain the variability in evolutionary pathways of parental care between taxa. Here, we review how parental decisions are regulated by physiological pathways and promote either parental effort or favour self-maintenance.

3.5.1 Molecular signals promoting parental effort

Vitellogenin is the precursor of most of the protein content of yolk in nearly all oviparous species. In insects, vitellogenin is produced from food and accumulated in the body, thereby directly linking resource availability to egg production (Page and Amdam 2007). In honeybees, vitellogenin levels fall when food resources are scarce, which in turn triggers foraging behaviour outside the colony as opposed to nursing behaviour (reviewed in Page and Amdam 2007). In insects, the vitellogenin signalling pathway is also linked to the juvenile hormone (JH) pathway (Page and Amdam 2007). JH signalling seems to link resource availability to vitellogenin secretion. For instance, in lubber grasshoppers (*Romalea microptera*) a threshold of food availability activates JH synthesis, which then stimulates vitellogenin production and oogenesis (Fronstin and Hatle 2008).

Leptin (or leptin-like proteins) controls food intake and immune response in vertebrates (reviewed in Henson and Castracane 2003; Otero et al. 2005). Leptin is produced mostly by adipocytes or lipogenic organs. Therefore, it has been suggested that reproduction is triggered when fat stores are enough large, and hence, the circulating levels of leptin exceed a certain signalling threshold (Henson and Castracane 2003). For instance, in great tits an artificial increase of leptin levels stimulated females to lay a second clutch (Lõhmus and Bjorklund 2009). Furthermore, female Siberian hamsters (*Phodopus sungorus*) whose circulating leptin levels were experimentally increased had lower rates of infanticide and produced more pups than controls (French et al. 2009). Thus, the availability of resources might even stimulate some parental behaviour by means of leptin signalling.

Prolactin also promotes and maintains incubation, gestation, and offspring care in vertebrates (Freeman et al. 2000). In birds, an experimental reduction of circulating prolactin inhibits incubation and also leads to brood desertion, whereas an experimental increase favours incubatory and protective behaviour (reviewed in Angelier and Chastel 2009). Prolactin levels are negatively controlled by glucocorticoids (next section). Treatment with glucocorticoids decreases prolactin levels in the blood of birds and rodents (Angelier and Chastel 2009). High fat reserves maintain glucocorticod secretion at low rates and hence prolactin-controlled behaviour may continue (Wingfield and Sapolsky 2003; Angelier and Chastel 2009; Spée et al. 2010). As in the case of vitellogenin and JH, these studies suggest that several hormones are simultaneously engaged in linking resource availability and parental care.

3.5.2 Pathways inhibiting parental effort

Some physiological changes as a result of short- or long-term stressors cause a redirection of resources to short-term vital processes and impair or threaten homeostasis, but can also inhibit parental investment. Such changes are usually known as the 'stress response' (see also McEwen and Wingfield 2003). In invertebrates, stress responses are modulated by HSPs. In fruit flies, for example, high levels of HSP70 (one of the major HSP proteins) increase longevity but reduce egg quality (i.e. hatching success; reviewed in Sorensen et al. 2003).

In vertebrates, the hypothalamic–pituitary–adrenal (HPA) axis induces a release of glucocorticoids into the blood a few minutes after exposure to a stressor (acute stress response; e.g. Wingfield and Sapolsky 2003). Baseline (low) glucocorticoid levels are nonetheless required for normal metabolism (Wingfield and Sapolsky 2003). The HPA response promotes the reallocation of resources from energy consuming systems (immunity, reproduction, etc.) to short-term survival (Wingfield and Sapolsky 2003). In addition to environmental (including social) stimuli, glucocorticoid secretion is triggered when lipid stores are exhausted and proteins from muscles and other tissues are catabolized to produce energy (e.g. Spée et al. 2010). In those circumstances, glucocorticoids stimulate glucogenesis and accelerate protein breakdown, thus optimizing energy production

(Challet et al. 1995) but also leading to clutch or brood desertion (Wingfield and Sapolsky 2003; Spée et al. 2010). Stress hormones may also favour reproductive effort under some circumstances (e.g. Bonier et al. 2009 and references therein). For example, high glucocorticoid levels intensify behaviours such as nest defence or foraging during reproduction (Bonier et al. 2009), while in mammals glucocorticoid-mediated immunosuppression prevents immune-induced damage to the foetus (e.g. Medina et al. 1993).

In birds, the glucocorticoid threshold that promotes desertion seems to depend on the reproductive value of current and future reproduction. House sparrows (*Passer domesticus*) raising experimentally enlarged broods reduced their acute (glucocorticoid) stress response in comparison with parents with reduced broods (Lendvai et al. 2007). A comparative analysis of 64 bird species showed that species with a higher value of the current brood compared to future breeding mounted a weaker corticosterone response during acute stress probably to enable successful breeding and maximize fitness; interestingly, females in species with more female-biased parental care also had weaker corticosterone responses (Bokony et al. 2009). A decrease in stress response during reproduction may not only be produced by inhibiting glucocorticoid secretion, but also by altering levels of glucocorticoid protein carriers in the blood or by blocking the glucocorticoid action on target tissues (central nervous or reproductive systems; Wingfield and Sapolsky 2003).

Finally, the hypothalamic–pituitary–gonadal (HPG) axis induces a release of sexual steroids that trigger mating behaviour at the expense of parental care, thereby playing a role in the trade-off between time devoted to care vs. time devoted to alternative mating opportunities (Section 3.3). It is well known from avian and mammalian studies that circulating testosterone levels decline when males start the parental care period, and experiments have shown that an increase in testosterone levels in males dramatically inhibits parental care (e.g. Adkins-Regan 2005; McGlothlin et al. 2007). In dark-eyed juncos (*Junco hyemalis*), the males that respond to aggression by producing higher testosterone levels are also those that contribute less to care, suggesting that the testosterone release pathway may even constitute a constraint for the evolution of parental care (McGlothlin et al. 2007).

3.6 Final remarks

In this chapter we have summarized important advances in the understanding of the mechanisms underlying the costs and benefits of parental care. Research in this area has increased enormously since the publication of Clutton-Brock's (1991) seminal book. Recurrent problems however remain unsolved. Research into costly mechanisms has provided new insights into the role of specific metabolites and oxidative stress (Section 3.4) and into control systems (Section 3.5), although empirical support for the relationship between mechanisms and fitness is still weak in many cases. This issue is well illustrated by Table 3.1. Although all the analysed physiological costs can be intuitively associated with fitness costs, few studies have examined exactly how parental care ('PC' in Table 3.1) induces both physiological and fitness costs ('FC') (i.e. PC→FC column in Table 3.1: 29 cases from 24 studies). Rarer still are studies that at the same time assess the positive correlation between physiological costs ('PHC' in Table 3.1) and fitness costs in the same dataset (i.e. PHC→FC column: 16 cases in 15 studies). Moreover, most work only assesses fitness proxies over a short period of time. In conclusion, evidence for connections between mechanisms and fitness is still quite weak with the strongest support from experiments on body mass loss in birds (see Table 3.1), although even these studies may not necessarily imply long-term costs, as the mass changes may be the result of tissue remodelling (e.g. as occurs in capital breeders; Speakman 2008).

Despite these difficulties, advances resulting from the study of phenotypic correlations and, above all, from manipulative experiments, have opened up new perspectives, for instance, revealing the importance of specific macro- and micronutrients in parental care. Another question is on the obligate costs of parental care, some of which are derived from resource acquisition rather than from the allocation of limited resources in competitive trade-offs. Obligate costs would include damage such as feather deterioration in hole-nesting birds, injuries caused when defending offspring, specific diet choice, or physiological damage (for example, damage revealed by increased levels of the repairing HSPs) induced by stress. Nevertheless, perhaps the best example is that of oxidative damage, since it may depend not only on limited resources such as antioxidants but also on metabolism and cell respiration; in other words damage is intimately related to a simple increase in metabolic activity (Metcalfe and Alonso-Alvarez 2010). Thus, in relation to mechanisms, the debate has so far been centred on resource allocation models (Leroi 2001; Barnes and Partridge 2003), but, as already mentioned, many cost and benefits have different currencies (Section 3.2). State models may help us to understand optimal parental decisions, that is the internal milieu of the individual determines its decisions when facing external challenges (Clark and Mangel 2000). More empirical demonstrations manipulating the state of individuals and assessing their consequences on parental care are still needed.

Another often neglected aspect concerns the costs of parental actions in relation to context-dependent benefits. For example, the allocation of maternal hormones and immunoglobulins to the egg or embryo seems to be a form of care that is cost-free for the parents, and their effects on offspring depend on when they will act and on their interaction with the particular environment at that time (Groothuis et al. 2005; Boulinier and Staszewski 2008). The study of individual states and context-dependence are promising areas for future empirical approaches since they may also serve to detect subtle hidden costs for parents.

One question meriting further attention is the possibility that parents inflict a certain level of stress on their offspring to favour their fitness. For instance, it has been recently shown that a moderate reduction in the food intake of yellow-legged gull (*Larus michaellis*) nestlings reduces oxidative damage during development (Noguera et al. 2011). This could be explained as a 'hormetic effect'. The 'hormetic model' proposes that fitness returns may describe a positive quadratic relationship with levels of a stressor, intermediate levels producing the highest fitness returns (reviewed in Costantini et al. 2010). The best known examples of a hormetic response are those induced by heat stress in insects (see HSPs) and dietary restriction in vertebrates. Few studies have ever addressed the hormetic response in the context of parental care (Noguera et al. 2011) and we ultimately still need to know whether or not parents can actively inflict stress on their offspring and what the impact of such potential stress-inducing strategies might be for offspring fitness.

The study of the control of signalling systems has opened new avenues of research, but more studies are required that link environmental cues with specific control systems that mediate the benefits and costs of parental care. The opportunities to manipulate care by offspring, by partners, or by reproductive parasites will also depend on the control systems of parental decisions, including transduction and molecular signalling pathways. Constraints in these regulatory mechanisms need to be explored since they may explain parental decisions and exactly why parents may make non optimal decisions (McNamara and Houston 2009). Moreover, we need to know whether the physiological signalling pathways that activate parental behaviour—thereby controlling decisions and resource allocation—are modulating costs or whether they are costly *per se* (Lessells 2008).

To conclude, studies of both phenotypic correlations and manipulative experiments have led to many relevant advances. However, approaches based on quantitative genetics or selection experiments aimed at disentangling genetic architecture (Reznick 1992) are still restricted to analyses of life-history traits (e.g. Roff 2002) and little work has been conducted on the mechanisms underpinning these traits to date (see Kim et al. 2010). The challenge for the future is to address this deficiency.

References

Adkins-Regan, E. (2005). *Hormones and Animal Social Behavior.* Princeton University Press, Princeton, NJ.

Alisauskas, R. T. and Ankney, C. D. (1994). Nutrition of breeding female ruddy ducks: the role of nutrient reserves. *Condor* 96, 878–97.

Alonso-Alvarez, C., Bertrand, S., Devevey, G., Prost, J., Faivre, B., and Sorci, G. (2004). Increased susceptibility to oxidative stress as a proximate cost of reproduction. *Ecology Letters* 7, 363–8.

Alonso-Alvarez, C., Bertrand, S., Devevey, G., et al. (2006). An experimental manipulation of life-history trajectories and resistance to oxidative stress. *Evolution* 60, 1913–24.

Angelier, F. and Chastel, O. (2009). General and comparative endocrinology stress, prolactin and parental investment in birds: A review. *General and Comparative Endocrinology* 163, 142–8.

Ardia, D. R., Schat, K., and Winkler, D. W. (2003). Reproductive effort reduces long-term immune function in breeding tree swallows (*Tachycineta bicolor*). *Proceedings of the Royal Society B* 270, 1679–83.

Arnold, K. E., Ramsay, S L., Donaldson, C., and Adam, A. (2007). Parental prey selection affects risk-taking behaviour and spatial learning in avian offspring. *Proceedings of the Royal Society B* 274, 2563–9.

Aron, S., Keller, L., and Passera, L. (2001). Role of resource availability on sex, caste and reproductive allocation ratios in the Argentine ant *Linepithema humile*. *Journal of Animal Ecology* 70, 831–9.

Badyaev, A. V. (2009). Evolutionary significance of phenotypic accommodation in novel environments: an empirical test of the Baldwin effect. *Philosophical Transactions of the Royal Society B* 364, 1125–41.

Baeza, J. A. and Fernandez, M. (2002). Active brood care in *Cancer setosus* (Crustacea: Decapoda): the relationship between female behaviour, embryo oxygen consumption and the cost of brooding. *Functional Ecology* 16, 241–51.

Balshine-Earn, S. and Earn, D. J. D. (1998). On the evolutionary pathway of parental care in mouth-brooding cichlid fishes. *Proceedings of the Royal Society B* 265, 2217–22.

Barnes, A. I. and Partridge, L. (2003). Costing reproduction. *Animal Behaviour* 66, 199–204.

Blanckenhorn, W. (1994). Fitness consequences of alternative life histories in water striders, *Aquarius remigis* (Heteroptera: Gerridae). *Oecologia* 97, 354–65.

Blount, J. D., Metcalfe, N. B., Arnold, K. E., Surai, P. F., and Monaghan, P. (2006). Effects of neonatal nutrition on adult reproduction in a passerine bird. *Ibis* 148, 509–14.

Bókony, V., Lendvai, A. Z., Liker, A., Angelier, F., Wingfield, J. C., and Chastel, O. (2009). Stress response and the value of reproduction: are birds prudent parents? *American Naturalist* 173, 589–98.

Bonier, F., Martin, P. R., Moore, I. T., and Wingfield, J. C. (2009). Do baseline glucocorticoids predict fitness? *Trends in Ecology and Evolution* 24, 634–42.

Bonilla, E., Medina-Leendertz, S., Villalobos, V., Molero, L., and Bohórquez, A. (2006). Paraquat-induced oxidative stress in Drosophila melanogaster: effects of melatonin, glutathione, serotonin, minocycline, lipoic acid and ascorbic acid. *Neurochemical Research* 31, 1425–32.

Bortolotti, G. R., Negro, J. J., Surai, P. F., and Prieto, P. (2003). Carotenoids in eggs and plasma of red-legged partridges: effects of diet and reproductive output. *Physiological and Biochemical Zoology* 76, 367–74.

Boulinier, T. and Staszewski, V. (2008). Maternal transfer of anti-bodies: raising immuno-ecology issues. *Trends in Ecology and Evolution* 23, 282–8.

Bourgeon, S., Le Maho, Y., and Raclot, T. (2009). Proximate and ultimate mechanisms underlying immunosuppression during the incubation fast in female eiders: roles of triiodothyronine and corticosterone. *General and Comparative Endocrinology* 163, 77–82.

Breuner, C. W., Patterson, S. H., and Hahn, T. P. (2008). In search of relationships between the acute adrenocortical response and fitness. *General and Comparative Endocrinology*, 157 288–95.

Brown, C. and Laland, K. N. (2001). Social learning and life skills training for hatchery reared fish. *Journal of Fish Biology* 59, 471–93.

Brown, J. L., Morales, V., and Summers, K. (2010). A key ecological trait drove the evolution of biparental care and monogamy in an amphibian. *American Naturalist* 175, 436–46.

Challet, E., Le Maho, Y., Robin, J.-P., Malan, A., and Cherel, Y. (1995). Involvement of corticosterone in the fasting-induced rise in protein utilization and locomotor activity. *Pharmacology, Biochemistry and Behaviour* 50, 405–12.

Cherel Y., Hobson K. A., and Weimerskirch H. (2005). Using stable isotopes to study resource acquisition and allocation in procellariiform seabirds. *Oecologia* 145, 533–40.

Clark, C. W. and Mangel, M. (2000). *Dynamic State Variable Models in Ecology*. Oxford University Press, Oxford, UK.

Clutton-Brock, T. H. (1991). *The Evolution of Parental Care*. Princeton University Press, Princeton.

Costantini, D., Metcalfe, N. B., and Monaghan, P. (2010). Ecological processes in a hormetic framework. *Ecology Letters* 13, 1435–47.

Coste, A., Sirard, J. C., Johansen, K., Cohen, J., and Kraehenbuhl, J. P. (2000). Nasal immunization of mice with virus-like particles protects offspring against rotavirus diarrhea. *Journal of Virology* 74, 8966–71.

Cox, R. M. and Calsbeek, R. (2010). Severe costs of reproduction persist in Anolis lizards despite the evolution of a single-egg clutch. *Evolution* 64, 1321–30.

Cox, R. M., Parker, E. U., Cheney, D. M., Liebl, A. L., Martin, L. B., and Calsbeek, R. (2010). Experimental evidence for physiological costs underlying the trade-off between reproduction and survival. *Functional Ecology* 24, 1262–9.

Curio, E. (1993). Proximate and developmental aspects of antipredator behavior. *Advances in the Study of Behavior* 22, 135–238.

Davies, N. B. and Welbergen, J. A. (2009). Social transmission of a host defense against cuckoo parasitism. *Science* 324, 1318–20.

Deeming, D. C. (2004). *Reptilian Incubation: Environment, Evolution and Behaviour*. Nottingham University Press, Nottingham.

Deerenberg, C., Pen, I., Dijkstra, C., Arkies, B. J., Visser, G. H., and Daan, S. (1995). Parental energy expenditure in relation to manipulated brood size in the European kestrel. *Zoology, Analysis of Complex Systems* 99, 38–47.

Descamps, S., Gilchrist, H. G., Joel, B., Buttler, E. I., and Forbes, M. R. (2009). Costs of reproduction in a long-lived bird: large clutch size is associated with low survival in the presence of a highly virulent disease. *Biology Letters* 5, 278–81.

Dijkstra, C., Bult, A., Bijlsma, S., and Daan, S. (1990). Brood size manipulations in the kestrel (*Falco tinnunculus*): effects on offspring and parent survival. *Journal of Animal* 59, 269–85.

Doughty, P. and Shine, R. (1998). Reproductive energy allocation and long-term energy stores in a viviparous lizard (*Eulamprus tympanum*). *Ecology* 79, 1073–83.

Drent, R. and Daan, S. (1980). The prudent parent. *Ardea* 68, 225–52.

Edward, D. A. and Chapman, T. (2011). Mechanisms underlaying reproductive trade-offs: costs of reproduction. In T. Flatt and A. Heyland eds. *Mechanisms of Life History Evolution*, Oxford University Press, Oxford.

Fedorka, K. M., Zuk, M., and Mousseau, T. A. (2004). Immune suppression and the cost of reproduction in the ground cricket, *Allonemobius socius*. *Evolution* 58, 2478–85.

Finkel, T. and Holbrook, N. J. (2000). Oxidants, oxidative stress and the biology of aging. *Nature* 408, 239–47.

Freeman, M. E., Kanyicska, B., Lerant, A., and Nagy, G. (2000). Prolactin: structure, function, and regulation of secretion. *Physiological Reviews* 80, 1523–631.

Freitak, D., Heckel D. G., and Vogel, H. (2009). Dietary-dependent trans-generational immune priming in an insect herbivore. *Proceedings of the Royal Society B* 276, 2617–24.

French, S. S., DeNardo, D. F., and Moore, M. C. (2007). Trade-offs between the reproductive and immune systems: facultative responses to resources or obligate responses to reproduction? *American Naturalist* 170, 79–89.

French, S. S., Greives, T. J., Zysling, D. A., Chester, E. M., and Demas, G. E. (2009). Leptin increases maternal investment. *Proceedings of the Royal Society B* 276, 4003–11.

Fronstin, R. B. and Hatle, J. D. (2008). A cumulative feeding threshold required for vitellogenesis can be obviated with juvenile hormone treatment in lubber grasshoppers. *Journal of Experimental Biology* 211, 79–85.

Ghalambor, C. K., Reznick, D. N., and Walker, J. A. (2004). Constraints on adaptive evolution: the functional trade-off between reproduction and fast-start swimming performance in the Trinidadian guppy (*Poecilia reticulada*). American Naturalist 164, 38–50.

Giacomello, E., Marchini, D., and Rasotto, M. B. (2006). A male sexually dimorphic trait provides antimicrobials to eggs in blenny fish. *Biology letters* 2, 330–3.

Giesing, E., Suski, C. D., Warner, R. E., and Bell, A. M. (2011). Female sticklebacks transfer information via eggs: Effects of maternal experience with predators on offspring. *Proceedings of the Royal Society B* 278, 1753–9.

Gil, D. (2008). Hormones in avian eggs: Physiology, ecology and behavior. *Advances in the Study of Behavior* 38, 337–98.

Gittleman, J. L. and Thompson, S. D. (1988). Energy allocation in mammalian reproduction. *American Zoologist*, 28, 863–75.

Golet, G. H., Irons, D. B., and Estes, J. A. (1998). Survival costs of chick rearing in black-legged kittiwakes. *Journal of Animal Ecology* 67, 827–41.

Golet, G. H., Schmutz, J. A., Irons, D. B., and Estes, J. A. (2004). Determinants of reproductive costs in the long-lived black-legged kittiwake: a multiyear experiment. *Ecological Monographs* 74, 353–72.

Grafen, A. (1988). On the uses of data on lifetime reproductive success. In T. H. Clutton-Brock ed. *Reproductive Success*, pp. 454–71. University of Chicago Press, Chicago.

Grandison, R. C., Piper, M. D. W., and Partridge, L. (2009). Amino-acid imbalance explains extension of lifespan by dietary restriction in *Drosophila*. *Nature* 462, 1061–4.

Green, B. S. and McCormick, M. I. (2005). O_2 replenishment to fish nests: males adjust brood care to ambient conditions and brood development. *Behavioral Ecology* 16, 389–97.

Grindstaff, J. L., Brodie, E. D., and Ketterson, E. D. (2003). Immune function across generations: integrating mechanism and evolutionary process in maternal antibody transmission. *Proceedings of the Royal Society B* 270, 2309–19.

Groothuis, T. G. G., Müller, W., von Engelhardt, N., Carere, C., and Eising, C. (2005). Maternal hormones as a tool to adjust offspring phenotype in avian species. *Neuroscience and Biobehavioral Reviews* 29, 329–52.

Hanson, L. A. (1998). Breast feeding provides passive and likely long lasting active immunity. *Annals of Allergy, Asthma and Immunology* 81, 523–37.

Hanssen, S. A., Hasselquist, D., Folstad, I., and Erikstad, K. E. (2005). Cost of reproduction in a long-lived bird: incubation effort reduces immune function and future reproduction. *Proceedings of the Royal Society B* 272, 1039–46.

Harshman, L. and Zera, A. J. (2007). The cost of reproduction: the devil in the details. *Trends in Ecology and Evolution* 22, 80–6.

Hau, M. and Wingfield, J. C. (2011). Hormonally-regulated trade-offs: Evolutionary variability and phenotypic plasticity in testosterone signaling pathways. In T. Flatt and A. Heyland eds. *Mechanisms of Life History Evolution*. Oxford University Press, Oxford.

Henson, M. C. and Castracane, V. D. (2003). *Leptin and Reproduction*. M. C. Henson and V. D. Castracane, eds. Kluwer Academic/Plenum Publishers, New York.

Hinde, C. A., Johnstone, R. A., and Kilner, R. M. (2010). Parent-offspring conflict and coadaptation. *Science*, 327, 1373–6.

Ho, D. H. and Burggren, W. W. (2010). Epigenetics and transgenerational transfer: a physiological perspective. *The Journal of Experimental Biology* 213, 3–16.

Hoover, J. P. and Reetz, M. J. (2006). Brood parasitism increases provisioning rate, and reduces offspring recruitment and adult return rates, in a cowbird host. *Oecologia* 149, 165–73.

Hoppitt, W. J. E., Brown, G. R., Kendal, R., et al. (2008). Lessons from animal teaching. *Trends in Ecology and Evolution* 23, 486–93.

Houston, A. I. and McNamara, J. M. (1999). *Models of Adaptive Behaviour*. Cambridge University Press, Cambridge.

Huestis, D. L. and Marshall, J. L. (2006). Interaction between maternal effects and temperature affects diapause occurrence in the cricket *Allonemobius socius*. *Oecologia* 146, 513–20.

Jacobsen, K.-O., Erikstad, K. E., and Saether, B.-E. (1995). An experimental study of the costs of reproduction in the kittiwake *Rissa Tridactyla*. *Ecology* 76, 1636–42.

Jones, J. C. and Oldroyd, B. P. (2007). Nest thermoregulation in social insects. *Advances in Insect Physiology* 33, 153–91.

Kalmbach, E., Griffiths, R., Crane, J. E., and Furness, R. W. (2004). Effects of experimentally increased egg production on female body condition and laying dates in the great skua *Stercorarius skua*. *Journal of Avian Biology* 35, 501–14.

Kim S-.Y., Velando, A., Sorci, G., and Alonso-Alvarez, C. (2010). Genetic correlation between resistance to oxidative stress and reproductive life span in a bird species. *Evolution* 64, 852–7.

Knouft, J. H., Page, L. M., and Plewa, M. J. (2003). Antimicrobial egg cleaning by the fringed darter (Perciformes: Percidae: *Etheostoma crossopterum*): implications of a novel component of parental care in fishes. *Proceedings of the Royal Society of London B* 270, 2405–11.

Knowles, S. C. L., Nakagawa, S., and Sheldon, B. C. (2009). Elevated reproductive effort increases blood parasitaemia and decreases immune function in birds: a meta-regression approach. *Functional Ecology* 23, 405–15.

Komatsu, M., O'Loughlin, P. M., Bruce, B., Yoshizawa, H., Tanakam, K., and Murakami, C. (2006). A gastric-brooding asteroid, *Smilasterias multipara*. *Zoological Science* 23, 699–705.

Landwer, A. J. (1994). Manipulation of egg production reveals costs of reproduction in the tree lizard (*Urosaurus ornatus*). *Oecologia* 100, 243–9.

Law, A. J., Pei, Q., Walker, M., et al. (2009). Early parental deprivation in the marmoset monkey produces long-term changes in hippocampal expression of genes involved in synaptic plasticity and implicated in mood disorder. *Neuropsychopharmacology* 34, 1381–94.

Lee, K. P., Simpson, S. J., Clissold, F. J., et al. (2008). Lifespan and reproduction in *Drosophila*: new insights from nutritional geometry. *Proceedings of the National Academy of Science USA* 105, 2498–503.

Lendvai, A. Z., Giraudeau, M., and Chastel, O. (2007). Reproduction and modulation of the stress response: an experimental test in the house sparrow. *Proceedings of the Royal Society B* 274, 391–7.

Leroi, A. M. (2001). Molecular signals versus the Loi de Balancement. *Trends in Ecology and Evolution* 16, 24–9.

Lessells, C. M. (1986). Brood size in Canada geese: a manipulation experiment. *Journal of Animal Ecology* 55, 669–89.

Lessells, C.(K.)M. (2008). Neuroendocrine control of life histories: what do we need to know to understand the evolution of phenotypic plasticity? *Philosophical Transactions of the Royal Society B* 363, 1589–98.

Li, D. and Jackson, R. R. (2003). A predator's preference for egg-carrying prey: a novel cost of parental care. *Behavioral Ecology and Sociobiology* 55, 129–36.

Lõhmus, M. and Björklund, M. (2009). Leptin affects life history decisions in a passerine bird: a field experiment. *PloS ONE* 4, e4602.

Lyon, B. E. and Eadie, J. M. (2008). Conspecific brood parasitism in birds: a life-history perspective. *Annual Review of Ecology, Evolution, and Systematics* 39, 343–63.

Madsen, T. and Shine, R. (1993). Costs of reproduction in a population of European adders. *Oecologia* 94, 488–95.

Magee, S. E., Neff, B. D., and Knapp, R. (2006). Plasma levels of androgens and cortisol in relation to breeding behavior in parental male bluegill sunfish, Lepomis macrochirus. *Hormones and Behavior* 49, 598–609.

Marshall, D. J. (2008). Transgenerational plasticity in the sea: context-dependent maternal effects across the life history. *Ecology* 89, 418–27.

Mateo, J. M. and Holmes, W. G. (1997). Development of alarm-call responses in Belding's ground squirrels: The role of dams. *Animal Behaviour* 54, 509–24.

McEwen, B. and Wingfield, J. C. (2003). The concept of allostasis in biology and biomedicine. *Hormones and Behavior* 43, 2–15.

McGlothlin, J. W., Jawor, J. M., and Ketterson, E. D. (2007). Natural variation in a testosterone-mediated trade-off between mating effort and parental effort. *American Naturalist* 170, 864–75.

McGraw, K. J., Adkins-Regan, E., and Parker, R. S. (2005). Maternally derived carotenoid pigments affect offspring survival, sex-ratio, and sexual attractiveness in a colorful songbird. *Naturwissenschaften* 92, 375–80.

McNamara, J. M. and Houston, A. I. (1986). The common currency for behavioral decisions. *American Naturalist* 127, 358–78.

McNamara, J. M. and Houston A. I. (2009). Integrating function and mechanism. *Trends in Ecology and Evolution* 24, 670–5.

Meaney, M. J. (2001). Maternal care, gene expression, and the transmission of individual differences in stress reactivity across generations. *Annual Review of Neuroscience* 24, 1161–92.

Medina, K. L., Smithson, G., and Kincade, P. W. (1993). Suppression of B lymphopoiesis during normal pregnancy. *The Journal of Experimental Medicine* 178, 1507–15.

Mërila, J. and Hemborg, C. (2000). Fitness and feather wear in the Collared Flycatcher *Ficedula albicollis*. *Journal of Avian Biology* 31, 504–10.

Merino, S., Moreno, J., Tomás, G., et al. (2006). Effects of parental effort on blood stress protein HSP60 and immunoglobulins in female blue tits: a brood size manipulation experiment. *Journal of Animal Ecology* 75, 1147–53.

Metcalfe, N. B. and Alonso-Alvarez, C. (2010). Oxidative stress as a life-history constraint: the role of reactive oxygen species in shaping phenotypes from conception to death. *Functional Ecology* 24, 984–96.

Miles, D. B., Sinervo, B., and Frankino, W. A. (2000). Reproductive burden, locomotor performance, and the cost of reproduction in free ranging lizards. *Evolution* 54, 1386–95.

Monaghan, P. (2008). Early growth conditions, phenotypic development and environmental change. *Philosophical transactions of the Royal Society B* 363, 1635–45.

Moore, I. T. and Hopkins, W. A. (2009). Interactions and trade-offs among physiological determinants of performance and reproductive success. *Integrative and Comparative Biology* 49, 441–51.

Morales, J., Velando, A., and Torres, R. (2009). Fecundity limits attractiveness when pigments are scarce. *Behavioral Ecology* 20, 117–23.

Mousseau, T. A. and Fox, C. W. (1998). *Maternal Effects as Adaptations*. Oxford University Press, Oxford, UK.

Myatt, L. and Cui, X. L. (2004). Oxidative stress in the placenta. *Histochemistry and Cell Biology* 122, 369–82.

Nilsson, J. Å. and Svensson, E. (1996). The cost of reproduction: a new link between current reproductive effort and future reproductive success. *Proceedings of the Royal Society B* 263, 711–14.

Noguera, J. C., Lores, M., Alonso-Álvarez, C., and Velando, A. (2011). Thrifty development: early-life diet restriction reduces oxidative damage during later growth. *Functional Ecology* 25, 1144–53.

Nordling, D., Andersson, M., Zohari, S., and Lars, G. (1998). Reproductive effort reduces specific immune response and parasite resistance. *Proceedings of the Royal Society B* 265, 1291–8.

Nur, N. (1984). The consequences of brood size for breeding blue tits. I. Adult survival, weight change and the cost of reproduction. *Journal of Animal Ecology* 53, 479–96.

Nur, N. (1988). The consequences of brood size for breeding blue tits. III. Measuring the cost of reproduction: survival, future fecundity, and differential dispersal. *Evolution* 42, 351–62.

Olsson, M., Shine, R., and Bak-Olsson, E. (2000). Locomotor impairment of gravid lizards: is the burden physical or physiological? *Journal of Evolutionary Biology* 13, 263–8.

Östlund-Nilsson, S. and Nilsson, G. E. (2004). Breathing with a mouth full of eggs: respiratory consequences of mouthbrooding in cardinalfish. *Proceedings of the Royal Society B* 271, 1015–22.

Otero, M., Lago, R., and Lago, F., et al. (2005). Leptin, from fat to inflammation: old questions and new insights. *FEBS Letters* 579, 295–301.

Page, R. E. and Amdam, G. V. (2007). The making of a social insect: developmental architectures of social design. *BioEssays* 29, 334–43.

Parker, G. A. and Maynard Smith, J. (1990). Optimality theory in evolutionary biology. *Nature* 348, 27–33.

Peluc, S. I., Sillett, T. S., Rotenberry, J. T., and Ghalambor, C. K. (2008). Adaptive phenotypic plasticity in an island songbird exposed to a novel predation risk. *Behavioral Ecology* 19, 830–5.

Pérez-Rodríguez, L. (2009). Carotenoids in evolutionary ecology: re-evaluating the antioxidant role. *Bioessays* 31, 1116–26.

Perry, C. and Roitberg, D. (2006). Trophic egg laying: hypotheses and tests. *Oikos* 112, 706–14.

Pigliucci, M. (2001). *Phenotypic Plasticity*. John Hopkins University Press, Baltimore, MD.

Pike, T. W., Blount, J. D., Lindström, J., and Metcalfe, N. B. (2007). Dietary carotenoid availability influences a male's ability to provide parental care. *Behavioral Ecology* 18, 1100–5.

Reguera, P. and Gomendio, M. (1999). Predation costs associated with parental care in the golden egg bug *Phyllomorpha laciniata* (Heteroptera: Coreidae). *Behavioral Ecology* 10, 541–4.

Reid, W. V. (1987). The cost of reproduction in the glaucous-winged gull. *Oecologia* 74, 458–67.

Reznick, D. (1992). Measuring the costs of reproduction. *Trends in Ecology and Evolution*, 7 42–5.

Roff, D. A. (2000). Trade-offs between growth and reproduction: an analysis of the quantitative genetic evidence. *Journal of Evolutionary Biology* 13, 434–45.

Roff, D. A. (2002). *Life History Evolution*. Sinauer Associates, Sunderland, MA.

Roth, O., Joop, G., Eggert, H., et al. (2010). Paternally derived immune priming for offspring in the red flour beetle, *Tribolium castaneum*. *Journal of Animal Ecology* 79, 403–13.

Sabat, A. M. (1994). Costs and benefits of parental effort in a brood-guarding fish (*Ambloplites rupestris*, Centrarchidae). *Behavioral Ecology* 5, 195–201.

Salmon, A. B., Marx, D. B., and Harshman, L. G. (2001). A cost of reproduction in *Drosophila melanogaster*: stress susceptibility. *Evolution* 55, 1600–8.

Salomon, M., Mayntz, D., Toft, S., and Lubin, Y. (2011). Maternal nutrition affects offspring performance via maternal care in a subsocial spider. *Behavioral Ecology and Sociobiology* 65, 1191–1202.

Scharf, I., Bauerfeind, S. S., Blanckenhorn, W. U., and Schaefer, M. A. (2010). Effects of maternal and offspring environmental conditions on growth, development and diapause in latitudinal yellow dung fly populations. *Climate Research* 43, 115–25.

Schradin, C. and Anzenberger, G. (2001). Costs of infant carrying in common marmosets, *Callithrix jacchus*: an experimental analysis. *Animal Behaviour* 62, 289–95.

Schradin, C., Schneider, C., and Yuen, C. H. (2009). Age at puberty in male African striped mice: the impact of food, population density and the presence of the father. *Functional Ecology* 23, 1004–13.

Shaffer, L. and Formanowicz, D. R. Jr (1996). A cost of viviparity and parental care in scorpions: reduced sprint speed and behavioural compensation. *Animal Behaviour* 51, 1017–24.

Simmons, L. W. (2011). Allocation of maternal- and ejaculate-derived proteins to reproduction in female crickets, *Teleogryllus oceanicus*. *Journal of Evolutionary Biology* 24, 132–8.

Sorensen, J. G., Kristensen, T. N., and Loeschcke, V. (2003). The evolutionary and ecological role of heat shock proteins. *Ecology Letters* 6, 1025–37.

Speakman, J. R. (2001). *Body Composition Analysis of Animals: A Handbook of Non-destructive Methods*. Cambridge University Press, Cambridge.

Speakman, J. R. (2008). The physiological costs of reproduction in small mammals. *Philosophical Transactions of the Royal Society of London B* 363, 375–98.

Spée, M., Beaulieu, M., Dervaux, A., Chastel, O., Le Maho, Y., and Raclot, T. (2010). Should I stay or should I go? Hormonal control of nest abandonment in a long-lived bird, the Adélie penguin. *Hormones and Behavior* 58, 762–8.

Stearns, S. C. (1992). *The Evolution of Life Histories*. Oxford University Press, Oxford, UK.

Stjernman, M., Råberg, L., and Nilsson, J.-A. (2004). Survival costs of reproduction in the blue tit (*Parus caeruleus*): a role for blood parasites? *Proceedings of the Royal Society B*, 271, 2387–94.

Storm, J. J. and Lima, S. L. (2010). Mothers forewarn off- spring about predators: a transgenerational maternal effect on behaviour. *American Naturalist* 175, 382–90.

Surai, P. F. (2002). *Natural Antioxidants in Avian Nutrition and Reproduction*. Nottingham University Press, Nottingham.

Suzuki, S., Kitamura, M., and Matsubayashi, K. (2005). Matriphagy in the hump earwig, *Anechura harmandi* (Dermaptera: Forficulidae), increases the survival rates of the offspring. *Journal of Ethology* 23, 211–13.

Svensson, I. (1988). Reproductive costs in two sex-role reversed pipefish species (Syngnathidae). *Journal of Animal Ecology* 57, 929–42.

Tallamy, D. W. (2005). Egg dumping in insects. *Annual Review of Entomology* 50, 347–70.

Tallamy, D. W. and Denno, R. F. (1982). Life-history trade-offs in *Gargaphia solani* (Hemiptera, Tingidae)—the cost of reproduction. *Ecology* 63, 616–20.

ten Cate, C., Lea, R., Ballintijn, M., and Sharp, P. (1993). Brood size affects behavior, interclutch interval, LH levels, and weight in ring dove (*Streptopelia risoria*) breeding pairs. *Hormones and Behavior* 27, 539–50.

Tinbergen, J. M. and Verhulst, S. (2000). A fixed energetic ceiling to parental effort in the great tit? *Journal of Animal Ecology* 69, 323–34.

Török, J., Hegyi, G., Tóth, L., and Könczey, R. (2004). Unpredictable food supply modifies costs of reproduction and hampers individual optimization. *Oecologia* 141, 432–43.

Trivers, R. L. (1972). Parental investment and sexual selection. In B. Campbell ed. *Sexual Selection and the Descent of Man, 1871–1971*, pp. 136–79. Aldine, Chicago.

Trumbo, S. T. (2007). Defending young biparentally: female risk-taking with and without a male in the burying beetle, *Nicrophorus pustulatus*. *Behavioral Ecology and Sociobiology* 61, 1717–23.

Tyler, M. J., Shearman, D. J. C., Franco, R., O'Brien, P., Seamark, R. F., and Kelly, R. (1983). Inhibition of gastric-acid secretion in the gastric brooding frog, *Rheobatrachus-silus*. *Science* 220, 609–10.

Tyndale, S., Letcher, R., Heath, J., and Heath, D. (2008). Why are salmon eggs red? Egg carotenoids and early life survival of Chinook salmon (*Oncorhynchus tshawytscha*). *Evolutionary Ecology Research* 10, 1187–99.

Uller, T. (2008). Developmental plasticity and the evolution of parental effects. *Trends in Ecology and Evolution* 23, 432–8.

Upreti, K. and Misro, M. (2002). Evaluation of oxidative stress and enzymatic antioxidant activity in brain during pregnancy and lactation in rats. *Health and Population-Perspectives and Issues* 25, 105–12.

van Noordwijk, A. J. and de Jong, G. (1986). Acquisition and allocation of resources: their influence on variation in life history tactics. *American Naturalist* 128, 137–42.

Veasey, J. S., Houston, D. C., and Metcalfe, N. B. (2001). A hidden cost of reproduction: the trade-off between clutch size and escape take-off speed in female zebra finches. *Journal of Animal Ecology* 70, 20–4.

Wake, M. H. and Dickie, R. (1998). Oviduct structure and function and reproductive modes in amphibians. *Journal of Experimental Zoology* 282, 477–506.

Wang, Y., Salmon, A. B., and Harshman, L. G. (2001). A cost of reproduction: oxidative stress susceptivility is associated with increased egg production in *Drosophila melanogaster*. *Experimental Gerontology* 36, 1349–59.

Weaver, I. C. G., Cervoni, N., Champagne, F. A., et al. (2004). Epigenetic programming by maternal behaviour. *Nature Neuroscience* 7, 847–54.

West-Eberhard, M. (2003). Developmental Plasticity and Evolution. Oxford University Press, Oxford, UK.

Wiersma, P., Selman, C., Speakman, J. R., and Verhulst, S. (2004). Birds sacrifice oxidative protection for reproduction. *Proceedings of the Royal Society of London B* 271, S360–3.

Wingfield, J. C. and Sapolsky, R. M. (2003). Reproduction and resistance to stress: when and how. *Journal of Neuroendocrinology* 15, 711–24.

Witter, M. S. and Cuthill, I. C. (1993). The ecological costs of avian fat storage. *Philosophical transactions of the Royal Society of London B* 340, 73–92.

CHAPTER 4

Patterns of parental care in vertebrates

Sigal Balshine

4.1 Introduction

Understanding parental care behaviour has remained a core research area in evolutionary behavioural ecology. Why the interest with parental behaviour? Many human cultures have a strong focus on children and parenting and our extensive and prolonged care for our own young undoubtedly contributes to our fascination with parental care in other animals. Also by studying parental care, behavioural scientists can gain a useful window on the social dynamics of family groups, providing insights on sexual conflict (Chapter 9), parent–offspring conflict (Chapter 7), sibling rivalry (Chapter 8), and kin-mediated cooperation (Emlen 1994, 1997). The strong link between parental effort and mating effort (and sexual selection patterns in general) has also driven the ever-growing interest in parental behaviour (Williams 1966; Trivers 1972; Emlen and Oring 1977; Kokko and Jennions 2008).

Parental care varies widely between species. While costly parental feeding of offspring is a near hallmark feature of birds and mammals, many species of fishes, amphibians, and reptiles also provide care for young by simply but vigorously guarding young against predators. Whether it is the mother or the father that defends the brood also varies widely between species. Understanding the key ecological factors selecting for care and explaining the plethora of parental care forms across different taxa has remained an enduring challenge in evolutionary behavioural ecology (Lack 1968; Gross 2005; Kvarnemo 2010). Empirical research has shown that providing care benefits parents by increasing offspring survival and increasing their reproductive success. However, parental care also has three potential costs: 1) a decrease in survival, 2) decreased growth and associated fecundity reduction, and 3) fewer remating opportunities (Gross and Sargent 1985). Scientists have used these costs and benefits to better understand when care will evolve and which sex will provide care (Maynard-Smith 1977; Balshine-Earn and Earn 1997; Houston et al. 2005). Vertebrate groups that show great variation in care types, like cichlid fishes and shorebirds, have been particular useful models in the search for a better understanding of the evolution of parental behaviour.

Since Clutton-Brock (1991) published his encyclopedic bible on parental care, two new technological advances (both molecular) have helped to further invigorate parental care research. First, phylogenetically based, comparative studies are now commonly employed in the study of parental care. The molecular revolution has facilitated the wide scale availability of molecular phylogenies, and analysis that links behaviour to these (Goodwin et al. 1998). These phylogenetic studies have strongly augmented and guided the more traditional behavioural ecology approach of experimental manipulation (Wright and Cuthill 1989). Second, the growth of genomic and bioinformatic studies has facilitated investigations into the divergence or conservation of genes, gene networks, and gene regulation across species or genera that share similar behaviour. Both these new directions have been made possible because of the rapid expansion of molecular data and because of impressive computational improvements that have facilitated large-scale database creation and analysis. In general, the molecular revolution is providing deep insights

The Evolution of Parental Care. First Edition. Edited by Nick J. Royle, Per T. Smiseth, and Mathias Kölliker.

into the evolutionary and physiological mechanisms underlying parental behaviour.

In the next section, I provide an overview of the forms of care that are generally observed across fishes, amphibians, reptiles, birds, and mammals and examine the factors that are thought to have selected for the evolution of viviparity and lactation. Then I discuss the evolution of male care, female care, and biparental care as well as the evolutionary transitions between these care states. I conclude the chapter with a brief overview of the patterns of parental care in humans. Whenever possible, I highlight new comparative and molecular studies in order to shed light on the evolution and maintenance of parental care and link care patterns across the different species and classes of vertebrates.

4.2 Forms of care

Some researchers prefer to use the term 'parental care' to refer only to only post-mating behaviours (such as care of eggs, larvae, or young after fertilization; see Chapter 1). However, some behaviours that occur before or during mating, such as nest building, egg provisioning, provisioning of the female with nuptial gifts or courtship feeding, are often still regarded as parental care as they lead to higher offspring survival. Here I describe the various forms of parental care in its broader sense.

4.2.1 Preparation of the physical rearing environment

The simplest form of parental care is the preparation of a territory to receive eggs or young (Fig. 4.1). Species differ in terms of whether they merely occupy an existing structure to receive young, or they modify these structures, or even create new structures *de novo*. Regardless, the preparation and construction of a nest, den, cavity, or burrow constitutes a basic but important aspect of parental behaviour that strongly influences the survival probabilities of young (Clutton-Brock 1991). These structures provide insulation and protection for young from adverse environmental conditions (such as low temperatures, rain, or desiccation) and may prevent predation. However, building such structures can be costly to the parent in terms of energy, time, and predation risk (Gauthier and Thomas 1993). Such costs are better born by some individuals than others, and hence the nest or burrow has become much more than the place where young are looked after and can be thought of as an extension of the individual's phenotype (Dawkins 1982; Soler et al. 1998). Females in many species have been shown to prefer males that build big, elaborate or particularly well-constructed nests (Soler et al. 1998; Östlund-Nilsson 2001). For example, male penduline tits *(Auriparus flaviceps)* build complex domed nests and these nests attract females, and males with the largest nests are most likely to mate, to mate earlier, and to have partners that faithfully look after young (Grubbauer and Hoi 1996; Szentirmai et al. 2005, Fig. 4.1a). Female choice is based on nest size even when male quality and habitat quality was held constant (Grubbauer and Hoi 1996). It turns out that larger nests have better insulation capacity, thereby reducing temperature fluctuations and promoting embryonic development (Grubbauer and Hoi 1996).

In mammals, den or burrow building may have evolved for purposes other than the rearing of young (Fig. 4.1b). These year round structures that provide insulation and protection are extremely commonly used among carnivores, rodents, insectivores, and lagomorph, and are especially common in species with an intense predation pressure (Birks et al. 2005). For example, some species of deer mice (genus *Peromyscus*) build complex burrows that contain nest chambers, specific holes used as an entrance, and several long escape tunnels that help minimize predation. The same burrow is used for both sleeping and looking after young. Recently behavioural experiments have mapped burrowing behaviour onto a *Peromyscus* molecular phylogeny showing differences between species in the shape and frequency of burrow making, and also revealing that species with complex burrows with many escape tunnels evolved from species with simpler burrowing behaviour (Weber and Hoekstra 2009).

Birds are without a doubt the master builders of the vertebrate world, using a diverse range of nesting materials and building nests of many different shapes and sizes. Most bird species build or mod-

Figure 4.1 Preparing for eggs and young. (a) A male Pendeline tit, *(Auriparus flaviceps)* with a domed shaped nests; females prefer large, well constructed nests. Photo credit: Rene van Dijk. (b) An eastern cottontail rabbit burrow with kits. Photo credit: courtesy of the Forest Preserve District of DuPage County. (c) A southern Masked weaver bird (*Ploceus velatus*) with its flask shaped nests. Photo credit: Chris Eason, courtesy of Wikimedia Commons. (d) The mud nest of a cliff swallow (*Petrochelidon pyrrhonota*). Photo credit: Mike Ross (MRoss46011@aol.com). (e) A male and female cichlid fish from Lake Tanganyika (*Cyathopharynx furcifer*) spawning in a crater nest. Photo credit: Robert Allen. (f) A line drawing of a male three-spined stickleback (*Gasterosteus aculeatus*) observing a female inspecting his nest. © Denis Barbulat 2011 used under license from Shutterstock.com.

ify a nest in which they deposit and incubate the eggs. Some species build open nests (e.g. American robins, *Turdus migratorius*), others excavate a cavity (e.g. downy woodpecker, *Picoides pubescens*), while some bird species simply take over the cavities made by other species lining the chamber with grass, moss, feathers, and hair to cushion the eggs inside (e.g. burrowing owls, *Athene cunicularia*; Martin and Li 1992). African weaver birds (Ploceidae) build dramatic nests that look like flasks hanging by woven loops from branches of trees. Each nest contains hundreds of strands of grass, twigs, or leaf fibres woven tightly together, and each nest contains a long downward facing, narrow entrance tunnel (Hansell 2000; Fig. 4.1c). Cliff swallows (*Petrochelidon pyrrhonota*) too build complicated nests that are made of mud mixed with bird saliva and are shaped like pottery jugs (Fig. 4.1d). A comparative phylogenetic analysis of the entire swallow family reveals that cavity nesting and nesting in mud-made structures evolved from simple burrowing (Winkler and Sheldon 1993; Sheldon et al. 2005). While most bird nests are short seasonal structures that are constructed to protect eggs, a few bird species like some of the larger raptors and megapods and some weavers have nests that last for many years and can be metres in diameter and height (Stone 1989). For example, nests of the colonial social weavers (*Philetairus socius*) not only last long but are also huge, often covering an entire tree. These multi-chambered colony-wide nests are honeycombed in shape and serve as a breeding area for hundreds of breeding pairs of weavers (Bartholomew et al. 1976).

Some reptiles, amphibians, and fish construct nests. Female iguanas (e.g. *Iguana iguana*) and snakes (e.g. *Pituophis melanoleucus*) can spend a considerable amount of time (many days) as well as substantial energy excavating a burrow in hard compact soils into which they deposit fertilized eggs (Doody et al. 2009). Similarly turtles and crocodiles make burrows in which they hide their eggs. (Shine 1988). Nest building has arisen a number of times in frogs. For example, in *Hyla boan*, the nest building gladiator frogs of Brazil, males build deep nests out of sand or clay within which the fertilized eggs undergo early development (Martins et al. 1998). Other frogs are also known to make floating bubble or foam nests on the surface of ponds, streams, or the axils of terrestrial bromeliads (Haddad and Prado 2005). Fish are also known to build nests. While simple pits or (redds) are dug by many fishes, like female salmon, other species like the Lake Tanganyika's cichlid *Cyathopharynx furcifer*, and three-spined sticklebacks, *Gasterosteus aculeatus*, construct remarkable nests. *Cyathopharynx furcifer* digs a crater shaped nest up to 2 metres in diameter and sticklebacks weave elaborate nests of plant material carefully glued together with a special kidney glycoprotein secretion known as spiggin (Balshine and Sloman 2011; Fig. 4.1e and 4.1f).

Although the costs and benefits of building these structures in specific ways have not yet been fully elucidated or manipulated experimentally, the watershed of new molecular phylogenies now available are providing insights on the evolutionary pathways and trajectories for nest building. These studies are shedding light on which nest building behaviours and nest types were ancestral and which are derived. In general, these studies show that the evolutionary trajectory towards fewer, higher quality offspring has been associated with increased selection for extensive pre-natal nest preparation. This makes sense because larger offspring represent 1) a greater lure for predators and 2) a greater proportion of a parent's lifetime reproductive output, that is too valuable to leave to chance.

4.2.2 Defence of offspring

Typically species that significantly modify the substrate on which they lay eggs (nest, cavity, and burrow builders) will also vigorously defend their young against predation. Animals that defend young usually do so in territories around their nests. Many species also protect and defend their young by keeping them in or attached to the parent's body (Fig. 4.2a). For example developing young can be kept in the parents' mouths (e.g. marine catfishes and many cichlid fishes), stomachs (e.g. myobatrachid frogs), ventral pouches (e.g. marsupials, seahorses, and pipefishes), embedded in skin (e.g. American banjo catfish and seadrag-

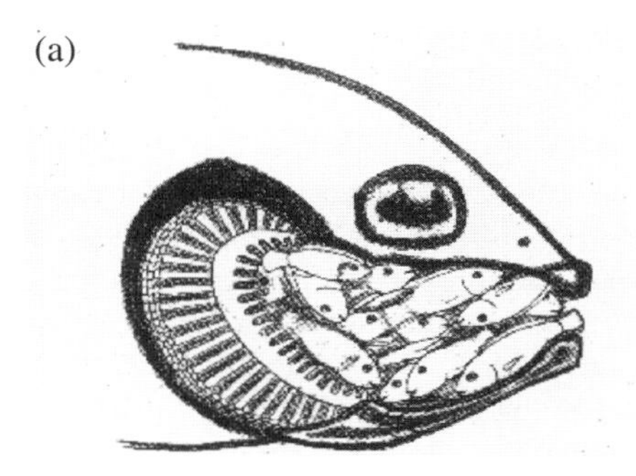

Figure 4.2 Post hatching care. (a) A line drawing of a cichlid fish mouth brooding its young. (b) A robin feeding young at the nest © Richard Stade. (c) A parent Bewick swan socially assisting its cygnets by ensuring they have access to feeding sites © Susan Marsh-Rollo.

ons), or inside the gills (e.g. cave fish). Most commonly, young remain inside the females reproductive tract. Live bearing or viviparity (see below) has evolved 21–22 times across fishes (Goodwin et al. 2002), at least once in amphibians, 102–115 times in reptiles, and 1–2 times in mammals (Reynolds et al. 2002). Internal brooding is an extremely effective method of protection because the only way a predator can capture or kill the young is to injure or kill the parent or force the parent to eject or drop its young.

4.2.3 Provisioning

4.2.3.1 Provisioning of gametes

All females provision their eggs with nutrient rich yolk stores but the degree of egg provisioning varies dramatically within and across species. Eggs vary wildly across species in terms of their size, yolk, albumen hormones, and nutrient composition. Among birds, the African ostrich (*Strutihio camelus*) has the largest eggs (21 cm in length and 1.4 kg), while the vervain hummingbird (*Mellisuga minima*) of Jamaica has the smallest (1 cm in length and 0.375 g) but each egg represents 16% of the female's total body mass (Bird 2004). Eggs vary not only between species but also among females within a species or population. In general, young that hatch from larger eggs have higher probability of survival especially if born in challenging environments (Nager and van Noordwijk, 1992, but see Christians 2002). But not all females make eggs of the same size, and egg size is influenced by many factors, including clutch size, the mother's phenotypic quality, environmental conditions such as food availability and density, as well as the predictability of the environmental conditions (Smith and Fretwell 1974; Christians 2002; Kindsvater et al. 2011). Parental care itself also appears to have coevolved with egg size; species that provide intense care for longer periods tend to have only a few, large eggs, while species that do not provide care or that provide less intense care for more young tend to produce smaller eggs (Shine 1988; Sargent et al. 1987; but see Summers et al. 2007).

It is often assumed that the costs of provisioning and protection of young are much greater than the costs of egg production. However, a careful study on lesser black-backed gulls (*Larus fuscus*) revealed that even a small increase in the cost of making an extra egg can have substantial impacts on how much energy parents will have to rear offspring (Monaghan et al. 1998). Even so, some researchers still do not accept that gamete provisioning is a truly a form of parental care, preferring to use the term parental care to refer only to behaviours that follow fertilization (see also Chapter 1). Once the retention of embryos within the female reproductive tract (termed viviparity, see below) evolved, additional provisioning or nourishment of developing embryos was possible beyond gamete provision.

4.2.3.2 Provisioning inside the parent and the evolution of live birth

In viviparous animals, the embryo develops within the mother's reproductive tract and the mother gives birth to live young, as opposed to the young developing in an egg outside of the mother's reproductive tract (known as oviparity). Viviparity can be as simple as embryo retention until hatching, or as complex as provision of nutrients by either direct absorption or by a specialized placental blood-vessel link (Reynolds et al. 2002). Viviparity has evolved among all vertebrate groups other than birds. Viviparity has the benefits of increasing offspring survival, but carries associated costs of reducing fecundity and mobility and increasing metabolic demands due to carrying offspring within the female (Wourms and Lombardi 1992; Qualls and Shine 1995). It has been argued that viviparity evolved 1) as an adaptation to cold and other rigorous climates (Tinkle and Gibbons 1977) or 2) as a way to deal with vastly unpredictable environments (Wootton 1990).

Across vertebrates, viviparity appears to have evolved independently 132 times (Blackburn 1995; Reynolds et al. 2002). A direct exchange of nutrients between mother and offspring via placenta-like structures is probably the most efficient way to provision developing offspring. However, viviparity provides offspring with direct access to maternal physiology, and selection may act on offspring to develop mechanisms for extracting resources more effectively from parents. Therefore, the placenta has been viewed as the battleground site where mother and offspring may fight over control of nutrients and the allocation of resources (Haig 1993).

Viviparity is found in all mammal species except for the five monotremes:, the platypus (Ornithorhynchidae) and the four species of spiny anteaters (Tachyglossidae), which all lay a single egg. The length of time that a female mammal will carry an embryo varies enormously from 12 days in the American opossum (*Didelphis virginiana*) to 660 days in African bush elephant (*Loxodonta africana*) (Hayssen 1993). Placental mammals have long pregnancies, followed by a relatively short lactation period (see below). In contrast, marsupials have much shorter pregnancies, followed by an extended period of lactation. Viviparity has been observed in 20% of the world's reptiles. It has arisen in many different reptile families, but is especially common among lizards and snakes (evolving some 102–115 times in the squamate reptiles; Shine 1985; Reynolds et al. 2002). The retention of developing embryos with maternal provisioning (live-bearing) has evolved in all three orders of Amphibians. Live bearing remains rare in frogs and toads (Anura) and in salamanders and newts (Urodela) (less than 1% of species), but is very common in the caecilian amphibians (occurring in 3 of 9 families and about 75% of all species; Wake 1993). It is thought to have evolved from egg laying and to represent a single evolutionary origin (Wilkinson and Nussbaum 1998). Viviparity has evolved from egg laying 21–22 times across all fishes (Dulvy and Reynolds 1997), and 12 times in teleost fishes (Goodwin et al. 2002). Interestingly, viviparity is the most common form of reproduction among sharks and rays (Elasmobranches). Viviparous fish species have larger offspring than egg laying fish but surprisingly do not have fewer young (Goodwin et al. 2002). Perhaps the most famous and odd case of viviparous fish is observed in seahorses and pipefish. In these fishes, females lay the eggs in the male's enclosed brood pouch within which the eggs are fertilized, and then aerated and nourished for several weeks (Kvarnemo 2010). The male seahorses eventually give birth via a series of forward and backwards muscular contortions to one young at a time (Vincent and Sadler 1995).

4.2.3.3 *Provisioning offspring outside the parent's body*

Many researchers have argued that the most energetically costly of parental behaviours is the feeding of newly hatched or born young (Fig. 4.2b; Drent and Daan 1980; but see Nager 2006). Food supplementation studies across different taxa have amply demonstrated that increased food availability results in young that emerge earlier, grow better, and have higher survival rates (Martin 1987; Christians 2002). In many small bird species, parents make more than 500 return feeding trips to the nest each day (Norberg 1981). All female mammals feed young with milk. Although hooded seal pups (*Cystophora cristata*) nurse for only 4 days, their mothers fast during this period and transfer an astonishing 8 kg of milk each day. In just four days, pups drinking this high fat (60%) milk manage to double their body mass (from 22 kg at birth to 45 kg at weaning). This species breeds on ice floes that often break up, and these unstable ecological conditions are thought to have selected for such intensive lactation (Boness and Bowen 1996).

Female mammals secrete milk from their mammary glands. How and why did lactation evolve in mammals? Although the duration of lactation varies wildly across mammals, ranging from 4 days in hooded seals to nearly three years in chimpanzees, *Pan toglodytes* (Gittleman and Thompson 1988; Hayssen 1993), recent comparative genomics and transcriptomics studies have revealed that all three mammalian lineages share highly conserved milk protein genes known as caseins (Lefèvre et al. 2010). The highly conserved nature of these genes suggests that it is likely that the origins of lactation, and the mammary gland itself, predate the common ancestor of living mammals. Molecular and fossil evidence suggests that the first mammal-like-reptiles called therapsids appeared at the end of the Triassic or the beginning of the Jurassic (166–240 million years ago). Along with the ability to lactate, therapsids possessed many mammalian traits such as endothermy, hair, and large brains (Hayssen 1993).

The mammary gland is thought to have evolved from a sweat or skin gland and the nipple from an associated hair follicle. Four major theories have been proposed to explain why these original secreting skin cells evolved into modern day mammary glands: 1) to keep the parchment-like eggs of early mammals moist (Oftedal 2002); 2) to provide extra nutrients to young (Hayssen 1993); 3) to keep offspring free from infection and provide immunological protection (Vorbach et al. 2006); and 4) to reduce juvenile mortality by maintaining a close mother–offspring contact (Hayssen 1993). These hypotheses are not necessarily mutually exclusive. Living monotremes, like the platypus, still produce parchment-shelled eggs and feed young milk that is secreted onto a patch of skin not a nipple. Lactation reduces the importance of provisioning the offspring with nutrients for growth through additional allocation of yolk to eggs. Indeed, the egg has been completely abandoned in the marsupial and placental mammals in favour of the placenta (Oftedal 2002).

Only female mammals lactate the young. Given that young fed from maternal energy reserves are well buffered from environmental fluctuations in food supply, a number of researchers have questioned why male lactation has not evolved? In fact, males have been shown to produce small amount of milk in two species of bats; *Dyacopterus spadecius* from Malaysia and *Pteropus capistrastus* from Papua New Guinea (Francis et al. 1994). Physiological barriers to male lactation exist including 1) the need for androgen suppression at puberty so that aromatase can orchestrate mammary gland development, and 2) the need for a change in the estrogen to progesterone ratio which in turn influences prolactin release and milk letdown (Daly 1979; Kunz and Hoskens 2009). While these proximate barriers clearly can be and have been surmounted in two species of male bats, functional lactation is unlikely to have been selected for in male mammals because of the high costs to males of associating with young via lost mating opportunities and paternity uncertainty (Kunz and Hoskens 2009).

Although rare, parental feeding of young among fish and amphibian species has also been reported. The young of the cichlid fish *Symphysodon discus* ingest the epidermal mucus from their parents' body (Buckley et al. 2010). Similarly, both male and female parents of the Central American convict cichlid, *Cichlasoma nigrofasciatum*, carefully lift up fallen leaves for their young providing them with benthic prey underneath the leaf litter. In Dendrobates frogs from Central America, females feed tadpoles unfertilized trophic eggs (Brust 1993).

4.2.3.4 Nutritionally independent young and social support

The most long-lasting parental care behaviour found in vertebrates is undoubtedly the continued support provided for nutritionally independent young. This type of care is typically found only in long-lived social vertebrates. Parents can continue to help their offspring and influence their fitness by providing them access to good feeding areas (Bewick swans, *Cygnus bewickii*; Scott 1980 Fig. 4.2c), by helping them acquire and defend a territory of their own (tree-toed sloths; Montgomery and Sunquist 1978), teaching skills, and by preventing conspecific attacks (Engh et al. 2000). In cercopithecine primates, such as Japanese macaques (*Macaca fuscata*), as well as in the spotted hyaena (*Crocuta crocuta*), offspring often remain for their entire lives in the same social group as

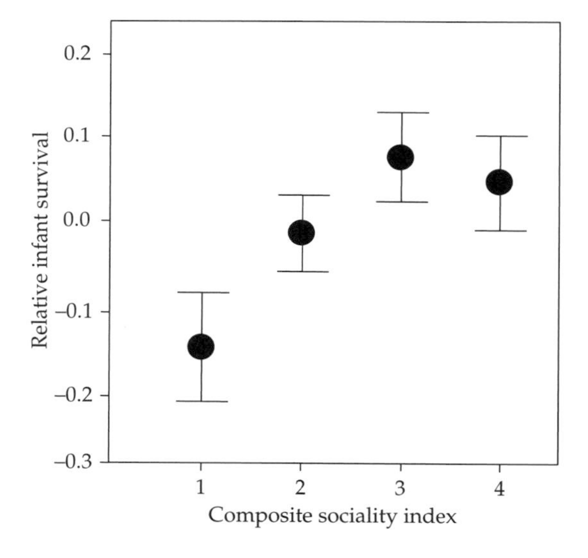

Figure 4.3 Redrawn from Silk et al. 2003, which is based on a 16 year study of savannah baboons in Kenya. In this study there was a strong positive effect of sociality on infant survival. The percentage of surviving infants increased with the mother's composite sociality score. The composite sociality score was based on three separate measures: 1) the time spent within 5 m of another adult conspecifics, 2) how much the mother was groomed by other adults in the group, and 3) how much time she spent grooming other adults. Females with high scores were considered more socially integrated than the average female and those with low scores less socially connected. Infant survival was calculated as the proportion of that female's infants that survived to 1 year of age. The main period of strong infant dependency is 1 year. Redrawn with permission from AAAS.

their mothers, and social rank is inherited. Based on a 16-year study on savannah baboons (*Papio cynocephalus*), in Amboseli, Kenya, Silk et al. (2003) showed that socially integrated females that had many living female relatives in the group were more likely to have their infants survive than were females that scored low on the sociality index (Fig. 4.3). The fitness effects of sociality via infant survival were independent of the effects of dominance rank, and environmental conditions. In some long-lived social vertebrates, known as cooperative breeders, some members of a social group forgo reproduction and help raise offspring (Emlen 1994). Such helpers may derive direct and indirect fitness benefits from living in the social group (Wong and Balshine 2010). For example, in the Tanganyikan cichlid *Neolamprologus pulcher*, sexually mature young continue to be vigorously guarded by parents or the dominant breeders of the social group (Wong and Balshine 2010).

4.3 Transitions in care

Our understanding of parental care evolution has been greatly enhanced by reconstructing the historical transitions in patterns of care across taxa and in particular by considering the variation in which sex provides parental care.

4.3.1 Parental care in fishes

Fishes provide care in a diverse fashion ranging from simple hiding of eggs, to guarding young in elaborately prepared structures or in/on the parent's body, and some species even feed young (Balshine and Sloman 2011; Fig. 4.1e and 4.1f). However most fish species do not provide any post-fertilization parental care. Only about 30% of the 500 known fish families show some type of parental care. Most often (in 78% of all cases) care is provided by only one parent (Gross and Sargent 1985; Reynolds et al. 2002) and male care (50–84%) is much more common than female care. Biparental care is the least common form of care in fishes (Mank et al. 2005). In some species, such as Galilee St. Peter's fish (*Sarotherodon galileaus*) and the brown bullhead (*Ameiurus nebulosus*), patterns of care are labile and paternal, maternal, and biparental care all co-exist (Blumer 1979; Balshine-Earn 1995). Based on a recent family level supertree for all ray-finned fishes, Mank et al. (2005) showed that male-only care has emerged at least 22 times (always within lineages with external fertilization), that biparental care arose at least 4 times, and that female-only care evolved independently at least 16 times. There have been at least 13 transitions to internal fertilization (and viviparity) all of which are associated with female care. The correlation between the mode of fertilization and the pattern of parental care suggests that there are two distinct pathways to male versus female care in fishes with the mode of fertilization (external versus internal) being the diverging starting point (Mank et al. 2005; Fig. 4.4a).

It is perhaps surprising, given that there are over 400 different fish families, that only one comparative analysis has investigated patterns of parental care within fish families (Goodwin et al. 1998). This study shows that among cichlid fishes there have been 21–30 changes from biparental to female only care, but that there have many fewer transitions in the other direction. This study supports the traditional route for parental care evolution, according to which male care evolves from none, that biparental care evolves from male care, and that female care evolves from biparental care (Gittleman 1981).

4.3.2 Parental care in amphibians

Most amphibian species abandon their eggs after laying them, but a few species display amazing parental care strategies. Some kind of parental behaviour is observed in 6–15% of the approximately 5000 anuran species (with male care being ancestral) and in 20% of around 500 salamander species (Summers et al. 2006; Summers and Earn 1999; Wells 2007). In total, parental care is thought to have evolved at least 41 times independently in this taxonomic class (Summers et al. 2006; Brown et al. 2010; K. Summers personal communication; Fig. 4.4b). The forms of parental care in amphibians are extraordinarily varied, including behaviours such as guarding the developing eggs, and carrying of eggs and tadpoles on the parents backs, on their hind legs, in dorsal pouches, vocal sacs, and even in the stomach (Corben et al. 1974). For example, male *Rhinoderma darwini* frogs carry eggs and young in their vocal sac until they have developed into

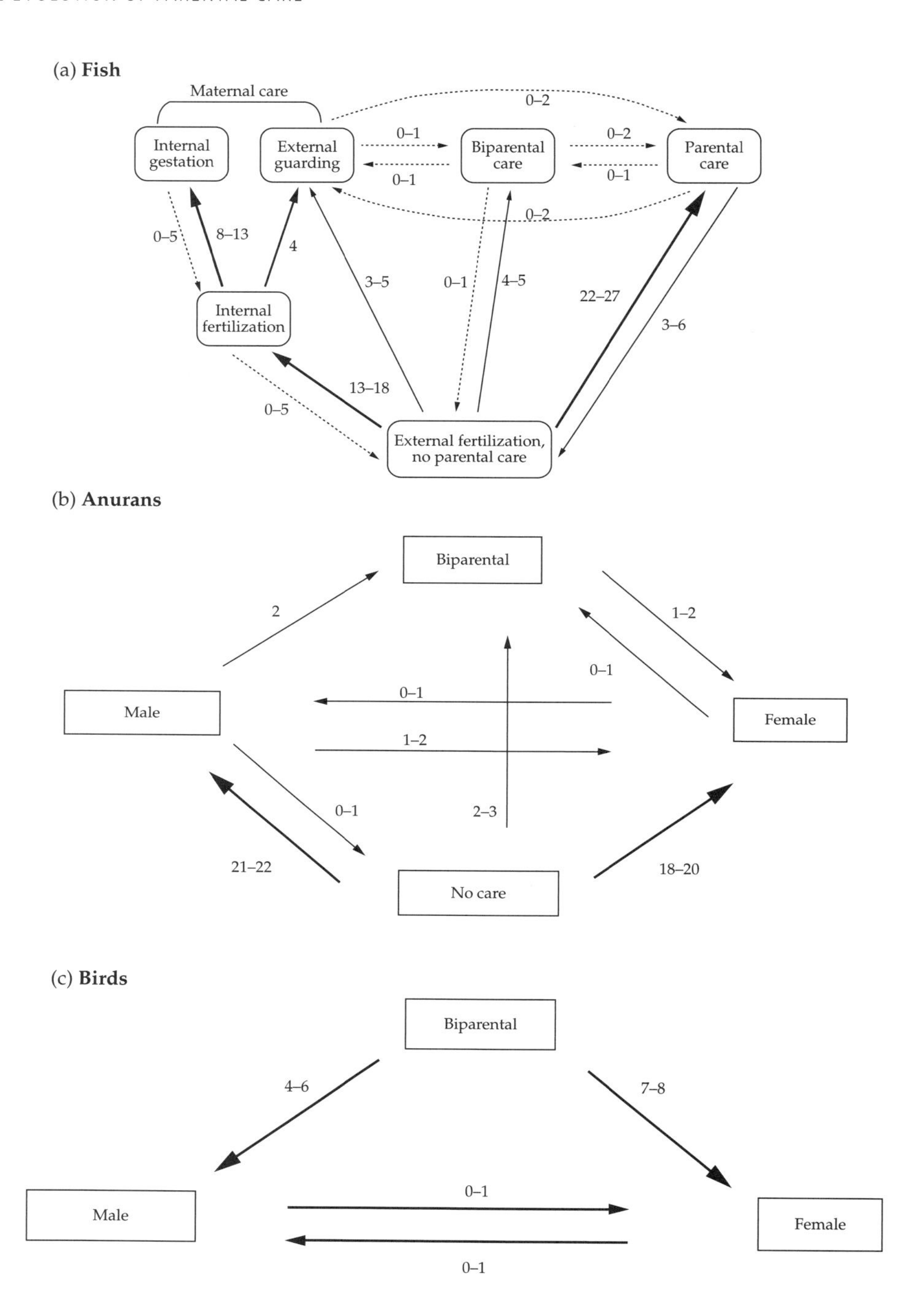

Figure 4.4 The most recent and well-accepted evolutionary models for transitions among parental care states in (a) fishes (taken from an analysis based on data for 228 families of ray-finned fishes Mank et al, 2005); (b) amphibians (based on estimates from Reynolds et al. 2002, Summers et al. 2006, and Brown et al. 2010) and (c) birds (taken from Reynolds et al. 2002). There are four possible states of parental care: no care, male, female, or biparental. The solid arrows show the likely direction of evolution among states. The number by each arrow refers to the minimum and maximum number of possible transitions. The dashed arrows refer to the selective factors promoting transitions in care. The percentage of families in each state is shown, and families including species in more than one state are counted more than once.

adults (Lutz 1947). In *Dendrobates pumlico*, another frog species from Central America, both parents will transport their young on their backs from one small water body to another and the females feed the growing tadpoles daily with unfertilized trophic eggs (Weygoldt 1987). As in fishes, patterns of parental care are varied, including male-only, female-only care and biparental care (Gross and Shine 1981; Crump 1996).

Care among amphibians is most common in tropical species where high predation rates may have forced parents to smaller water bodies that are relative predator free (Magnusson and Hero 1991), but where there is likely to be greater fluctuations in water levels, temperature, oxygen levels, and food availability (Wells 2007). Parental care could ameliorate such harsh fluctuating conditions because parents can move young around to better sites as the environmental conditions deteriorate (Bickford 2004). The size of the breeding pool size is associated with the evolution of parental care in frogs (Brown et al. 2010), and the most intensive form of parental care observed in frogs, feeding of trophic eggs to tadpoles, evolved in concert with the use of extremely small breeding pools and biparental care (Summers and Earn 1999; Brown et al. 2010).

4.3.3 Parental care in reptiles

Like fishes and amphibians the most common pattern in reptiles is the complete absence of care. Maternal care occurs only in about 1% of oviparous lizards and 3% of oviparous snakes (Shine 1988; De Fraipont et al. 1996). Care is usually limited to nest guarding, though some pythons coil around clutches and generate heat to incubate the eggs by means of shivering (Shine 1988). However maternal care is widespread in crocodilians, with the females of all 8 species guarding nests and young (Ferguson 1985; Lang 1987). Care by the male alone has never been recorded in a reptile (Shine 1988). There is currently no robust or comprehensive estimate for the number of times that care has evolved in reptiles.

4.3.4 Parental care in birds

In most bird species (90–95%) both parents look after the young chicks, while in 4–8% of bird species females alone look after young, and in 1–2% males alone do so (Silver et al. 1985; Cockburn 2006). Typically birds provide parental care by building a nest, incubating eggs, and then defending and feeding the chicks. Why do males and females birds commonly provide joint care? Males tend to provide care when the fitness derived via offspring survival is greater than benefits of abandoning young to seek out new mates (Székely et al. 1999). In many species, if males do not help raise young, some or even all the young chicks perish (Reynolds and Székely 1997). In many bird species, chicks are completely helpless at hatching, requiring constant feeding and warmth in order to grow and develop. Male and female birds are equally capable of building a nest, incubating, and feeding young (Ketterson and Nolan 1994). There are little anatomical or physiological sex specific specializations (in contrast to mammals) that would predispose one sex to provide parental care over the other sex. Simple biparental care with identical care roles, which is common in birds, may ensure that each parent can replace the other should the one die or leave (Oring 1982).

Given how common biparental care is across birds, scientists have long assumed that it is the most primitive form of care, and have concentrated their efforts in explaining how male or female care could have possibly evolved from this ancestral state (Lack 1968; Emlen 1994; Emlen and Oring 1977; Oring 1982; Székely and Reynolds 1995; Owens 2002). However, more recently, researchers have argued that male-only care in the form of egg guarding is the most likely ancestral form of care in birds and that biparental and female care are derived from it (Wesołowski 1994; Varricchio et al. 2008). Initially, there was a strong rejection of the suggestion that male-only care evolved first (Burley and Johnson 2002; Tullberg et al. 2002), but in recent years the idea has received substantial support from molecular, taxonomical, and paleontological studies. First, the fossil record shows that the clutch volume to adult body mass of three theropod dinosaurs (considered to be either closely related to birds or the direct ancestors of birds) matches closely that of birds with paternal care (Varricchio et al. 2008). Second, in the most primitive of all living birds known collectively as paleognaths (e.g. ratities and tinamous; Harshman et al. 2008, Phillips et al. 2010), all but 2 of the 60 species in this group have male-only care (Handford and Mares 1985). The proponents of the 'male-care-evolved-first' hypothesis argue that care first evolved in males and not in females because 1) females would have been energetically constrained by producing large eggs, 2) care would have significantly decreased future female fecundity, and 3), as in fishes, territorial males could combine egg protection and the attraction of additional mates, thereby lowering the costs of paternal care to males (Ah King et al. 2005). Biparental care probably evolved from male-only care due to harsh environmental conditions favouring the constant presence of one parent for incubation or protection. Such simple biparental care would in turn provide the platform from which role specialization, uniparental double-clutching, and then male-only or female-only care emerged as environmental conditions became more benign (Wesołowski 1994, 2004). Although a formal analysis of transitions in care across bird species has yet to be conducted, a preliminary analysis based on a partly resolved tree suggests that of eight independent transitions towards female care all but one occur through a biparental care intermediate step (Reynolds et al. 2002; Fig. 4.4c). The origins of care patterns in bird continue to be an area of great excitement, stimulating lots of empirical and theoretical work.

4.3.5 Parental care in mammals

In mammals, females always provide care and they usually do so alone or as part of a kin group. In monotremes, females lay and incubate a single egg in the female's abdominal pouch (echidnas, Tachyglossidae) or in a burrow (platypus, Ornithorhynchidae). Female monotremes do not possess nipples, but instead their milk oozes out of their skin and young lick milk from the milk-soaked fur (Brawand et al. 2008). Female marsupials provide care for their extremely altricial young within a pouch or skin fold that contains a mammary gland to which the offspring remains permanently attached as the teat swells in its mouth (Long 1969). Although marsupials have very short gestational periods (4–5 weeks), the young are nursed for nearly a year (Russell 1982). In placental mammals, there is a wide range of parental care by females after birth. In some species such as the guinea pig, *Cavia porcellus*, the young are extremely precocial, young are active soon after birth, able to feed themselves, and do not require a parent to keep warm (Laurien-Kehnen and Trillmich 2003). Other species, such as most other rodents, cats, and dogs, have highly dependent young that need to be warmed, fed and protected.

Males assist in care in only 9–10% of mammalian genera, including many primate, carnivore, and rodent species (Kleiman and Malcolm 1981). The general assumption is that female care among mammals is primitive and that biparental care is derived (Reichard and Boesch 2003). When males provide care, they typically carry, feed, warm, and guard the young against predators. In Siberian hamsters, *Phadopus campbelli*, males assist in the female's delivery, clear the nostrils to open the pups' airways, and lick and clean the pups of membranes immediately after birth (Jones and Wynne-Edwards 2000). In the California mouse, *Peromyscus californicus*, the removal of the male results in lower offspring survival (Cantoni and Brown 1997). The decreased survival is not a result of a reduction in care but due to the presence of infanticidal intruders who try to mate with the mother. Although biparental care is certainly rare in mammals, it appears to have evolved from female care 9 times and to have been lost 3 times (Reynolds et al. 2002). Males will care more or less depending on the costs of lost mating opportunities, and reduced mobility or foraging success caused by looking after young (Woodroffe and Vincent 1994).

4.4 Parental care in humans

Compared with other mammals, *Homo sapiens* provide intensive and long lasting post-natal parental care for a relatively small number of offspring. Human mothers provide nutrition and protection during the 9 months of pregnancy and supply milk from their own reserves during infancy, and both parents usually continue to support their offspring for their entire lifespan. There are at least three distinctive characteristics of human parental care behaviour that is rarely observed in other mammals and that require explanation: 1) the exceptionally long period of parental care, 2) the considerable amount of male care, and 3) kin support in rearing young (Hill and Kaplan 1999).

4.4.1 Exceptionally long parental care duration

Humans look after their offspring through infancy, childhood, puberty, and often well after sexual maturity. This represents an unusually long period of dependence, even among primates (Hill and Kaplan 1999). For example, human parents continue to provide financial and emotional support for their offspring, even after their offspring have become parents in their own right. What event in hominid evolution selected for this long parental care period? The fossil record suggests that over the last 4 million years, brain volume has increased threefold and this change is associated with a doubling in the developmental period (Alexander 1979). Dunbar (1993) proposed that large brains were necessary early in hominid evolution to deal with the complexities of social life and in support of this idea he demonstrated that brain size covaries with group size among non-human primates. The long developmental period that arose with large brains would have enabled sufficient time to learn how to deal with the complexities of social living (coalition and cooperation) and such skills would have be necessary to control access to resources and to coordinate competition with other groups (Dunbar 2000). Parenting by both mother and father would have supported this long developmental period and selected for the efficient acquisition of social skills and competences (Geary and Flinn 2001).

4.4.2 Male care and support

Although male care is extremely rare among mammals, male humans care for young in every culture studied to date (Marlowe 2000). Men provide social protection and material resources to their wives and children (Marlowe 2000). In some cultures, men spend time holding and babysitting their children. For example, Hadza men from Tanzania, protect their offspring by remaining close to them for 12%

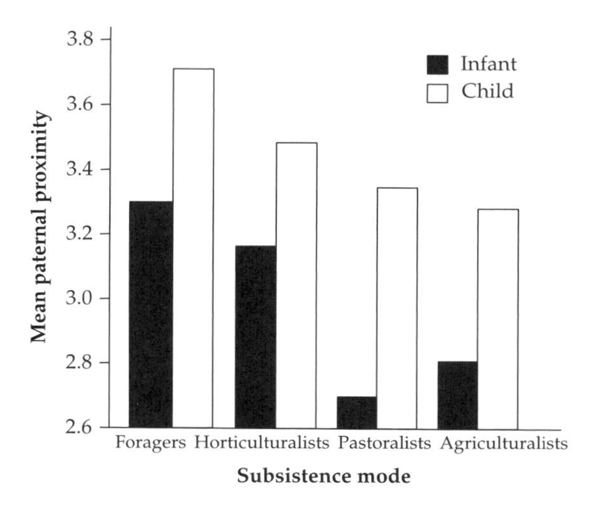

Figure 4.5 Data of the average father–offspring proximity according to major subsistence mode, based on an original sample of 186 Standard Cross Cultural societies, and redrawn from Marlowe 2000. Hunter-gatherers or foragers (n = 42) practice no agriculture; horticulturalists (n = 70) get the majority of their diet from agriculture; pastoralist (n = 17) acquire the majority of their diet from domesticated animals; and agriculturists (n = 57) practice intense forms of agriculture using irrigation, fertilization, and plows. Used with permission from Elsevier.

of the daytime and 100% of time of the nighttime (Marlowe 1999). The degree of male involvement with children varies widely with ecological and social circumstances. For example, men in foraging societies tend to provide the most child care, interacting closely and regularly with their infants for much longer periods than men in agriculturist, and pastoralist societies (Marlowe 2000; Fig. 4.5).

Why do male humans always provide some type of care? Two main hypotheses have been put forth: 1) paternal care in humans was selected for by the need for male provisioning especially during lactation or gestation periods when women could not hunt efficiently (Washburn and Lancaster 1968; Deacon 1997); 2) paternal care was selected as biproduct of mate guarding (van Rhijn 1991; Hawkes 2004). A number of authors have argued that men may end up providing parental care as a consequence of protecting a mate against harassment from other males (Hawkes 2004). The parental care via mate guarding hypothesis has received more support from the available ethnology and human behavioural ecology data than the provisioning hypothesis (Kelly 1995). Male provisioning and defence of offspring appears to have had more to do with attracting mates and competing with other men over mates rather than providing care for children (Hill and Kaplan 1988). For example, a hunter's family rarely receives more food than the rest of the village or social group (Hill and Kaplan 1988). Hence, male hunting and defence are effectively a kind of public good, as these characteristics do not only help the man's wife and his offspring but benefit the whole social group (Hawkes and Bliege Bird 2002). Great hunters and fighters develop reputations and hence gain more mates. The costly signals of male quality demonstrated via hunting capacity appear to more strongly line up with status and mating access than with male parental provisioning *per se* (Hawkes and Bliege Bird 2002). In essence, what appears to be a form of male parental care may in fact be represent a form of male mating effort.

4.4.3 Support from constellations of kin

While women and their children clearly form the nucleus of a family, human families are typically embedded in wider kin networks that provide a considerable amount of assistance in the provisioning, protection, and socialization of children (Hrdy 2009). Humans are unusual among mammals in their strong reliance on extended kin assistance for rearing offspring. The kin that help rear offspring are often an older, non-reproductive sibling or an older relative like a grandmother or an aunt (Hawkes et al. 1998; Shanley and Kirkwood 2001). These relatives operate as a 'helper at the nest' significantly increasing reproductive success of their breeding relatives (Hrdy 2009). It has been hypothesized that menopause (the cessation of ovarian function and female fertility, that occurs between the ages of 40–60) is an adaptation to extend post-reproductive lifespan and increases investment by older females (Hawkes 2003; Lahdenpera et al. 2004; but see Tang et al. 1996). Older females with extensive experience of both social and physical environments can increase the fitness of their children and grandchildren by transferring valuable social survival skills and connections and teaching them how to better compete and negotiate social environments within a kin network (Hawkes 2003).

4.4.4 Humans as a study system for parental care

Although humans are unique in their long parental care duration and in the degree to which kin help raise offspring, humans, like other species, allocate their investment in offspring judiciously considering factors such as parentage, offspring quality, and parental resources. These patterns are best demonstrated in studies of the cross-cultural variation in parental neglect, offspring abandonment, and infanticide (Daly and Wilson 1988). Collectively they show that parentage, offspring quality, and the availability of parental resources all influence care in our own species. Parentage matters: human parents spend the most money on genetic children from current relationships and least on their stepchildren, especially those from relationships that have already ended (Anderson et al. 2007). It turns out that having a step-parent is the most powerful epidemiological risk factor for child abuse, suggesting that parental investment is strongly influenced by relatedness (Davis and Daly 1997; Westneat and Sherman 1993; but see Temrin et al. 2000 and Daly and Wilson 2001). Among the Aché Indians, a traditional hunter-gather tribe from Paraguay, children who have lost their natural fathers have a significantly increased risk of dying before the age of 15 compared to children whose fathers remain around (Hill and Kaplan 1988). These children are apparently commonly killed by adult men from within their social group who do not want to provide for young that are not their own (Hill and Kaplan 1988). Offspring quality matters: although very much a strategy of last resort, parents will sometimes abandon or even kill their own offspring. The frequency of infanticide increases if the offspring is seriously ill, has major birth defects, severe physical deformities, and hence poor probable future fitness (Daly and Wilson 1988). Parental resources matter: when the social and physical costs of raising a child are too high, humans across different societies will commit infanticide or simply abandon offspring (Daly and Wilson 1988). In this way infanticide can be seen as a strategic allocation decision occurring more regularly when a second baby is born soon after the last one or when there are too few resources to raise a child. In Canada, France, and England, infanticide is more common among unwed mothers (Daly and Wilson 1988). Interestingly, while the practice of twin killing is rare (1%) in societies where mothers regularly get help from their female relatives in rearing children, it is a fairly common practice (43%) in societies where mothers have to carry the bulk of the parenthood burden alone (Granzberg 1973).

These cross-cultural studies of infanticide demonstrate further that environments can pose differential costs associated with parental provisioning (Daly and Wilson 1984). For example, in agricultural and pastoral societies in African, infanticide rates are extremely low (Hrdy et al. 1994). In many of these societies, women are calorically self-sufficient, direct paternal investment is small, and the temporary fostering or more permanent adoption of children to kin is extremely common (Hrdy et al. 1994; Marlowe 2000). In contrast, in New Guinean and Amazonian hunter-gather tribal societies, infanticide rates are much higher (12–38%). In these societies adoption and fostering is not a viable option because when foraging conditions worsen everyone in the group faces the same deteriorating foraging conditions (Hill and Kaplan 1988). Also, in these societies, men invest a great deal in care, and children are heavily dependent on male input as the lack of paternal support increases the risk for infanticide (Hill and Kaplan 1988).

In summary, in humans the importance of group solidarity and cooperation, in the face of intense intergroup competition has apparently selected for a very long care duration, as well as group and biparental care of offspring. In addition, the evolution of menopause and the presence of non-reproductive helpers with little incentive to disperse has dramatically changed the cost–benefits ratio for providing care in humans.

4.5 Concluding remarks

The most obvious conclusion to draw from this review is that parental care is highly dynamic and variable. Although, we have a fairly solid understanding of factors that determine whether a parent will or will not provide care, we still need to develop a firm theoretical foundation to understand the variation that exists in parental care form. Why, for example, do some species look after their young in a nest while other species do so on or in the parental body? Several promising developments (all comparative and based on molecular revolution) are already significantly advancing our understanding of the evolution of parental care. From a theoretical prospective, the explosion of recent studies on sexual conflict over the extent of care for young will undoubtedly shed light on the generality of how life history and ecological variables interact to mediate levels and types of parental behaviour. More experiments and broad-scaled phylogenetically based comparative ansly-ses are needed to explore if particular forms or types of care are associated with high or low levels of sexual conflict over care. Further progress is highly likely if we continue to combine broad-scale approaches and the levels of analysis on which we investigate parental care (i.e. adaptive function, proximate mechanisms, evolutionary history, and development).

Acknowledgements

I thank Mart Gross and Tim Clutton-Brock for introducing me to the amazing world of parental care research, to Per Terje Smiseth, Lotta Kvarnemo, and Matt Bell for helpful comments on the manuscript. I also thank Susan Marsh-Rollo for her help with the references and figures, and David, Arielle, and Maya Earn for teaching me about the benefits of parental care first hand. Support for this research was provided via the Canada Research Chairs Program and a Discovery Grant from the Natural Sciences and Engineering Research Council of Canada.

References

Ah-King, M., Kvarnemo, C., and Tullberg, B. (2005). The influence of territoriality and mating system on the evolution of male care: a study on fish. *Journal of Evolutionary Biology* 18, 371–82.

Alexander, R. D. (1979). *Darwinism and Human Affairs*. University of Washington Press, Seattle.

Anderson, K. G., Kaplan, H., and Lancaster, J. B. (2007). Confidence of paternity, divorce, and investment in children by Albuquerque men. *Evolution and Human Behavior* 28, 1–10.

Balshine, S. and Sloman, K. (2011). Parental care in fishes. In A. P. Farrell, *Encyclopedia of Fish Physiology: From Genome to Environment*, San Diego: Academic Press, San Diego.

Balshine-Earn, S. (1995). The costs of parental care in Galilee St. Peter's Fish *Sarotherodon galilaeus*. *Animal Behaviour* 50, 1–7.

Balshine-Earn, S. and Earn, D. J. D. (1997). An evolutionary model of parental care in St. Peter's fish. *Journal of Theoretical Biology* 184, 423–31.

Bartholomew, G. A., White, F. N., and Howell, T. R. (1976). The thermal significance of the nest of the sociable weaver *Philetairus socius*: summer observations. *Ibis* 118, 402–11.

Bickford, D. P. (2004). Differential parental care behaviors of arboreal and terrestrial microhylid frogs from Papua New Guinea. *Behavioral Ecology and Sociobiology* 55, 402–9.

Bird, D. M. (2004). *The Bird Almanac: A Guide to Essential Facts and Figures of the World's Birds*. Firefly Books, Richmond Hill.

Birks, J., Messenger, J. E., and Halliwell, E. C. (2005). Diversity of den sites used by pine martens Martes martes: a response to the scarcity of arboreal cavities? *Mammal Review* 35, 313–20.

Blackburn, D. G. (1995). Saltationist and punctuated equilibrium-models for the evolution of viviparity and placentation. *Journal of Theoretical Biology* 174, 199–216.

Blumer, L. S. (1979). Male parental care in the bony fishes. *Quarterly Review of Biology* 54, 149–61.

Boness, D. J. and Bowen, W. D. (1996). The evolution of maternal care in pinnipeds. *Bioscience* 46, 645–54.

Brawand, D., Wahli, W., and Kaessmann, H. (2008). Loss of egg yolk genes in mammals and the origin of lactation and placentation. *PLOS Biology* 6, 507–17.

Brown, J. L., Morales, V., and Summers, K. (2010). A key ecological trait drove the evolution of biparental care and monogamy in an amphibian. *American Naturalist* 175, 436–46.

Brust, D. G. (1993). Maternal brood care by *Dendrobates pumilio*: a frog that feeds its young. *Journal of Herpetology* 27, 96–8.

Buckley, J., Maunder, R. J., Foey, A., Pearce, J., Val, A. L., and Sloman, K. A. (2010). Biparental mucus feeding: a unique example of parental care in an Amazonian cichlid. *Journal of Experimental Biology* 213, 3787–95.

Burley, N. T. and Johnson, K. (2002). The evolution of avian parental care. *Philosophical Transactions of the Royal Society of London, Series B* 357, 241–50.

Cantoni, D. and Brown, R. E. (1997). Paternal investment and reproductive success in the California mouse, *Peromyscus californicus*. *Animal Behaviour* 54, 377–86.

Christians, J. K. (2002). Avian egg size: variation within species and inflexibility within individuals. *Biological Reviews* 77, 1–26.

Clutton-Brock, T. H. (1991). *The Evolution of Parental Care*. Princeton University Press, Princeton.

Cockburn, A. (2006). Prevalence of different modes of parental care in birds. *Proceedings of the Royal Society, Series B* 273, 1375–83.

Corben, C. J., Ingram, G. J., and Tyler, M. J. (1974). Gastric brooding: unique form of parental care in an Australian frog. *Science* 186, 946–7.

Crump, M. L. (1996). Parental care among the Amphibia. *Advances in the Study of Behavior* 25, 109–44.

Daly, M. (1979). Why don't male mammals lactate? *Journal of Theoretical Biology* 78, 325–45.

Daly, M. and Wilson, M. (eds.) (1984). *A Sociobiological Analysis of Human Infanticide*, pp. 487–502. Aldine, New York.

Daly, M. and Wilson, M. (1988). The Darwinian psychology of discriminative parental solicitude. *Nebraska Symposium on Motivation* 35, 91–144.

Daly, M. and Wilson, M. (2001). An assessment of some proposed exceptions to the phenomena of nepotistic discrimination against stepchildren. *Annales Zoologici Fennici* 38, 287–96.

Davis, J. N. and Daly, M. (1997). Evolutionary theory and the human family. *Quarterly Review of Biology* 72, 407–35.

Dawkins, R. (1982). *The Extended Phenotype*. Freeman, Oxford.

De Fraipont, M., Clobert, J., and Barbault, R. (1996). The evolution of oviparity with egg guarding and viviparity in lizards and snakes: a phylogenetic analysis. *Evolution*, 50, 391–400.

Deacon, T. (1997). *The Symbolic Species*. W. W. Norton, New York.

Doody, J. S., Freedberg, S., and Keogh, J. S. (2009). Communal egg-laying in reptiles and amphibians: evolutionary patterns and hypotheses. *Quarterly Review of Biology*, 84, 229–52.

Drent, R. and Daan, S. (1980). The prudent parent: energetic adjustments in avian breeding. *Ardea* 68, 225–52.

Dulvy, N. K. and Reynolds, J. D. (1997). Evolutionary transitions among egg-laying, live-bearing and maternal inputs in sharks and rays. *Proceedings of the Royal Society, Series B* 264, 1309–15.

Dunbar, R. I. M. (1993). Coevolution of neocortex size, group size and language in humans. *Behavior and Brain Science* 16, 681–735.

Dunbar, R. I. M. (2000). On the origin of the human mind. In P. Carruthers and A. Chamberlain, eds. *The Evolution of Mind*, pp. 238–53. Cambridge University Press, Cambridge.

Emlen, S. T. (1994). Benefits, constraints and the evolution of the family. *Trends in Ecology and Evolution* 9, 282–5.

Emlen, S. T. (1997). Predicting family dynamics in social vertebrates. In J. R. Krebs and N. B. Davies, eds. *Behavioural Ecology: An Evolutionary Approach*, 4th edition, pp. 228–53. Blackwell, Oxford.

Emlen, S. T. and Oring, L. W. (1977). Ecology, sexual selection, and evolution of mating systems. *Science* 197, 215–23.

Engh, A. L., Esch, K., Smale, L., and Holekamp, K. E. (2000). Mechanisms of maternal rank 'inheritance' in the spotted hyaena, *Crocuta crocuta*. *Animal Behaviour* 60, 323–32.

Ferguson, M. W. J. (1985). The reproductive biology and embryology of the crocodilians. In C. Gans, F. S. Billet, and P. F.A Maderson, eds. *Biology of the Reptilia*, pp. 330–491. John Wiley, New York.

Francis, C. M., Anthony, E. L. P., and Brunton, J. A. (1994). Lactation in male fruit bats. *Nature* 367, 691–2.

Gauthier, M. and Thomas, D. W. (1993). Nest site selection and cost of nest building by cliff swallows (*Hirundo pyrrhonota*). *Canadian Journal of Zoology* 71, 1120–3.

Geary, D. C. (2000). Evolution and proximate expression of human paternal investment. *Psychological Bulletin* 126, 55–77.

Gittleman, J. L. (1981). The phylogeny of parental care in fishes. *Animal Behaviour* 29, 936–41.

Gittleman, J. L. and Thompson, S. D. (1988). Energy allocation in mammalian reproduction. *American Zoologist* 28, 863–75.

Goodwin, N. B., Balshine-Earn, S., and Reynolds, J. D. (1998). Evolutionary transitions in parental care in cichlid fish. *Proceedings of the Royal Society of London, Series B* 265, 2265–72.

Goodwin, N. B., Dulvy, N. K., and Reynolds, J. D. (2002). Life-history correlates of the evolution of live bearing in fishes. *Philosophical Transactions of the Royal Society of London, Series B* 357, 259–67.

Granzberg, G. (1973). Twin infanticide—a cross-cultural test of a materialistic explanation. *Ethos* 1, 405–12.

Gross, M. R. (2005). The evolution of parental care. *Quarterly Review of Biology* 80, 37–45.

Gross, M. R. and Sargent, R. C. (1985). The evolution of male and female parental care in fishes. *American Zoologist* 25, 807–22.

Gross, M. R. and Shine, R. (1981). Parental care and mode of fertilization in ectothermic vertebrates. *Evolution* 35, 775–93.

Grubbauer, P. and Hoi, H. (1996). Female penduline tits (*Remiz pendulinus*) choosing high quality nests benefit by decreased incubation effort and increased hatching success. *Ecoscience* 3, 274–9.

Haddad, C. F. B. and Prado, C. P. A. (2005). Reproductive modes in frogs and their unexpected diversity in the Atlantic Forest of Brazil. *BioScience* 55, 207–17.

Haig, D. (1993). Genetic conflicts in human pregnancy. *Quarterly Review of Biology* 68, 495–532.

Handford, P. and Mares, M. (1985). The mating systems of ratites and tinamous—an evolutionary perspective. *Biological Journal of the Linnean Society* 25, 77–104.

Hansell MH. (2000). *Bird Nests and Construction Behaviour*. Cambridge University Press, Cambridge.

Harshman, J., Braun, E. L., Braun, M. J., Huddleston, C. J., and Bowie, R. C. K. (2008). Phylogenomic evidence for multiple losses of flight in ratite birds. *Proceedings of the National Academy of Sciences of the United States of America* 105, 13462–7.

Hawkes, K. (2003). Grandmothers and the evolution of human longevity. *American Journal of Human Biology* 15, 380–400.

Hawkes, K. (2004). Mating, parenting, and the evolution of human pair bonds. In B. Chapais and C. Berman, eds. *Kinship and Behavior in Primates*, pp. 443–74. Oxford University Press, Oxford.

Hawkes, K. and Bird, R. B. (2002). Showing off, handicap signaling, and the evolution of men's work. *Evolutionary Anthropology* 11, 58–67.

Hawkes, K., O'Connell, J. F., Jones, N. G. B., Alvarez, H., and Charnov, E. L. (1998). Grandmothering, menopause, and the evolution of human life histories. *Proceedings of the National Academy of Sciences of the United States of America* 95, 1336–9.

Hayssen, V. (1993). Empirical and theoretical constraints on the evolution of lactation. *Journal of Dairy Science* 76, 3213–33.

Hill, K. and Kaplan, H. (1988). Tradeoffs in male and female reproductive strategies among the Ache: Part 1. In L. Betzig, M. B. Mulder, and P. Turke, eds. *Human Reproductive Behaviour: A Darwinian Perspective*, pp. 277–89. Cambridge University Press, Cambridge.

Hill, K. and Kaplan, H. (1999). Life history traits in humans: theory and empirical studies. *Annual Review of Anthropology* 28, 397–430.

Houston, A. I., Székely, T., and McNamara, J. M. (2005). Conflict between parents over care. *Trends in Ecology and Evolution* 20, 33–8.

Hrdy, S. (2009). *Mothers and Others: The Evolutionary Origins of Mutual Understanding*. Harvard University Press, Cambridge, Mass.

Hrdy, S., Janson, C., and van Schaik, C. (1994). Infanticide: let's not throw out the baby with the bath water. *Evolutionary Anthropology* 3, 151–4.

Jones, J. S. and Wynne-Edwards, K. E. (2000). Paternal hamsters mechanically assist the delivery, consume amniotic fluid and placenta, remove fetal membranes, and provide parental care during the birth process. *Hormones and Behavior* 37, 116–25.

Kelly, R. L. (1995). *The Foraging Spectrum: Diversity in Hunter-Gatherer Lifeways*. Smithsonian Institution-Press, Washington.

Ketterson, E. D. and Nolan, V. (1994). Male parental behavior in birds. *Annual Review of Ecology and Systematics* 25, 601–28.

Kindsvater, H. K., Bonsall, M. B., and Alonzo, S. H. (2011). Survival costs of reproduction predict age-dependent variation in maternal investment *Journal of Evolutionary Biology* 24, 2230–40.

Kleiman, D. G. and Malcolm, J. R. (1981). The evolution of male parental investment in mammals. In D. J. Gubernick and P. H. Klopfer, eds. *Parental Care in Mammals*, pp. 347–87. Plenum, New York.

Kokko, H. and Jennions, M. D. (2008). Parental investment, sexual selection and sex ratios. *Journal of Evolutionary Biology* 21, 919–48.

Kunz, T. H. and Hosken, D. J. (2009). Male lactation: why, why not and is it care? *Trends in Ecology and Evolution* 24, 80–5.

Kvarnemo, C. (2010). Parental care. In D. F. Westneat and C. W. Fox, eds. *Evolutionary Behavioral Ecology*. Oxford University Press, Oxford.

Lack, D. (1968). *Ecological Adaptions for Breeding in Birds*. Methuen, London.

Lahdenpera, M., Lummaa, V., Helle, S., Tremblay, M., and Russell, A. (2004). Fitness benefits of prolonged post-reproductive lifespan in women. *Nature* 428, 178–81.

Lang, J. W. (1987). Crocodilian behaviour: implications for management. pp. 273–94. In G. J. W. Webb, S. C. Manolis, and P. J. Whitehead, eds. Wildlife management: crocodiles and alligators. Beatty, Surrey.

Laurien-Kehnen, C. and Trillmich, F. (2003). Lactation performance of guinea pigs (*Cavia porcellus*) does not respond to experimental manipulation of pup demands. *Behavioral Ecology and Sociobiology* 53, 145–52.

Lefèvre, C. M., Menzies, K., Sharp, J. A., and Nicholas, K. R. (2010). Comparative genomics and transcriptomics of lactation. In P. Pontarotti, ed. *Evolutionary Biology—Concepts, Molecular and Morphological Evolution*, pp. 115–32. Springer, Berlin.

Long, C. A. (1969). The origin and evolution of mammary glands. *BioScience* 19, 519–23.

Lutz, B. (1947). Trends towards non-aquatic and direct development in frogs. *Copeia* 4, 242–52.

Magnusson, W. E. and Hero, J. M. (1991). Predation and the evolution of complex oviposition behavior in amazon rain-forest frogs. *Oecologia* 86, 310–18.

Mank, J. E., Promislow, D. E. L., and Avise, J. C. (2005). Phylogenetic perspectives in the evolution of parental care in ray-finned fishes. *Evolution* 59, 1570–8.

Marlowe, F. (1999). Showoffs or providers?: The parenting effort of Hadza men. *Evolution and Human Behavior* 20, 391–404.

Marlowe, F. (2000). Paternal investment and the human mating system. *Behavioural Processes* 51, 45–61.

Martin, T. E. (1987). Food as a limit on breeding birds: a life- history perspective. *Annual Review of Ecology and Systematics* 18, 453–87.

Martin, T. E. and Li, P. (1992). Life history traits of open- versus cavity-nesting birds. *Ecology* 73, 579–92.

Martins, M., Pombal, J. P., and Haddad, C. F. B. (1998). Escalated aggressive behaviour and facultative parental care in the nest building gladiator frog, *Hyla faber*. *Amphibia-Reptilia* 19, 65–73.

Maynard Smith, J. (1977). Parental investment: a prospective analysis. *Animal Behaviour* 25, 1–9.

Monaghan, P., Nager, R., and Houston, D. (1998). The price of eggs: increased investment in egg production reduces the offspring rearing capacity of parents. *Proceedings of the Royal Society of London, Series B* 265, 1731–5.

Montgomery, G. G. and Sunquist, M. E. (1978). Habitat selection and use by two-toed and three-toed sloths. In G. G. Montgomery, eds. *The Ecology of Arboreal Folivores*, pp. 329–59. Smithsonian Institution Press, Washington, DC.

Nager, R. G. (2006). The challenges of making eggs. *Ardea* 94, 323–46.

Nager, R. G. and van Noordwijk, A. J. (1992). Energetic limitation in the egg-laying period of great tits. *Proceedings of the Royal Society of London, Series B* 249, 259–63.

Norberg, R. A. (1981). Temporary weight decrease in breeding birds may result in more fledged young. *American Naturalist* 118, 838–50.

Oftedal, O. T. (2002). The origin of lactation as a water source for parchment-shelled eggs. *Journal of Mammary Gland Biology and Neoplasia* 7, 253–66.

Oring, L. W. (1982). Avian mating systems. In D. S. Farner, J. R. King, and K. C. Parkes, eds. *Avian Biology*, pp. 1–92. Academic Press, New York.

Östlund-Nilsson, S. (2001). Fifteen-spined stickleback (*Spinachia spinachia*) females prefer males with more secretional threads in their nests: an honest-condition display by males. *Behavioral Ecology and Sociobiology* 50, 263–9.

Owens, I. P. F. (2002). Male-only care and classical polyandry in birds: phylogeny, ecology and sex differences in remating opportunities. *Philosophical Transactions of the Royal Society of London, Series B* 357, 283–93.

Phillips, M. J., Gibb, G. C., Crimp, E. A., and Penny, D. (2010). Tinamous and Moa flock together: Mitochondrial genome sequence analysis reveals independent losses of flight among Ratites. *Systematic Biology* 59, 90–107.

Qualls, C. P. and Shine, R. (1995). Maternal body-volume as a constraint on reproductive output in lizards—evidence from the evolution of viviparity. *Oecologia* 103, 73–8.

Reichard, U. H. and Boesch, C. (2003). *Monogamy: Mating Strategies and Partnerships in Birds, Humans and other Mammals*. Cambridge University Press, Cambridge.

Reynolds, J. D., Goodwin, N. B., and Freckleton, R. P. (2002). Evolutionary transitions in parental care and live bearing in vertebrates. *Philosophical Transactions of the Royal Society, Series B* 357, 269–81.

Reynolds, J. D. and Székely, T. (1997). The evolution of parental care in shorebirds: life histories, ecology, and sexual selection. *Behavioral Ecology* 8, 126–34.

Russell, E. M. (1982). Patterns of parental care and parental investment in marsupials. *Biological Reviews* 57, 423–86.

Sargent, R. G., Taylor, P. D., and Gross, M. R. (1987). Parental care and the evolution of egg size in fishes. *American Naturalist* 129, 32–46.

Scott, D. K. (1980). Functional aspects of prolonged parental care in Bewick's swans. *Animal Behaviour* 28, 938–52.

Shanley, D. P. and Kirkwood, T. B. L. (2001). Evolution of the human menopause. *Bioessays* 23, 282–7.

Sheldon, F. H., Whittingham, L. A., Moyle, R. G., Slikas, B., and Winkler, D. W. (2005). Phylogeny of swallows (Aves: Hirundinidae) estimated from nuclear and mitochondrial DNA sequences. *Molecular Phylogenetics and Evolution* 35, 254–70.

Shine, R. (1985). The evolution of viviparity in reptiles: an ecological analysis. In C. Gans and F. Billet, eds. *Biology of the Reptilia: Development*, pp. 605–94. John Wiley, New York.

Shine, R. (1988). Parental care in reptiles. In C. Gans and R. B. Huey, eds. *Biology of the Reptilia*, pp. 275–330. Liss, New York.

Silk, J. B., Alberts, S. C., and Altmann, J. (2003). Social bonds of female baboons enhance infant survival. *Science* 302, 1231–4.

Silver, R., Andrews, H., and Ball, G. F. (1985). Parental care in an ecological perspective: A quantitative analysis of avian subfamilies. *American Zoologist* 25, 807–22.

Smith, C. C. and Fretwell, S. D. (1974). The optimal balance between size and number of offspring. *American Naturalist* 108, 499–506.

Soler, J. J., Møller, A. P., and Soler, M. (1998). Nest building, sexual selection and parental investment. *Evolutionary Ecology* 12, 427–41.

Stone, T. (1989). Origins and environmental significance of shell and earth mounds in Northern Australia. *Archaeology in Oceania* 24, 59–64.

Summers, K. and Earn, D. J. D. (1999). The cost of polygyny and the evolution of female care in poison frogs. *Biological Journal of the Linnean Society* 66, 515–38.

Summers, K., McKeon, C. S., and Heying, H. (2006). The evolution of parental care and egg size: a comparative analysis in frogs. *Proceedings of the Royal Society, Series B* 273, 687–92.

Summers, K., McKeon, C. S., Heying, H., Hall, J., and Patrick, W. (2007). Social and environmental influences on egg size evolution in frogs. *Journal of Zoology, London* 271, 225–32.

Székely, T. and Reynolds, J. D. (1995). Evolutionary transitions in parental care in shorebirds. *Proceedings of the Royal Society of London, Series B* 262, 57–64.

Székely, T., Cuthill, I. C., and Kis, J. (1999). Brood desertion in Kentish plover: sex differences in remating opportunities. *Behavioral Ecology* 10, 185–90.

Szentirmai, I., Komdeur, J., and Székely, T. (2005). What makes a nest-building male successful? Male behavior and female care in penduline tits. *Behavioral Ecology* 16, 994–1000.

Tang, M. X., Jacobs, D., Stern, Y., Marder, K., and Schofield, P. (1996). Effect of oestrogen during menopause on risk and age at onset of Alzheimer's disease. *Lancet* 348, 429–32.

Temrin, H., Buchmayer, S., and Enquist, M. (2000). Step-parents and infanticide: new data contradict evolutionary predictions. *Proceedings of the Royal Society of London, Series B* 267, 943–5.

Tinkle, D. W. and Gibbons, J. W. (1977). The distribution and evolution of viviparity in reptiles. *Miscellaneous Publications Museum of Zoology, University of Michigan*, 154, 1–55.

Trivers, R. L. (1972). Parental investment and sexual selection. In B. Campbell, ed. *Sexual Selection and the Descent of Man, 1871–1971*, pp. 136–79. Aldine, Chicago.

Tullberg, B. S., Ah-King, M., and Temrin, H. (2002). Phylogenetic reconstruction of parental-care systems in the ancestors of birds. *Philosophical Transactions of the Royal Society, Series B* 357, 251–7.

van Rhijn, J. G. (1991). Mate guarding as a key factor in the evolution of parental care in birds. *Animal Behaviour* 41, 963–70.

Varricchio, D. J., Moore, J. R., Erickson, G. M., Norell, M. A., Jackson, F. D., and Borkowski, J. J. (2008). Avian paternal care had dinosaur origin. *Science* 322, 1826–8.

Vincent, A. C. J. and Sadler, L. M. (1995). Faithful pair bonds in wild seahorses, *Hippocampus whitei*. *Animal Behaviour* 50, 1557–69.

Vorbach, C., Capecchi, M. R., and Penninger, J. M. (2006). Evolution of the mammary gland from the innate immune system? *Bioessays* 28, 606–16.

Wake, M. H. (1993). Evolution of oviductal gestation in amphibians. *Journal of Experimental Zoology* 266, 394–413.

Washburn, S. L. and Lancaster, C. L. (1968). The evolution of hunting. In R. B. Lee and I. DeVore eds. *Man the Hunter*, pp. 293–303. Aldine, Chicago.

Weber, J. N. and Hoekstra, H. E. (2009). The evolution of burrowing behaviour in deer mice (genus *Peromyscus*). *Animal Behaviour* 77, 603–9.

Wells, K. (2007). *The Ecology and Behavior of Amphibians*. University of Chicago Press, Chicago.

Wesołowski, T. (1994). On the origin of parental care and the early evolution of male and female parental-roles in birds. *American Naturalist* 143, 39–58.

Wesołowski, T. (2004). The origin of parental care in birds: a reassessment. *Behavioral Ecology* 15, 520–3.

Westneat, D. F. and Sherman, P. W. (1993). Parentage and the evolution of parental behavior. *Behavioral Ecology* 4, 66–77.

Weygoldt, P. (1987). Evolution of parental care in dart poison frogs (Amphibia: Anura: Dendrobatidae). *Zeitschrift für Zoologisch, Systematik und Evolutionsforschung* 25, 51–67.

Wilkinson, M. and Nussbaum, R. A. (1998). Caecilian viviparity and amniote origins. *Journal of Natural History* 32, 1403–9.

Williams, G. C. (1966). *Adaption and Natural Selection*. Princeton University Press, Princeton.

Winkler, D. W. and Sheldon, F. H. (1993). Evolution of nest construction in swallows (Hirundinidae)—a molecular phylogenetic perspective. *Proceedings of the National Academy of Sciences of the United States of America* 90, 5705–7.

Wong, M. and Balshine, S. (2010). The evolution of cooperative breeding in the African cichlid fish, *Neolamprologus pulcher*. *Biological Reviews* 86, 511–30.

Woodroffe, R. and Vincent, A. (1994). Mothers little helpers—patterns of male care in mammals. *Trends in Ecology and Evolution* 9, 294–7.

Wootton, R. J. (1990). *Ecology of Teleost Fishes*. Chapman and Hall, London.

Wourms, J. P. and Lombardi, J. (1992). Reflections on the evolution of piscine viviparity. *American Zoologist* 32, 276–93.

Wright, J. and Cuthill, I. (1989). Manipulation of sex-differences in parental care. *Behavioral Ecology and Sociobiology* 25, 171–81.

CHAPTER 5

Patterns of parental care in invertebrates

Stephen T. Trumbo

5.1 Introduction

The tremendous diversity of social behaviour among the invertebrates is an asset and a challenge. There is richness in both the number of phylogenetic lineages that have evolved extended parental care and the forms of care provided by parents. Fifty families of insects in more than a dozen orders have evolved parental care (Costa 2006), and sixteen separate lineages of arthropods exhibit paternal care (Tallamy 2001; Nazareth and Machado 2009). There are three principal avenues for progress in evolutionary biology: new theory to test, new technology for measurement, and new subject matter. This last avenue is still wide open for students of parental care in invertebrates. Among birds and mammals, extended parental care is universal (Chapter 4) and most forms of care have been well studied. On the other hand, new species and new forms of care in the invertebrates are being discovered yearly, permitting a creative interplay between inductive and deductive approaches. The non-eusocial invertebrates are where we will find the subject matter for understanding the origins of parental care, transitions between types of care, and manipulable systems for testing theories. Broad phylogenetic comparisons will permit testing of the ecological factors that favour different social solutions, and also whether convergent parental behaviour is built upon convergent physiological mechanisms. More narrow comparisons will reveal how closely related species can take widely divergent social paths. For studies of single species, invertebrates offer many pragmatic advantages over vertebrates. First, most invertebrates live in a very different sensory world than humans. While this imposes barriers to our understanding, it also allows manipulations, such as observation under red light, that have minimal effect on experimental subjects. Second, the smaller size and shorter lives of many invertebrates make it practical to follow large numbers of subjects over their lifetime, often in the laboratory. Third, many invertebrates are also suitable subjects for selection experiments and genetic analysis, which will eventually allow an understanding of how interactions between genes and developmental environments produce the tremendous variation in parental care and social behaviour.

According to broad definitions of parental care, care includes all parental traits that enhance offspring fitness (Chapter 1). For the purpose of discussing the origins and transitions of extended care in the present chapter, consideration is confined primarily to post-fertilization traits that increase offspring fitness, beyond the temporary housing and passage of the fertilized egg within the female. Viviparity and ovoviviparity will therefore be discussed, but the amount of yolk in an egg and selection of an oviposition site will not. After an overview of the forms of parental care, the origins, transitions, and loss of parental care will be discussed, including male versus female care. Lastly, the microbiology of care is salient for appreciating the complexity of social invertebrates and needs to be integrated into our understanding of parental care. The physiology of care is not treated here because this topic was covered in a prior review (Trumbo 2002).

The Evolution of Parental Care. First Edition. Edited by Nick J. Royle, Per T. Smiseth, and Mathias Kölliker.

5.2 Forms of care

The evolution of parental care in many independent lineages of invertebrates has resulted in diverse forms of care (Chapter 1). An exhaustive review of all forms of care in invertebrates is not feasible. Although this chapter uses a slightly different scheme to characterize the diversity of forms of care, the categories used here can easily be fitted into the more general scheme used in Chapter 1. I start with two basic forms of parental care, the use of trophic eggs and sedentary protection of eggs. I then discuss more complex forms, many of which were built upon the initial parent–egg association.

5.2.1 Trophic eggs

Trophic eggs are food provisions that do not require direct maternal–offspring contact, and either may be the sole form of care or part of a diverse suite of parental behaviours. Reduction of sibling cannibalism may have been an important ecological need at the origin of trophic egg production. In some species, juveniles cannibalize viable eggs or opportunistically consume inviable eggs before leaving the oviposition site. In such species, the origin of trophic eggs may be preceded by kin selection on cannibalistic young to discriminate between viable and inviable eggs. To assist young in making the correct choice of which eggs to consume, maternally derived cues could be incorporated into trophic eggs that initially mimicked the kin-selected discrimination cues employed by juveniles. Production of trophic eggs might reduce cannibalism, thereby enhancing direct fitness of parents and indirect fitness of offspring (Perry and Roitberg 2006). Because accurate discrimination between eggs may not increase the direct fitness of young, the use of trophic eggs may not necessarily have originated as a form of parental care.

Although there is debate in particular species about whether a consumed egg was indeed 'trophic', there are also clear cases of use of trophic eggs. True trophic eggs should be easy to distinguish from non-trophic eggs, should be available at the appropriate time for developing young, and should provide a nutritional benefit not easily supplied by the mother (West and Alexander 1963). In the wood-feeding passalid beetle *Cylindrocaulus patalis*, trophic eggs are paler, softer, and with a less complex chorion compared to viable eggs. If third instar larvae stridulate, they are fed a trophic egg by the mother (Ento et al. 2008). In the burrower bug *Canthophorus niveimarginatus*, females produce some trophic eggs at the same time as they produce viable eggs, but they also lay additional trophic eggs after the nymphs hatch (Filippi et al. 2009). In burrower bugs, trophic eggs can provide food for nymphs while the mother forages for seeds away from the nest (Hironaka et al. 2005). In each of these systems, production of trophic eggs is part of a suite of parental behaviours. Investigation of systems with only trophic egg provisioning will provide insight into its origin and allow tests of the reduction of cannibalism hypothesis.

5.2.2 Attending eggs and offspring

After a female oviposits in a selected location, a tendency to linger near the clutch might offer modest protection from predators or parasitoids. Protection requires clumping of eggs (as opposed to scattering in the environment), a behaviour that may have preceded extended care due to the enhancement of an aposematic effect, facilitation of feeding among siblings, or oviposition near a patchy food source. Subsequent to the evolution of post-ovipositional care, young may be under selection to aggregate to facilitate care. At its origin, protection of young may evolve without specialized parental behaviour. Tallamy and Schaefer (1997) have pointed out that maternal defensive behaviours in the lace bug *Gargaphia solani* are similar to those used in conspecific interactions, and were likely to have been co-opted from the former without a long period of evolutionary modification. Over time, defensive behaviour can become increasingly complex. In the treehopper *Umbonia crassicornus*, mothers tilt their elongated pronotum, and fan and buzz potential threats, protecting their young until adulthood. In the presence of a predator, offspring produce synchronous vibrations that inform the mother on which side of the aggregation the threat is more imminent (Ramaswamy and Cocroft 2009). Not all predators, however, can be deterred by active defense. Female spider mites (*Stigmaeopsis*) employ a form of misdirection where they produce numerous 'void' nests in addition to the true nest that contains an egg mass. Void nests appear to reduce predator search efficiency, and ultimately reduce predator motivation (Saito et al. 2008).

5.2.3 Protection and facilitating feeding of mobile young

Mobile young feeding out in the open are vulnerable to predators and parasitoids (Tallamy and Wood 1986). In some species, the protective mother simply follows her young around. For example, in the acanthosomatid bug *Elasmucha dorsalis*, protection lasts for over two months while the nymphs feed on fruits or flowers (Kûdo et al. 1989). The presence of the mother can also facilitate cooperative behaviour among juveniles, such as in scorpions where groups of young are able to subdue much larger prey than young feeding alone (Mahsberg 2001). The benefits of mass action or byproduct mutualisms among aggregating juveniles may exceed the costs of heightened predation only in the presence of a defending parent. Mothers can also play a more direct role in facilitating feeding. Mothers of the treehopper *Umbonia crassicornus* make a series of spiral slits down the stem of the host plant where their young aggregate to feed. Parental females stand nearby, not only threatening potential predators, but stroking wandering nymphs (Wood 1976). Such 'herding' behaviour has evolved convergently in folivorous beetles (Windsor and Choe 1994), as well as in fungus feeding beetles of the genus *Pselaphacus* (Preston-Mafham and Preston-Mafham 1993). In this latter group, the mother shepherds her larvae from fungus to fungus, first sitting over them and then walking off, some young underneath and some straggling behind in a line, perhaps following a pheromone trail. Some mothers appear to lead larvae out to feeding sites each day and back to a hiding place each night, permitting a rapid development that may be as short as four days.

5.2.4 Brooding behaviour and viviparity

An alternative form of protection is to carry eggs or young either internally or externally. Brood pouches are common among non-insect invertebrates. In an exceptional case, mothers of the tailless whipscorpion *Phrynus marginemaculatus* protect their young for at least 11 months. Others, such as fathers of all species of sea spiders, simply carry eggs on ovigerous legs (Barreto and Avise 2008), as does the mother in the only squid species known to have post-spawning care (Seibel et al. 2005). Viviparous scorpion mothers extend care by assisting in birth and then transferring young to her back. During transfer, young that have difficulty separating from their birth membranes are more likely to be cannibalized by the mother, salvaging nutrients from offspring unlikely to thrive (Mahsberg 2001).

Leeches provide diverse forms of care, ranging from the production of a protective cocoon for eggs, via brooding of eggs and young, to feeding of young (Kutschera and Wirtz 2001). In the hermaphroditic leech *Helobdella papillornata* the parent carries young on its venter for up to 60 days and provisions them with gastropods. Paez et al. (2004) argue that although parental care in this sit-and-wait predator is costly, it may be less so than it would be in leeches that actively hunt. Cockroaches are also known for carrying young in a variety of ways. Oviparous species carry an ootheca externally, false ovoviviparous species retract the ootheca into the abdomen, true ovoviviparous cockroaches carry eggs and then the young in a brood sac without producing an ootheca, and viviparous species nourish young internally with secretions (Nalepa and Bell 1997). In the ovoviviparous *Thorax porcellana*, care is extended when hatching nymphs ride in a specialized compartment on the dorsum (Fig. 5.1) and obtain liquid nourishment through pores of the mother by using their mandibles to pierce the mother's cuticle (Bhoopathy 1998). These specialized mandibular 'teeth' are lost at a later instar when young no longer associate with the mother.

In aquatic environments, oxygen availability can be a fundamental constraint on the number of embryos that can be brooded and there can be considerable costs to making oxygen available to the centre of a large embryo mass (Fernández et al. 2000). Smith (1997) has related how the back-brooding male giant water bug must keep eggs moist to prevent desiccation, yet periodically push

Figure 5.1 (a) Centipede (believed to be *Scolopendra subspinus*) wrapped around her eggs (Scott Camazine); (b) a webspinner (*Partenembia reclusa*) mother emerging from underneath her silk nest (Janice Edgerly-Rooks); (c) a male giant water bug (*Belostoma flumineum*) carrying eggs of two females who oviposited close in time (Scott Kight); (d) the cockroach *Thorax porcellana* carrying nymphs in a specialized compartment on her dorsum beneath the carapace (Natasha Mhatre).

the eggs above the water surface to increase oxygen availability. Where paternal care is so essential, it is predicted that male sexual advertisement might reflect his parental ability (Kelly and Alonzo 2009). The belostomatids would be an attractive model for this question because the pumping sexual display of males shares many features in common with the vigorous oxygenating behaviour during care. The ancestral pattern within Belostomatidae of ovipositing eggs on emergent vegetation adequately met the oxygen demand, but necessitated that the male imbibe and regurgitate water onto the eggs, or else stand above the egg cluster and dribble water down toward the clutch (Ichikawa 1988). These extraordinary adaptations for aerating and hydrating eggs may be necessitated by selection for large eggs that must rely on passive diffusion of oxygen (Smith 1997).

5.2.5 Nest building and burrowing

The ability to create and maintain a long-term favourable microenvironment for eggs or young (nesting) is a foundation for many of the remarkable social behaviours found in invertebrates. Webspinners build a branched silk tunnel system that not only protects against predators, but provides a favourable physical environment by elevating humidity and reducing temperature (Edgerly 1997). Enhancing offspring fitness by environmental buffering is most evident in extreme environments such as the desert where burrow systems make the habitat tolerable (Rasa 1998). A remarkable example of transition to a niche widely divergent from the ancestral state occurs in the crab *Metopaulias depressus*, which maintains a nursery in water-holding epiphytic bromeliads. During a 9-week period the mother removes debris that reduces substrates for oxygen-consuming decomposers, and adds shells to increase the availability of calcium carbonate (Diesel and Schuh 1993).

5.2.6 Food provisioning

More complex parent–offspring interactions can develop subsequent to the evolution of nesting. In addition to trophic eggs (see Section 5.2.1), there are four basic types of food provisioning based on a nest: 1) mass provisioning for each offspring, 2) mass provisioning with a large resource (e.g. carrion) for multiple offspring, 3) nesting within the food source that is modified to serve as shelter by the parent (e.g. a log or tree), and 4) progressive provisioning of young. In a classic work, Evans (1958) outlined a series of increasingly more complex nesting behaviours among wasps, from mass provisioning of a single food item for each larva, to progressive provisioning of smaller items. Diversity of forms of provisioning behaviour is also seen in spiders. The simplest form, and perhaps the origin of provisioning in spiders, might have occurred when a mother tolerates young for an extended period near a site where she stores food (Buskirk 1981). Progressive provisioning occurs in many social spiders in which parents and sometimes juveniles cooperate to capture numerous prey, and in some cases (e.g. *Anelosimus studiosus*) the young stay until maturity (Jones and Parker 2002). The spider *Stegodyphys lineatus* regurgitates repeatedly to young and then makes the ultimate sacrifice, allowing the young to feed upon her body (Salomon et al. 2011).

Some dung- and carrion-feeders are mass provisioners of a sort. The nutrient-rich resource typically requires ongoing maintenance to control microbial competitors and ward off potential usurpers (Hanski and Cambefort 1991). Dung feeders match the available food resource to brood number by partitioning the resource for each larva, while burying beetles adjust the total number of young to match the size of the indivisible carrion resource (Wilson and Fudge 1984). Wood feeders typically live inside their food resource, which also provides shelter. After egg-laying, parents can progressively extend tunnel systems to meet the demands of growing young who feed off wood, fecal material, or fungi that use wood as a substrate (Kirkendall et al. 1997; Schuster and Schuster 1997).

For many types of food provisioning, there is the potential for feedback from young to alter subsequent parental care. Mass provisioners such as burying beetles and dung beetles, and progressive provisioners such as burrower bugs and earwigs, adjust their parental behaviour to cues from offspring. Overlap of parents and offspring and parent–offspring communication can be co-opted for the elaboration of more complex social behavior. For example, permanent-social behaviour is seen in progressively provisioning spiders (Avilés 1997), eusociality evolved from progressive provisioning Hymenoptera, and eusociality from wood-feeding in bark beetles (Kent and Simpson 1992).

The most intimate parent–offspring interactions are observed when parents feed young by regurgitation (stomodeal trophallaxis). Regurgitation allows parents to soften food, add digestive enzymes, and transfer symbionts. Parents may respond to cues from offspring as in the earwig *Forficula auricularia*, which increases provisioning to high quality offspring (Mas et al. 2009). Regurgitation behaviour may be especially critical for altricial young with poorly developed sensory abilities and unsclerotized mouthparts, as found in the woodroach *Salganea* (Nalepa et al. 2008). In the burying beetle *Nicrophorus vespilloides*, parents respond to the hunger level of offspring, while offspring increase begging in response to a chemical cue from the parent (Smiseth et al. 2010). Mutual adjustment of provisioning and begging behavior by parent and offspring suggests a complex co-evolution in which conflicts over the amount of care are resolved.

5.3 Origins and transitions of parental care

5.3.1 Factors promoting care

Students of evolutionary biology are taught very early that an understanding of the maintenance of an adaptation does not explain its origin. The origin of parental care has not attracted as much attention as other issues (Clutton-Brock 1991), such as which sex provides care, transitions between types of care, or the origin of eusociality from 'presocial' behavior. The universality of extended care among birds and mammals, and their origins in a distant reptilian past, explains some of the neglect. Biologists studying ectotherms, in which different forms of care have evolved many times independently, have shown more interest in this issue, stimulated by the opportunity for comparative work (Reynolds et al. 2002). Using dung beetles, Halffter and Edmonds (1982) outlined an increasing complexity of parental provisioning across extant species that is suggestive of a past progressive evolutionary sequence. The diversity of parental care within some groups (e.g. no care, maternal care, paternal care, amphisexual care in harvestmen) show promise of uncovering the ecological tipping points that nudge different species toward one evolutionary pathway or the other (Machado and Macías-Ordóñez 2007). Theory and broad phylogenetic analyses, long employed for topics such as which sex provides care, have been a relatively recent approach for understanding the origins of care.

The first modern attempts to explain the origin of parental care among invertebrates emphasized ecological pressures. The prime movers of care were proposed to be stable and structured environments (K selection for parental investment), harsh environments (care as a buffer allowing expansion of the realized niche), use of rich, ephemeral resources (enhanced competition often leading to biparental cooperation), and intense predation pressure on eggs (Wilson 1975). There are two practical problems with using this framework. Most care-giving species, such as the bromeliad crab described earlier, exhibit multiple forms of care, making it difficult to determine which adaptation and which prime mover was important at the origin of care. Secondly, while most insects face at least one of these ecological challenges, extended parental care is rare (about 1% of insect species, Costa 2006). Among other groups of invertebrates, however, parental care is quite common.

Barriers to care among insects are thought to exist both because of a lack of clear benefits of care and the substantial hurdle of evolving necessary innovations. The insect ovipositor allows eggs to be dispersed in the smallest hiding places, and the remarkable ability of the insect egg to allow gas exchange while retaining even smaller molecules of water, greatly reduces the benefits of care for many species (Zeh et al. 1989). Even if the potential benefits of parental care were substantial, small-bodied, short-lived insects with limited means to modify the environment or deter predators might have difficulty evolving effective care-giving (Zeh and Smith 1985; Tallamy and Wood 1986). Only an innovation such as the modification of an ovipositor into a sting or the evolution of burrowing legs or silk production might provide a pathway through the adaptive landscape to reach parental care.

A second ecological approach to the origin of parental care emphasizes food resources, particularly their persistence, dispersion, and nutritional value (Tallamy and Wood 1986). Folivores, which often feed out in the open, tend to provide care in the form of protection against parasitoids and small predators. Detritivores, which typically feed on dispersed sources, either must forego feeding during care (e.g. earwigs) or carry their young with them (e.g. cockroaches). The use of protein-rich resources such as carrion and dung, or the use of sheltering wood may select for biparental guarding of the resource, which in turn may set the stage for protection of young (carrion and dung) or nutritional assistance (wood). Insect predators feed on unpredictable resources and sacrifice foraging opportunities to provide care. Tallamy (2001) noted that six of the seven insect lineages that have evolved male-only care have a predatory lifestyle, suggesting that maternal care would entail high fecundity costs. The association between a predatory habit and the absence of maternal care, however, does not hold as well for non-insects. The possession of effective defence mechanisms such as venoms has perhaps promoted extended maternal care among predatory centipedes (Fig. 5.1), spiders, and scorpions (Costa 2006).

A third ecological approach proposes that a primary barrier to the evolution of parental care is the cost of future reproduction (Trivers 1972; Tallamy and Schaefer 1997). Parental care is not only an adaptation that can solve an ecological problem, but may incur substantial fitness costs to parents. Among species with indeterminate growth such as fish and many marine invertebrates, parental care will subtract resources that otherwise could be allocated toward adult growth. Because fecundity is often related to body size by positive allometry, care will have high costs in species with the potential for large body size. This may explain the association of parental care and small body size in many ectotherms with indeterminate growth (Strathmann and Strathmann 1982).

Consideration of life-history trade-offs, feeding ecology, and Wilson's prime movers provides a plausible explanation of why some insects exhibit parental care. Left unanswered are the questions why other insects exploiting the same resources in the same habitats do not provide care, and why some invertebrate groups (spiders, scorpions, pseudoscorpions, centipedes, octopuses) are primarily parental (Costa 2006)? There are thousands of species of invertebrates that use small vertebrate carrion as a resource, some facultatively, some obligatory. Only the burying beetles (65 species) and some of the tropical scarabs have evolved parental defence of the carcass from competitors, and provisioning of food to larvae. Other carrion specialists have taken diverse parental and non-parental evolutionary routes. The sarcophagid flies employ an alternative form of parental care, bypassing an external egg stage and depositing larvae directly on the resource. Many other dipterans simply maximize speed, locating and ovipositing on a carcass within minutes of availability, while other non-parental invertebrates are latecomers and consume the leftovers. One predisposition for subsocial behavior occurs when organisms have been selected to modify their microenvironment, a process that increases the variance in environmental quality. Creating highly favourable environmental space selects for adults to remain in such spaces, and also provides an advantageous place for eggs and young to develop (Nowak et al. 2010). Such different approaches to the same resource, leading to alternative forms of parental care or to the absence of extended care, are only to be understood by overlaying life-history and ecological approaches on a phylogenetic framework.

Phylogeny will also be necessary to understand why parental care is present or absent in non-insect invertebrates. While most insects use an ovipositor to selectively locate and protect their eggs without extended care, many groups employ viviparity (scorpions), brood pouches (pseudoscorpions, tailless whipscorpions, amphipods, isopods), or egg cases (spiders) to protect young. Viviparity has also evolved numerous times among diverse marine invertebrates and in selected groups of insects such as dipterans (Reynolds et al. 2002). To understand why flies should evolve viviparity, but rarely other forms of extended parental care, will require a thorough understanding of the physiology of female reproductive systems and embryo development. Difficult, but rewarding work in evo-devo may be the only path to understand phylogenetic biases in the intensity and forms of parental care.

Gillespie and McClintock (2007) and Poulin et al. (2002) argue that historical biogeography is also needed to understand patterns of parental care. Among echinoderms, more species exhibit brooding behaviour in the Antarctica than near the Equator. While the difficult environment has been proposed to be a prime mover for care, in this case an evolutionary as well as an ecological time scale must be considered. Poulin et al. (2002) suggest that the extinction of phytoplankton during periodic shifts to colder climates may have eliminated the food source for planktotrophic echinoderms, favouring non-broadcasting species that brood their young. Brooding species are thought to have greater opportunity to buffer environmental conditions for young, allowing young to subsist (indirectly) on the food sources of their mothers.

As opposed to the extrinsic hypotheses for the origin of parental care, an intrinsic factor, haplodiploidy, is thought to predispose some groups toward care, specifically maternal care, especially when inbreeding is prevalent (Linksvayer and Wade 2005). Although haplodiploidy is well known to be associated with eusociality, it appears to be more strongly linked to subsociality (Alexander et al. 1991). Subsocial behaviour in mites and ticks is found only among haplodiploids (Saito 1997), and also in haplodiploid bark beetles (Kirkendall et al. 1997), thrips (Crespi and Mound 1997), and of course, Hymenoptera. Reeve (1993) proposed the protected invasion hypothesis to explain why the loss of dominant alleles for care and allocare to genetic drift is rarer in females of haplodiploid species. Analysis of this hypothesis has focused almost exclusively on the evolution of eusociality and awaits testing by students of subsocial behaviour.

5.3.2 Male versus female care

From the earliest days of the field, sociobiology has provided novel insights into conflicts among individuals, including conflict between males and females over mating and parenting (Trivers 1972). In this section, invertebrate models that appear particularly favourable for addressing the questions of which sex provides care, why exclusive male care has evolved, and how transitions between the various parenting systems might occur, are highlighted. In the next section, cooperation and conflicts between males and females in biparental systems are discussed.

Attractive systems for understanding the origin of male versus female care are those, like the *Rhinocoris* assassin bugs or the harvestmen, that exhibit a diversity of parental behaviour despite living in similar habitat, having a similar feeding ecology, and having the same mode of fertilization. Using a model that does not restrict parents to caring for one brood at a time, Manica and Johnstone (2004) demonstrate that female care can be favoured when the time to produce a second batch of eggs (processing time) is long and population density is low as in *Rhinocoris carmelita*, and paternal care is favoured when processing time is short and population density is high, as in *R. tristis*. Encounter rates with available females is a critical factor in allowing males to care for multiple broods, and in the evolution of exclusive male care (Gilbert et al. 2010). As a caveat, we cannot know whether the life-history and ecological conditions required by the model were present at the origin of care. Even greater diversity of care-giving (no care, maternal care, paternal care, amphisexual care) is found among harvestmen (Buzatto and Machado 2009; Nazareth and Machado 2010). Paternal care in the harvestman *Pseudopucrolia* seems to be best explained by sexual selection, as guarding males are more attractive to mates than non-guarders, so much so that males will initially guard unrelated eggs.

Exclusive male care has evolved in at least 16 independent lineages among the arthropods. As with *Rhinocoris* and harvestmen, most cases are explained by territoriality that initially allowed males to provide a benefit to young (protection) at low cost, while still attracting additional mates. This pathway to polygyny (Paternal Care Polygyny) is distinct from other forms of polygyny (Resource Defence, Female Defence, Lek) in that females are attracted to males that have initiated care-giving for another brood. Even with exclusive male care, there can be sharp conflicts between the interests of males and females. In *R. tristis*, eggs are better protected under leaves, but many males care out in the open on stems where they are more likely to attract other females. Presumably this divergent caring behaviour is maintained by the trade-off for males between less effective care for more eggs and more effective care for fewer eggs, and by the high costs for females of searching for caring males under leaves (Gilbert and Manica 2009). In other cases, exclusive male care may be a solution enhancing the reproductive potential of both sexes. In the polychaete worm *Neanthes arenaceodentata* the transfer of parental costs from the female to the male may directly benefit the male by allowing the female to put all of her resources into eggs. The female produces extremely large eggs for polychaetes, and dies soon after mating, after which she is often eaten by the male who obtains more energy for brooding (Schroeder and Hermans 1975).

The paternal care polygyny hypothesis for uniparental male care does not explain why females would oviposit where another female has placed a clutch. Two ideas for the origin of this behaviour have been proposed. Tallamy (related in Costa 2006, p. 35) suggests that multiple females might initially oviposit in the same location for reasons unrelated to care such as to facilitate feeding of her offspring, to enhance an aposematic display, or to dilute predation pressure. If so, males might be selected to intercept females directly before oviposition by defending the site against rival males. Trumbo (1996) offered an alternative possibility that females previously selected for dumping eggs in the clutch of another female could transfer that behaviour instead toward a territorial male. A male with a clutch might be more attractive to a female looking to dump eggs. This has the result of females copying the mate choice of prior females, establishing the potential for polygyny right at the origin of paternal care (Ridley 1978).

A second type of exclusive male care, found in sea spiders (Barreto and Avise 2008) and giant water bugs (Smith 1997), occurs when the male carries eggs. In these circumstances care is likely to be more costly and it is less clear whether the potential for polygyny is important. The lack of suitable oviposition sites and high fecundity costs of maternal care may have been the driving forces for male care (Tallamy 2001). Although there may be some potential for polygyny in the giant water bugs, it does not appear to be great. The number of eggs a male can carry is limited by his backspace and it may be inefficient for males to carry eggs oviposited at widely different times. When a male is carrying a partial brood, for a short time he may accept a partial brood of a second female (see Fig. 5.1). Males, however, may remove a partial brood from a first mate to make room for a full complement from a second female, suggesting that simply adding eggs from multiple females is not always a good option (Kruse 1990; Kight et al. 2000). The back-brooding belostomatids are thought to have evolved from territorial males that protected and hydrated eggs that females oviposited on emergent vegetation, as occurs in extant *Lethocerus* in which the potential for polygyny may have been greater (Ichikawa 1988). An exemplary comparative study integrating the ecology, phylogeny and physiology, of care in these groups can be found in Smith (1997).

The evolutionary pathways leading to exclusive male care or exclusive maternal care are not well understood. Models of biparental care suggest that biparental care can be unstable when either parent can desert, forcing the other parent into a 'cruel bind' where it either must provide care or experience no reproductive success (Trivers 1972). Investigators of invertebrate parental care, however, have not proposed such a pathway for exclusive male care, exclusive female care, or asymmetric biparental care. In three insect systems with biparental or amphisexual care, we know the response when one parent deserts or is experimentally removed. The responses indicate that one parent could easily manipulate the other into providing more parental care, but this seems to be rarely done. In the reduviid, *Rhinocoris tristis*, males typically provide all parental care. It was noticed, however, that if the male were removed, the female parent would return and provide care until young hatch, acting just as aggressively as a male parent. Beal and Tallamy (2006) argue that female care is rare in nature, however, because the male rarely deserts and males will guard unrelated eggs if given the opportunity. Male care is likely maintained by the ability to attract mates, rather than by female parenting decisions.

In the harvestmen *Serracutisoma proximum*, which shows female uniparental care, females guard eggs within a male's superterritory. If one of a male's females is temporarily removed, the male will guard the clutch for 2–9 days (Buzatto and Machado 2009). Although males will not guard for the entire period of egg development (37 days), the male's response suggests that females could take considerable advantage of the male's willingness to care for moderate intervals. Females will occasionally do so for several hours during cold periods, but could be expected to exploit the male's response for longer periods of foraging during warmer periods. Females, however, seem to fail to do so.

In burying beetles (*Nicrophorus* spp.) with biparental care, the female typically provides care until the larvae disperse from the carcass (10–15 days), while the male leaves several days earlier. If the female dies or is removed, the male compensates by providing more care and staying until larvae disperse, achieving the same degree of reproductive success as the female (Trumbo 1991; Fetherston et al. 1994). Females, however, are almost never known to exploit this paternal response by leaving early. Female behaviour may be constrained because she has to completely abandon her brood to induce the male to provide more care. If she simply reduces her level of care while remaining in the nest, the male does not compensate (Suzuki and Nagano 2009). Paternal care can lessen the lifetime costs of care for the female (Jenkins et al. 2000), but there is no invertebrate system in which this has been hypothesized to be the route to the complete abandonment of care by the female, as proposed for birds (Emlen and Oring 1977). In each of these three systems (*Rhinocoris*, *Serracutisoma*, *Nicrophorus*), the parent that normally provides more care seems to do so primarily because of its own costs and benefits.

5.3.3 Biparental care

While competition for a resource or the necessity of two parents to construct a nest may explain the origin of biparental care, it does not explain how extended cooperative biparental associations can be stable through evolutionary time. The sources of conflict that might destabilize biparental care are now well modelled (Chapters 6 and 9). Both parents profit from high productivity of the parental pair, but each parent does better by having its partner bear the costs of care, allowing resources to be saved for future reproduction (Trivers 1972). An understanding of conflict and cooperation at multiple levels in primitively eusocial insects might provide a framework for also understanding biparental care. Groups with more cooperative members can achieve higher group output (number of reproductively capable offspring) than competing groups with less cooperative members. Within the group, however, conflicts exist because individuals can attempt to shift costs to group members. An additional source of conflict occurs over genetic representation in offspring produced. In species with biparental care, intraspecific brood parasitism by females that mate with the paired male, and sperm storage by the paired female, place the male and female in direct conflict. Despite these sources of conflict, both primitively eusocial and biparental groups persist. Stabilizing features of social units include an alignment of genetic interests, mass action by a larger group, policing, insurance against death of a group member, division of labour with or without task specialization, and a lack of future reproductive opportunities (Oster and Wilson 1987; Keller and Reeve 1999). Among invertebrates, biparental care is usually associated with a nest (Eickwort 1981), a structured environment that can facilitate selection for division of labour, and a valuable resource that must be protected.

There are some tasks that a single parent cannot finish within the requisite time for successful brood production (necessity of mass action). Time constraints appear to be important for many invertebrates that take advantage of transient reproductive opportunities in severe habitats. In the desert isopod *Hemilepistus reamuri*, both parents fashion a burrow system to moderate temperature and increase microhabitat humidity (Linsenmair 1987). Similar constraints occur for the tenebrionid beetle *Parastizopus armaticeps* in which the male takes over burrow building from the female while the female begins foraging. When food is made available experimentally, the female has more time to assist in working on the burrow, a critical feature of reproductive success in the desert (Rasa 1998; Rasa 1999).

Biparental care occurs in log-inhabiting cockroaches in which slow-growing young are dependent on parents for an extended period (Nalepa et al. 2008). Reproductive attempts may be abandoned if one parent dies early in the nesting cycle, suggesting the combined effort from two parents is necessary to prepare the nest. Abandonment of reproduction with the death or desertion of a partner suggests the potential for a form of policing in which continued cooperation depends on a minimal level of partner presence and effort. Female *Trypoxylon* wasps will not forage for the brood unless a male is present to guard the nest. Such female decision rules would select for the male to make his presence known. Communication might promote stability of biparental care by coordinating and monitoring activity. Acoustic and chemical communication between partners is varied and complex in the passalid beetle *Odontotaenius disjunctus* (Schuster and Schuster 1997) and in the burying beetle *Nicrophorus vespilloides* (Steiger et al. 2009). Disruption of acoustic communication in *N. mexicanus* prevents coordination between partners (Huerta and Halffter 1992). Repeated interaction of just two individuals in a confined area such as a nest provides an available mechanism for both monitoring partner effort and to increase confidence of parentage.

Parents are expected to be more likely to provide care to related than unrelated young. House et al. (2008) explored the genetics of repeated mating in burying beetles and its effect on paternity. They found that males are under strong selective pressure to mate repeatedly to increase paternity while females are not under strong selection to refuse copulations because the costs are small. The sexual dynamic results in numerous, brief copulations in the first 24 h on a carcass before oviposition begins, with males siring $>90\%$ of the brood (Müller and Eggert 1989). Although species with male care are generally expected to have high paternity, the effect of variation in the confidence of parentage on parental behaviour (Chapter 11) has received little attention in invertebrates. The burying beetles provide an attractive system for manipulating expectation of parentage because of frequent visits to the nests by intruders of either sex.

If young need at least one parent to survive until independence, then the presence of a second parent can act as insurance should one parent disappear or die. If insurance is important for biparental care, then analysis might reveal that one parent performs most parental tasks, and the second parent only becomes active if the first parent is gone (or the second parent switches to tasks normally performed by the first parent). Removal of one parent has demonstrated this effect in several insects with biparental care, demonstrating the compensating plasticity of parental behaviour, and perhaps also giving the false impression that specialization is limited. In the composting beetle *Cephalodesmius armiger*, the female will take over the male's tasks if he is removed (Dalgleish and Elgar 2005). In burying beetles paired males will do less feeding of young or nest maintenance than the female but will almost completely compensate if the mate is removed (Rauter and Moore 2004; Smiseth and Moore 2004).

Division of labour is a key component of cooperation in social insects. Division of labour can occur without task specialization, where individuals take turns performing a task or when two non-specialists work at different locations. Benefits are realized by reducing inter-task travel time, by performing two tasks simultaneously, and by coordinating activity. In burrow builders, it is common for one parent to work (or guard) outside the burrow or near the entrance while the second parent works underground (Monteith and Storey 1981; Linsenmair 1987; Hunt and Simmons 2002). The presence of biparental care despite the ability of a single parent to complete all tasks suggests that some benefit is achieved even without specialization.

Detailed behavioural analyses of biparental invertebrates have uncovered sex-role specialization of parental tasks. In the dung beetle, *Canthon cyanellus*, the male pushes and the female pulls the first dung ball toward the burrow. The male then excavates under the dung ball that the female covers with soil. The female shapes the dung ball into a pear shape while the male gathers additional dung balls. The female is also specialized to produce antifungal secretions from sternal glands (Favila and Díaz 1996). In the bark beetles, a male makes a nuptial chamber, guards the entrance, and removes frass and debris while the female constructs egg galleries (Kirkendall et al. 1997). A similar degree of specialization is found in diverse dung beetle species (Monteith and Storey 1981), ambrosia beetles (Kirkendall et al. 1997), and burying beetles (Walling et al. 2008; Cotter and Kilner 2010).

To maximize efficiencies from specialization, an individual with a tendency to perform a task more frequently should also be superior in performing that task. A correlation between superior performance and biased behavioural tendency has been demonstrated most convincingly in the burying beetles where males have both a greater tendency to guard and are superior defenders (reviewed in Trumbo 2006). Biparental care may be unstable if both parents can perform all tasks equally and either one will fully compensate for a reduction in the amount of care provided by its partner (Houston et al. 2005). Incomplete compensation, on the other hand, is thought to promote extended biparental care. Many models of biparental conflict assume that each parent is a phenocopy of the other in regards to care-giving, rather than viewing the biparental association as a potential synergism (Motro 1994). With specialization, incomplete compensation is inherent because the non-specialist cannot perform the neglected tasks equivalently (Trumbo 2006).

An interesting class of specialization occurs when a conspecific threatens a nest. Paradoxically, sexual conflict over protection of the nest and offspring can promote extended biparental associations. A conspecific may attempt to expel the same-sex rival and pair with the partner, committing infanticide as part of the takeover (Trumbo 2006; King and Fashing 2007). The costs of a takeover are much greater for the same-sex resident. In both the burying beetle, *Nicrophorus orbicollis* and the passalid beetle *Odontotaenius disjunctus* adults with young are more aggressive toward the same-sex intruder (Valenzuela-González 1986), suggesting that each parent must stay to protect its reproductive interests, promoting an extended association (Trumbo 2006). The potential for competitive takeovers may be the primary reason that high value resources are thought to be a driver of biparental care (Wilson 1975).

Specialization of parental tasks may evolve without a long history of selection for specialization (Lessells 1999). Key aspects of specialization may be present from the origin, beginning with differences in behavioural tendencies of the two sexes, as hypothesized for female co-foundresses in social insects (Fewell and Page 1999). In many biparental lineages, male care has been hypothesized to have evolved from guarding of the female or a critical resource (Alcock 1975). Defensive behaviour is reported as a male-biased behaviour in many biparental species (Kirkendall et al. 1997; Rasa 1999). After plasticity of parental behaviour has evolved, specialization can be maintained by differences in response thresholds for stimuli eliciting parental behaviour (see a similar rationale for specialization among social insect workers, Robinson and Page 1989). Small differences in thresholds of response can result in exaggerated differences in behavioural repertoires as the more sensitive individual reduces the task-inducing stimuli that would eventually trigger a response in the less sensitive individual. Considerable task specialization is facilitated while allowing rapid plasticity of behaviour when, for example, one parent is absent and task-related stimuli abruptly change. Flexibility of behaviour, even among task specialists, can be important to resolve conflict in species with costly parental care (Royle et al. 2010).

Two well developed biparental systems where specialization has not been reported occur in the desert isopod *Hemilepistus reaumuri* and in *Cryptocercus* woodroaches (Linsenmair 1987; Nalepa and Bell 1997). Although further study might reveal specialization, it should be noted that both of these species are semelparous. When parents have limited opportunity for re-pairing then their reproductive interests are closely aligned, reducing conflict and selecting for extended parent–offspring associations. The lack of reproductive opportunities likely contributes to the high level of care in these species, as it may also do toward the end of reproductive life in the iteroparous species discussed earlier.

5.3.4 The loss of parental care

Parental care encompasses a set of co-adapted traits integrating adult and offspring behaviour that could be considered to be an advance in sociality (Wilson 1975). Once evolved, why would a species discard such apparent progress in social behaviour? And even if a changing environment reduced the benefits from care, correlated changes in eggs and young that occurred during the evolution of care might leave young helpless in the absence of parents. Experiments that remove parents routinely produce devastating mortality of offspring due to predators, fungal attack, or desiccation (reviewed in Tallamy and Wood 1986; Trumbo 1996). Eggs of parental species often lack antimicrobials, egg cases, toxins, and other protections commonly found in non-parental species (Zeh et al. 1989). Eberhard (1975) suggested that some care-giving species may be in a parental 'trap' that constrains the loss of care. In the pentatomid *Antiteuchus tripterus*, the guarding mother can repel generalist predators, but increases the vulnerability to a specialist parasitoid that uses her as a cue to find her clutch. While the abandonment of parental care might appear to benefit the mother, Eberhard (1975) hypothesized that clustered eggs and thin egg shells that co-evolve with parental care may now prevent the evolution of a non-parental lifestyle.

In contrast to the parental trap (Eberhard 1975) and social advancement (Wilson 1975) perspectives, which suggest that care would rarely be lost as an adaptation, Tallamy and Schaefer (1997) argue that parental care may have been common among basal groups of invertebrates but has been repeatedly lost because of the high costs of care. The primary costs of care are a reduction in lifetime fecundity and increased vulnerability to predators while giving care. Tallamy and Schaefer (1997) hypothesize that parental care is plesiomorphic within certain clades of hemipterans and that it often is ecologically less successful than a non-parental lifestyle. By example, the pentatomids are thought to be derived from a ground-nesting ancestor with an ecology similar to extant cydnid bugs (Filippi et al. 2009). Parental care was retained, or occasionally re-evolved, when host plant seasonality enforced semelparity, or when care took on additional functions such as feeding. Comparative analysis and greater knowledge of the natural history of caregivers will be necessary to evaluate these contrasting perspectives on the evolution and de-evolution of parental care. In one such analysis of the Membracinae, Lin et al. (2004) suggest that although parental care can evolve and be lost as an adaptation, it may not be as evolutionarily labile as proposed by Tallamy and Schaefer (1997).

While there is considerable debate about the likelihood of loss of derived social traits such as parental care and eusociality, recent work suggests that male care (Reynolds et al. 2002) and complex forms of care may be more resistant to loss. A phylogenetic analysis of care in ray-finned fish suggests that viviparity has evolved numerous times, but has never been lost (Mank et al. 2005). The presence of viviparity in several diverse groups of invertebrate would make good tests for the generality of this finding. The co-evolution of provisioning behaviour and competition among offspring may also make parental care resistant to loss. Gardner and Smiseth (2011) modelled how provisioning is expected to lead to both the choice of safer nest sites by parents and to more sibling competition (provisioning being more difficult to share than guarding). Sibling competition, in turn, is expected to lead to greater provisioning and parental attendance, inhibiting the loss of care. Phylogenetic tests to examine whether guarding (such as occurs in most insect folivores) is more easily lost as an adaptation compared to more complex forms of care, remain to be done.

All investigators agree that parental care can be readily lost when care can be off-loaded onto another female or to a different species. The vulnerability of eggs and young is not an issue when there are alternative care-givers and the female achieves greater lifetime fecundity. Egg dumping into the nest of a care-giver is best documented in hemipterans (Tallamy 2005). Two females of the North American treehopper *Publilia concava* will sometimes produce a common clutch that is almost always (98%) abandoned by at least one female (Zink 2005). Non-guarding females are more likely to start a second brood (Zink 2003). The lace bug *Gargaphia solani* also dumps eggs facultatively, dumping eggs when a guarding female is available, and caring for her own eggs when this is not the case (Tallamy and Horton 1990). These condition-dependent behaviours are mediated by juvenile hormone that promotes oogenesis and abandonment at the expense of guarding (Tallamy et al. 2002). Egg dumping might be beneficial for the recipient because dumpers typically lay their eggs on the periphery of the clutch where mortality is higher. Defences against egg dumping in lace bugs are therefore not well developed. Tallamy (2005) contrasts this behaviour with egg dumping in burying beetles (Müller et al. 1990) which rear young that are food-limited on small carcasses. Parasitism has clear costs for the care-giver, and potential brood parasites are vigourously attacked, occasionally fatally.

The costs of maternal care also could be transferred to others through mutualisms with other species. There are many examples of ant–treehopper and ant–lycaenid mutualisms, some involving parental and some non-parental species. The clearest case of shifting the costs of care from care-giving treehoppers to ant mutualists is *Publilia reticulata* (Bristow 1983). The presence of ants is the apparent cue for the mother to abandon her brood, her protection now superfluous.

5.4 Microbiology of care

The microbiology of care in invertebrates is a rapidly expanding subfield of microbial ecology. Close familial associations within a nest create hygienic problems for all care-giving species. Many organisms also use microbes as digestive symbionts. Some invertebrates face additional challenges, however, when species mass provision their nest with food that needs to be preserved for significant periods of time. Microbes also can be sources of nutrition because of the lesser caloric requirements of small-bodied ectotherms and the ability of dexterous mouthparts to manipulate small fungal cultures. Vertical transmission of starter fungal cultures, digestive symbionts, and microbes aiding in food preservation are facilitated by parent–offspring contact. It is rapidly becoming appreciated that an important parental task is management of the microbial community to minimize costs and maximize benefits associated with interactions with microbes.

A simple hygienic mechanism is to separate living areas from refuse areas, a behaviour that has been recorded in a burrowing cricket, bark beetles, and even the moth ear mite that has a separate defecation chamber within a bat's ear cavity (Treat 1958; West and Alexander 1963; Kirkendall et al. 1997). When a separate refuse area is not possible, webspinners encase debris in silk to isolate it from the family (Edgerly 1997).

Grooming of eggs, typically thought to be an antimicrobial strategy, has been noted in nearly a dozen groups and may be universal among earwigs and parental centipedes (reviewed in Costa 2006). In most cases it is not clear if the defence is chemical application or mechanical removal. Some centipedes and earwigs extend grooming to young (Lamb 1976). Grooming of eggs has been speculated to be necessary in species that oviposit in organic substrates (Costa 2006). The burrower bug, *Sehirus cinctus*, is one of the few hemipterans to nest in the soil and has considerable egg grooming (Sites and McPherson 1982). Non-parental invertebrates, however, frequently oviposit in organic substrates such as rotting wood, carrion, and dung and achieve high hatching rates, so the particular selective pressures that led to egg grooming in species with parental care are not obvious. Parents can also combat microbes indirectly by eliminating substrates on which microbes grow. Male bark beetles *Ips* spp. remove frass from tunnel systems while patrolling galleries for predators or conspecific males (Robertson 1998). One of the more intricate antimicrobial strategies occurs in amphipods of the genus *Phronima* (Hirose et al. 2005). This group feeds on tunicates and uses the tunicate barrel as a nursery. *Phronima* eats the animal tissue of the tunicate, but leaves the gelatinous matrix intact. After grazing on the epidermis, tunicate cuticular layers regenerate a living layer that protects the nursery from microbes.

The coevolution of care and egg properties has taken two distinct pathways among parental insects. In some care-giving species, the ancestral egg adaptations have been retained, allowing the parent to oviposit away from the nest (e.g. burying beetles, Pukowski 1933) or away from active adult areas (e.g. bark beetles, Kirkendall et al. 1997). Displacement of eggs keeps them out of harm's way, either from normal adult activity or from ovicidal competitors (Scott 1997). Other care-giving species keep eggs close by and groom them. In these species, experimental removal of the parent typically results in eggs succumbing to fungal attack or desiccation, indicating that adaptations typical of non-parental species have been lost. Detailed comparative study of egg anatomy of non-parental species, parental species that groom eggs, and parental species that do not groom eggs would be enlightening. One could hypothesize substantial costs of antimicrobial and antidesiccant adaptations for eggs, and that parents of some species pay a smaller cost by egg attendance.

European beewolf (*Philanthus triangulum*) mothers provide antifungal protection through to the pupal stage. Females transfer an inoculum of bacteria (*Streptomyces philantii*) from glands located on the antennae to the brood cell where an egg is oviposited (Kaltenpoth et al. 2005). These bacteria produce antibiotics that protect the larvae against fungal attack. The larvae later apply the bacteria to the silk of their cocoon within which they will pupate in the soil. Wood-feeding spruce beetles *Dendroctonus rufipennis* need to maintain an entire gallery system clear of antagonistic fungi. Adults exude oral secretions that inhibit growth of invading fungi. Filtered-sterilized secretions do not inhibit fungal growth, providing evidence that bacteria within the secretions are responsible for the antifungal activity (Cardoza et al. 2011).

Mass provisioning, whether incorporating dung or carrion into a nest, or hunting for prey to stockpile for young, presents considerable microbial problems. In addition to preventing spoilage, antimicrobial application can reduce cues that might lead to detection of the nest and food by competitors. The European beewolf brings paralysed (but not dead) prey to its nest, and then licks the body surface of the prey to apply protective hydrocarbons while it makes a brood cell (Strohm and Linsenmair 2001). Both the licking behaviour and the use of paralysed prey help to prevent spoilage. The scarab beetle *Canthon cyanellus* applies an antibiotic from its sternal glands to dung balls prepared for young (Pluot-Sigwalt 1988). Similarly, the burying beetle *Nicrophorus vespilloides* applies anal secretions by rubbing the tip of its abdomen back and forth across a prepared small vertebrate carcass (Rozen et al. 2008; Cotter and Kilner 2010). It remains to be investigated whether these treatments should be regarded as sterilization of the resource, or the use of one type of microbe to limit growth of another, more harmful, microbe.

To exploit cellulose-rich resources such as wood or leaves, invertebrates benefit from digestive symbionts. Parental care provides a conduit for symbionts to be passed to offspring either by trophallaxis as in the woodroach *Cryptocercus* (Nalepa and Bell 1997), or by consumption of faeces as in the passalid beetle *Odontotaenius disjunctus* (Schuster and Schuster 1997). Among subsocial wood feeders, the passalid beetles have one of the shortest maturation times of young despite their soft mouthparts, benefiting from pre-digestion of wood by microbes in frass outside the body (an 'external rumen') (Halffter 1991; Schuster and Schuster 1997). Similar external digestion occurs in the composting beetle *Cephalodesmius armiger*, which macerates leaves, mixes them with faeces, and produces an artificial 'dung' ball for offspring (Monteith and Storey 1981). Another wood feeder, the stag beetle *Dorcas rectus*, consumes both wood and wood-rotting fungi, utilizing specialized abdominal pouches, mycangia, to transport fungal cultures (Tanahashi et al. 2010). Mutualist bacteria may produce chemicals that limit growth of competing, parasitic fungi. Antibiotic treatment reduced, but did not stop growth in larvae, suggesting that although bacteria may aid digestion, they are not absolutely necessary. A fascinating but little understood form of invertebrate–microbe symbiosis occurs in shield bugs of the family Plataspidae. In these species, mycetomes, specialized digestive pouches, are packed with bacteria that can be passed from mother to her eggs. In many species, mothers smear a microbial secretion over eggs, while in *Coptosoma scutellatum* the mother produces eggs that contain a packet of bacteria at the base. Remarkably, after the nymph hatches it inoculates itself, which is essential for its survival, by siphoning bacteria from the storehouse within the egg (Schneider 1940 in Costa 2006). Both nutritional and hygienic hypotheses have been generated for these taxon specific symbioses. This system seems ripe for molecular approaches that will permit a detailed exploration of the numbers and kinds of microbes involved.

Yet more complex associations with fungi are found in insects that actively manage the substrate on which fungi grow. Although several additional groups, including thrips and webspinners, are known to consume fungal mycelia and/or spores, true cultivation of fungi is thought to have evolved only among Macrotermininae termites, leaf-cutting ants, and ambrosia beetles (Mueller et al. 2005). These agriculturalists maintain the proper microenvironment to grow colonies of symbiotic fungi, alter the growth form of fungi by selective consumption, control the growth of competing fungi both by removal and the use of antibiotic-producing bacterial symbionts, and carry starter cultures to begin a new colony (De Fine Licht et al. 2005; Little et al. 2006; Scott et al. 2008). In some bark beetles, growth can be so luxurious that the crop must be harvested simply to prevent overgrowth within the tunnel system. There are a number of factors that promote mutualisms between host and symbiont (Leigh 2010). At the origin of host–symbiont associations there was presumably considerable genetic diversity among possible symbionts. Hosts that selected and vertically transmitted more cooperative strains likely would have produced more offspring than hosts working with less cooperative strains. The cultivation of fungal gardens is a rapidly growing area of research that best exemplifies the changing view of microbes as either commensals or competitors, to one where microbial interactions are actively managed by social invertebrates. For further information, readers are directed to the reviews by Currie (2001), Mueller et al. (2005), and Aanen (2006).

Acknowledgements

I thank Glauco Machado, Per Smiseth, and Hugh Drummond for thoughtful comments on the writing that improved this chapter. Christine Nalepa, Scott Kight, Lisa Filippi, and Janice Edgerly-Rooks shared useful information that was unpublished. Janice Edgerly-Rooks, Natasha Mhatre, Scott Kight, and Scott Camazine kindly provided images for Fig 5.1. Lastly, James Costa's *The Other Insect Societies* (2006) was a labour saving resource that was appreciated increasingly as the chapter was being written.

References

Aanen, D. (2006). As you reap, so shall you sow: coupling of harvesting and inoculating stabilizes the mutualism between termites and fungi. *Biology Letters* 2, 209–12.

Alcock, J. (1975). Territorial behavior by males of *Philanthus multimaculatus* (Hymenoptera: Sphecidae) with a review of male territoriality in male sphecids. *Animal Behaviour* 23, 889–95.

Alexander, R., Noonan, K. M., and Crespi, B. J. (1991). The evolution of eusociality. In P. W. Sherman, J. U. M. Jarvis, and R. Alexander, eds. *The Biology of the Naked Mole Rats*, pp. 3–44, Princeton University Press, Princeton.

Avilés, L. (1997). Causes and consequences of cooperation and permanent-sociality in spiders. In J. Choe and B. Crespi, eds. *The Evolution of Social Behavior in Insects and Arachnids*, pp. 476–98. Cambridge University Press, Cambridge.

Barreto, F. S. and Avise, J. C. (2008). Polygynandry and sexual size dimorphism in the sea spider *Ammothea hilgendorfi* (Pycnogonida: Ammotheidae), a marine arthropod with brood-carrying males. *Molecular Ecology* 17, 4164–75.

Beal, C. A. and Tallamy, D. W. (2006). A new record of amphisexual care in an insect with exclusive paternal care: *Rhynocoris tristis* (Heteroptera: Reduviidae). *Journal of Ethology* 24, 305–307.

Bhoopathy, S. (1998). Incidence of parental care in the cockroach *Thorax porcellana* (Saravas) (Blaberidae: Blattaria). *Current Science* 74, 248–51.

Bristow, C. M. (1983). Treehoppers transfer parental care to ants: a new benefit of mutualism. *Science* 220, 532–3.

Buskirk, R. E. (1981). Sociality in the arachnids. In H. Herrman, ed. *Social Insects, Volume II*, pp. 282–367. Academic Press, New York.

Buzatto, B. A. and Machado, G. (2009). Amphisexual care in *Acutisoma proximum* (Arachnida, Opiliones), a Neotropical harvestman with exclusive maternal care. *Insectes Sociaux* 56, 106–8.

Cardoza, Y. J., Klepzig, K. D., and Raffa, K. F. (2011). Bacteria in oral secretions of an endophytic insect inhibit antagonistic fungi. *Ecological Entomology* 31, 636–45.

Clutton-Brock, T. H. (1991). *The Evolution of Parental Care* Princeton University Press, Princeton.

Costa, J. T. (2006). *The Other Insect Societies* Harvard University Press, Cambridge, Mass.

Cotter, S. C. and Kilner, R. M. (2010). Sexual division of antibacterial resource defence in breeding burying beetles, *Nicrophorus vespilloides*. *Journal of Animal Ecology* 79, 35–43.

Crespi, B. J. and Mound, L. A. (1997). Ecology and evolution of social behavior among Australian gall thrips and their allies. In J. Choe and B. Crespi, eds. *The Evolution of Social Behavior in Insects and Arachnids*, Cambridge University Press, Cambridge.

Currie, C. (2001). A community of ants, fungi, and bacteria: a multilateral approach to studying symbiosis. *Annual Review of Microbiology* 55, 357–80.

Dalgleish, E. A. and Elgar, M. A. (2005). Breeding ecology of the rainforest dung beetle *Cephalodesmius armiger* (Scarabaeidae) in Tooloom National Park. *Australian Journal of Zoology* 53, 95–102.

De Fine Licht, H. H., Andersen, A., and Aanen, D. K. (2005). *Termitomyces* sp. associated with the termite *Macrotermes natalensis* has a heterothallic mating system and multinucleate cells. *Mycological Research* 309, 314–18.

Diesel, R. and Schuh, M. (1993). Maternal care in the bromeliad crab, *Metopaulias depressus* (Decapoda): maintaining oxygen, pH and calcium levels optimal for the larvae. *Behavioral Ecology and Sociobiology* 32, 11–15.

Eberhard, W. G. (1975). The ecology and behavior of a subsocial pentatomid bug and two scelionid wasps: strategy and counterstrategy in a host and its parasitoids. *Smithsonian Contributions in Zoology* 205, 1–39.

Edgerly, J. S. (1997). Life beneath silk walls: a review of the primitively social Embiidina. In J. Choe and B. Crespi, eds. *The Evolution of Social Behavior in Insects and Arachnids*, pp. 14–25. Cambridge University Press, Cambridge.

Eickwort, A. C. (1981). Presocial insects. In H. R. Hermann, ed. *Social Insects, Volume II*, pp. 199–280. Academic Press, New York.

Emlen, S. T. and Oring, L. W. (1977). Ecology, sexual selection, and the evolution of mating systems. *Science* 197, 215–23.

Ento, K., Araya, K., and Kûdo, S.-I. (2008). Trophic egg provisioning in a passalid beetle (Coleoptera). *European Journal of Entomology* 105, 99–104.

Evans, H. E. (1958). The evolution of social life in wasps. *Proceedings of the 10th International Congress of Entomology* 2, 449–57.

Favila, M. E. and Díaz, A. (1996). *Canthon cyanellus cyanellus* LeConte (Coleoptera: Scarabaeidae) makes a nest in the field with several brood balls. *Coleopterists Bulletin* 50, 52–60.

Fernández, M., Bock, C., and Pörtner, H.-O. (2000). The cost of being a caring mother: the ignored factor in the reproduction of marine invertebrates. *Ecology Letters* 3, 487–94.

Fetherston, I. A., Scott, M. P., and Traniello, J. F. A. (1994). Behavioural compensation for mate loss in the burying beetle *Nicrophorus orbicollis*. *Animal Behaviour* 47, 777–85.

Fewell, J. H. and Page, R. E. (1999). The emergence of division of labour in forced associations of normally solitary ant queens. *Evolutionary Ecology Research* 1, 537–48.

Filippi, L., Baba, N., Inadomi, K., Yanagi, T., Hironaka, M., and Nomakuchi, S. (2009). Pre- and post-hatch trophic egg production in the subsocial burrower bug, *Canthophorus niveimarginatus* (Heteroptera: Cydnidae). *Naturwissenschaften* 96, 201–11.

Gardner, A. and Smiseth, P. T. (2011). Evolution of parental care driven by mutual reinforcement of parental food provisioning and sibling competition. *Proceedings of the Royal Society, Series B* 278, 196–203.

Gilbert, J. and Manica, A. (2009). Brood conspicuousness and clutch viability in male-caring assassin bugs (*Rhinocoris tristis*). *Ecological Entomology* 34, 176–82.

Gilbert, J. D. J., Thomas, L. K., and Manica, A. (2010). Quantifying the benefits and costs of parental care in assassin bugs. *Ecological Entomology* 35, 639–51.

Gillespie, J. M. and McClintock, J. B. (2007). Brooding in echinoderms: how can modern experimental techniques add to our historical perspective? *Journal of Experimental Marine Biology and Ecology* 342, 191–201.

Halffter, G. (1991). Feeding, bisexual cooperation and subsocial behavior in three groups of Coleoptera. In M. Zunino, X. Belles, and M. Blas, eds. *Advances in Coleopterology*, pp. 263–80. Veracruz, Mexico.

Halffter, G. and Edmonds, W. D. (1982). *The Nesting Behavior of Dung Beetles (Scarabaeinae): An Ecological and Evolutive Approach*, Instituto de Ecologia, Mexico City.

Hanski, I. and Cambefort, Y. (1991). *Dung Beetle Ecology*, Princeton University Press, Princeton.

Hironaka, M., Nomakuchi, S., Iwakuma, S., and Filippi, L. (2005). Trophic egg production in a subsocial shield bug, *Parastrachia japonensis* Scott (Heteroptera: Parastrachiidae), and its functional value. *Ethology* 111, 1089–102.

Hirose, E., Aoki, M. N., and Nishikawa, J. (2005). Still alive? Fine structure of the barrels made by *Phronima* (Crustacea: Amphipoda) *Journal of the Marine Biological Association of the United Kingdom*, 85, 1435–9.

House, C. M., Evans, G. M. V., Smiseth, P. T., Stamper, C. E., Walling, C. A., and Moore, A. J. (2008). The evolution of repeated mating in the burying beetle, *Nicrophorus vespilloides*. *Evolution* 62, 2004–14.

Houston, A. I., Székely, T., and McNamara, J. M. (2005). Conflict between parents over care. *Trends in Ecology and Evolution* 20, 33–8.

Huerta, C. and Halffter, G. (1992). Inhibition of stridulation in *Nicrophorus* (Coleoptera: Silphidae): consequences for reproduction. *Elytron* 6, 151–7.

Hunt, J. and Simmons, L. W. (2002). Behavioural dynamics of biparental care in the dung beetle *Onthophagus taurus*. *Animal Behaviour* 64, 65–75.

Ichikawa, N. (1988). Male brooding behaviour of the giant water bug, *Lethocerus deyrollei* Vuillefroy (Hemiptera: Belostomatidae). *Journal of Ethology* 6, 121–7.

Jenkins, E. V., Morris, C., and Blackman, S. (2000). Delayed benefits of paternal care in the burying beetle *Nicrophorus vespilloides*. *Animal Behaviour* 60, 443–51.

Jones, T. C. and Parker, P. G. (2002). Delayed juvenile dispersal benefits both mother and offspring in the cooperative spider *Anelosimus studiosus* (Araneae: Theriidae). *Behavioral Ecology* 13, 142–8.

Kaltenpoth, M., Göttler, W., Herzner, G., and Strohm, E. (2005). Symbiotic bacteria protect wasp larvae from fungal infestation. *Current Biology* 15, 475–9.

Keller, L. and Reeve, H. K. (1999). Dynamics of conflict within insect societies. In L. Keller, ed. *Levels of Selection in Evolution*, pp. 153–75. Princeton University Press, Princeton.

Kelly, N. B. and Alonzo, S. H. (2009). Will male advertisement be a reliable indicator of paternal care, if offspring survival depends on male care? *Proceedings of the Royal Society, Series B* 276, 3175–83.

Kent, D. and Simpson, J. A. (1992). Eusociality in the beetle *Austroplatypus incompertus* (Coleoptera:Curculionidae). *Naturwissenschaften* 79, 86–7.

Kight, S. L., Batino, M., and Zhang, Z. (2000). Temperature-dependent parental investment in the giant waterbug *Belostoma flumineum* (Heteroptera: Belostomatidae). *Annals of the Entomological Society of America* 93, 340–2.

King, A. and Fashing, N. (2007). Infanticidal behavior in the subsocial beetle *Odontotaenius disjunctus* (Illiger) (Coleoptera: Passalidae). *Journal of Insect Behavior* 20, 527–36.

Kirkendall, L. R., Kent, D. S., and Raffa, K. A. (1997). Interactions among males, females and offspring in bark and ambrosia beetles: the significance of living in tunnels for the evolution of social behavior. In J. Choe and B. Crespi, eds. *The Evolution of Social Behavior in Insects and Arachnids*, pp. 181–215. Cambridge University Press, Cambridge.

Kruse, K. C. (1990). Male backspace availability in the giant waterbug (*Belostoma flumineum*). *Behavioral Ecology and Sociobiology* 26, 281–9.

Kûdo, S.-I., Satô, M., and Ôhara, M. (1989). Prolonged maternal care in *Elasmucha dorsalis* (Heteroptera: Acanthosomatidae). *Journal of Ethology* 7, 75–81.

Kutschera, U. and Wirtz, P. (2001). The evolution of parental care in freshwater leeches. *Theory in Biosciences* 120, 115–37.

Lamb, R. J. (1976). Parental behaviour in the Dermaptera with special reference to *Forficula auricularia* (Dermaptera: Forficulidae). *Canadian Entomologist* 108, 609–19.

Leigh, E. G. (2010). The evolution of mutualism. *Journal of Evolutionary Biology* 23, 2507–28.

Lessells, C. M. (1999). Sexual conflict in animals. In L. Keller, ed. *Levels of Selection in Evolution*, pp. 77–99. Princeton University Press, Princeton.

Lin, C.-P., Danforth, B. N., and Wood, T. K. (2004). Molecular phylogenetics and evolution of maternal care in Membracinae treehoppers. *Systematic Biology* 53, 400–21.

Linksvayer, T. A. and Wade, M. J. (2005). The evolutionary origin and elaboration of sociality in the aculeate hymenoptera: maternal effects, sib-social effects, and heterochrony. *Quarterly Review of Biology* 80, 317–36.

Linsenmair, K. E. (1987). Kin recognition in subsocial arthropods, in particular in the desert isopod *Hemilepistus reamuri*. In D. J. C. Fletcher and C. D. Michener, eds. *Kin Recognition in Animals*, pp. 121–208. Wiley, New York.

Little, A. E. F., Murakami, T., Mueller, U. G., and Currie, C. R. (2006). Defending against parasites: fungus-growing ants combine specialized behaviours and microbial symbionts to protect their fungus gardens. *Biology Letters* 2, 12–16.

Machado, G. and Macías-Ordóñez, R. (2007). Reproduction. In R. Pinto Da Rocha, G. Machado, and G. Giribet, eds. *The Biology of Opiliones*, pp. 414–54. Harvard University Press, Cambridge, Mass.

Mahsberg, D. (2001). Brood care and social behavior. In P. Brownwell and G. Polis, eds. *Scorpion Biology and Research*, pp. 257–77. Oxford University Press, New York.

Manica, A. and Johnstone, R. A. (2004). The evolution of paternal care with overlapping broods. *American Naturalist* 164, 517–30.

Mank, J. E., Promislow, D. E. L., and Avise, J. C. (2005). Phylogenetic perspectives in the evolution of parental care in ray-finned fishes. *Evolution* 59, 1570–8.

Mas, F., Haynes, K. F., and Kölliker, M. (2009). A chemical signal of offspring quality affects maternal care in a social insect. *Proceedings of the Royal Society, Series B* 276, 2847–53.

Monteith, G. H. and Storey, R. I. (1981). The biology of *Cephalodesmius* a genus of dung beetles which synthesizes 'dung' from plant material (Coleoptera: Scarabaeidae: Scarabaeinae). *Memoirs of the Queensland Museum* 20, 253–71.

Motro, U. (1994). Evolutionary and continuous stability in asymmetric games with continuous strategy sets—the parental investment conflict as an example. *American Naturalist* 144, 229–41.

Mueller, U. G., Gerardo, N. M., Aanen, D. K., Six, D. L., and Schultz, T. R. (2005). The evolution of agriculture in insects. *Annual Review of Ecology and Systematics* 36, 563–95.

Müller, J. K. and Eggert, A.-K. (1989). Paternity assurance by 'helpful' males: adaptations to sperm competition in burying beetles. *Behavioral Ecology and Sociobiology* 24, 245–9.

Müller, J. K., Eggert, A.-K., and Dressel, J. (1990). Intraspecific brood parasitism in the burying beetle, *Necrophorus vespilloides* (Coleoptera: Silphidae). *Animal Behaviour* 40, 491–9.

Nalepa, C. A. and Bell, W. J. (1997). Postovulation parental investment and parental care in cockroaches. In J. Choe and B. Crepi, eds. *The Evolution of Social Behaviour in Insects and Arachnids*, pp. 26–46. Cambridge University Press, Cambridge.

Nalepa, C. A., Maekawa, K., Shimada, K., Saito, Y., Arellano, C., and Matsumoto, T. (2008). Altricial development in subsocial wood-feeding cockroaches. *Zoological Science* 25, 1190–8.

Nazareth, T. M. and Machado, G. (2009). Reproductive behavior of *Chavesincola inexpectabilis* (Opiliones, Gonyleptidae) with description of a new and independently evolved case of paternal care in harvestmen. *Journal of Arachnology* 37, 127–34.

Nazareth, T. M. and Machado, G. (2010). Mating system and exclusive postzygotic paternal care in a Neotropical harvestman (Arachnida: Opiliones). *Animal Behaviour* 79, 547–54.

Nowak, M. A., Tarnita, C. E., and Wilson, E. O. (2010). The evolution of eusociality. *Nature* 466, 1057–62.

Oster, G. F. and Wilson, E. O. (1987). *Caste and Ecology in Social Insects* Princeton University Press, Princeton.

Paez, D., Govedich, F. R., Bain, B. A., Kellett, M., and Burd, M. (2004). Costs of parental care on hunting behaviour of *Helobdella papillornata* (Euhirudinea: Glossiphoniidae). *Hydrobiologia* 519, 185–8.

Perry, J. C. and Roitberg, B. D. (2006). Trophic egg laying: hypotheses and tests. *Oikos* 112, 706–14.

Pluot-Sigwalt, D. (1988). Données sur l'activité et le rôle de quelques glandes tégumentaires, sternales, pygidiales et autres, chez deux espèces de *Canthon* (Col., Scarabaeidae). *Bulletin de la Société Entomologique de France* 93, 89–98.

Poulin, E., Palma, A. T., and Feral, J. P. (2002). Evolutionary versus ecological success in Antarctic benthic invertebrates. *Trends in Ecology & Evolution* 17, 218–22.

Preston-Mafham, R. and Preston-Mafham, K. (1993). *The Encyclopedia of Land Invertebrate Behaviour* MIT Press, Cambridge, Mass.

Pukowski, E. (1933). Ökologische Untersuchungen an *Necrophorus* F. *Zeitschrift für Morphologie und Ökologie der Tiere* 27, 518–86.

Ramaswamy, K. and Cocroft, R. B. (2009). Collective signals in treehopper broods provide predator localization cues to the defending mother. *Animal Behaviour* 78, 697–704.

Rasa, O. A. E. (1998). Biparental investment and reproductive success in a subsocial desert beetle: the role of maternal effort. *Behavioral Ecology and Sociobiology* 43, 105–13.

Rasa, O. A. E. (1999). Division of labour and extended parenting in a desert tenebrionid beetle. *Ethology* 105, 37–56.

Rauter, C. M. and Moore, A. J. (2004). Time constraints and trade-offs among parental care behaviours: effects of brood size, sex and loss of mate. *Animal Behaviour* 68, 695–702.

Reeve, H. K. (1993). Haplodiploidy, eusociality and absence of male parental and alloparental care in Hymenoptera: a unifying genetic hypothesis distinct from kin selection theory. *Philosophical Transactions of the Royal Society, Series B* 342, 335–52.

Reynolds, J. D., Goodwin, N. B., and Freckleton, R. P. (2002). Evolutionary transitions in parental care and live bearing in vertebrates. *Philosophical Transactions of the Royal Society, Series B* 357, 269–81.

Ridley, M. (1978). Paternal care. *Animal Behaviour* 26, 904–32.

Robertson, I. C. (1998). Paternal care enhances male reproductive success in pine engraver beetles. *Animal Behaviour* 56, 595–602.

Robinson, G. E. and Page, R. E. (1989). Genetic determination of nectar foraging, pollen foraging, and nest-site scouting in honey bee colonies. *Behavioral Ecology and Sociobiology* 24, 317–23.

Royle, N. J., Schuett, W., and Dall, S. R. X. (2010). Behavioral consistency and the resolution of sexual conflict over parental investment. *Behavioral Ecology* 21, 1125–30.

Rozen, D. E., Engelmoer, D. J. P., and Smiseth, P. T. (2008). Antimicrobial strategies in burying beetles breeding on carrion. *Proceedings of the National Academy of Sciences* 105, 17890–5.

Saito, Y. (1997). Sociality and kin selection in Acari. In J. Choe and B. Crespi, eds. *The Evolution of Social Behavior in Insects and Arachnids*, pp. 443–457. Cambridge University Press, Cambridge.

Saito, Y., Chittenden, A. R., Mori, K., Ito, K., and Yamauchi, A. (2008). An overlooked side effect of nest-scattering behavior to decrease predation risk (Acari: Tetranychidae, Stigmaeidae). *Behavioral Ecology and Sociobiology* 63, 33–42.

Salomon, M., Mayntz, D., Toft, S., and Lubin, Y. (2011). Maternal nutrition affects offspring performance via maternal care in a subsocial spider. *Behavioral Ecology and Sociobiology* 65, 1191–202.

Schneider, G. (1940). Beiträge zur kenntnis der symbiontischen einrichtungen der heteropteren. *Zeitschrift für Morphologie und Ökologie der Tiere* 36, 595–644.

Schroeder, P. C. and Hermans, C. (1975). Annelida: Polychaeta. In A. C. Giese and J. S. Pearse, eds. *Reproduction of Marine Invertebrates: Volume 3, Annelida and Echiurans*, pp. 1–213. Academic Press, New York.

Schuster, J. C. and Schuster, L. B. (1997). The evolution of social behavior in Passalidae (Coleoptera). In J. Choe and B. Crespi, eds. *The Evolution of Social Behavior in Insects and Arachnids*, pp. 260–9. Cambridge University Press, Cambridge.

Scott, J. J., Oh, D.-C., Yuceer, M. C., Klepzig, K. D., Clardy, J., and Currie, C. R. (2008). Bacterial protection of beetle-fungus mutualism. *Science* 322, 63.

Scott, M. P. (1997). Reproductive dominance and differential ovicide in the communally breeding burying beetle *Nicrophorus tomentosus*. *Behavioral Ecology and Sociobiology* 40, 313–20.

Seibel, B. A., Robinson, B. H., and Haddock, S. H. D. (2005). Post-spawning egg care by a squid. *Nature* 438, 929.

Sites, R. W. and McPherson, J. E. (1982). Life history and laboratory rearing of *Sehirus cinctus cinctus* (Hemiptera: Cydnidae), with description of immature stages. *Annals of the Entomological Society of America* 75, 210–15.

Smiseth, P. T., Andrews, C., Brown, E., and Prentice, P. M. (2010). Chemical stimuli from parents trigger larval begging in burying beetles. *Behavioral Ecology* 21, 526–31.

Smiseth, P. T. and Moore, A. J. (2004). Behavioral dynamics between caring males and females in a beetle with facultative biparental care. *Behavioral Ecology* 15, 621–8.

Smith, R. L. (1997). Evolution of paternal care in the giant water bugs (Heteroptera: Belostomatidae). In J. Choe and B. J. Crespi, eds. *The Evolution of Social Behavior in Insects and Arachnids*, pp. 116–49. Cambridge University Press, Cambridge.

Steiger, S., Whitlow, S., Peschke, K., and Müller, J. K. (2009). Surface chemicals inform about sex and breeding status in the biparental burying beetle *Nicrophorus vespilloides*. *Ethology* 115, 178–85.

Strathmann, R. R. and Strathmann, M. F. (1982). The relationship between adult size and brooding in marine invertebrates. *American Naturalist* 119, 91–101.

Strohm, E. and Linsenmair, E. (2001). Females of the European beewolf preserve their honeybee prey against competing fungi. *Ecological Entomology* 26, 198–203.

Suzuki, S. and Nagano, M. (2009). To compensate or not? Caring parents respond differentially to mate removal and mate handicapping in the burying beetle, *Nicrophorus quadripunctatus*. *Ethology* 115, 1–6.

Tallamy, D. W. (2001). Evolution of exclusive paternal care in arthropods. *Annual Review of Entomology* 46, 139–65.

Tallamy, D. W. (2005). Egg dumping in insects. *Annual Review of Entomology* 50, 347–70.

Tallamy, D. W. and Horton, L. A. (1990). Costs and benefits of the egg-dumping alternatives in *Gargaphia* lace bugs (Hemiptera: Tingidae). *Animal Behaviour* 39, 352–9.

Tallamy, D. W., Monaco, E. L. and Pesek, J. D. (2002). Hormonal control of egg dumping and guarding in the lace bug, *Gargaphia solani* (Hemiptera: Tingidae). *Journal of Insect Behavior* 15, 467–75.

Tallamy, D. W. and Schaefer, C. (1997). Maternal care in the Hemiptera: ancestry, alternatives, and current adaptive value. In J. C. Choe and B. J. Crespi, eds. *The Evolution of Social Behavior in Insects and Arachnids*, pp. 94–115. Cambridge University Press, Cambridge.

Tallamy, D. W. and Wood, T. K. (1986). Convergence patterns in subsocial insects. *Annual Review of Entomology* 31, 369–90.

Tanahashi, M., Kubota, K., Matsushita, N., and Togashi, K. (2010). Discovery of mycangia and the associated xylose-fermenting yeasts in stag beetles (Coleoptera: Lucanidae). *Naturwissenschaften* 97, 311–17.

Treat, A. (1958). Social organization in the moth ear mite. *Proceedings of the 10th International Congress of Entomology* 2, 475–80.

Trivers, R. L. (1972). Parental investment and sexual selection. In B. Campbell, ed. *Sexual Selection and the Descent of Man*, pp. 136–79. Aldine, Chicago.

Trumbo, S. T. (1991). Reproductive benefits and the duration of paternal care in a biparental burying beetle, *Necrophorus orbicollis*. *Behaviour* 117, 82–105.

Trumbo, S. T. (1996). Parental care in invertebrates. *Advances in the Study of Behavior* 25, 3–51.

Trumbo, S. T. (2002). Hormonal regulation of parental care in insects. In D. W. Pfaff, A. P. Arnold, A. M. Etgen, S. E. Fahrbach, R. L. Moss, and R. R. Rubin, eds. *Hormones, Brain and Behavior*, pp. 115–39. Academic Press, New York.

Trumbo, S. T. (2006). Infanticide, sexual selection and task specialization in a biparental burying beetle. *Animal Behaviour* 72, 1159–67.

Valenzuela-González, J. (1986). Territorial behavior of the subsocial beetle *Heliscus tropicus* under laboratory conditions (Coleoptera: Passalidae). *Folia Entomológica Mexicana* 70, 53–63.

Walling, C. A., Stamper, C. E., Smiseth, P. T., and Moore, A. J. (2008). The quantitative genetics of sex differences in parenting. *Proceedings of the National Academy of Sciences of the United States of America* 105, 18430–5.

West, M. J. and Alexander, R. D. (1963). Sub-social behavior in a burrowing cricket *Anurogryllus muticus* (De Geer). *Ohio Journal of Science* 63, 19–24.

Wilson, D. S. and Fudge, J. (1984). Burying beetles: intraspecific interactions and reproductive success in the field. *Ecological Entomology* 9, 195–203.

Wilson, E. O. (1975). *Sociobiology: The New Synthesis* Harvard University Press, Cambridge, Mass.

Windsor, D. M. and Choe, J. C. (1994). Origins of parental care in chrysomelid beetles. In P. Jolivet, M. Cox and E. Petitpierre, eds. *Novel Aspects of the Biology of Chrysomelidae*, pp. 111–17. Kluwer, Dordrecht.

Wood, T. K. (1976). Alarm behavior of brooding female *Umbonia crassicornis* (Homoptera: Membracidae). *Annals of the Entomological Society of America* 69, 340–4.

Zeh, D. W. and Smith, R. L. (1985). Paternal investment by terrestrial arthropods. *American Zoologist* 25, 785–805.

Zeh, D. W., Zeh, J. A., and Smith, R. L. (1989). Ovipositors, amnions and eggshell architecture in the diversification of terrestrial arthropods. *Quarterly Review of Biology* 64, 147–68.

Zink, A. G. (2003). Quantifying the costs and benefits of parental care in female treehoppers. *Behavioral Ecology* 14, 687–93.

Zink, A. G. (2005). The dynamics of brood desertion among communally breeding females in the treehopper, *Publilia concava*. *Behavioral Ecology and Sociobiology* 58, 466–73.

CHAPTER 6

Sex differences in parental care

Hanna Kokko and Michael D. Jennions

6.1 Introduction

Why can we sometimes successfully distinguish males from females at a glance, even in a species we have never seen before (Fig. 6.1)? One answer is that we make an often biologically justified assumption that females are the main care-givers.

Figure 6.1 Which parent is the male and which is the female? (a,b) Chacma baboon (*Papio Ursinus*); (c,d) Leaden flycatcher (*Myiagra rubecula*); (e,f) earwig. Answer: Males are on the right hand side. (Photo credits: (a,b) Peter Henzi; (c,d) Geoffrey Dabb; (e,f) Joel Meunier.)

In about 90% of mammals, males fertilize eggs and thereafter provide no parental care (Chapter 4). Likewise, in arthropods that do provide parental care there is a very strong bias towards female-only care. Biparental care is rare in this taxon, and estimates of the number of times male-only care has evolved range from 'at least eight' (Tallamy 2000) to sixteen (Chapter 5). Although new discoveries of male care continue to be made (e.g. Requena et al. 2010) as a whole, male care appears rare in this enormous taxonomic group. In reptiles, parental care is uncommon: eggs are usual buried and abandoned (Chapter 4). If egg guarding occurs, however, it is almost always by females, alongside a few cases of biparental care in crocodilians (Reynolds et al. 2002). In birds, the sex bias in parental care is weaker. There is female-only care in 8% of species and biparental care in 90% (9% with and 81% without helpers) (Cockburn 2006). In biparental birds there is, however, still a propensity for females to spend more time building nests and incubating eggs (Schwagmeyer et al. 1999; Møller and Cuervo 2000).

The list above suggests that guessing the sex of an individual based on how much it cares for its young is a reasonable rule of thumb. It works poorly, however, in those taxa that do not show female-biased care (Fig. 6.2). Amphibians show fantastic variation in both the form and sexual division of parental care (Chapter 4). In some species females lay trophic eggs to feed tadpoles, or brood frogs inside their stomachs, or carry developing frogs in fleshy capsules on their backs. Males sometimes guard tadpoles, build canals to move tadpoles between temporary pools, or transport them to new sites (Wells 2007). There are even species where males carry froglets on their backs (Bickford 2002). Biparental care is rare in anurans, but male-only

The Evolution of Parental Care. First Edition. Edited by Nick J. Royle, Per T. Smiseth, and Mathias Kölliker.

Figure 6.2 Which parent is the male and which is the female? (a,b) Kentish Plover *(Charadrius alexandrinus)*; (c,d) Dendrobatid frog (*Hyloaxalus nexipus*). Answer: Males are on the right hand side in these 'sex-role reversed' species. (Photo credits: (a) Sándor Kovács; (b) Monif Alrashidi (Farasan Islands); (c,d) Jason Lee Brown)

and female-only care has evolved equally often (Reynolds et al. 2002; Summers et al. 2006).

In teleost fish, one can more reliably use caring to identify the sexes, but the pattern is reversed (Chapter 4): male-only care occurs in nine times as many genera as female-only care (Reynolds et al. 2002). Excluding livebearers, phylogenetic studies of evolutionary transitions suggest that female-only care is a more recently derived state, arising almost exclusively from biparental care (e.g. cichlids: Klett and Meyer 2002; Gonzalez-Voyer et al. 2008; ray-finned fish: Mank et al. 2005).

The stereotype for non-biologists is that females are always the primary care-givers. Although this is probably a product of naïvely extrapolating from the norm in most human societies, our brief taxonomic review reveals a strong trend towards female-biased care in nature. This is true whether we consider the absolute number of species where female care predominates (given invertebrate biodiversity) or the number of independent evolutionary transitions towards greater female than male care. Here our goal is to explain why females tend to provide more care, and what, on the other hand, maintains such fantastic diversity in the sexual division of parental duties.

6.2 Why does an individual's sex predict its behaviour?

A predictable relationship between an individual's sex and breeding behaviour is most parsimoniously explained if there is something inherent to being a male or a female that applies across all taxa. We can immediately eliminate the mechanism of sex determination as a common denominator as it varies greatly among taxa. Some taxa have sex chromosomes, but females are either the heterogametic (e.g. ZW in birds and butterflies) or homogametic (e.g. XX in placental mammals) sex. In haplodiploid species such as ants and wasps, males develop from unfertilized eggs and females from fertilized eggs. Still other species exhibit temperature-dependent sex determination. The sex determination mechanism can even differ among populations of the same species, as in the snow skink *Niveoscincus ocellatus* which has temperature-dependent sex determination at low altitudes and and genotypic sex determination in highlands (Pen et al. 2010).

How then do we decide to which sex an individual belongs? Why can we confidently state that pregnant seahorses are always male when nobody knows the exact mechanism of sex determination in this taxon? The answer is that there is an agreement by convention: individuals producing the smaller of two gamete types—sperm or pollen—are males, and those producing larger gametes—eggs or ovules—are females. This classification usually works unambiguously because, when gamete size varies, it generally shows strong bimodality with one very large and one very small gamete fusing to form a zygote (i.e. anisogamy). In species where mating types are indistinguishable based on gamete size the terminology of two sexes—males and females—is abandoned. Instead we refer to sex between isogamous mating strains, usually labelled + and − (e.g. many fungi, algae).

The question of why there is a tendency towards female-biased care can therefore be rephrased as: why is producing a smaller gamete associated with providing less post-zygotic care? To answer this question it is important to understand why anisogamy evolves. The reasons why males are sparing in provisioning each gamete with resources have surprisingly much in common with the reasons why males might be less willing to subsequently invest in additional parental care. Gametic resources that are available to the zygote after fertilization represent the minimal possible level of parental investment (see Chapter 1). A study of the evolution of gamete size can therefore also be considered to be a study of the evolution of parental care.

6.3 The first sex difference in parental care: anisogamy

There are several competing explanations for the evolution of anisogamy (expertly reviewed by Lessells et al. 2009), but the most common one invokes the economics of parental investment and gamete production. Lower investment in each gamete allows a parent to produce more gametes on the same budget, similar to the general principle of a trade-off between offspring size and number (Smith and Fretwell 1974). For a parent, the optimal offspring size maximizes the number of surviving offspring. This optimum is predicted by size-dependent offspring survival trading-off against parental fecundity. As a consequence an intermediate level of fecundity is expected to be favoured.

Optimality models for offspring size in egg-laying species assume that only one parent determines size at birth: egg size predicts offspring size, as sperm are too tiny to make a difference. This *a priori* assumption is, however, inappropriate when considering the evolution of anisogamy itself. Here the gamete size of both parents is free to evolve.

A gamete must survive until it finds a compatible gamete to form a zygote. Thereafter, zygote survival depends on how both gametes influence offspring size. Small gametes are not necessarily destined to give rise to offspring with poor survival prospects if they fuse with larger gametes. In effect, males are selected to produce small gametes (sperm) to 'parasitize' greater parental provisioning of gametes (eggs) by females.

Consider an isogamous situation with + and − mating types. Initially mutants of one mating type that produce smaller that average gametes (a proto-male) will suffer a decline in offspring survival because this slightly reduces zygote size. This is more than compensated for, however, by the mutants' higher fertilization success, simply because proto-males produce more gametes. In turn, increased likelihood of fertilization by a smaller gamete selects for greater gamete size in the other mating type (proto-females) to compensate for the expected reduction in zygote size. Models that make plausible assumptions about the size-number trade-off and size-dependent gamete and zygote survival show that the resultant disruptive selection readily drives the evolution of anisogamy (Parker et al. 1972; Bulmer and Parker 2002).

Interestingly, classic models of anisogamy do not explicitly model one consequence of altering gamete size: a gamete might never become part of a zygote. If one sex makes many more gametes than the other then most of these are doomed as a result of the *Fisher condition* (Houston and McNamara 2005). R. A. Fisher insightfully noted that since each offspring in a sexual, diploid species has one genetic parent of each sex, the total reproductive output of each sex must be identical at the population level (Fisher 1930). Fisher used this mathematical necessity to explain the ubiquity of a 1:1 offspring sex ratio: if one sex is overproduced then the other, rarer sex has a higher per capita reproductive success, and parents are selected to produce this sex. A stable outcome only exists when parents invest equally into sons and daughters. This translates to a 1:1 offspring sex ratio if the cost of producing sons and daughters is identical.

In diploid species the Fisher condition is a reminder that the number of successful sperm must equal the number of successful eggs. But we never observe a 1:1 ratio of eggs to sperm. Most sperm are wasted. At first sight this makes the profitability of 'gametic parasitism' by males seem less lucrative. If most sperm never find an egg to parasitize, why does this not select for males that produce fewer, larger gametes? A recent model for gamete size evolution that explicitly derives survival over time shows that isogamy can indeed be maintained when gametes of both sexes are 'reasonably likely' to become a zygote (Lehtonen and Kokko 2011a). Anisogamy only evolves when there is a deviation from this 'reasonable likelihood' either in the direction of a high encounter rate such that proto-sperm deplete proto-eggs and many proto-sperm are outcompeted, or in the direction of a low encounter rate so that *both* gamete types struggle to find each other.

Why do these opposing scenarios both lead to anisogamy? What matters is the relative strength of selection on proto-males to increase the number of their own gametes that will fuse with compatible gametes, compared with selection to elevate zygote survival (by increased investment through larger gametes). The trade-off between these two benefits determines the net reproductive success of proto-males.

In a high encounter scenario, such as a broadcast spawner living at high density, the greater abundance of proto-sperm than proto-eggs makes them function like tickets in a raffle where a limited number of prizes are guaranteed (the eggs are there). The only way to do better than other participants is to have more tickets. (In this admittedly unusual raffle each participant has a piece of paper which he can cut up to make tickets of any size, and larger tickets then increase the value of any prize that is won.) Succeeding in this initial stage of competition—sperm competition—is a better predictor of male reproductive success than elevating the survival of any resulting zygotes. Increased zygote survival would, of course, be beneficial, but it is difficult for a male to elevate this component of fitness economically. To increase the amount of parental provisioning a male has to put extra energy into *every* sperm because *he can not predict in advance which of his 'tickets', if any, will win a prize.* Most of his parental investment would be wasted because so few sperm locate an egg. We will later return to this argument because it is as relevant for post-zygotic male care when paternity is uncertain as it is for pre-zygotic provisioning of gametes.

Why does the opposite scenario of very low gamete density also select for anisogamy? Selection on a proto-male to make his gametes locate eggs before rivals do is now replaced by the task of locating an egg in the first place. Producing numerous sperm is now advantageous because it raises the likelihood that at least some succeed in locating an egg. The analogy with a raffle with a guaranteed prize is now invalid because some prizes are never awarded. Perhaps a better analogy is putting up many 'Lost Pet' notices in the hope that one will be read by whoever has found your missing cat. Females are no longer victims of parasitism: they too benefit from males producing numerous, albeit small, sperm because some of their eggs would otherwise go unfertilized.

The largest absolute number of fertilizations might occur if both sexes produced small gametes. Eggs do not evolve to be small, however, because this would overly compromise the survival prospects of the resulting zygotes. There is even the potential for selection for bigger eggs if these are easier for sperm to locate (Levitan 2010).

In summary, the most ancient form of parental investment is the gametic provisioning that determines zygote size. Male investment per gamete is reduced because it trades off with efficiently locating unfertilized eggs. The likely returns of parental investment per gamete are low if most sperm have difficulties in finding eggs to fertilize: either because unfertilized eggs are rare (low density), or because rivals compete to fertilize eggs first (high density).

6.4 What happened next? Sex roles in post-zygotic parental care

When eggs and sperm of a broadcast spawner meet to form a zygote far away from parents, the female parent has by definition provided the greater parental investment. This book would not exist, however, if the story ended there. Post-zygotic parental investment expressed in many forms of parental care has repeatedly evolved in both external and internal fertilizers (Chapters 4 and 5). Parental care requires that at least one parent can locate zygotes formed from its gametes. In internal fertilizers, the zygote is initially surrounded by parental tissue (e.g. the brood pouch of a male seahorse, or the reproductive tract of an inseminated female guppy). Whichever sex provides this tissue is clearly predisposed to continue to provide care. Livebearer females could be said to provide more post-zygotic care than males simply because they are in the position to transfer additional energy resources to developing offspring (fish: Pollux et al. 2009), possibly also leading to greater post-birth provisioning (e.g. lactation in mammals).

However, explanations based on the ability of a parent to locate its young do not predict female care as easily as appears at first sight. Seahorses (Sygnathidae) are a reminder that 'pregnancy', or brooding in general, is not *a priori* a female trait. Either sex could take sole responsibility for carrying and/or guarding young. Nothing in principle prevents males from doing this, as illustrated by Belastomatid water bugs where a female cements fertilized eggs onto the male's back (Inada et al. 2011).

Even in cases where the female carries young internally or externally, there is often no obvious barrier to later male paternal investment. For example, male mammals have not lost the physiological and morphological machinery required for lactation (Wynne-Edwards and Reburn 2008), so why do they not make use of it (Kunz and Hosken 2009)? Similarly, when fertilization is external and eggs are laid onto a substrate, both parents are present when zygotes form. The ability of a parent to locate its young clearly affects whether care is provided, but this only shifts the question to 'why are there sex differences in the ease of location'? For example, why does internal fertilization more often occur inside females than males?

One approach to explain sex biased care is to treat fundamental sex-specific constraints (e.g. female gestation) as a given. We can then focus on current environmental or social factors that influence the likelihood that each parent encounters its offspring. For example, male care might be related to whether there is sufficient food for females to remain on male territories during pregnancy. This approach can help to account for special cases such as sex role reversal in taxa where males provide the bulk of care. We are, however, primarily interested in working from first principles, rather than invoking current contingencies. Instead of assuming that the female has traits that force it to stay with zygotes, we want to ask why egg-producers (females) evolve to stay with their young more often than sperm-producers (males) in most, but not all, taxa?

Any explanation for sex-biased care must account for potential sexual conflict (Chapter 9). Unless there is lifelong monogamy, each parent is likely to do better if it made the other bear the full costs of caring. In the case of short-term stability of biparental care, a reduction in care by one parent ought to be only partially compensated for by the other (Harrison et al. 2009). If fully compensated for, one parent could desert without its fitness declining. Most behavioural studies of parental care that estimate the pay-offs from caring and deserting assume that both sexes will adaptively respond to each other in an attempt to maximize fitness. Measuring behavioural responses thereby allows us to predict the underlying direction of fitness changes when a deserting parent's fitness is affected by the subsequent decisions of its deserted mate (e.g. Jennions and Polakow 2001). Over an evolutionary time frame, however, short-term behavioural responses must be combined with asking how the *mean* behaviour of each sex affects the fitness of the other sex, until an evolutionary stable state is reached where individuals of neither sex improve their fitness by behaving differently (e.g. Kokko 1999).

We will now provide a non-mathematical account of recent models that account for the evolution of female-biased care. We start by explaining what factors select *against* male care. We then discuss factors that limit the fitness gains males can achieve by caring less. Together these help to explain why male-only or biparental care can sometimes evolve while a general female bias towards greater care-giving still persists.

6.5 Uncertain parentage reduces male care

Earlier we stated that anisogamy results in sperm vastly outnumbering eggs. There are two assumptions behind this statement. First, that there are not many more female than male adults currently trying to breed. Second, that each sex invests a similar amount of its budget into gamete production. If, for example, there are very few males releasing sperm, or each male only produces a few sperm, then sperm might not outnumber eggs. Selection driving anisogamy results in an enormous size difference between eggs and sperm, however, so such deviations would require a massive bias of the breeding sex ratio towards females and/or far greater female than male investment into gametes. Although female reproductive investment into gametes can greatly exceed that of a male (counting gametic biomass, Hayward and Gillooly 2011), sperm tend to be so much smaller than eggs that the assumption is robust: typically numerous sperm compete to locate eggs before their rivals, rather than the reverse.

While the overabundance of sperm relative to eggs is a relatively robust pattern seen in all species, sperm competition is less ubiquitous. Monogamy can eliminate sperm competition, as can low adult density in broadcast spawners (Levitan 2010). Exceptions aside though, given the numerical abundance of sperm, it is likely that sperm from several males are seeking out the same egg shortly before fertilization. This creates uncertainty as to which male has sired a zygote. This uncertainty compromises a male's ability to locate his own young, both when fertilization is external and when it is internal and females mate multiply (which is common, Slatyer et al. 2011). In contrast, the relative rarity of eggs usually makes it easy for a female to avoid mistaking the eggs of another female for her own. Exceptions include egg dumping (some birds and insects lay eggs in others' nests, Petrie and Møller 1991; Tallamy 2005) which, in extreme cases, causes mistakes in identifying offspring that are exploited by other species (brood parasites, e.g. cuckoos; Chapter 13). A female's ability to locate her own young is also compromised by communal spawning. Even so, the numerical superiority of sperm to eggs creates a clear sexual asymmetry: males are far more often uncertain about a zygote's parentage than are females.

Note that sticking to the convention 'uncertain parentage' (usually phrased as 'uncertain paternity') should not prevent us from realizing that it is actually the *certainty* of the decline in mean relatedness to putative offspring that matters. If, for example, three males ejaculate when a female spawns, the average male reproductive success is a third of a clutch. Across species, or between the sexes, higher uncertainty of parentage decreases the benefit of caring. Sperm competition makes it harder for a male to identify his young, even if they reside in the vicinity.

Although the above principle is simple, the relationship between confidence of paternity and male care is deceptively complex in the literature (Chapter 11). This is because much work concerns phenotypic plasticity of individuals (i.e. adaptive shifts in male care over his lifetime). As a result, it is easy to question the pivotal role of uncertain parentage. Before dealing with potential sources of confusion, we would like to reassure the reader by asking a few simple questions. If the only benefit of caring is to increase the survival of one's own offspring, should a male with no paternity in a brood provide care? (Answer: no.) Is the maximum benefit gained from caring linearly related to the proportion of offspring in a brood that are your own? (Answer: yes.) So, if species vary in the mean uncertainty of paternity (certainty of siring fewer offspring per brood), and nothing else, is the maximum benefit of male care lower when uncertainty is greater? (Answer: yes.) So, if females have a higher certainty of parentage than males, is the maximum benefit of caring greater for females? (Answer: yes.)

So, why is there confusion about the relationship between the certainty of paternity and male care? There are at least three reasons.

6.5.1 Why exactly does paternity matter?

It is easy to fall into the trap of arguing that paternity should not affect parental care given that lifetime fitness depends on balancing current and future reproduction. If paternity is likely to be higher in a subsequent breeding attempt, then selection might favour 'saving' energy to invest more into one's own survival and future care. But if a given male consistently has the same average likelihood of paternity (be it low or high) then his current and future reproductive success in a brood are devalued by the same amount due to lost paternity. This seemingly implies that average paternity will not influence the mean level of male care across species (Maynard Smith 1977).

The flaw in the argument is neglecting to consider that someone had to sire the extra-pair young (Queller 1997; Houston and McNamara 2002). A species cannot have low paternity for all its males without many males gaining significant fitness through extra-pair young. Extra-pair mating opportunities elsewhere yield offspring that are part of a male's future reproductive success. This is why low paternity, even if it is consistent across broods, does not have equivalent effects on a male's current and future reproduction (Houston and McNamara 2002). Males that reduce care in their current brood can redirect investment into gaining additional extra-pair matings in the near or far future.

Within-species studies of phenotypic plasticity show that below average paternity can reduce parental care (e.g. Neff 2003) or, on the flipside, increase offspring cannibalism (Gray et al. 2007). Investigating the paternity–care correlation *across* males can, however, sometimes be misleading. For example, unattractive males might provide the best care despite low paternity because their attempts to gain fitness via the extra-pair route are particularly unprofitable. Ideally, one should investigate how care changes when a male's current share of paternity differs from his future expectations (which might be the population average, or an expectation based on his own life-history). This can be achieved by experimentally manipulating a male's perceived share of paternity (e.g. Neff 2003). Another line of evidence comes from phylogenetic studies. Across bird species, paternal care is low when extra-pair paternity is high (Møller and Birkhead 1993; Møller and Cuervo 2000). This is an encouraging sign that paternity has a strong influence on male care, because mean paternity is affected by other traits that could conceivably have the reverse effect on male care, thereby negating the predicted pattern (Houston and McNamara 2002).

6.5.2 Behavioural and evolutionary time scales are not equivalent

Ignoring differences in how selection acts over behavioural and evolutionary time scales generates confusion. Average paternity influences male care decisions because males possess 'evolutionary knowledge' about their expected paternity, based on past selection to provide a level of care that maximizes fitness. Simultaneously, a male's likelihood of siring a given offspring can differ from the average male in the current generation, or even his own expected lifetime average. This makes it dangerous to test how a factor such as paternity affects care by assuming that behavioural or other phenotypic responses will mimic the predicted evolutionary response to selection. Specifically, how does one interpret a case where there is no behavioural response to low paternity? Does it mean that lower paternity does not select for reduced care? Or that selection for adaptive phenotypic plasticity has been insufficient, perhaps because the relevant paternity cues or ability to detect them do not exist?

A failure to distinguish between selection acting over behavioural and evolutionary time scales can easily lead to claims that females will pursue mating strategies that promote the evolution of desirable male traits. For example, Morton et al. (2010) argued that 'genetic monogamy [...] may be a female tactic that reduces the likelihood of males evolving counter-adaptations to female desertion'. Although it seems self-evident that selection can not favour female traits that reduce undesirable evolutionary outcome for other females in the future (except as a fortuitous byproduct), this type of reasoning is still common (Kokko and Jennions 2010).

The accuracy of proximate cues to detect deviations from prevailing parentage certainty is pertinent to the outcome of sexual conflict over care (Kokko 1999; Mauck et al. 1999). Consider a species with biparental care where extra-pair matings sometimes occur. If a female could signal to her mate that he has paternity that is above the population average over recent evolutionary time she might induce him to provide more care (Shellman-Reeve and Reeve 2000). Such a signalling system is, however, unlikely to be stable because all females benefit from convincing their social mate that he has high paternity.

If males fail to detect lower than average paternity then, *within* each generation, polyandrous females are not penalized by reduced male care (Kokko 1999). Selection on male parenting decisions will now only depend on accumulated 'evolutionary knowledge' of average paternity in the population. Consequently, small benefits—or correlated selection between males and females (Forstmeier et al. 2011)—are enough to favour extra-pair mating (Slatyer et al. 2011), and this makes paternity decline *over evolutionary time*. Updated *evolutionary knowledge* of average paternity will then decrease mean male care over successive generations.

If females evolve to be sufficiently promiscuous and become better at biasing paternity towards extra-pair males (e.g. Pryke et al. 2010), one possible evolutionary outcome is female-only care (Fig. 6.3). The resulting decline in mean female breeding success can be interpreted as a 'tragedy of the commons' (Kokko 1999; Rankin et al. 2007). Females evolve to lose a valuable resource (male care) because no female within any generation pays the price for her own actions.

6.5.3 Traits that protect paternity can co-evolve with care

The co-evolution between paternity and parental care has an extra twist when traits that protect paternity also enhance offspring survival, or when sexual selection for traits that increase male care simultaneously increases paternity (Kvarnemo 2006). If such 'magic' traits exist, male care will evolve more readily than if responses to changes in paternity arise from factors uncorrelated with male care itself.

Protecting paternity involves defence of unfertilized eggs (e.g. pre-copulatory mate guarding), while enhancing offspring survival often depends on processes that occur after zygote formation. If care evolves as a correlated response to protecting paternity then the relevant traits must bridge this temporal gap. Kvarnemo (2006) lists a range of intriguing possibilities as to how this might occur. For example, nuptial gifts or courtship feeding can increase sperm uptake by females (increasingly a male's likely parentage) while also providing more nutrients for offspring. Similarly, a male fish with a well-constructed, more defendable nest might have greater paternity and offer better protection to young. The ideas of Kvarnemo (2006) are worthy of further study, not least because any factor that keeps parents near putative young is a prerequisite for the future provision of care.

6.6 Don't bother caring till the going gets tough: the OSR

Although the Fisher condition usually keeps the primary sex ratio near 1:1, there is no selection to maintain sex ratio equality after parental investment ends (West 2009). Dramatic differences in mortality between the sexes are possible. How then does the adult sex ratio (ASR: proportion of males among adults) affect parental care? This is an important topic as recent theoretical work reveals some counterintuitive outcomes. A persistent theme in the literature has been the assertion that whichever sex experiences greater difficulties in finding mates will benefit more from increasing its mating effort (Trivers 1972; Dawkins 1989; Clutton-Brock and Parker 1992). This is often read to imply that this sex will gain less by providing more parental care.

The operational sex ratio (OSR), the ratio of sexually available males to females (Emlen and Oring 1977) measures the sex-specific difficulty of finding mates. The OSR takes the ASR as its baseline, but modifies it by removing individuals who are currently unavailable as mates—for example caring for young or replenishing resources required to mate and breed again. These individuals are spending 'time out' (Clutton-Brock and Parker 1992; Parker and Simmons 1996) from the mating pool, while available individuals are spending 'time in'. The relationship between the OSR and ASR is not

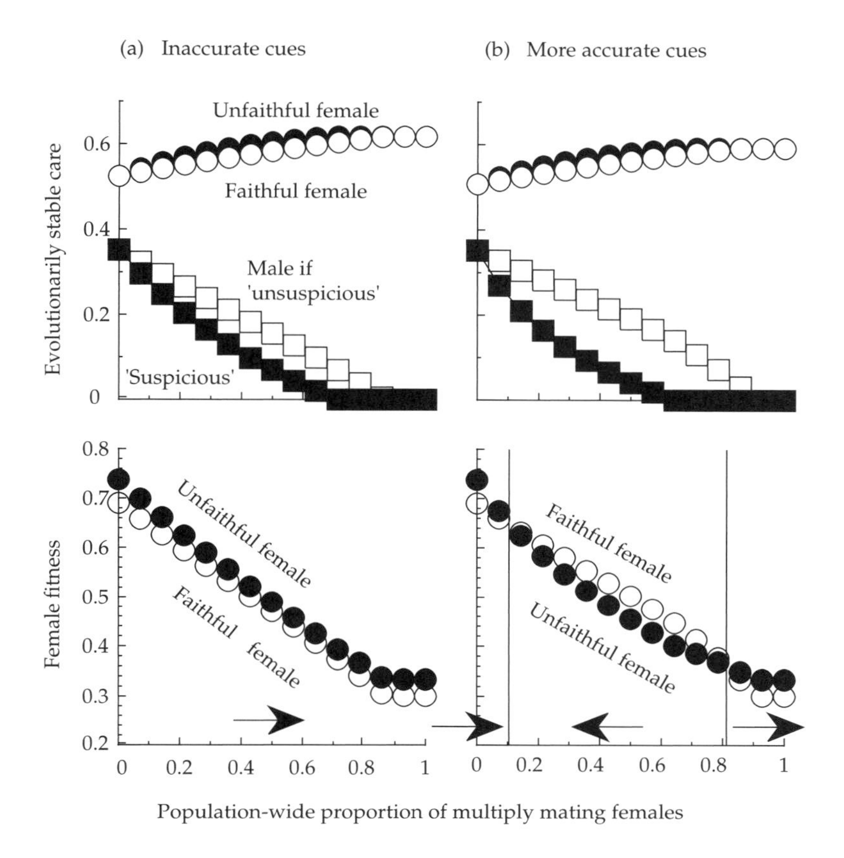

Figure 6.3 Predictions of evolutionarily stable care effort by females (circles) and males (squares) when females can either mate with their social mate ('faithful female') or multiply ('unfaithful female'), and males can to some extent detect female mating behaviour but this is either relatively inaccurate (a) or relatively accurate (b). Evolutionary changes occur along the x axis, behavioural responses along the y axis. Males possess evolutionary knowledge of average female behaviour, and multiple mating by females selects against male care (which always declines with the proportion of multiply mating females). Whether multiple mating is selected against or not in females, however, depends on the fitness difference between unfaithful and faithful females (vertical differences between circles in the female fitness graph). Males paired to unfaithful females tend to be suspicious but some of them remain unsuspicious. Males paired to faithful females tend to be unsuspicious but some of them become erroneously suspicious. When the accuracy of cues is low (case (a)), errors in such judgements are common. Consequently males are not selected to put much weight on behavioural evidence, and the care effort between the different types of males remains largely similar at each point in time. The mating system fails to penalize unfaithful females sufficiently to prevent multiple mating from spreading. The horizontal arrows, depicting female evolution over time, show that the proportion of multiply mating females approaches 1, and that male care is entirely lost from the system. Selection works against female multiple mating only if females in each generation lose a large amount of male care by being unfaithful. This occurs in the middle region of (b), indicated by the arrow pointing left, but not elsewhere. Even (b) will evolve towards more multiple mating if the population starts out from a point where multiple mating is, for any reason, already very common (rightmost arrow). Figure is modified from Kokko (1999).

necessarily linear because parental care differences influence the former but not the latter. (Sexual differences in caring in conjunction with different mortality rates for 'time in' and 'time out' activities will, however, affect each sex's mortality rate thereby changing the ASR; see Section 6.7.)

If mate-finding difficulties always favour greater mating effort at the expense of reduced parental care, any initial bias in the OSR will be self-reinforcing until only the rarer sex in the OSR provides parental care. According to this logic, self-reinforcement happens because if sex A cares slightly more than sex B this increases sex A's 'time out'. Sex A consequently becomes rare in the mating pool. Sex B responds to the intensified competition for matings by investing in traits that increase the likelihood of mating, thereby diverting resources away from caring. Reduced care by sex B, in turn, selects for greater care by sex A to ensure offspring survival. This makes sex A ever rarer in the mating pool, and so on until eventually sex B provides no care.

There is an obvious empirical problem with any account that suggests extreme sexual divergence in care is inevitable: biparental care exists. Some cases can be explained by the benefits of synergy. Care models predict, rather unsurprisingly, that biparental care is more likely if two parents perform parental duties more efficiently than either parent caring alone (Kokko and Johnstone 2002). The real problem, however, is that the self-reinforcement process relies on an effect of the OSR that is mathematically problematic. Numerous models show that an overabundance of males selects for *more* male care (Yamamura and Tsuji 1993; McNamara et al. 2000; Houston and McNamara 2002; Kokko and Jennions 2008). Similar conclusions have been reached when modelling other traits that increase male fitness by remaining with his current mate. For example, selection for mate guarding to protect paternity, rather than seeking out other females, is stronger in more male-biased populations (Carroll and Corneli 1995; Fromhage et al. 2008). Indeed, it should be obvious that a male is more likely to desert a current female or his offspring when there are *more*, not fewer, reproductive opportunities available elsewhere. This occurs when the OSR is female-biased.

It is now worth recalling an insight from thinking about gamete size evolution. Gamete size can show self-reinforcing divergence due to the greater abundance of sperm than eggs. Why then does male abundance (in the OSR) not inevitably lead to a similar divergence towards an ever more male-biased OSR? Instead, models predict that difficulty securing additional matings makes a male more likely to care. The solution lies with the seemingly trivial fact that a male is only in a position to provide care once he has mated and his potential offspring exist. We noted earlier that males are not selected to better provision their gametes because '*the male cannot predict in advance which of his "tickets", if any, will win the prize*'.

Contrast this with the evolution of post-zygotic male care, and assume that the level of paternity is p (between 0 and 1). How does the expected future number of reproductive opportunities with new females influence whether a male should continue to care to improve offspring survival, or desert and seek a new mate? A male in a male-biased population experiences a trade-off between investing in offspring *that already exist* and investing in gaining future offspring that *might never exist*. Caring for existing offspring offers an assured fitness return that remains unchanged (albeit discounted by p) as future mating opportunities become scarcer, while the return from future offspring is weighted by the likelihood of securing a mating. The likelihood is clearly lower when the number of competitors per female increases. This is why, for the same paternity certainty p, a male gains relatively more by improving the survival of his existing young when securing additional matings is more difficult. A male-biased OSR therefore selects for increased male care. Although some current paternal investment is wasted if parentage p is less than 100%, a reasonable p still allows a male to direct paternal care towards existing genetic descendants. In contrast, this is impossible at the pre-zygotic stage as a male cannot direct material investment into those few sperm that will end up forming zygotes.

In sum, a numerical bias in gametes (more sperm than eggs) tends to create sperm competition, which reduces the certainty of paternity and makes it hard for males to locate their own young and preferentially direct care towards them. It does not follow, however, that male and female care levels will diverge completely because negative frequency-dependent selection (Fisher condition) begins to operate as soon as care differences arise. The moment there is intense pre-copulatory competition for mates (male-biased OSR), greater post-zygotic male care is favoured simply because attempts to find new females are now less profitable. Although not always strong enough to prevent divergence, there is a logical relationship between one sex caring for longer and the opposite sex experiencing difficulties in securing a mate. The resultant selection on the less caring sex to do something else to raise fitness—such as provide more care—can explain biparental care even when synergistic benefits of care are absent (Kokko and Jennions 2008).

6.7 Orwell was right, not all animals are equal: sexual selection and the adult sex ratio

If uncertainty of parentage and biased OSRs were the only factors driving parental care, we would predict that female multiple mating or group spawning would reduce male care, which is then countered by frequency-dependent selection as male mating opportunities decline. By analogy, differences in the costs of making sons and daughters select for a biased primary sex ratio, but the inevitable frequency-dependent selection on the sex ratio (again due to the Fisher condition) prevents one sex from becoming completely absent. If the analogy is taken at face value, this makes it difficult to explain uniparental care. What is missing from the explanation is that not all individuals of a given sex experience competition equally. The Fisher condition only dictates that individuals of the more numerous sex in the OSR mate *on average* less often than those of the rarer sex. This does not preclude *some* individuals having a reliably higher mating rate.

Sexual selection acts on traits that increase success under competing for mates (or, more accurately, for access to fertilizable gametes). Successful individuals spend less time in the mating pool than is the average time for their sex. Crucially, only individuals that mate have their propensity to provide care exposed to selection. Those that never have offspring are never in a position to decide how to apportion care to offspring.

So should a mated individual care or desert? If sexual selection is strong, mating *per se* indicates that an individual bears traits that will yield above average mating success in the future. These individuals, by virtue of their superiority, can partially escape the Fisherian frequency-dependent selection that dictates a lower *average* mating rate for individuals of their sex. Consequently, desertion becomes more profitable for the average *mated* individual of the sexually selected sex who is expected to repeat his (or her) success with new mates.

As the relative strength of sexual selection increases, the mating system can shift from biparental care with a mild sex bias to uniparental care (for numerical examples see Kokko and Jennions 2008; Lehtonen and Kokko 2011b). The Fisher condition never disappears, but with sufficiently strong sexual selection, it no longer constrains *mated* individuals of the more numerous (less caring) sex in the mating pool to have a lower mating rate than that of the other sex.

Given the central role of sexual selection in determining parental care it is important that we can empirically measure the strength of sexual selection. We can then contrast the mating rates of favoured individuals with those experienced by the average individual of that sex. The OSR is directly related to the average mating rate, but less clearly to the former. It is tempting to assume that sexual selection is always stronger on the more common sex in the mating pool (Emlen and Oring 1977), but this is false because mating difficulties are not synonymous with sexual selection. A biased OSR lowers the mean mating success per time unit for the more common sex, but it does not follow that mating success has automatically begun to depend more strongly on any particular trait (Fitze and Le Galliard 2008; Kokko and Jennions 2008; Klug et al. 2010; Jennions et al. 2012). What defines sexual selection is a causal link between specific traits and the ability to fertilize gametes. Such causality cannot be directly shown by documenting variance in mating success or be inferred from a biased OSR. This is important, because unless there are sexually selected traits that increase the bearer's mating success, a mated individual cannot assume it will have above average mating success after deserting its current young. Without repeatability in the likelihood of mating no individual can avoid the full force of Fisherian frequency-dependence.

The relationship between the OSR and sexual selection has to be determined empirically, but surprisingly few studies have formally quantified this relationship (Fitze and Le Galliard 2008) or its proxies (Weir et al. 2011). This is despite the fundamental importance of the relationship in predicting sex biases in care. If sexual selection rapidly intensifies as the OSR becomes biased, uniparental care is far more likely because a mild initial bias in care becomes self-reinforcing (Lehtonen and Kokko 2011b). Successful individuals of the more common sex in the mating pool will comprise an ever smaller subset of that sex. They will be unaffected by the increasing bias in the OSR that lowers the mating rate for the average individual of their sex, and selection will still favour desertion of young to seek out additional matings. If, however, sexual selection saturates or declines at a high OSR (e.g. because it is then less economical for a male to monopolize matings due to increased interactions with competitors, Klug et al. 2010) then the subset of successful males becomes larger. The *per capita* decline in mean mating rate with an increasingly biased OSR more strongly affects mated males and thereby selects for greater male care (Lehtonen and Kokko 2011b).

One of the most interesting feedbacks between sexual selection, sex ratios, and parental care could operate through differences in the mortality costs of caring and competing. The ASR sets the baseline for the OSR. The OSR is the ASR corrected for the difference in care by each sex that determines their 'time out' of the mating pool. Anything that influences the ASR will therefore affect the balance of the forces that determine the level of male and female care. But what if sex differences in care directly affected the ASR? Parental care and competition for matings are often dangerous activities (Liker and Székely 2005) and/or require initial investment in traits that affect mortality (Moore and Wilson 2002). If caring is a more life-threatening activity than mate searching then whichever sex cares more will become rarer in the population. This, all else being equal, will—perhaps counterintuitively—select for the opposite sex to care more (despite the mortality costs) as future mating opportunities become scarcer. In contrast, if seeking out mates is the more dangerous activity then the sex that spends less time caring and more time in the mating pool will become rarer in the ASR. All else being equal, the surviving members of this sex then have a relatively greater number of mating opportunities so they are not selected to care more.

The outcome of a scenario where the more caring sex is rare is reminiscent of demographic patterns familiar to ornithologists. Although avian ASRs are notoriously difficult to measure and inferences are often based on non-territorial male 'floaters' (e.g. Kosztolányi et al. 2011), it appears broadly true that the ASR is often male-biased in biparentally caring birds (Donald 2007). This fits well with the idea that the Fisher condition makes males less optimistic about securing matings elsewhere when males are common (and extra-pair paternity can conceivably explain the slightly greater female investment per brood in biparental birds). Intriguingly, bird species that lack male care tend to have a female-biased ASR (Donald 2007), which resembles the typical situation in mammals. In most mammals male care is absent and there is often very high male mortality (usually attributed to male investment in sexually selected traits and immune system suppression; Moore and Wilson 2002) creating a female-biased ASR. Males—especially those who have already proven their ability to gain matings by siring young—can therefore be relatively optimistic about their future mating success, and are unlikely to benefit as much by caring for young. Even though the lack of male care might, depending on the ASR, lead to a strongly male-biased OSR, the Fisher condition reminds us that each baby is sired by a living male. The mating optimism of males is therefore appropriate despite intense competition for mates due to a male-biased OSR (e.g. in lekking antelopes males who mate are likely to mate again, while those who do not are not in the position to provide care).

6.8 Conclusions

We have sketched an outline of why females care more than males but, as every biologist knows, historic contingencies can result in taxon-specific traits that seemingly challenge general rules. When observed outcomes disagree with theoretical predictions it is often prudent to check whether the underlying assumptions hold rather than automatically assuming that the theory is fatally flawed. The common occurrence of male-only care in fish is illustrative (see also Chapter 2). We have assumed a trade-off between caring and remating. Although valid in many taxa it does not seem to hold in fish because parental care involves activities like nest defence and egg fanning that can be performed almost equally easily for a few or many offspring (Stiver and Alonzo 2009). Males are therefore able to receive eggs from additional mates and females appear to be willing to lay eggs with already mated males because it carries no cost or because the presence of eggs signals that the male is a good parent and/or because the *per capita* risk of predation/cannibalism declines when a male guards more eggs. Little or no increase in parental costs when caring for many zygotes can also explain some other counterintuitive results. For example, despite our argument that uncertainty of paternity lowers care, male care increases with lower paternity in the ocellated wrasse *Symphodus ocellatus* (Alonzo and Heckman 2010). This makes sense, however, if the male is caring for an absolutely greater number of sired offspring, and the costs of care are not directly proportion to the number of young guarded.

Of course, there are still taxa where it is a challenge to understand why only one sex cares, and why it is a given sex. For example, many frogs have rather similar breeding systems and comparable ecological requirements so why, within the same genus, do only males provide care in some species and females in others? Likewise, mouthbrooding in fish seems to place limits on mating rates, so why has male-only care still evolved? Can these differences be explained by 'tipping points' within our general framework? For example, do small ecological changes produce large shifts in the strength of sexual selection, the degree of multiple mating, or the ASR? Or do these differences depend on taxon-specific traits? For example, in some species of fish (Kraak and Groothuis 1994) and insects (Tallamy 2000) females actually prefer males already providing care. Or are there important general factors that are missing from current theoretical models?

It will be rewarding to empirically test the model we have outlined. This is a challenge because—as we have noted—predictions are made about evolutionary responses that are not necessarily mimicked when making comparisons between parents in different environments (i.e. adaptive plasticity). This places limitations on the extent to which behavioural studies of extant variation in care can test the model. Cross-species studies are more powerful and there have been some valiant attempts to conduct phylogenetic tests to determine whether transitions between major patterns of parental care can be predicted based on the intensity of sexual selection (e.g. Gonzalez-Voyer et al. 2008; Olson et al. 2009). Of course, phylogenetic studies can only report correlations so they need to be interpreted with caution when inferring causality.

Ideally, researchers could run experimental evolution studies akin to recent ones in which lines are assigned to monogamy and polyandry treatments and trait evolution is then monitored (e.g. Simmons and Garcia-Gonzalez 2008), but with a focus on parental care. One could, for example, establish selection lines in which the ASR is either heavily male-biased or female-biased each generation. The obvious problem is, however, identifying a suitable species with a sufficiently short generation time and readily measurable care—perhaps dung or burying beetles? More pragmatically one could rely on 'natural experiments' and investigate differences among populations in key model parameters (e.g. ASR, mean paternity) to test whether the predicted divergence in parental care exists.

As an interesting example, we recently noted a dataset on variance in male and female mating success in various human populations classified as polygynous, serially monogamous, or monogamous (Brown et al. 2009). The level of male care per female is presumably lowest in the polygyny and highest in monogamy. If the mean reproductive success of males and females was correctly estimated then, given the Fisher condition, the inverse

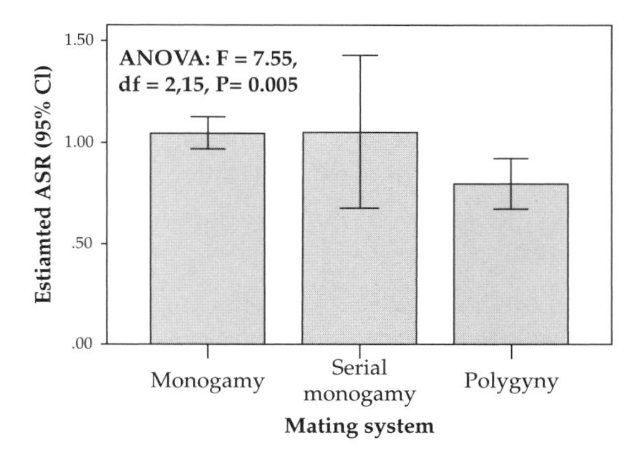

Figure 6.4 Estimated adult sex ratio (ASR) based on the ratio of mean reproductive success (number of live births) for females and males for populations assigned to three mating system types. The Fisher condition applies so males and females must have the same total reproductive output. Hence any difference in mean reproductive success has to be due to a biased ASR. The data is from Brown et al. (2009) who compiled data from 18 populations. Note that when Brown et al. classified a population as having two mating systems (polygyny and one other) we classified it as polygyny. Sample sizes are 6, 3, and 9.

of this ratio is a measure of the ASR. The social systems differed significantly in their estimated ASR (Fig. 6.4). Despite no overlap in the populations considered, the pattern in Brown et al.'s dataset is essentially identical to an older analysis by Ember (1974), who likewise noted that polygynous human societies appear to have more males per female than non-polygynous ones.

Finally, we end with a quirk of natural history to remind us that nature always challenges grand theories unified by common themes. We have made much of the Fisher condition—one mother and one father. Even this 'fact' is, however, violated in at least one diploid, sexually reproducing organism—and a primate no less. As with most Callitrichid primates, the marmoset *Callithrix kuhlii* lives in polyandrous groups with two males caring for offspring. Fascinatingly, they are sometimes chimeras because siblings exchange cells in utero. One offspring can have two genetic fathers! From an evolutionary perspective, it is important that the germ line tissue is also chimeric. Either father's genes can therefore be transmitted to the next generation (Ross et al. 2007). Amazingly, the more chimeric offspring were, the more often they were carried by both males, presumably because offspring scent profiles matched both fathers. This should remind us how easily the limits of our imagination can constrain our ability to consider all evolutionarily relevant factors. That said, we are still confident that paternity, sexual selection, and adult sex ratios will continue to predict general patterns of parental care.

References

Alonzo, S. H. and Heckman, K. L. (2010). The unexpected but understandable dynamics of mating, paternity and paternal care in the ocellated wrasse. *Proceedings of the Royal Society London, Series B* 277, 115–22.

Bickford, D. (2002). Animal behaviour—male parenting of New Guinea froglets. *Nature* 418, 601–2.

Brown, G. R., Laland, K. N., and Borgerhoff-Mulder, M. (2009). Bateman's principle and human sex roles. *Trends in Ecology and Evolution* 24, 297–304.

Bulmer, M. G. and Parker, G. A. (2002). The evolution of anisogamy: a game-theoretic approach. *Proceedings of the Royal Society London, Series B* 269, 2381–8.

Carroll, S. P. and Corneli, P. S. (1995). Divergence in male mating tactics between two populations of the soapberry bug: II. Genetic change and the evolution of a plastic reaction norm in a variable social environment. *Behavioural Ecology* 6, 46–56.

Clutton-Brock, T. H. and Parker, G. A. (1992). Potential reproductive rates and the operation of sexual selection. *Quarterly Review of Biology* 67, 437–56.

Cockburn, A. (2006). Prevalence of different modes of parental care in birds. *Proceedings of the Royal Society London, Series B* 273, 1375–83.

Dawkins, R. (1989). *The Selfish Gene*, 2nd ed. Oxford University Press, Oxford.

Donald, P. F. (2007). Adult sex ratios in wild bird populations. *Ibis* 149, 671–92.

Ember, M. (1974). Warfare, sex ratio, and polygyny. *Ethnology* 13, 197–206.

Emlen, S. T. and Oring, L. W. (1977). Ecology, sexual selection, and the evolution of mating systems. *Science* 197, 215–23.

Fisher, R. A. (1930). *The Genetical Theory of Natural Selection*. Oxford University Press, Oxford.

Fitze, P. S. and Le Galliard, J.-F. (2008). Operational sex ratio, sexual conflict and the intensity of sexual selection. *Ecology Letters* 11, 432–9.

Forstmeier, W., Martin, K., Bolund, E., Schielzeth, H., and Kempenaers, B. (2011). Female extrapair mating behavior can evolve via indirect selection on males. *Proceedings of the National Academy of Sciences* U.S.A. 108, 10608–13.

Fromhage, L., McNamara, J. M., and Houston, A. I. (2008). A model for the evolutionary maintenance of monogyny in spiders. *Journal of Theoretical Biology*, 250, 524–31.

Gonzalez-Voyer, A., Fitzpatrick, J. L., and Kolm, N. (2008). Sexual selection determines parental care patterns in cichlid fishes. *Evolution*, 62, 2015–26.

Gray, S. M., Dill, L. M., and McKinnon, J. S. (2007). Cuckoldry incites cannibalism: male fish turn to cannibalism when perceived certainty of paternity decreases. *American Naturalist* 169, 258–63.

Harrison, F., Barta, Z., Cuthill, I., and Székely, T. (2009). How is sexual conflict over parental care resolved? A meta-analysis. *Journal of Evolutionary Biology* 22, 1800–12.

Hayward, A. and Gillooly, J. F. (2011). The cost of sex: quantifying energetic investment in gamete production by males and females. *PLoS ONE* 6, e16557.

Houston, A. I. and McNamara, J. M. (2002). A self-consistent approach to paternity and parental effort. *Philosophical Transactions of the Royal Society of London* B 357, 351–62.

Houston, A, I. and McNamara, J. M. (2005). John Maynard Smith and the importance of consistency in evolutionary game theory. *Biology and Philosophy* 20, 933–950.

Inada, K., Kitade, O., and Morino, H. (2011). Paternity analysis in an egg-carrying aquatic insect *Appasus major* (Hemiptera: Belostomatidae) using microsatellite DNA markers. *Entomological Science* 14, 43–8.

Jennions, M. D. and Polakow, D. A. (2001). The effect of partial brood loss on male desertion in a cichlid fish: an experimental test. *Behavioural Ecology* 12, 84–92.

Jennions, M. D., Kokko, H., and Klug, H. (2012). The opportunity to be misled in studies of sexual selection. *Journal of Evolutionary Biology* 25: 591–598.

Klett, V. and Meyer, A. (2002). What, if anything, is a Tilapia? Mitochondrial ND2 phylogeny of Tilapiines and the evolution of parental care systems in the African cichlid fishes. *Molecular Biology and Evolution* 19, 865–83.

Klug, H., Heuschele, J., Jennions, M. D., and Kokko, H. (2010). The mismeasurement of sexual selection. *Journal of Evolutionary Biology,* 23, 447–462.

Kokko, H. (1999). Cuckoldry and the stability of biparental care. *Ecology Letters* 2, 247–55.

Kokko, H. and Jennions, M. D. (2008). Parental investment, sexual selection and sex ratios. *Journal of Evolutionary Biology* 21, 919–48.

Kokko, H. and Jennions, M. D. (2010). Behavioural ecology: ways to raise tadpoles. *Nature* 464, 990–1.

Kokko, H. and Johnstone, R. A. (2002). Why is mutual mate choice not the norm? Operational sex ratios, sex roles, and the evolution of sexually dimorphic and monomorphic signalling. *Philosophical Transactions of the Royal Society of London Series B* 357, 319–30.

Kosztolányi, A., Barta, Z., Küpper, C., and Székely, T. (2011). Persistence of an extreme male-biased adult sex ratio in a natural population of polyandrous bird. *Journal of Evolutionary Biology* 24, 1842–6.

Kraak, S. B. M. and Groothuis, T. G. G. (1994). Female preference for nests with eggs is based on the presence of the eggs themselves. *Behaviour* 131, 189–206.

Kunz, T. H. and Hosken, D. J. (2009). Male lactation: why, why not and is it care? *Trends in Ecology and Evolution* 24, 80–5.

Kvarnemo, C. (2006). Evolution and maintenance of male care: is increased paternity a neglected benefit of care? *Behavioural Ecology* 17, 144–8.

Lehtonen, J. and Kokko, H. (2011a). Two roads to two sexes: unifying gamete competition and gamete limitation in a single model of anisogamy evolution. *Behavioural Ecology and Sociobiology* 65, 445–59.

Lehtonen, J. and Kokko, H. (2011b). Positive feedback and alternative stable states in inbreeding, cooperation, sex roles and other evolutionary processes. *Philosophical Transactions of the Royal Society of London* B 367, 211–221.

Lessells, C. M., Snook, R. R., and Hosken, D. J. (2009). The evolutionary origin and maintenance of sperm: selection for a small, motile gamete mating type. In T. R. Birkhead, D. J. Hosken, D. J., and S. Pitnick, eds. *Sperm Biology: An Evolutionary Perspective*, pp. 43–67. Academic Press, New York.

Levitan, D. R. (2010). Sexual selection in external fertilizers. In D. F. Westneat and C. W. Fox, eds. *Evolutionary Behavioral Ecology*, pp. 365–78. Oxford University Press, Oxford.

Liker, A. and Székely, T. (2005). Mortality costs of sexual selection and parental care in natural populations of birds. *Evolution* 59, 890–7.

Mank, J. E., Promislow, D. E. L., and Avise, J. C. (2005). Phylogenetic perspectives in the evolution of parental care in ray-finned fishes. *Evolution* 59, 1570–8.

Mauck, R. A., Marschall, E. A., and Parker, P. G. (1999). Adult survival and imperfect assessment of parentage: effects on male parenting decisions.*American Naturalist* 154, 99–109.

Maynard Smith, J. (1977). Parental investment: a prospective analysis. *Animal Behaviour* 25, 1–9.

McNamara, J. M., Székely, T., Webb, J. N., and Houston, A. I. (2000). A dynamic game-theoretic model of parental care. *Journal of Theoretical Biology* 205, 605–23.

Møller, A. P. and Birkhead, T. R. (1993). Certainty of paternity covaries with paternal care in birds. *Behavioural Ecology and Sociobiology* 33, 261–8

Møller, A. P. and Cuervo, J. J. (2000). The evolution of paternity and paternal care in birds. *Behavioural Ecology* 11, 472–85.

Moore, S. L. and Wilson, K. (2002). Parasites as a viability cost of sexual selection in natural populations of mammals. *Science* 297, 2015–18.

Morton, E. S., Stutchbury, B. J. M., and Chiver, I. (2010). Parental conflict and brood desertion by females in blue-headed vireos. *Behavioural Ecology and Sociobiology* 64, 947–54.

Neff, B. D. (2003). Decisions about parental care in response to perceived paternity. *Nature* 422, 716–19.

Olson, V. A., Webb, T. J., Freckleton, R. P., and Székely, T. (2009). Are parental care trade-offs in shorebirds driven by parental investment or sexual selection? *Journal of Evolutionary Biology* 22, 672–82.

Parker, G. A., Baker, R. R., and Smith, V. G. F. (1972). The origin and evolution of gamete dimorphism and the male-female phenomenon. *Journal of Theoretical Biology* 36, 181–98.

Parker, G. A. and Simmons, L. W. (1996). Parental investment and the control of sexual selection: predicting the direction of sexual competition. *Proceedings of the Royal Society London, Series B* 263, 315–21.

Pen, I., Uller, T., Feldmeyer, B., Harts, A., While, G. M., and Wapstra, E. (2010). Climate-driven population divergence in sex-determining systems. *Nature* 468, 436–8.

Petrie, M. and Møller, A. P. (1991). Laying eggs in others' nests: intraspecific brood parasitism in birds. *Trends in Ecology and Evolution* 6, 315–20.

Pollux, B. J. A., Pires, M. N., Banet, I. A., and Reznick, D. N. (2009). Evolution of placentas in the fish family Poeciliidae: an empirical study of macroevolution. *Ann. Rev. Ecol. Syst.* 40, 271–89.

Pryke, S. R., Rollins, L. A., and Griffith, S. C. (2010). Females use multiple mating and genetically loaded sperm competition to target compatible genes. *Science* 329, 964–7.

Queller, D. C. (1997). Why do females care more than males? *Proceedings of the Royal Society of London B* 264, 1555–7.

Rankin, D. J., Bargum, K., and Kokko, H. (2007). The tragedy of the commons in evolutionary biology. *Trends in Ecology and Evolution* 22, 643–51.

Requena, G. S., Nazareth, T. M., Schwertner, C. F., and Machado, G. (2010). First cases of exclusive paternal care in stink bugs (Hemiptera: Pentatomidae). *Zoologia* 27, 1018–21.

Reynolds, J. D., Goodwin, N. B., and Freckleton, R. P. (2002). Evolutionary transitions in parental care and live bearing in vertebrates. *Philosophical Transactions of the Royal Society of London B* 357, 269–81.

Ross, C. N., French, J. A., and Orti, G. (2007). Germ-line chimerism and paternal care in marmosets (*Callithrix kuhlii*). *Proceedings of the National Academy of Sciences* U.S.A. 104, 6278–82.

Schwagmeyer, P. L., St Clair, R. C., Moodie, J. D., Lamey, T. C., Schnell, G. D., and Moodie, M. N. (1999). Species differences in male parental care in birds: a reexamination of correlates with paternity. *Auk* 116, 487–503.

Shellman-Reeve, J. and Reeve, H. K. (2000). Extra-pair paternity as the results of reproductive transactions between paired mates. *Proceedings of the Royal Society of London B* 267, 2543–6.doi:10.1098/rspb.2000.1318

Simmons, L. W. and Garcia-Gonzalez, F. (2008). Evolutionary reduction in testes size and competitive fertilization success in response to the experimental removal of sexual selection in dung beetles. *Evolution* 62, 2580–91.

Slatyer, R., Mautz, B., Backwell, P. R. Y., and Jennions, M. D. (2011). Estimating genetic benefits of polyandry from experimental studies: a meta-analysis. *Biological Reviews of the Cambridge Philosophical Society* 87, 1–33.

Smith, C. C. and Fretwell, S. D. (1974). Optimal balance between size and number of offspring.*American Naturalist* 108, 499–506.

Stiver, K. A. and Alonzo, S. H. (2009). Parental and mating effort: is there necessarily a trade-off? *Ethology* 115, 1101–26.

Summers, K., McKeon, C. S., and Heying, H. (2006). The evolution of parental care and egg size: a comparative analysis in frogs. *Proceedings of the Royal Society London B* 273, 687–92.

Tallamy, D. W. (2000). Sexual selection and the evolution of exclusive paternal care in arthropods. *Animal Behaviour* 60, 559–67.

Tallamy, D. W. (2005). Egg dumping in insects. *Annual Reviews of Entomology* 50, 347–70.

Trivers, R. (1972). Parental investment and sexual selection. In B. Campbell, ed. *Sexual Selection and the Descent of Man 1871–1971*, pp. 139–79. Aldine Press, Chicago.

Weir, L. K., Grant, J. W. A., and Hutchings, J. A. (2011). The influence of operational sex ratio on the intensity of competition for mates. *American Naturalist* 177, 167–76.

Wells, K. D. (2007). *The Ecology and Behaviour of Amphibians*. University of Chicago Press, Chicago.

West, S. A. (2009). *Sex Allocation*. Princeton University Press, Princeton.

Wynne-Edwards, K. E. and Reburn, C. J. (2008). Behavioral endocrinology of mammalian fatherhood. *Trends in Ecology and Evolution* 15, 464–8.

Yamamura, N. and Tsuji, N. (1993). Parental care as a game. *Journal of Evolutionary Biology* 6, 103–27.

SECTION II

Conflict and Cooperation in Parental Care

CHAPTER 7

Parent–offspring conflict

Rebecca M. Kilner and Camilla A. Hinde

7.1 Introduction

Disputes between parents and their young might seem easy enough to spot in everyday human life, but the notion of a general, evolutionary conflict between offspring and their parents has proved surprisingly slippery (reviewed by Godfray 1995). Nevertheless, today, almost four decades since the concept was first proposed by Trivers (1974), parent–offspring conflict has theoretically robust foundations (reviewed by Godfray 1995; Godfray and Johnstone 2000) and there is diverse evidence that it is a significant selective force in nature (although perhaps not always in the ways initially assumed, reviewed by Kilner and Hinde 2008). In a recent review, we showed how information warfare lies at the heart of parent–offspring conflict in many instances (Kilner and Hinde 2008). Our aim in this chapter is to attempt to reconstruct the evolutionary consequences of this conflict for traits in offspring and their parents, focusing relatively little on signalling this time. We start with a quick recap of basic conflict theory before outlining diverse empirical evidence for an evolutionary conflict between parents and their young. Next we consider how conflict might link pairs of traits in parents and their offspring, showing how co-evolution between the two parties becomes focused on these particular characters. We conclude with a discussion about the outcomes of parent–offspring conflict: does it always end in a stable equilibrium between parents and their young, as predicted by the many 'resolution' models of conflict? Or is instability widespread, with parents and offspring frequently alternating in who gains the upperhand?

7.2 The theory of parent–offspring conflict

The concept of parent–offspring conflict was outlined by Robert Trivers in 1974, when he applied the revolutionary gene-level analyses of social evolution pioneered by Hamilton (1963, 1964a, 1964b) to his own ideas about parental investment (Trivers 1972). Trivers (1974) developed his idea of parent–offspring conflict by imagining the interactions between a single offspring and its mother, framing the conflict as a battle over the division of resources between the current offspring and its future sibling. This sort of parent–offspring conflict is often referred to as 'interbrood conflict', to distinguish it from 'intrabrood conflict' in which offspring enter into conflict with parents over the division of resources among members of the current brood (Macnair and Parker 1979; Mock and Parker 1997).

Interbrood and intrabrood conflict each arise because asymmetries in relatedness generate contrasting optimal levels of investment for parents and young, and these are illustrated in Fig. 7.1. (For ease of comparison between the two forms of conflict, here we describe parent–offspring conflict over the benefits of investment to parents and offspring, although in his original formulation, Trivers (1974) focused on conflict over costs.) For example, in a monogamous sexually reproducing family (where the relatedness of parents and offspring is 0.5), individual offspring value the benefits of parental investment twice as highly as their parents (Lazurus and Inglis 1986). The optimal level of investment from the offspring's perspective thus exceeds that of its parents, and the disparity between the two optima generates conflict (Fig. 7.1).

The Evolution of Parental Care. First Edition. Edited by Nick J. Royle, Per T. Smiseth, and Mathias Kölliker.

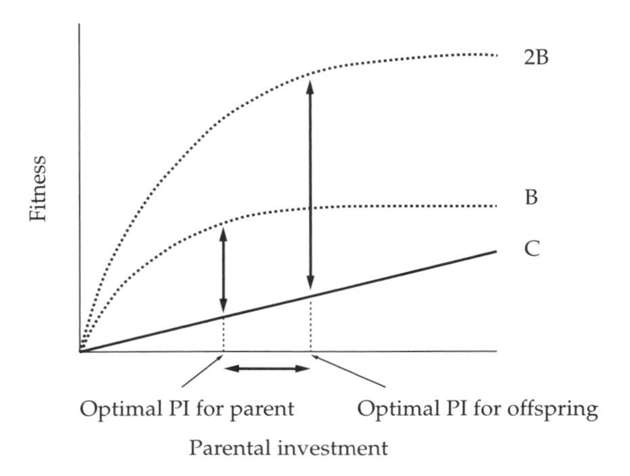

Figure 7.1 Parent–offspring conflict, after Figure 1b in Lazurus and Inglis 1986, redrawn from Kilner and Hinde 2008. 'C' shows the fitness costs experienced by parents and offspring as a consequence of parental investment; 'B' denotes the fitness benefits gained by parents from the provision of parental investment; '2B' shows the fitness benefits experienced by offspring as their parents supply investment (assuming a monogamous mating system in a diploid species). The optimal levels of investment for each party can be found at the point at which they experience greatest benefit for least cost. The horizontal arrow indicates the disparity between optima, which is the source of parent–offspring conflict. Redrawn with permission from Elsevier.

The same reasoning applies whether we are contemplating interbrood conflict over the total level of parental investment supplied to current young, or intrabrood conflict over the division of investment within the current brood. Interbrood conflict selects offspring to develop mechanisms that influence the total amount of parental investment supplied, by forcing parents to supply resources that they would prefer to withhold for reproduction in the future. Interbrood conflict therefore potentially influences the evolution of interactions between offspring and parents. Intrabrood conflict, however, selects offspring to distort the equitable distribution of resources among the brood favoured by parents (all else being equal) that are equally related to each of their current young. Young animals may skew investment towards themselves and away from their siblings (in whom they have a lower genetic stake) either by appealing directly to parents or by outcompeting rival siblings (Chapter 8). Intrabrood conflict can therefore influence the evolution of interactions among offspring as well as those between parents and their young.

At this point, we should resolve some conceptual niggles that often follow the outline of the theory in its simplest terms. In some respects, it may seem counterintuitive that a conflict of interest can ever arise between parents and their young. After all, offspring surely constitute their parent's fitness, so how can parents be in conflict with their own fitness? To see past this obstacle involves thinking about natural selection on genes for strategies rather than on individuals, just as we do when thinking about other sorts of kin-selected altruism. Conflict between parents and offspring may seem impossible when considered at the level of the individual but, taking the gene's-eye view, we can see that conflict between selfish strategies is perfectly plausible. What about the problem that one day offspring will themselves become parents and fall victim to the very same ploy they used as offspring to gain an advantage over their parents (Alexander 1974)? Again, gene-level thinking provides the answer. It is perfectly possible for the optimal strategy for offspring to be quite different from the optimal strategy for parents (Dawkins 1976), even when both strategies are expressed by the same individual. This was shown to be the case in the formal population genetic models of parent–offspring conflict developed by Geoff Parker and others (Macnair and Parker 1978; Parker and Macnair 1978; Stamps et al. 1978; Macnair and Parker 1979; Metcalf et al. 1979). These studies, and others subsequently (reviewed by Godfray 1995; Mock and Parker 1997; Godfray and Johnstone 2000; Parker et al. 2002), have placed the notion of parent–offspring conflict on very robust theoretical foundations.

7.3 Evidence of parent–offspring conflict: battlegrounds for antagonistic interests

7.3.1 Behavioural disputes are not evidence of evolutionary conflict

Now that we have a clearer understanding of parent–offspring conflict theory, we can move on to assess the evidence for its existence in nature. Demonstrating an evolutionary conflict of interests between parents and their young is harder than it might at first appear. As Mock and Forbes (1992) have pointed out, the bleats, whines, squawks, and fights that are characteristic of many family disputes are not themselves evidence of evolutionary conflict and should instead be referred to as 'evolutionary squabbles'. (We prefer the term 'behavioural disputes' to be absolutely clear that such observations imply nothing about the fitness of either party.) For example, acts of aggression between young, which were traditionally considered to serve the evolutionary interests of the dominant offspring, and to run counter to those of the mother (e.g. O'Connor 1978), may actually promote maternal fitness after all, providing a useful mechanism for adaptive brood reduction (Mock and Parker 1997) or even adaptive infanticide (Trillmich and Wolf 2008).

In short, to find evidence of parent–offspring conflict, it is not sufficient to witness a fight. Specifically, there must be evidence that 1) optimal parental investment levels differ between parents and offspring and 2) shifting the usual allocation of parental investment causes a simultaneous increase in fitness to one party and a decrease in fitness to the other. Criterion 1) provides evidence of what Godfray (1995) has termed 'the battleground' for parent–offspring conflict. Criterion 2) provides evidence of the antagonistic fitness interests of offspring and parents within this battleground. As we now show, evidence from at least three different contexts fulfils both of these criteria.

7.3.2 Parent–offspring conflict in natural vertebrate populations

Recent analyses of data collected from natural vertebrate populations have used quantitative genetic techniques to measure selection on mothers and offspring and in so doing have revealed good evidence for the criteria for conflict specified above. Wilson et al. (2005) measured the direction of selection acting on maternal and offspring components of individual fitness by using long-term data collected from free-living Soay sheep *Ovis aries*, with one analysis focusing particularly on litter size. They found that selection operating through maternal lifetime reproductive success favours a greater litter size than selection operating through offspring lifetime reproductive success: in other words mothers are selected to produce larger litters while offspring favour fewer offspring per reproductive bout. Thus 1) the optimal litter size differs for mothers and their young and 2) an increase in litter size will promote maternal fitness whilst simultaneously reducing offspring fitness. Similarly, Janzen and Warner (2009) quantified the direction of selection acting on egg size in turtles, from both the mother's and the offspring's perspective, by using laboratory incubators to measure the hatchability of eggs taken from natural populations of three different species. In general, they found that selection on offspring favoured a larger egg size than selection acting on mothers. So clearly 1) the optimal egg size differs for mothers and their offspring and 2) an increase in egg size will promote offspring fitness whilst at the same time reducing maternal fitness.

7.3.3 IGF-II and other examples of genomic imprinting

The antagonistic interests of genes expressed in parents and offspring are manifest in quite a different way in mammals. For the majority of genes in these species, gene expression is not contingent on whether they are inherited from the mother or the father. By contrast, for the so-called 'imprinted genes' their pattern of expression depends on which parent transmitted them (reviewed by Haig 2004): some are only expressed if they are inherited maternally, some only if they are inherited paternally. Haig and Westoby (1989) hypothesize that this curious pattern of genetic inheritance has evolved as a consequence of parent–offspring conflict. More specifically, they argue, it results from conflict between maternally and paternally inherited genes within the offspring over the provision of maternal investment. In species where females bear young by more than one father, paternally derived genes are expected to demand more maternal investment than those which have been inherited from the mother herself because, unlike the maternally derived genes, the paternally derived genes experience none of the associated costs of maternal care (Haig and Westoby 1989).

The *Igf2* and *Igf2r* genes in mice beautifully support Haig and Westoby's (1989) idea. *Igf2* is paternally imprinted and encodes IGF-II, an insulin-related polypeptide that plays a key role in extracting resources from the mother during pregnancy. When the paternal allele of *Igf2* is experimentally inactivated, offspring are 60% their normal size at birth. Inactivation of the maternal allele has no such effect on birth weight (Haig 1997). Counteracting the effects of *Igf2* is the maternally imprinted *Igf2r*, which encodes a receptor that acts as a sink for IGF-II, thus reducing its influence on resource transfer from mother to offspring (Haig and Graham 1991). When the maternal allele of this gene is inactivated, offspring are 20–30% larger than normal at birth, while inactivation of the paternal allele leaves birth weight unchanged (Haig 1997). The functioning alleles of *Igf2* and *Igf2r* thus balance the resources given to the offspring against the resources that are retained by the mother and potentially given to other young (Haig and Graham 1991), and their expression generates the contrasting optima for mothers and their offspring (and fathers). The paternally imprinted *Igf2* promotes the offspring's fitness at the expense of maternal fitness, while the maternally imprinted *Igf2r* promotes maternal fitness at the expense of the offspring's fitness (and that of their father).

Although there is clear evidence that parent–offspring conflict has caused the evolution of genomic imprinting of *Igf2* and *Igf2r*, and other loci (Haig 2000), it should be said that the evidence is less clearcut for other imprinted genes and the generality of Haig and Westoby's (1989) idea therefore remains to determined (Haig 2004).

7.3.4 Sex ratio wars in the social insects

The sex ratio wars of the social Hymenoptera provide further evidence of an evolutionary conflict of interest, this time between the queen and her worker offspring. The conflict centres on the proportion of male versus female reproductives that should be produced (Trivers and Hare 1976; Bourke and Franks 1995; Mock and Parker 1997; Sündstrom and Boomsma 2001; Ratnieks et al. 2006). In the Hymenoptera, sex is determined by haplodiploidy with fertilized eggs yielding daughters and unfertilized eggs producing sons. In the social Hymenoptera, queens are equally related to sons and daughters ($r = 0.5$) but, in colonies with one queen who is singly mated, the workers are three times as related to their sisters ($r = 0.75$) as their brothers ($r = 0.25$). This quirk of natural history offers the rare opportunity to quantify distinct optima theoretically from both the maternal and offspring's perspectives. Whereas queens are selected to produce an even ratio of daughters versus sons (1:1), workers are selected to favour a female-biased sex ratio of 3:1 (Trivers and Hare 1976). Furthermore, each party can act to promote their own fitness, whilst simultaneously reducing the fitness of the other. Queens can adjust the sex ratio in their favour by carefully allocating sperm to half the eggs they lay, while workers can bias sex allocation in their favour by selectively destroying male eggs (Ratnieks et al. 2006). For example, in bumblebees *Bombus terrestris*, sex ratios are closer to the queen's optimum than to the workers' (Ratnieks et al. 2006). By contrast, in the wood ant *Formica truncorum*, colonies with singly-mated queens show female-biased sex ratios that are close to the worker's optimum (Ratnieks et al. 2006). The key point here is that, once again, the two criteria for parent–offspring conflict are fulfilled.

7.4 Co-evolution of traits in offspring and their parents

Evidence from a diverse range of species thus nicely delimits the battlegrounds within which selection acts in opposing directions on genes expressed in parents and their young, supporting the theory of an evolutionary conflict of interest between the two parties. Given the existence of parent–offspring conflict, how does evolution then proceed? As we show in this section, one key consequence is that conflict drives the co-evolution of pairs of reciprocally acting traits in offspring and their parents: these are traits in offspring that are selected by parents and that influence parental behaviour, and traits in parents that are selected by offspring and that influence offspring behaviour. Furthermore, the nature of the co-evolutionary trajectory between parent and offspring strategies depends critically on the ecology of the species in question.

7.4.1 The co-evolution of supply and demand

In general, parent–offspring conflict causes reciprocally acting traits in parents and their young to co-evolve. In principle, this could happen for any pair of traits influencing parent-offspring interactions where the fitness consequences are antagonistic for each party (a 'conflictor locus' in the offspring versus a 'suppressor locus' in parents, to generalize Parker and Macnair's (1979) terminology). However, it has been especially well-studied for interactions over provisioning in which offspring use begging behaviour to demand food, and in which parents use begging behaviour to determine their rate of brood provisioning. In this section, we therefore focus specifically on this particular pair of traits (see Fig. 7.2 for two examples).

Offspring frequently have private information about their quality or condition, which parents are keen to know so as to allocate resources at levels close to their optimum (Grodzinski and Lotem 2007). Trivers (1974) recognized that offspring could exploit this asymmetry in information to their advantage through exaggerated advertising of their true condition, so as to procure more resources than is optimal for parents to supply. Thus the co-evolution of parental supply and offspring demand is set in motion, with parents selecting the intensity with which offspring demand investment, and with offspring selecting the generosity of their parents through the amount of investment they offer in response (Parker and Macnair 1979; Hussell 1988; Kölliker 2003). Quantitative genetics models more specifically predict the co-adaptation of genes determining offspring solicitation behaviour and genes influencing provisioning behaviour, ultimately leading to a genetic correlation between the two (see Wolf and Brodie 1998; Kölliker et al. 2005; Chapter 16).

But in which direction, exactly, does co-evolution proceed? For example, are ever more exaggerated demands by offspring met with increasing sales resistance in parents, thereby eventually establishing a negative genetic correlation between offspring and parental traits? Or are increasingly intense solicitation displays rewarded by parents with ever increasing levels of investment, eventually generating a positive correlation between traits in parents and offspring? This is the sort of question that interests the quantitative geneticists who study family life, and the co-adaptations that arise as a result (see Chapter 16 for greater discussion of this topic). Nevertheless, the answer strongly depends on which party currently has the upperhand in controlling levels of brood provisioning (Kölliker et al. 2005; Smiseth et al. 2008; Hinde et al. 2010), something of greater interest to the behavioural ecologists who study parent–offspring conflict (Smiseth et al. 2008; Hinde et al. 2010).

Figure 7.2 Two examples of offspring demand and parental supply. (a) A burying beetle larva soliciting food from a provisioning parent. Photo credit O. Krüger. (b) A canary nestling begging for food. Photo credit F. Trabanco.

When parents control provisioning (i.e. they currently set the ceiling on the total resources allocated to the current brood), and provide more care to offspring that beg more intensely, it causes selection to act more strongly on traits in offspring than in parents (Kölliker et al. 2005; Hinde et al. 2010). We might expect parental control of provisioning in species where environmental food distribution sets a limit on how frequently parents can provision their young, leaving offspring relatively powerless to nudge brood provisioning rates higher. A situation like this might arise in seabirds, for example, which commonly travel long distances to collect prey, or in seed eating birds where adults spend a long time at each feeding patch filling their crop, thus visiting the nest relatively infrequently. It may also occur when the rate of provisioning is limited by maternal physiology, which may be the case in lactating mammals, or paternal physiology, if provisioning is genetically determined (e.g. Dor and Lotem 2010), or in species where the resources available for offspring nourishment are fixed and determined before birth, as in the burying beetle where the food available for developing young is dependent on the size of the carcass breeding resource (e.g. Bartlett and Ashworth 1988). Parental life-history could also enforce parental control of provisioning, for example if species are long-lived and must defend resources for future reproduction from the demands of current young (Kölliker et al. 2005; Thorogood et al. 2011).

Regardless of the precise ecological (or physiological) conditions that lead to parental control of provisioning, the outcome is the same: under these circumstances, a positive genetic correlation between parent and offspring traits is predicted (Kölliker et al. 2005). The empirical evidence is broadly consistent with this prediction: in mice *Mus musculus* (Hager and Johnstone 2003; Curley et al. 2004), earwigs *Forficula auricularia* (Mas et al. 2009), burying beetles *Nicrophorus vespilloides* (Lock et al. 2004; Lock et al. 2007), great tits *Parus major* (Kölliker et al. 2000), and canaries *Serinus canaria* (Hinde et al. 2010) a positive correlation between offspring demand and parental supply is indeed observed. In other species, for example those which produce one offspring per breeding attempt, or which have short lives and typically have just one or two bouts of reproduction, offspring currently exert greater control over the total levels of care supplied by parents. Offspring control of provisioning causes selection to act more strongly on traits in the parent than on traits on the young (Hinde et al. 2010). Under these circumstances, a negative genetic correlation is predicted (Kölliker et al. 2005). Again, the evidence is largely consistent with the prediction: in sheep, macaques *Macaca mulatta* (Kölliker et al. 2005) and burrower bugs *Sehirus cincta* (Agrawal et al. 2001) there is a negative correlation between offspring demands and parental supply (although, as with the positive correlations described above, it is unclear whether these correlations are genetic).

7.4.2 Plasticity

As is evident from the preceding section, quantitative genetic models can offer powerful insights into the ways in which parent and offspring traits co-evolve. Nevertheless, as is always the case in biology, the real world is likely to be more complex and interesting than the world described by theory alone. For a start, it is probably incorrect to imagine a dichotomy between parental and offspring control over provisioning (Royle et al. 2002; Hinde and Kilner 2007). Instead, the balance of power between the two parties is more likely to lie on a continuum (Royle et al. 2002; Hinde and Kilner 2007) and probably varies among families with parental quality and changing ecological conditions. In Hihi *Notiomystis cincta*, for example, the extent to which nestlings can influence provisioning at the nest declines considerably if their parents are likely to breed again that year (Thorogood et al. 2011). Parent and offspring traits will continue to co-evolve against this changing backdrop but unless one party consistently controls provisioning, presumably a purely genetic correlation between offspring demand and parental supply cannot arise. Perhaps this explains the lack of such a correlation in the house sparrow *Passer domesticus* (Dor and Lotem 2010), even though parental provisioning behaviour and offspring begging behaviour are each heritable in this species (Dor and Lotem 2009, 2010).

Furthermore, we know that environmental effects generate considerable variation in both the intensity with which offspring demand food (e.g. Kedar et al. 2000; Grodzinski and Lotem 2007) and the rate at which parents provision. Any correlations between parental and offspring behaviour are therefore unlikely to be exclusively genetic (Kölliker et al. 2005; Smiseth et al. 2008). For example, offspring begging intensity varies with nestling age, need, and condition (reviewed by Kilner and Johnstone 1997) while the rate at which parents provision their young is contingent on their brood's begging behaviour and the behaviour of their partner (e.g. Hinde and Kilner 2007), amongst other things. Rather than finding genetic correlations between parent and offspring behaviours, we should instead expect to see behavioural reaction norms becoming linked as the two parties interact (Smiseth et al. 2008). In behavioural ecology parlance, this equates to the behavioural negotiation of rules for offspring demand and parental supply (McNamara et al. 1999; Smiseth et al. 2008; for examples of negotiated behaviour see Hinde 2006; Johnstone and Hinde 2006; Hinde and Kilner 2007).

Moreover, plasticity is not confined to variation on a behavioural time scale, such as within a single breeding attempt. Parents also vary between breeding attempts in the investment they supply to their young and their offspring likewise vary in the strength of their demands (e.g. Lock et al. 2007; Hinde et al. 2009, 2010). Recent evidence suggests that in these 'interbrood' cases, reaction norms for parents and offspring are hormonally determined and maternal effects can link the two sets of reaction norms before birth for the period of offspring dependence (Hinde et al. 2009, 2010). For example, in canaries, mothers respond to nestling begging by supplying more food (Kilner 1995). However, the amount of food mothers actually supply for a given level of begging (i.e. the extent of maternal generosity) varies from female to female and is greater when maternal androgen levels are higher. Nestling begging intensity likewise increases with increasing food deprivation (Kilner 1995) but the intensity of demand for a given level of hunger varies considerably between chicks, being greater when testosterone levels are higher (Buchanan et al. 2007). Despite all this variation, brood begging intensity is positively correlated with the extent of maternal generosity across families (Hinde et al. 2009). Together, these results suggest that reaction norms in maternal supply are linked with reaction norms in offspring demand through (unknown) maternal substances in the egg (Hinde et al. 2010).

It is plausible, though not proven, that hormonally determined reaction norms might similarly determine offspring demand and parental supply in burying beetles *Nicrophorus*. In *N. orbicollis*, juvenile hormone (JH) levels rise rapidly at the onset of maternal care, suggesting that this hormone plays a key role in regulating the extent of maternal generosity (Scott and Panaitof 2004). JH also determines the intensity of offspring demands in *N. vespilloides*, with greater doses increasing larval begging intensity (Crook et al. 2008). Finally, experiments with *N. vespilloides* have established a positive correlation between maternal generosity and larval begging (Lock et al. 2004), which may arise through maternal effects in the egg (Lock et al. 2007).

Has all this interbrood plasticity in both parent and offspring behaviours arisen through co-evolution? It is tempting to speculate that plasticity is a by-product of co-evolution in changing world, where ecological circumstances might otherwise randomly give one party the upperhand. A sudden glut in resources, for example, could make it easier for offspring to bring investment levels closer to their optimum while times of leanness could make it more straightforward for parents to impose investment at their optimum. Perhaps plasticity buffers each party against the vagaries of fluctuations in the environment, while the reaction norms in each party continue to co-evolve?

7.5 What is the outcome of co-evolution between offspring and parents?

Although quantitative genetic techniques can give us powerful insights into the nature and direction of coevolutionary trajectory between parent and offspring traits, we must turn to the phenotypic theoretical models from behavioural ecology to learn more about the eventual outcomes of parent–offspring conflict (Smiseth et al. 2008). Will offspring and their parents tussle forever over the provision of investment? Or does co-evolution eventually conclude with a stable outcome, perhaps with one party emerging as the outright winner?

Before we can attempt to answer these questions, we need to be clear about our definitions of 'stable' and 'unstable' outcomes. In the discussion that follows, we consider the outcome to be unstable if it is possible to see variation in ecological time, with some parents sometimes investing closer to their optimum (parents 'win') and other parents investing closer to the offspring's optimum (offspring 'win'). By contrast, if the outcome is evolutionarily stable then investment is constant with respect to the position of the parent and offspring's optimal investment levels. It is important to note that by following these definitions, we cannot infer much about conflict outcome by observing variation in either parent or offspring behaviour alone: a stable outcome is possible (in the sense that investment is constantly fixed with respect to the parent and offspring optima) even when there is plasticity in the behaviours shown by parents and their young. Thus, to draw conclusions about conflict stability we really need to have measures of parent and offspring fitness.

7.5.1 Unstable outcomes

An evolutionarily unstable outcome to parent–offspring conflict was shown theoretically in one of the earliest analyses of the genetic conflict between parents and offspring (Parker and Macnair 1979). In this model, parental behaviour is determined by one of two alleles: one that responds to offspring demands by supplying more food and one that ignores solicitation altogether. Parents gain by showing some sensitivity to begging because investment is then allocated prudently (e.g. Grodzinksi and Lotem 2007) but suffer by responding to begging because then they are exploited by their offspring. Offspring, on the other hand, gain resources by begging but do so at some cost to their fitness. Offspring behaviour is also determined by one of two alleles: one that demands investment at optimal levels for the parent and one that demands more than is optimal for the parent to supply. With these assumptions in place, simulations show that there are no pairs of parent and offspring behaviours that yield an evolutionarily stable outcome. Instead, under some conditions, cycling of alternate parental and offspring behaviours ensues (Parker and Macnair 1979): parental sensitivity to begging selects offspring which demand more than is optimal for parents to supply. This, in turn, selects parents that ignore offspring solicitation altogether, which favours offspring who seek investment at levels optimal for the parent, which again favours parents that are sensitive to their young. And so on. Although there is a known genetic basis for diverse parent and offspring behaviours (e.g. Haig and Westoby 1989; Kölliker and Richner 2001; Dor and Lotem 2009, 2010), it is not yet known how their frequencies change over evolutionary time. Whether stable limit cycles like this actually occur in nature therefore remains to be determined. Nevertheless, other lines of evidence suggest that there are instances of parent–offspring conflict where the outcome is unstable over evolutionary time in different ways.

7.5.1.1 *Co-evolution between foetus and mother*

David Haig (1993) has identified numerous pairs of potentially conflicting traits in mammalian mothers and their offspring *in utero*. Genomic analyses certainly confirm that there has been rapid, escalating co-evolution within some of these pairings (Summers and Crespi 2005; Crespi and Summers 2006). But to find indications of an unstable outcome of co-evolution between mother and foetus we must focus specifically on the blood sugar wars waged between mothers and their unborn young. After a meal a mother and foetus must decide how to share the resulting increase in blood sugar (Haig 1993). The mother can keep most of the blood sugar if she quickly reduces her blood sugar levels after eating, by secreting high levels of insulin. The foetus counteracts these maternal tactics by secreting human placental lactogen (hPL), which increases the mother's resistance to insulin and so keeps blood sugar levels elevated for longer, thereby allowing the foetus to take a greater portion. In response, the mother swamps the effect of hPL by increasing her levels of insulin production, so escalating a hormonal battle with the unborn baby over levels of glucose in the blood (Haig 1993). Haig (1993) interprets these observations as evidence of co-evolutionary warfare in which selection favours offspring that produce high levels of placental hormones, but favours mothers who become increasingly resistant to the hormones' effect. Furthermore, he argues that some mothers clearly lose while their offspring gain: some mothers suffer from conditions such as diabetes in later life, yet give birth to heavy, thriving babies. Other women, however, seem to escape altogether the scars of blood sugar warfare, suffering no obvious long-term fitness costs at all while still delivering healthy babies (Haig 1993). These observations thus appear to fit our criteria for an unstable outcome to parent–offspring conflict because some mothers seemingly invest close to their optimum while others invest closer to their offspring's optimum.

7.5.1.2 Co-evolution between queens and workers

The sex ratio wars seen in Hymenopteran populations also exhibit unstable outcomes, in this case as a result of changing ecological conditions which can temporarily favour one party over the other (e.g. Chapuisat et al. 1997; Ratnieks et al. 2006). The particular outcome seems to depend on the opportunity for workers to bias the sex ratio and on the accuracy of the information available to workers about the sex of the developing young and their maternity and paternity. When workers are powerless, or deprived of this key information, sex ratios lie closer to the queen's optimum (reviewed by Ratnieks et al. 2006). This sort of unstable parent–offspring conflict thus closely resembles instances of unstable sexual conflict over mating, where the outcome is similarly variable and also influenced by ecology (e.g. Davies 1992; reviewed by Arnqvist and Rowe 2005).

7.5.2 Stable outcomes

7.5.2.1 Investment at the parent's optimum, with no interaction with offspring

In some cases, co-evolution between parents and their young has yet to get underway because there is seemingly no reciprocal conflicting locus in offspring (or at least, not one that is currently expressed). For example, parents may so completely overwhelm offspring with their physical dominance, and their control over the provision of care, that offspring are incapable of contesting investment decisions (Godfray 1995). For example, offspring cannot determine the extent to which their egg will be provisioned, nor the size of the family into which they will hatch, yet both of these will influence the level of investment the offspring obtains and hence its future fitness (e.g. de Kogel and Prijs 1996; Nager et al. 2000). In these circumstances, the likely result is a stable outcome with investment fixed at the parent's optimum. The clearest evidence for this comes from parent–offspring conflict over turtle egg provisioning, in which the conflict is resolved very close to the parent's optimum (Janzen and Warner 2009).

7.5.2.2 Parents interact directly with offspring

When offspring have the opportunity to contest parental investment decisions then it is less straightforward to see how the conflict will end and in whose favour it will be resolved. This is where the phenotypic models used by behavioural ecologists come into their own. Here, game theory is used to find pairs of behavioural strategies in parents and offspring that are evolutionarily stable, and to determine whether investment lies closer to the parent or offspring's optimum when these evolutionarily stable strategies are employed. A substantial theoretical literature of this sort now exists (reviewed by Godfray 1995; Mock and Parker 1997; Godfray and Johnstone 2000; Parker et al. 2002), much of it focusing on the role of stable offspring begging and parental provisioning strategies in resolving parent–offspring conflict.

One general finding is that the outcome of conflict depends critically on the way in which parental provisioning behaviours select offspring begging behaviours over evolutionary time (i.e. the effect of supply on demand (ESD; Hussell 1988; Parker et al. 2002) or Mechanism 1 (Mock and Parker 1997)). These co-evolutionary relationships are generally assumed at the outset in game theory models. For example, in the so-called scramble models of conflict resolution, championed by Geoff Parker and his colleagues (e.g. Macnair and Parker 1979; Parker and Macnair 1979; Parker 1985; Parker et al. 1989; Mock and Parker 1997; Parker et al. 2002) an increase in parental investment levels selects a reduction in begging levels over evolutionary time. In these models, individual offspring control the extent to which they are provisioned through scramble competition. Offspring pay a fitness cost when they compete for food, and by threatening excessive competition, effectively blackmail parents into supplying more food so that resources are not wasted on extravagant begging (Godfray 1995; Johnstone 1996). With a negative ESD function like this, in which greater parental investment selects reduced begging, a stable outcome is reached in which investment levels lie somewhere between parent and offspring optima (Mock and Parker 1997; Parker et al. 2002).

Put like this, Parker et al.'s game theory models fit pleasingly with the quantitative genetic approach to understanding parent–offspring interactions, described in Section 7.4 (see also Smiseth et al. 2008). Quantitative genetics theory provides a biological reason to link offspring control of provisioning and the negative ESD relationship that are otherwise rather unconnected arbitrary assumptions in the scramble models of conflict resolution (Mock and Parker 1997; Parker et al. 2002).

In the so-called signalling resolution models of parent–offspring conflict (Godfray 1991, 1995; Johnstone 1996, 1999) the ESD function is rather different (and it is harder to find as neat a correspondence with the theoretical quantitative genetic work). The aim of these models is to explain the paradoxical exuberance of short-range communication between offspring and their parents: why do young animals shout so loudly in their parent's ear when noisy behaviour like this draws the unwanted attention of potential predators and wastes energy? Here, parents control provisioning and the condition dependence of offspring begging behaviour is allowed to evolve. At equilibrium, offspring begging behaviour is costly and accurately signals need and parents use begging behaviour to allocate food in relation to need. Costly begging is essential to stabilize offspring–parent interactions and this explains the paradoxically flamboyant offspring begging behaviour. Critically, in these models the supply of investment does not change offspring begging behaviour on an evolutionary time scale: offspring beg at their ESS level irrespective of parental behaviour (Mock and Parker 1997; Parker et al. 2002). In other words, the slope of the ESD function is 0 (a scenario for which there is currently no equivalent quantitative genetic model). As a result, parents are able to supply investment at or close to their optimum (Godfray 1991).

Whether either of these theoretical treatments of conflict resolution actually describes events in the real world is still tantalizingly beyond our reach, even though empiricists can find systems that match some parts of each theoretical approach very well (reviewed by Kilner and Johnstone 1997; Royle et al. 2002; Kilner and Hinde 2008). Perhaps the clearest example of conflict apparently resolved close to the parent's optimum comes from experiments with captive canaries, although the mechanism for conflict resolution revealed by this work was not foreshadowed by theory.

In canaries, parents visit the nest relatively infrequently with food, and feed their young seed by regurgitation. Canary feeding ecology thus gives parents a high level of control over provisioning.

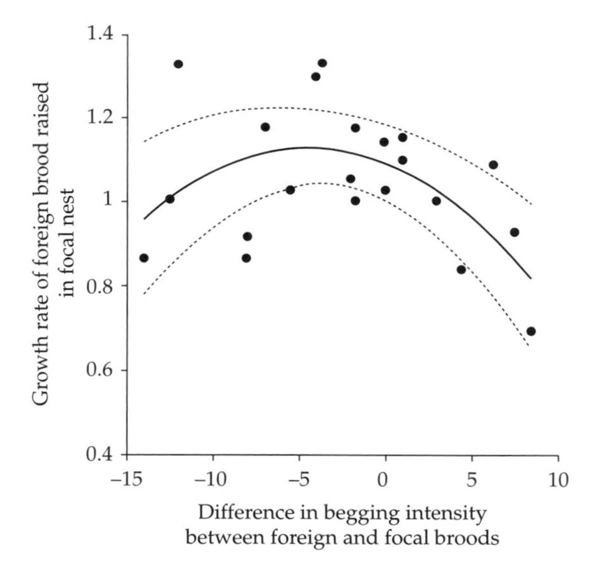

Figure 7.3 Regression plot showing the effect of a change in brood demand at the focal nest on the growth rate of the foreign brood. Each datapoint is collected from one pair (N = 21 pairs with begging data that reared one foreign brood and one brood of their own). The least squares regression lines are shown with 95% confidence intervals. This experiment shows that when the system of pre-natal signalling is disrupted experimentally by swapping young between nests, offspring that beg more than their foster mother's own chicks grow relatively slowly because they waste resources on unrewarded begging (redrawn from Hinde et al. 2010).

Experiments show that when nestlings in the brood vary in hunger, mothers control food allocation by carefully choosing which offspring to feed (Kilner 1995, 1997). When the brood is very hungry, mothers cede control of food allocation to their offspring, inciting sibling competition and rewarding only the most competitive offspring (Kilner 1995, 2002). So even when offspring divide food amongst themselves, it is done at the mother's bidding. The begging behaviour performed by young canaries is costly, because offspring induced to beg too vigorously for food then gain mass more slowly (Kilner 2001). The costs of begging, in conjunction with strong maternal control over provisioning, limit the offspring's capacity to procure resources at levels greater than the maternal optimum (Hinde et al. 2010). In cross-fostering experiments, foreign offspring whose begging intensities were naturally greater than those exhibited by natal broods were unable to use their greater demands to extract substantially more resources from their foster parents. Instead, they wasted energy on unrewarded begging and grew more slowly than the natal brood as a result (Fig. 7.3). Such wastefulness does not arise under natural conditions because offspring are careful to scale their demands to the extent of their parent's generosity (Hinde et al. 2009), probably using signals of maternal quality carried in the eggs (Hinde et al. 2010). In short, multiple lines of evidence show that mothers have the upper hand in determining the rate at which offspring are provisioned. They deploy signals in the egg before hatching to determine brood begging levels, which are enforced by costly begging coupled with strong maternal control over provisioning rates. Moreover, this system of maternal control appears resistant to fluctuations in environmental conditions: in times of plenty mothers simply increase their generosity, and their brood begs more vigorously, while under harsher conditions, mothers provide less food and brood demands are correspondingly less intense (Hinde et al. 2009). In this way, mothers are able to defend their future fecundity from the demands of their young (Hinde et al. 2010). Presumably this means they are able to provision at levels close to their optimum for generation after generation, although there is no direct evidence that this is the case.

7.6 Future directions

Historically, the field of parent–offspring conflict has developed through advances in theory, from the first population genetic models indicating that the concept of an evolutionary battleground between parents and their young was robust (Parker and Macnair 1978; Macnair and Parker 1978), to the first resolution models (Parker and Macnair 1979; Macnair Parker 1979; Parker 1985), to the signalling models that arguably sparked the field's renaissance by stimulating long overdue experimental work (Godfray 1991, 1995). Now it is time for empirical research to point the way for future work on parent–offspring conflict, and our chapter indicates three ways in which this might be accomplished. First, empirical work can identify otherwise hidden battlegrounds for parent–offspring conflict through sophisticated analyses of long-term datasets collected in the field (e.g. Wilson et al. 2005; Janzen and Warner 2009). In addition, new genomic datasets provide a 'genetic fossil record' of battles waged between parents and their young in evolutionary history (e.g. Summers and Crespi 2005; Crespi and Summers 2006).

Second, empirical work is beginning to take us away from the concept of universal conflict resolutions (Godfray 1995) by yielding evidence that unstable outcomes are equally possible. Consequently, a challenge for the future is to predict when parent–offspring conflict should have a stable resolution and when it should not (see Lessells 2006 for equivalent predictions concerning the outcome of sexual conflict). Comparative analyses of empirical data from diverse systems can lead the way here (e.g. see Lessells 2006).

Finally, empirical work is starting to bridge the divide between quantitative genetic analyses of parent–offspring co-adaptation and behavioural ecological analyses of conflict resolution, (Kölliker et al. 2000; Kölliker 2003; Kölliker et al. 2005; Smiseth et al. 2008; Hinde et al. 2010; Chapter 16). Future work here might fruitfully focus on the plasticity of parental and offspring strategies. These range from plasticity in parent and offspring strategies within a breeding attempt involving learning and signalling, by which strategies are negotiated in behavioural time (e.g. Smiseth et al. 2008); to plasticity between breeding attempts, hormonally induced in response to environmental cues (e.g. Hinde et al. 2009, 2010); to variation among individuals as a function of context or condition (e.g. Thorogood et al. 2011). In our view, the field should move beyond dealing in population averages, as previously encouraged by the resolution models for understanding parent–offspring conflict, and focus instead on accounting for individual variation in behaviour.

References

Agrawal, A. F., Brodie III, E. D., and Brown, J. (2001). Parent-offspring coadaptation and the dual genetic control of maternal care. *Science* 292, 1710–12.

Alexander, R. D. (1974). The evolution of social behaviour. *Annual Review of Ecology and Systematics* 5, 325–83.

Arnqvist, G. and Rowe, L. (2005). *Sexual Conflict*. Princeton University Press, Princeton.

Bartlett, J. and Ashworth, C. M. (1988). Brood size and fitness in *Nicrophorus vespilloides* (Coleoptera: Silphidae). *Behavioral Ecology and Sociobiology* 22, 429–34.

Bourke, A. F. G. and Franks, N. R. (1995). *Social Evolution in Ants*. Princeton University Press, Princeton.

Buchanan, K. L., Goldsmith, A. R., Hinde, C. A., Griffith, S. C., and Kilner, R. M. (2007). Does testosterone mediate the trade-off between nestling begging and growth in the canary (*Serinus canaria*)? *Hormones and Behavior* 52, 664–71.

Chapuisat, M., Sündstrom, L., and Keller, L. (1997). Sex-ratio regulation: The economics of fratricide in ants. *Proceedings of the Royal Society, Series B* 264, 1255–60.

Crespi, B. J. and Summers, K. (2006). Positive selection in the evolution of cancer. *Biological Reviews* 81, 407–24.

Crook, T. C., Flatt, T., and Smiseth, P. T. (2008). Hormonal modulation of larval begging and growth in the burying beetle *Nicrophorus vespilloides*. *Animal Behaviour* 75, 71–7.

Curley, J. P., Barton, S., Surani, A., and Keverne, E. B. (2004). Coadaptation in mother and infant regulated by a paternally expressed imprinted gene. *Proceedings of the Royal Society, Series B* 271, 1303–9.

Davies, N. B. (1992). *Dunnock Behaviour and Social Evolution*. Oxford University Press, Oxford.

Dawkins, R. (1976). *The Selfish Gene*. Oxford University Press, Oxford.

de Kogel, C. H. and Prijs, H. J. (1996). Effects of brood size manipulations on sexual attractiveness of offspring in the zebra finch. *Animal Behaviour* 51, 699–708.

Dor, R. and Lotem, A. (2009). Heritability of nestling begging intensity in the house sparrow (*Passer domesticus*). *Evolution* 63, 743–8.

Dor, R. and Lotem, A. (2010). Parental effort and response to nestling begging in the house sparrow: repeatability, heritability and parent-offspring co-evolution. *Journal of Evolutionary Biology* 23, 1605–12.

Godfray, H. C. J. (1991). Signalling of need by offspring to their parents. *Nature* 352, 328–30.

Godfray, H. C. J. (1995). Evolutionary theory of parent-offspring conflict. *Nature* 376, 133–8.

Godfray, H. C. J. and Johnstone, R. A. (2000). Begging and bleating: The evolution of parent-offspring signalling. *Philosophical Transactions of the Royal Society, Series B* 355, 1581–91.

Grodzinski, U. and Lotem, A. (2007). The adaptive value of parental responsiveness to nestling begging. *Proceedings of the Royal Society, Series B* 274, 2449–56.

Hager, R. and Johnstone, R. A. (2003). The genetic basis of family conflict resolution in mice. *Nature* 421, 533–5.

Haig, D. (1993). Genetic conflicts in human pregnancy. *Quarterly Reviews in Biology* 68, 495–532.

Haig, D. (1997). The social gene. In J. R. Krebs and N. B. Davies, eds. *Behavioural Ecology: An Evolutionary Approach*, pp. 284–304. Blackwell Science, Oxford.

Haig, D. (2000). The kinship theory of genomic imprinting. *Annual Review of Ecology and Systematics* 31, 9–32.

Haig, D. (2004). Genomic imprinting and kinship: How good is the evidence? *Annual Review of Genetics* 38, 553–85.

Haig, D. and Graham, C. (1991). Genomic imprinting and the strange case of the insulin-like growth factor-II receptor. *Cell* 64, 1045–6.

Haig, D. and Westoby, M. (1989). Parent-specific gene-expression and the triploid endosperm. *American Naturalist* 134, 147–55.

Hamilton, W. D. (1963). The evolution of altruistic behaviour. *American Naturalist* 97, 354–6.

Hamilton, W. D. (1964a). The genetical evolution of social behaviour I. *Journal of Theoretical Biology* 7, 1–16.

Hamilton, W. D. (1964b). The genetical evolution of social behaviour II. *Journal of Theoretical Biology* 7, 17–52.

Hinde, C. A. (2006). Negotiation over offspring care? A positive response to partner-provisioning rate in great tits. *Behavioral Ecology* 17, 6–12.

Johnstone, R. A. and Hinde, C. A. (2006). Negotiation over offspring care – how should parents respond to each other's efforts? *Behavioral Ecology* 17, 818–27.

Hinde, C. A. and Kilner, R. M. (2007). Negotiations within the family over the supply of parental care. *Proceedings of the Royal Society, Series B* 274, 53–61.

Hinde, C. A., Buchanan, K. L., and Kilner, R. M. (2009). Prenatal environmental effects match offspring begging to parental provisioning. *Proceedings of the Royal Society, Series B* 276, 2787–94.

Hinde, C. A., Johnstone, R. A., and Kilner, R. M. (2010). Parent-offspring conflict and coadaptation. *Science* 327, 1373–6.

Hussell, D. J. T. (1988). Supply and demand in tree swallow broods: a model of parent-offspring food-provisioning interactions in birds. *American Naturalist* 131, 175–202.

Janzen, F. J. and Warner, D. A. (2009). Parent-offspring conflict and selection on egg size in turtles. *Journal of Evolutionary Biology* 22, 2222–30.

Johnstone, R. A. (1996). Begging signals and parent-offspring conflict: Do parents always win? *Proceedings of the Royal Society, Series B* 263, 1677–81.

Johnstone, R. A. (1999). Signaling of need, sibling competition, and the cost of honesty. *Proceedings of the National Academy of Sciences of the United States of America* 96, 12644–9.

Kedar, H., Rodriguez-Girones, M. A., Yedvab, S., Winkler, D. W., and Lotem, A. (2000). Experimental evidence for offspring learning in parent-offspring communication. *Proceedings of the Royal Society, Series B* 267, 1723–7.

Kilner, R. (1995). When do canary parents respond to nestling signals of need? *Proceedings of the Royal Society, Series B* 260, 343–8.

Kilner, R. (1997). Mouth colour is a reliable signal of need in begging canary nestlings. *Proceedings of the Royal Society B* 264, 963–8.

Kilner, R. and Johnstone, R. A. (1997). Begging the question: are offspring solicitation behaviours signals of need? *Trends in Ecology and Evolution* 12, 11–15.

Kilner, R. M. (2001). A growth cost of begging in captive canary chicks. *Proceedings of the National Academy of Sciences of the United States of America* 98, 11394–8.

Kilner, R. M. (2002). Sex differences in canary (*Serinus canaria*) provisioning rules. *Behavioral Ecology and Sociobiology* 52, 400–7.

Kilner, R. M. and Hinde, C. A. (2008). Information warfare and parent-offspring conflict. *Advances in the Study of Behavior* 38, 283–336.

Kölliker, M. (2003). Estimating mechanisms and equilibria for offspring begging and parental provisioning *Proceedings of the Royal Society, Series B* 270, S110–13.

Kölliker, M., Brinkhof, M., Heeb, P., Fitze, P., and Richner, H. (2000). The quantative genetic basis of offspring solicitation and parental response in a passerine bird with parental care. *Proceedings of the Royal Society, Series B* 267, 2127–32.

Kölliker, M., Brodie, E. D., and Moore, A. J. (2005). The coadaptation of parental supply and offspring demand. *American Naturalist* 166, 506–16.

Kölliker, M. and Richner, H. (2001). Parent-offspring conflict and the genetics of offspring solicitation and parental response. *Animal Behaviour* 62, 395–407.

Lazurus, J. and Inglis, I. R. (1986). Shared and unshared parental investment, parent-offspring conflict and brood size. *Animal Behaviour* 34,1791–804.

Lessells, C. M. (2006). The evolutionary outcome of sexual conflict. *Philosophical Transactions of the Royal Society, Series B* 361, 301–17.

Lock, J. E., Smiseth, P. T., and Moore, A. J. (2004). Selection, inheritance, and the evolution of parent-offspring interactions. *American Naturalist* 164, 13–24.

Lock, J. E., Smiseth, P. T., Moore, P. J., and Moore, A. J. (2007). Coadaptation of prenatal and postnatal maternal effects. *American Naturalist* 170, 709–18.

Macnair, M. R. and Parker, G. A. (1978). Models of parent-offspring conflict. II Promiscuity. *Animal Behaviour* 26, 111–22.

Macnair, M. R. and Parker, G. A. (1979). Models of parent-offspring conflict. III Intra-brood conflict. *Animal Behaviour* 27, 1202–9.

Mas, F., Haynes, K. F., and Kölliker, M. (2009). A chemical signal of offspring quality affects maternal care in a social insect. *Proceedings of the Royal Society, Series B* 276, 2847–53.

McNamara, J., Gasson, C., and Houston, A. (1999). Incorporating rules for responding into evolutionary games. *Nature* 401, 368–71.

Metcalf, R. A., Stamps, J. A., and Krishnan, V. V. (1979). Parent-offspring conflict that is not limited by degree of kinship. *Journal of Theoretical Biology* 76, 99–107.

Mock, D. W. and Forbes, L. S. (1992). Parent offspring conflict: a case of arrested development. *Trends in Ecology and Evolution* 7, 409–13.

Mock, D. W. and Parker, G. A. (1997). *The Evolution of Sibling Rivalry*. Oxford University Press, Oxford.

Nager, R. G., Monaghan, P., and Houston, D. C. (2000). Within-clutch trade-offs between the number and quality of eggs: experimental manipulations in gulls. *Ecology* 81, 1339–50.

O'Connor, R. J. (1978). Brood reduction in birds: selection for infanticide, fratricide, and suicide? *Animal Behaviour* 26, 79–96.

Parker, G. A. (1985). Models of parent-offspring conflict. V. Effects of the behaviour of the two parents. *Animal Behaviour* 33, 519–33.

Parker, G. A. and Macnair, M. R. (1978). Models of parent-offspring conflict. I. Monogamy. *Animal Behaviour* 26, 97–110.

Parker, G. A. and Macnair, M. R. (1979). Models of parent-offspring conflict. IV. Suppression: evolutionary retaliation by the parent. *Animal Behaviour* 27, 1210–35.

Parker, G. A., Mock, D. W., and Lamey, T. C. (1989). How selfish should stronger sibs be? *American Naturalist* 133, 846–68.

Parker, G. A., Royle, N., and Hartley, I. (2002). Intrafamilial conflict and parental investment: A synthesis. *Philosophical Transactions of the Royal Society, Series B* 357, 295–307.

Ratnieks, F. L. W., Foster, K. R., and Wenseleers, T. (2006). Conflict resolution in insect societies. *Annual Review of Entomology* 51, 581–608.

Royle, N. J., Hartley, I. R., and Parker, G. A. (2002). Begging for control: When are offspring solicitation behaviours honest? *Trends in Ecology and Evolution* 17, 434–40.

Scott, M. P. and Panaitof, S. C. (2004). Social stimuli affect juvenile hormone during breeding in biparental burying beetles (Silphidae: *Nicrophorus*). *Hormones and Behavior* 45, 159–67.

Smiseth, P. T., Wright, J., and Kölliker, M. (2008). Parent-offspring conflict and co-adaptation: behavioural ecology meets quantitative genetics. *Proceedings of the Royal Society, Series B* 275, 1823–30.

Stamps, J. A., Metcalf, R. A., and Krishnan, V. V. (1978). A genetic analysis of parent-offspring conflict. *Behavioral Ecology and Sociobiology* 3, 369–92.

Summers, K. and Crespi, B. (2005). Cadherins in maternal-foetal interactions: Red queen with a green beard? *Proceedings of the Royal Society, Series B* 272, 643–9.

Sündstrom, L. and Boomsma, J. J. (2001). Conflicts and alliances in insect families. *Heredity* 86, 515–21.

Thorogood, R., Ewen, J. G., and Kilner, R. M. (2011). Sense and sensitivity: responsiveness to offspring signals varies with the parents' potential to breed again. *Proceedings of the Royal Society, Series B* 278, 2638–45.

Trillmich, F. and Wolf, J. B. W. (2008). Parent-offspring and sibling conflict in Galápagos fur seals and sea lions. *Behavioral Ecology and Sociobiology* 62, 363–75.

Trivers, R. L. (1972). Parental investment and sexual selection. In B. Campbell, ed. *Sexual Selection and the Descent of Man 1871–1971*, pp. 136–79. Aldine Press, Chicago.

Trivers, R. L. (1974). Parent-offspring conflict. *American Zoologist* 14, 249–64.

Trivers, R. L. and Hare, H. (1976). Haplodiploidy and the evolution of the social insects. *Science* 191, 249–64.

Wilson, A. J., Pilkington, J. G., Pemberton, J. M., Coltman, D. W., Overall, A. D. J., Byrne, K. A., and Kruuk, L. E. B. (2005). Selection on mothers and offspring: whose phenotype is it and does it matter? *Evolution* 59, 451–63.

Wolf, J. B. and Brodie, E. D. (1998). The coadaptation of parental and offspring characters. *Evolution* 52, 299–308.

CHAPTER 8

Sibling competition and cooperation over parental care

Alexandre Roulin and Amélie N. Dreiss

8.1 Introduction

Until the 1960s and 1970s, evolutionary biologists envisioned family interactions as harmonious, with parents maximizing the number of surviving offspring (Lack 1947). However, after the development of the theories of kin selection and parent–offspring conflict (Hamilton 1964; Trivers 1974), it became evident that family members might have conflicting interests concerning the allocation of parental resources and that such conflicts may be particularly violent between siblings. Sibling competition refers to rivalry between siblings over access to limited parental resources (Box 8.1). The cause of sibling rivalry stems from offspring demanding more resources from their parents than parents are willing to supply. This mis-match between supply and demand of resources is the outcome of three key life history strategies (Stearns 1992). First, because reproductive activities are costly, parents are selected to optimally allocate resources between the different reproductive events rather than to maximize effort at the current attempt. Second, parents face a trade-off between offspring number and quality, and hence they usually maximize their fitness by producing several medium-quality offspring rather than by producing fewer higher-quality offspring. Third, parents often create more offspring than they can rear to independence either because resources become scarcer than anticipated by the parents or because marginal offspring are created as an insurance against early failure of the core offspring (Forbes 1991). The limitation of resources leads to three forms of conflict between family members: siblings compete among each other to share resources (this chapter), offspring are in conflict with their parents over how much parents should invest in providing resources (Chapter 7), and in species with biparental care the mother and father are in conflict over how much effort each party should assume (Chapter 9).

The observation that even closely related individuals compete intensely for resources may seem counterintuitive at first sight. The Arabic proverb 'I against my brothers, my brothers and I against my cousins and I, my brothers, and my cousins against the strangers' perfectly illustrates that even though individuals often support close relatives, siblings may compete when confined in a restricted space where shared resources are limited. There is thus a trade-off between behaving altruistically towards relatives to derive indirect genetic benefits and competing with them to obtain direct material benefits (Mock and Parker 1997; West et al. 2002). This makes the study of sibling interactions challenging as such interactions range from cooperation to fierce competition (Drummond 2001; Roulin 2002), and hence it can help understand the evolution of selfishness and altruism among close relatives (Box 8.1). As emphasized by Mock and Parker (1997), Hamilton's kin selection theory not only sets the conditions promoting altruism but also specifies the evolutionary limits on selfish behaviour. Following Hamilton's rule an allele coding for altruism will spread in a population when the benefit of being altruistic multiplied by the coefficient of relatedness between the altruistic donor and its recipient exceeds its costs. Conversely, the inverse Hamilton's rule states that an allele coding for selfishness will spread if the benefits of being selfish exceed the costs to the victim multiplied by the coefficient of

The Evolution of Parental Care. First Edition. Edited by Nick J. Royle, Per T. Smiseth, and Mathias Kölliker.

Box 8.1 Definition of sibling competition and sibling cooperation

Sibling: one of two or more individuals having at least one parent in common. Full-sibs have two parents in common and half-sibs have one parent in common.

Sibling competition: interactions between siblings that increase the fitness of an individual offspring at an expense of the fitness of its siblings. The extent of this reduction in fitness due to sibling competition should be compared with the situation where siblings would not be interacting or by measuring the negative effect that one individual has upon its sibling(s) by consuming, or controlling access, to some limited resource.

Sibling cooperation: interactions between siblings that have a positive effect on the fitness of an individual offspring as well as the fitness of its siblings. Although cooperative acts can entail costs, the net benefit should be positive either directly (i.e. the actor of a cooperative act stands to gain material benefits in terms of extra resources or lower resource loss compared to the situation where cooperation would not have reduced the level of sibling competition) or indirectly by giving resources to siblings with whom they share genes.

Selfish behaviour: a behaviour that enhances the fitness of the individual expressing the behaviour at the expense of the individual that the behaviour is targeted at.

Altruistic behaviour: a behaviour enhancing the fitness of the individual that the behaviour is targeted at at the expense of the individual expressing the behaviour.

relatedness between the selfish individual and the victim (Mock and Parker 1997). Assuming that the propensity of being selfish is heritable, Alexander (1974) proposed that a conflictor gene coding for selfish behaviour in offspring would not spread, since selfish individuals would have reduced fitness as parents as a consequence of the selfish behaviour of their own offspring. However, because selfish offspring out-compete their non-selfish siblings, the 'conflictor gene' should spread because it provides fitness benefits early in life even at a cost paid at a later stage (Macnair and Parker 1979; Chapter 7).

The function of sibling competition and its evolutionary implications can be more complex than just a matter of competing for limited resources. Even though sibling competition may often reduce parental fitness by decreasing the number of offspring and their survival prospects through lethal fights or other deleterious interactions, sibling competition may also provide a strategic means by which parents enhance their own fitness (Simmons 1988). Indeed, if the amount of resources available for breeding is unpredictable, parents may produce more offspring than they can normally rear (Mock and Parker 1997) so that they can take advantage of years when resources are abundant, while at the same time needing a mechanism for eliminating surplus offspring in years when resources are scarce. Sibling competition provides a mechanism for producing high quality offspring while selectively eliminating individuals of low residual reproductive value (Simmons 1988). The elimination of some offspring through sibling competition, a process that is usually denoted 'brood reduction' in the ornithological literature, may release part of the resources from marginal to core offspring (Forbes 1991).

In this chapter, we first highlight theoretical and empirical aspects of the different forms of sibling competition and cooperation with a specific focus on aggressive and non-aggressive forms of competition. We then elaborate on the factors generating various forms of sibling interactions including parental overproduction of offspring, sibling dominance hierarchies, and the nature of food supply. We finally discuss future directions in the study of sibling competition and cooperation. The study of sibling interactions is important as it sheds light on the evolution of social interactions among siblings, including aggression and communication (Godfray 1995), and of various life-history traits such as growth rate (Royle et al. 1999). Although sibling competition for space, oxygen supply in aquatic clutches or light in plants also prevails in species without parental care, we will focus exclusively on competition for limited parental resources in this chapter.

8.2 Forms of sibling competition and cooperation

The forms and intensity of sibling competition differs between animal taxa, lifestages (i.e. before birth, soon after birth, or just before independence) and the forms of care parents provide for their offspring. As our chapter deals with sibling competition over parental care, and sibling competition can operate under all forms of care, we provide a schematic overview of forms of care and how they relate to sibling competition (Fig. 8.1). Mock and Parker (1997) already reviewed the diversity in sibling competition in various organisms and Hudson and Trillmich (2008) reviewed sibling competition in mammals. Although the study of sibling competition has also spread to insects and other invertebrates (e.g. Smiseth et al. 2007; Dobler and Kölliker 2010), most studies have been performed on birds reflecting a substantial taxonomic bias in our understanding of sibling competition. A search in the web of sciences with the key words 'sibling competition' and a taxonomic name yielded 289 studies in birds (85.8%), 31 in mammals (9.2%), 9 in insects (2.7%), 7 in fishes (2.1%), and 1 in reptiles (0.3%).

Within-family distribution of parental resources involves complex interactions between parents and offspring, with each party potentially having some degree of control over resource allocation. Competition can take place between contemporary siblings

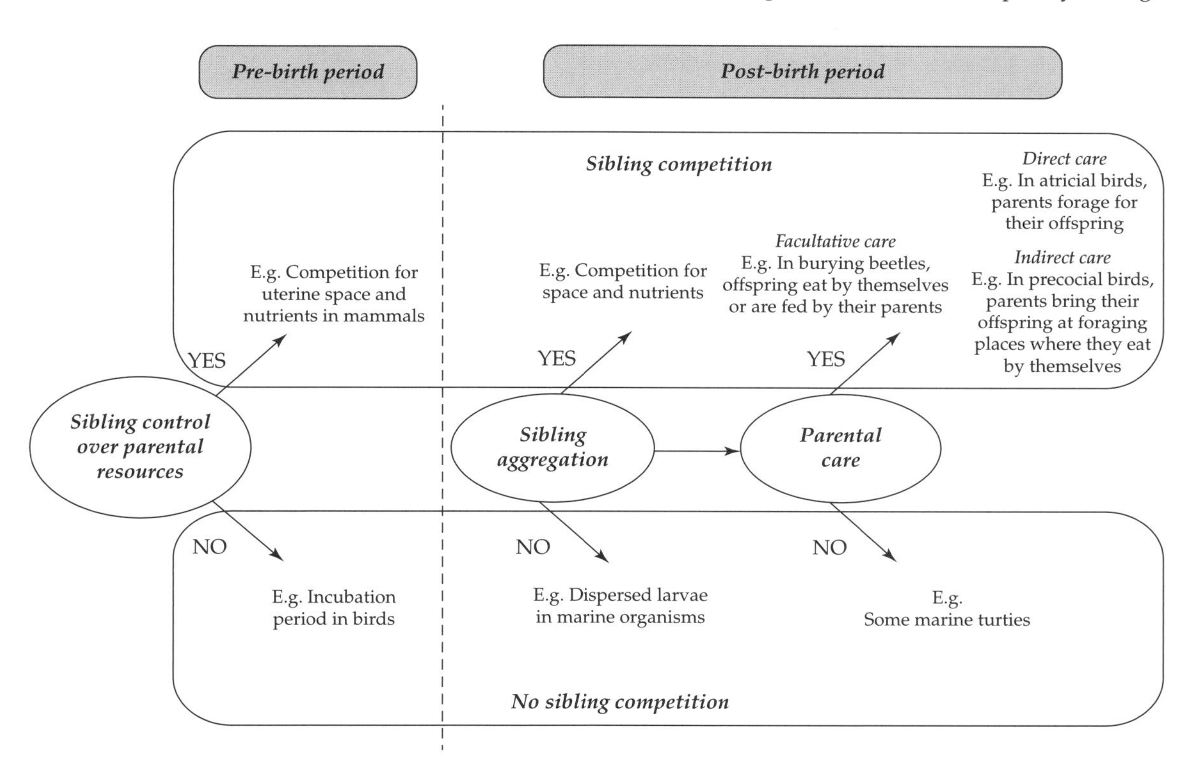

Figure 8.1 Overview of the major lifestages during which siblings compete over parental care. Before birth, sibling competition is absent unless the offspring can have some control over the amount of resources they receive from their parents, such as in mammals and viviparous lizards. After birth, sibling competition is absent if siblings are dispersed over a wide area preventing them interacting and hence competing. This situation is found for instance in some marine organisms. However, the most frequent situation is where siblings are aggregated, at least momentarily, and hence compete to extract limited resources. In many species, parents provide care either directly by feeding their offspring or indirectly if the offspring can feed by themselves but still requires their parents to find the best foraging places or to be protected against predators. Parental care can be facultative with the offspring obtaining resources from their parents or by themselves. In this situation, parental care is not strictly necessary but is still beneficial. The different forms of care are not mutually exclusive as in altricial species, parental care can be obligatory in the first days after birth and then facultative once the offspring have developed the necessary locomotory ability to find food. All these different situations affect the form and magnitude of sibling competition.

Box 8.2 Definitions of the different modes of sibling competition

Sub-lethal sibling competition: the role of fighting is trivial compared to other forms of competition including non-aggressive monopolization of resources, jostling for the position in the limited space where parents are more likely to deliver food, and scramble competition through the expression of various begging behaviour. Sub-lethal competition is over shares of resources and it may reduce the siblings' future prospects without actually killing them.

Lethal sibling competition: aggression or resource monopolization by some individuals leading to death of siblings. Frequent fighting can lead to the death of one or several individuals, usually the subordinate ones. Siblicide is 'obligate' when the dominant individual almost invariably kills its younger sibling(s). Siblicide is 'facultative' when the level of aggressiveness between siblings depends on some environmental factors such as food supply.

Sibling cannibalism: consumption of dead siblings. Cannibalism does not necessarily imply siblicide.

Brood reduction: elimination of part of the offspring in a brood when parents cannot raise all offspring often due to an unpredictable shortage in food supply. Mock and Parker (1997) have referred to the brood reduction hypothesis as the resource-tracking hypothesis. Offspring that parents can normally raise are called 'core offspring' and the supernumerary offspring that can be raised only if there are enough resources are denoted 'insurance', 'marginal', or 'runt' offspring. The occurrence of brood reduction gives parents a choice to adjust the size of the initial brood to current conditions and in doing so they choose the qualitatively best progeny ('progeny choice hypothesis').

Dominance: a consistent agonistic asymmetry between individuals. It can result from unequal fighting ability usually due to age differences between siblings or from trained winning and losing where some individuals do not contest their subordinate role even if in some cases they could physically reverse the dominance relationship.

Sibling negotiation: a signalling system between siblings that establishes the priority access to resources among siblings. Before parents come back to the nest with food, siblings signal to each other their need and hence willingness to compete at the parent's arrival. Upon parent's arrival each individual offspring competes to a level that depends on their own need and the need of their siblings as assessed from the sib–sib communication system. Typically, an individual will reduce its investment in competition if its sibling informed it of the intention to compete intensely.

(i.e. individuals resulting from the same reproductive attempt) or between individuals from different cohorts (i.e. siblings born in successive reproductive attempts) (Trillmich and Wolf 2008). Parents may *a priori* decide the absolute amount of resources to be invested in a given reproductive attempt and how these resources are divided among the different offspring. However, offspring may often actively control resource allocation by competing with their siblings and/or by communicating their need more efficiently to the parents. Sibling interactions can be relatively simple with individuals jostling for the favourable positions near to where the parents deliver food, begging for food from parents through loud calls or begging postures, or through the establishment of social dominance hierarchies (Drummond 2006). Other sibling interactions may be more complex, involving siblings communicating to each other their willingness to compete once parents are back with food (Roulin 2002). We distinguish between three major forms of social interactions among siblings depending on parental resources, namely lethal sibling competition, sub-lethal sibling competition, and sibling cooperation, in decreasing order of agonistic interaction (Box 8.2).

8.2.1 Sub-lethal sibling competition

In sub-lethal forms of sibling competition, the distribution of food among siblings is directly related either to a strict dominance hierarchy often based on size or to investment in non-aggressive forms of scramble competition for parental resources such as begging. The main interest of studying these two modes of sibling competition is to determine the circumstances under which a selfish individual is willing to take a larger share of the resources than it allows its siblings to take. Under scramble competition food acquisition is related to investment in sibling competition and asymmetry in resource holding potential. Under strict dominance hierarchy, in contrast, the strongest sibling decides its food acquisition and the amount of food received by others depends on the willingness of the despotic individuals to share resources with their subordinates (Mock and Parker 1997). Thus, a higher level of sibling competition is expected under strict dominance hierarchy than under scramble competition (Gonzalez-Voyer et al. 2007).

Strict 'dominance hierarchies' usually result from staggered birth, such as hatching asynchrony in birds (e.g. Viñuela 1999). Some mechanisms can reinforce dominance hierarchies. For instance, each individual has a history of victories and defeats against its competitors, which conditions its level of aggressiveness (if usually victorious) or submissiveness (if usually defeated) through mechanisms of learning (Drummond 2006). Each individual thus learns its exact position in the linear dominance hierarchy that prevails in its nest, allowing it to behave in the most profitable way when interacting with a specific sibling. Although siblings can compete for parental resources in non-violent way, the outcome of such contests can be detrimental when other individuals are better able to attract parental attention, and hence obtain a larger than fair share of the resources (Drummond 2006). Game theoretical models of strict dominance hierarchy among three siblings produced non-intuitive results (Mock and Parker 1997). The first two dominant individuals A and B achieve similar fitness, while the smallest individual C suffers disproportionally because the most dominant individual A competes mainly with sibling C rather than with sibling B as shown in cattle egrets (*Bubulcus ibis*; Mock and Lamey 1991). As a consequence, variation in the amount of parental resources affects mainly individuals A and C and to a lower extent individual B because when food becomes short, A monopolizes the share of C.

Under 'scramble competition' models, parents have no control over the allocation of resources. Instead, offspring control food allocation through scramble competition, and hence the way food is shared among siblings depends strongly on their begging levels. This implies that the offspring that begs at the highest level, and is therefore fed in priority, is not necessarily the hungriest individual since the cost of begging between offspring may depend on their competitive strength. Under scramble competition models, the intensity of sibling competition is predicted to depend on whether the cost associated with begging for food from parents is carried exclusively by the begging individual (e.g. when begging is energetically costly), or whether it is also carried by its siblings (e.g. when begging attracts predators) (Godfray and Parker 1992). Scramble competition to attract parental attention involves conspicuous vocalizations, the display of morphological traits such as colourful gapes in birds and other ornaments (Lyon et al. 1994), the release of chemical compounds and tactile behaviour in invertebrates (e.g. Smiseth and Moore 2002).

Although scramble competition models assume that food distribution is under offspring control, parents may control the provisioning rate and sometimes how food is shared among the offspring. Thus, the effect of parental provisioning on sibling competition will depend on the amount and quality of alternative food that can be processed by offspring without parental help (Smiseth and Moore 2002). Furthermore, physical interactions among siblings can reduce the parent's ability to feed specific offspring. One individual can supplant its siblings by jostling for the nest position where parents predictably deliver food, such as in the great tit (*Parus major*; Kölliker et al. 1998) or by chasing its parents more rapidly as in the banded mongoose (*Mungo mungo*; Gilchrist 2008). Without an experimental approach it is difficult to determine whether within-progeny food allocation is under parental or offspring control. This situation prevails because needy individuals may simultaneously produce conspicuous signals that are directed to the parents and compete against their siblings (Royle et al. 2002). Two studies that tested the relative importance of these two factors in the way food is shared among the progeny reported a more important role of physical sibling competition than offspring signalling towards parents in the canary (*Serinus canaria*: Kilner 1995) and in the great tit (Tanner et al. 2008). Thus, the outcome of sibling competition does not necessarily coincide with parental interest and this may select for parental behaviour to keep control, as much as possible, of within-progeny food allocation.

8.2.2 Lethal sibling competition

Theoretical developments of sibling aggressive interactions leading to siblicide were initially motivated by the study of 'brood reduction', the phenomenon where family size is reduced to better match the parental supply of resources to the number of offspring. The seminal paper of O'Connor (1978) explores the conditions under which selection would favour brood reduction through siblicide, filial infanticide, or suicide. Although this theoretical study arrived to the intuitive result that selection tends to favour siblicide by dominant rather than suicide by the subordinate, it was the first to envisage brood reduction not only as an adult adaptation but also as an offspring adaptation. His model also arrived at interesting conclusions regarding the selective value of brood reduction from the point of view of the parents, and the survivor and victim nestlings in relation to brood size and the age at which brood reduction should occur. A first important result from this model is that siblicide is favoured more readily than either filial infanticide or suicide. This indicates that the selective value of brood reduction is higher for the surviving nestlings than it is for their parents. This is particularly true for small broods and in species with long rearing periods. The reason for this is that the resources released by the death of one individual will be shared among fewer siblings in smaller broods than in larger ones and that the resources will be shaped over a longer period if the death occurs relatively early.

Aggressive sibling interactions based on physical or chemical weaponry are relatively frequent in a variety of vertebrate, invertebrate and plant taxa (e.g. Fox 1975; Krishnamurthy et al. 1997). Drummond (2006) reviewed various forms of aggressive sibling competition in birds (Table 8.1) and other animals. In spotted hyena (*Crocuta crocuta*) and several species of canids, dominance relationships between cubs become fixed after a long period of violent fights, threats, and submissive displays (White 2008). In pigs (*Sus scrofa*), felids and hyraxes young compete violently for exclusive access of particular teats. In several fish species, broodmates can be aggressive and they learn their position in the linear dominance hierarchy that becomes stable after a period of intense fights (Drummond 2006).

In cattle egrets, siblicide is triggered by food shortage, and the aggressive young that commit

Table 8.1 Various forms of aggressive relationships between nestling birds after Drummond (2006)

Dominance relationships	Definition	Examples
Aggression–submission	One individual aggressive and broodmates respond submissively with specific displays	Blue-footed booby *Sula nebouxii* Kittiwake *Rissa tridactyla* Osprey *Pandion haliaetus* Some grebes Some herons
Aggression–aggression	All broodmates are aggressive	Brown booby *Sula leucogaster* Imperial eagle *Aquila heliaca* White pelican *Pelecanus erythrorhynchos*
Aggression–resistance	All broodmates are aggressive but subordinates that are constantly aggressed limit themselves to retaliating to aggression	Cattle egret *Bubulcus ibis* Great egret *Egretta alba* Brown pelican *Pelecanus occidentalis*
Aggression–avoidance	The subordinate individual learns to avoid dominant siblings but does not display submissive behaviour	Oystercatcher *Haematopus bachmani* South Polar skua *Catharacta maccormicki*
Rotating dominance	Aggressive and submissive displays are used by each broodmate in turn	Crested ibis *Nipponia nippon*

siblicide benefit from the death of their sibling as more food becomes available to them (Creighton and Schnell 1996). The death of a sibling would make more resources available to the surviving siblings (e.g. Drummond et al. 2000) provided that parents do not reduce their reproductive investment following the reduction in family size. If parental provisioning decreases after brood reduction in such a way that the amount of food per surviving offspring is lower after than it was before the death of siblings, siblicide could be non-adaptive as suggested for the brown pelican (*Pelecanus occidentalis*: Ploger 1997). Thus, brood reduction due to siblicide is expected to be evolutionary stable only when offspring fitness is higher following parental readjustment of feeding effort (Forbes 1993). Brood reduction can occur even if interactions between siblings are non-aggressive. For example, the largest sibling may obtain a larger than equal fraction of the resources up to the point where their smaller siblings starve to death (Drummond 2006).

8.2.3 Sib–sib cooperation

In the context of family interactions, Hamilton's rule can be used to explain why siblings often are not overly aggressive or competitive, since an individual stands to gain indirect genetic benefits from related siblings (Mock and Parker 1997). So far, most studies on sibling interactions have demonstrated that siblings compete among each other over parental resources while there are much fewer reported instances where siblings cooperate to obtain parental resources. In the following, we review the various cooperative acts between siblings showing that siblings can sometimes behave peacefully.

Reported examples of cooperative acts among siblings are rather rare in offspring that are still dependent of parental resources, but such cases may be ignored as they have received less attention from evolutionary biologists than agonistic interactions. This includes events of individuals preening each other (i.e. allopreening) in the Mississippi kite (*Ictinia mississippiensis*: Botelho et al. 1993), feeding each other (i.e. allofeeding or food sharing) in the barn owl (*Tyto alba*: Marti 1989), huddling in mammals to improve thermoregulation (e.g. Forbes 2007), or forming coalitions with litter-mates against other unrelated juveniles in the spotted hyaena (Smale et al. 1995). Dominant blue-footed booby nestlings (*Sula nebouxii*) can moderate their own selfishness to allow their siblings to share food even during short-term food shortage (Anderson and Ricklefs 1995). In line with a game-theoretical model showing that siblings may help each other to induce their parents increasing feeding rate (Johnstone 2004; see also Forbes 2007), Mathevon and Charrier (2004) found that black-headed gull nestlings (*Larus ridibundus*) solicit food from their parents at a lower level in the presence of other siblings than when alone. By coordinating their begging behaviour, siblings may reduce their personal investment in this costly activity and increase their benefits as parents allocate more food to the brood when siblings coordinate their begging. A similar observation has been made in banded mongoose pups (Bell 2007). These two studies open up the possibility that siblings can cooperate to extract more resources from parents.

8.2.4 Sibling negotiation

Sibling competition is often energetically costly and can involve dangerous violent behaviours (see above). To reduce the cost of competition, siblings may negotiate among each other which individual will be given priority of access to the impending parental resources. When the outcome of sibling competition is predictable, siblings may be better off negotiating food resources instead of fighting desperately at any cost. Negotiation is therefore a form of cooperation to reduce the cost of sibling competition to obtain parental attention.

The development of the sibling negotiation hypothesis was motivated by the observation that barn owl nestlings vocalize not only when parents bring food, as in many other organisms with parental care, but also in the absence of parents: each individual nestling producing on average 1786 calls per night (Roulin 2002). Detailed observations suggest that the primary function of these calls is to communicate with siblings rather than with parents (Roulin 2002). When nestlings are old enough to swallow entire prey items or tear apart pieces of meat, parents return to the nest only to bring a single indivisible prey item that is consumed by a single offspring. Following the game-theoretical model of Johnstone and Roulin (2003), only the hungriest individual should compete for the impending food item and it should communicate its intention to its less hungry siblings who in turn should temporarily retreat from competing as the outcome is predictable. Experiments showed that when a competitor is hungry, and hence highly vocal, its siblings do indeed refrain from competing, thus giving the vocal individual easier access to food (Roulin 2002; Dreiss et al. 2010). Thus, sibling negotiating reduces the cost of sibling competition.

Two forms of cheating may occur in this sibling communication system. First, individuals may not signal their need but compete once a parent arrives with a food item. When individuals are not informed about their siblings' need, they would be expected to compete more intensely than when siblings engage in negotiation. For this reason, an individual that does not signal its need will face more intense competition from its siblings than an individual that has negotiated priority of access. Thus, this form of cheating may not be a viable solution if the benefit of reducing the level of sibling competition through negotiation is higher than the cost of negotiating. Note that if food becomes so short that sharing food with siblings may lead to starvation, individuals may stop negotiating. This event should not be considered as cheating but rather as a case where the absence of negotiation is the optimal strategy. Second, individuals may negotiate at a higher level than that predicted by their need as a means to deter their siblings from competing. However, the level of negotiating should correspond to the expected benefit of seeing siblings withdraw from a given contest because the costs of negotiating ensure honesty of the signal (Roulin 2002). In other words, given the benefits of negotiation, individuals are expected to minimize their costs, and hence individuals should not negotiate beyond an optimal level where the costs of negotiating are higher than the expected benefits.

There is evidence for sibling negotiation to resolve competition over the next indivisible food item delivered by a parent from the barn owl and the spotless starling (*Sturnus unicolor*: Bulmer et al. 2007). Experimental data showed that in the absence of parents nestling barn owls vocally refrain when the value of the next delivered prey item will be higher for its nest-mates (Roulin et al. 2000). In the black-headed gull (*Larus ridibundus*: Mathevon and Charrier 2004) and in meerkats (*Suricata suricatta*: Madden et al. 2009) researchers also observed that an individual offspring vocalizes less intensely in the presence of highly vocal siblings as predicted by the sibling negotiation hypothesis. In other species, however, begging in the absence of parents may be the result of nestlings responding to cues that were wrongly interpreted as the arrival of the parent (e.g. Dor et al. 2007). More data are required in a large range of organisms and consideration that sibling negotiation may not necessarily involve vocalizations but physical behaviour. Negotiation should occur in the prolonged absence of the parents in species in which food delivered by parents is indivisible, and when it is hard for one chick to monopolize access to resources (Johnstone and Roulin 2003).

8.3 Conditions promoting sibling competition

A key issue is to understand which ecological factors and life-history traits determine the level and the mode of sibling competition across species (Table 8.2). This was the goal of a comparative study of aggression in avian broods carried out by Gonzalez-Voyer et al. (2007). Provided that the species have the necessary weaponry to inflict injuries, that the potential victim cannot escape, and that the difference in age between siblings is pronounced, this study identified three life-history traits associated with aggressive competition. First, sibling aggression was more intense among species with small than large broods (as predicted by theoretical models of brood reduction; see Section 8.2.2 above), maybe because dominant individuals find it more difficult to impose their dominance by force in large broods (but see Drummond and Rodriguez 2009). Second, sibling aggression is more frequent and violent when parents deposit food on the nest floor rather than pass it directly from the parent's beak to the chick's beak. When food is deposited on the nest floor, it becomes accessible to all brood-

Table 8.2 Major factors that influence the incidence and intensity of sibling competition over parental care. The exact role of each factor is often unknown but should be considered in intra- and interspecific studies

Ecology	Food quantity, nest topography, parasitism, predation
Life-history	Length of rearing period of dependency to parents, hatching asynchrony (i.e. size and age difference between siblings), family size, altricial/precocial
Physiology	Growth rate, immunity
Morphology	Effective weaponry (e.g. pointed or sharp bill), sensory/motor skills
Genetics	Degree of relatedness between siblings
Parental behaviour	Feeding rate, feeding method (food size, mode of food deposition), divisibility of food (i.e. number of offspring fed per feeding visit), maternal effects

mates, thereby promoting intense competition over the food. This finding points out the modulating effect parental food provisioning behaviour can have on sibling competition. Finally, aggressiveness was more prevalent in species with long rearing periods, probably because longer cohabitation between siblings provides greater benefits of dominance (see Section 8.2.2). Whether subordinate siblings can perform submissive behaviour and avoid encounters, according to brood confinement and escape possibility, may also shape sibling competition (Drummond 2006). As discussed below, other related factors that promote sibling competition include a reduction in food supply, the presence of effective weapons, and asymmetries in resource holding potential and age between siblings.

8.3.1 Food amount

The amount of resources available is assumed to be the ultimate cause of conflict, but may also act as a proximate trigger of sibling competition. The latter is referred to as the food amount hypothesis, and has received empirical support from experimental manipulations in which the amount of food supplied by parents was found to influence offspring competitive behaviours (Mock et al. 1987). In facultatively siblicidal vertebrates, such as blue-footed booby, black-legged kittiwake (*Rissa tridactyla*), osprey (*Pandion haliaetus*) and black guillemot (*Cepphus grille*) the level of aggression between broodmates increases with food deprivation, whereas in obligately siblicidal species such as eagles food supply does not seem to regulate aggressiveness (Drummond 2001). In meerkats, where fights between pups are frequent but not fatal, the frequency of aggression is directly associated with food availability (Hodge et al. 2009). Similarly, when food becomes unpredictably scarce, for instance due to spell of rain, competition can become lethal as shown in black-legged kittiwakes (Braun and Hunt 1983).

8.3.2 Weapons

Violent interactions to resolve sibling contests over resources often involve effective weaponry that may inflict injuries to siblings. For example, piglets display sharp teeth used exclusively to displace siblings from maternal teats (Fraser and Thompson 1991), and the bill of egrets and eagles is used to beat siblings (Mock and Parker 1997). In egrets, chicks hatch asynchronously and the biggest individuals inflict serious injuries to their last-hatched siblings that frequently die as a consequence (Mock and Parker 1997).

8.3.3 Age difference between siblings

Size differences between siblings due to staggered birth can reduce the level of competition (Mock and Ploger 1987) because siblings reach their maximum food demand sequentially in time (Stenning 1995) or because the resulting competitive asymmetries would reduce the intensity of competitive interactions since low-ranking individuals cannot contest the dominance advantage of first-born siblings (Nathan et al. 2001). However, there might be an optimal level of birth asynchrony as the risk of starvation and/or violent death for the last-born individuals increases with age difference between siblings. For example, in black kites (*Milvus migrans*) aggression between siblings is more pronounced in broods with either reduced or increased levels of hatching synchrony (Viñuela 1999). The optimal level of birth asynchrony from the offspring' and parents' point of views may be difficult to achieve because birth asynchrony can evolve for other reasons than to modulate sibling competition (Amundsen and Slagsvold 1991). The situation may be particularly intricate in sexually dimorphic species where the offspring of the larger sex has a competitive advantage over offspring of the weaker sex. In some eagles, parents tend to give birth first to offspring of the weaker sex, in this case males, and then to give birth to offspring of the stronger sex, in this case females (Bortolotti 1986).

An interesting case of age difference between siblings is in species in which the young attain independence after their mother gives birth to the subsequent progeny. In such species, offspring of different cohorts may compete for parental resources. Although young of the first cohort will have a size advantage in the competition, the success in sibling competition can depend on the expected benefit mothers derive from investing in the young versus older offspring. For instance, in Galapagos fur seals (*Arctocephalus galapagoensis*) and sea lions (*Zalophus wollebaeki*) 23% of all pups are born while the mother is still nursing an older offspring. Although the older offspring has a competitive advantage over its younger sib, the mother often defends her younger offspring when attacked by the older sibling (Trillmich and Wolf 2008).

8.3.4 Parental manipulation of sibling competition

The difficulty in tracking resources that will become available to offspring after birth, and unexpected early failure of offspring, can select for the production of an optimistic family size and mechanisms that facilitate a subsequent reduction in family size should resources be less abundant than anticipated (Simmons 1988). This situation led to the concept of core offspring (usually the first-born offspring) that are likely to survive even if resources become scarce, and of marginal or insurance offspring (usually the last-born offspring) that have much lower survival expectancy except in situations where resources are abundant and/or where first-born siblings have died prematurely. In situations where insurance offspring are redundant, they can be eliminated by parental infanticide or sibling competition and may be even cannibalized as a means to recycle them as a food source (Forbes 1991). The effect of sibling competition on the elimination of insurance offspring can be direct through violent aggression from core offspring, or indirect through starvation if core offspring monopolize food resources (Drummond 2006). Whereas suicide is a theoretically plausible mechanism for reducing family size for the good of relatives, evidence of its existence remains elusive, which perhaps is not surprising given that the threshold for siblicide and infanticide is much lower than that for suicide (O'Connor 1978).

Even if food shortage is a major cause for sibling competition, its occurrence and intensity is not always associated with variation in food supply (Drummond 2001). In a variety of birds, including eagles and cranes, parents lay two eggs despite raising only one offspring (e.g. Miller 1973). If both eggs hatch, the older sibling will kill its younger sibling relatively soon after birth. This behaviour has puzzled evolutionary biologists because siblicide happens even in the absence of food shortages and where the extra egg does not seem to be laid as insurance against the failure of other eggs or the death of other nestlings for reasons other than siblicide (Cash and Evans 1986). In long-lived species, it might be particularly important for parents to produce offspring of high quality rather than to produce a greater number of offspring. An alternative hypothesis for the adaptive function of extra egg and obligate siblicide may then be that parents select the individual offspring with highest residual reproductive value (Simmons 1988; Jeon 2008). This hypothesis may explain why parents are often indifferent to violent behaviours displayed by their offspring (Drummond 1993). For instance, the parents interrupted in less than 1% of the 3000 fights recorded in a study on the great egret (*Egretta alba*: Mock and Parker 1997). Finally, senior nestlings in obligatory siblicide species kill their junior siblings very early in life, when the younger offspring may still be valuable as a form of insurance to the parents. The desperado sibling hypothesis proposed by Drummond (1993) postulates that the older nestling kills its younger sibling soon after birth before it becomes strong enough to behave violently. To test this hypothesis, junior chicks were experimentally allowed to survive up to an age at which they would usually be dead (on average within 6.5 days after hatching in boobies) (Drummond et al. 2003). To this end, junior brown boobies (*Sula leucogaster*) were temporarily fostered in nests of blue-footed boobies (*S. nebouxii*), a facultatively siblicide species. As predicted by the desperado sibling hypothesis, junior brown boobies were seven times more aggressive in nests of blue-footed booby than in nests of brown boobies.

8.3.5 Parental strategies to reduce sibling competition

Although O'Connor's (1978) model predicts that parents and dominant offspring share a similar interest in reducing family size when resources become scarce, it is often unclear whether lethal sibling competition is beneficial to parents (Drummond 1993). If there is conflict between parents and dominant offspring over the optimal level of violence, parents may try to suppress aggression actively (Cash and Evans 1986) by punishing aggressive offspring (White 2008) or by increasing the total amount of resources devoted to their progeny. In some species, simply the presence of parents appeases their offspring (Mock 1987). Although these parental strategies are plausible, there is still no clear demonstration that they evolved to reduce the level of aggression between offspring. For example, parental control over the level of violence may not be feasible if sibling aggression is not triggered by the amount of food and if offspring resume aggression once the parents are away foraging.

Parents may have evolved other ways to control the level of sibling competition by for example modifying the spacing of birth via hatching asynchrony to alter the competitive hierarchy within a brood (e.g. Viñuela 1999). For instance, in pied flycatchers (*Ficedula hypoleuca*), the body mass of surviving offspring increased with the degree of hatching asynchrony, although nestling mortality was higher (Slagsvold 1986). Another possible way by which parents may control the level of sibling competition is through differential allocation of maternal resources into each offspring before birth. Examples of such strategies may include differences in egg size in relation to the order in which eggs were laid in species with hatching asynchrony (Slagsvold et al. 1984) and sex ratio biases (Martyka et al. 2010; Chapter 10), both of which may have pronounced effect on sibling competition. In red-winged blackbirds (*Agelaius phoeniceus*), nestlings have a higher survival prospects in the early days of life if hatching from large than small eggs, an effect that was more pronounced in late- than early-hatched nestlings (Forbes and Wiebe 2010). In the European blackbird (*Turdus merula*), male eggs are larger than female eggs and egg size increases with the laying sequence in female eggs only (Martyka et al. 2010). Similarly, females can control birth order of male and female offspring if sibling competition has sex-specific long-term effects as shown in the scops owl (*Otus scops*) (Blanco et al. 2002). Because females are larger than males, and hence have a competitive advantage, hatching a male offspring early in the hatching sequence may increase the probability of producing at least one male in good condition.

Another possibility for parents to control the level of sibling competition is to adjust the degree of genetic relatedness between siblings by mating with a single male or with multiple males. In the invasive sessile marine gastropod (*Crepidula fornicata*), where siblings are reared together, a higher degree of relatedness is associated with lower variation in growth rate (Le Cam et al. 2009). Even if there is no parental care in this gastropod, this example raises the possibility that variation in relatedness may affect sibling competition (see also Briskie et al. 1994). Accordingly, in the barn swallow (*Hirundo rustica*), nestlings beg to a higher level when relatedness among siblings is reduced (e.g. Boncoraglio et al. 2009).

Parents can also influence sibling competition by showing favouritism towards particular types of offspring. For instance, the female parent may feed preferentially the smallest offspring while the male takes care of the largest offspring (Lessells 2002).

This pattern of within-brood food allocation, which is termed brood division, may reduce the level of sibling competition because only the offspring that are usually fed by the father compete when he brings food, whereas their other siblings compete when the mother brings food (Lessells 2002). A similar outcome can be achieved when parents deliver food from different locations in the nest, thereby forcing the offspring to compete for food delivered from either the mother or the father. The two parents can be viewed as two foraging patches differing in profitability due to sex-specific parental feeding rates and parental favouritism towards specific offspring (Kölliker et al. 1998). Fewer individuals therefore compete for the same patch of resources, reducing the overall level of sibling competition. For example, in the great tit, hungry nestlings position themselves close to the location within nests where mothers feed offspring (Tanner et al. 2008). These patterns generated by parental behaviour emphasize the importance of studying the different forms of behaviours used by offspring to compete against siblings for resources, leading to offspring specialization on some resources and feeding tactics. To sum up, parents can control the level of sibling competition by modulating their behaviour and life-history traits in many ways.

8.4 Conditions promoting sibling cooperation

Although kin selection theory proposes that a high degree of relatedness promotes the evolution of cooperation, the study of sibling interactions has primarily concerned agonistic behaviours (Mock and Parker 1997). Animals that are still dependent on parental resources have been rarely observed to help their siblings obtain resources but are often seen competing over parental attention. Two explanations can account for the lack of detailed studies on cooperative behaviour between young siblings. First, despite the fact that allopreening and allofeeding have been observed in young animals, suggesting that cooperation between siblings may be not uncommon, little effort has been made to carry out research on these behaviours. Second, sibling competition over parental resources may prevent the evolution of cooperative behaviour if the indirect genetic benefits gained from helping are lower than their costs (West et al. 2002). For instance, dominant offspring may behave selfishly to avoid their subordinate siblings becoming stronger (Drummond et al. 2003). In the following, we discuss the two major factors that may promote cooperation, namely genetic and ecological factors.

Sibling cooperation can be promoted by relatedness. For instance, siblicide and cannibalism is more often directed towards half-siblings than towards full-siblings (Pfennig 1997), and aggressiveness and contest for parental attention can be exacerbated by low relatedness due to multiple paternity (Boncoraglio et al. 2009). In some invertebrates such as the European earwig (*Forficula auricularia*), where there is a high risk that individuals from other families exploit the same pool of resources, young are often able to discriminate siblings from unrelated same-aged conspecifics (Dobler and Kölliker 2010). Yet, although full siblings should be more willing to spare each other because of the inclusive fitness cost of harming close relatives (Mock and Parker 1997), they often share more similar needs and life-history traits, which can intensify the level of discord. Increasing the level of genetic diversity in a family by multiple mating may thus reduce the intensity of sibling competition if reduced relatedness is associated with greater diversity in offspring needs or competitive behaviours (Le Cam et al. 2009; see also Aguirre and Marshall 2012). This point emphasizes the dual effect that relatedness may have on the level of sibling competition and cooperation; although increased relatedness is expected to reduce the degree of sibling competition through indirect fitness benefits, it may sometimes increase the degree of sibling competition if more closely related offspring are more similar with respect to their needs for parental resources and the competitive strategies they use to obtain them.

Sibling competition entails substantial costs to the offspring, and behaviours that temper these costs may therefore be expected to evolve. For instance, at the end of the rearing period crested ibis (*Nipponia nippon*) siblings ritualize aggression to prevent violent interactions (Li et al. 2004). From a theoretical point of view, siblings that share limited resources should be expected to make a decision about whether to refrain from competing with their siblings or to attempt to monopolize the resources at the expense of their siblings. An individual is expected to behave altruistically if the coefficient of relatedness r between competitors and the cost c of competing are high and if the value v of the resources and the probability p of monopolizing the resources are low (Johnstone and Roulin 2003). For example, to appreciate the role of p in promoting sibling cooperation, it can be useful to compare the models of Godfray (1995) and Johnstone and Roulin (2003). Godfray's model is applicable to situations where parents allocate food resources in direct proportion to the investment in begging of each offspring. Because offspring are rewarded for any investment in sibling competition at each parental feeding visit, it always pays to invest in competition (in this case, food is divisible and each offspring obtains some portion of the resources). Thus, in Godfray's model sibling competition will promote an escalation in agonistic interactions. In Johnstone and Roulin's model, the food provided by parents cannot be divided among the offspring, thus implying that a single offspring is fed per parental feeding visit. In this case, selection should favour the transfer of information between siblings regarding the extent to which each individual is hungry and willing to compete. The hungriest individual is expected to succeed in this enterprise if the asymmetry in need between siblings is high. When this is the case, the hungriest individual should signal its willingness to compete to deter the others from claiming the next food resources. This sibling signalling system, denoted sibling negotiation, should prevail unless the supply of food becomes short because, if this is the case, an individual offspring would no longer benefit from sharing the resources with its siblings.

8.5 Perspectives

This chapter has highlighted the most important issues regarding the ecology and evolution of sibling competition. The book on sibling rivalry by Mock and Parker (1997) and comparative studies such as the one by Gonzalez-Voyer et al. (2007) have pinpointed the most important factors that can account for interspecific variation in the level of sibling competition, such as the presence of weaponry, pronounced age asymmetry between siblings, long rearing period, small family size, and how parents distribute food. As pointed out in this chapter, one of the most understudied aspects of sibling interactions is the interplay between sibling competition and sibling cooperation. Although evolutionary biologists have investigated numerous aspects of the causes and consequences of sibling competition, a number of issues remain to be considered. In the following, we highlight three issues that have recently emerged and that may advance our understanding of the diversity of family life.

First, variation between individuals in the way they interact with siblings has been mainly studied in relation to an individual's sex and position in the within-brood age/size hierarchy (Mock and Parker 1997). There has been little interest in the effects of other intrinsic differences on the behaviour of competing siblings. For example, a consideration of personality differences between siblings may shed new light on our understanding of both the ecological implications of personality and on family interactions. Apart from examples in humans (Brody 1998), little evidence exists for the possibility that siblings differ in their inherent dispositions to act competitively and aggressively or to act cooperatively. Personality, defined as consistent interindividual variation in behaviour over time and across contexts, is frequently reported from mature animals (Réale et al. 2007), but there is currently no information on how parent and offspring personalities may modify the dynamics of family interactions, including sibling competitive interactions and the way parents distribute food among the progeny (Roulin et al. 2010; Royle et al. 2010). For example, Stamps (2007) proposed that interindividual differences in growth rates would favour the evolution of personality traits with fast-growing individuals being selected to take more risks in foraging than slow-growing conspecifics. Thus, fast-growing young may be willing to invest more effort in conspicuous begging signals that attract predators than slow-growing young, and such differences between siblings may affect the dynamics of sibling competition. For instance, different personalities may ontogenetically develop in human siblings as alternative strategies to attract parental attention

(Sulloway 2001). Thus, competition between siblings may lead to a niche specialization within the brood.

Second, little attention has been given to the long-term consequences of the competition occurring early in life. The development of personalities, reproductive choices, viability, and other life-history traits may be influenced by the outcome or intensity of sibling competition (e.g. growing up as dominant versus subordinate offspring), but there is currently little evidence concerning this suggestion (Drummond et al. 2003; Sanchez-Macouzet and Drummond 2011). It has been known for a long time that life-history traits, including survival and reproduction, depend on the environmental conditions experienced as offspring, such as stress and food supply (Stearns 1992). However, little is known about how physiological factors may influence the evolution of sibling competition (Table 8.2). For instance, investment in immunity due to high parasitism levels may affect competitive interactions by diverting resources to fight pathogens at the expanse of allocating them to competitive behaviours. Comparative analyses are needed to evaluate the role of various ecological factors and of interspecific variation in life-history traits and physiology on sibling competition.

Third, the dynamics of sibling interactions may help understand how sibling disputes (Chapter 7) are resolved behaviourally. Usually ecologists consider the final outcome of sibling interactions but neglect how they arrived to this outcome. For instance, siblings often adjust their begging behaviour to each other once parents are back with food, an outcome that can depend on complex interactions taking place in the prolonged absence of the parents (Dreiss et al. 2010). A last point of interest is the potential for post-conflict behaviours such as reconciliation, consolation, and empathy. For instance, adult ravens engage in reconciliation after a conflicting situation and do so more often when interacting with kin than with non-kin (Fraser and Bugnyar 2011). Reconciliation may reduce the intensity of sibling disputes or its negative consequences on psychology in animals with high cognitive ability. In the same vein, little is currently known as to whether individual siblings are able to form coalitions as a countermeasure against dominant siblings (but see Smale et al. 1995). In other words, complex forms of social behaviour coupled with the cognitive ability to memorize past social interactions may reveal unexpected forms of conflict resolution between siblings.

Acknowledgements

The Swiss National Science Foundation supported the present work (grant 31003A_120517 to A.R.). We are grateful to Per Smiseth, Mathias Kölliker, Nick Royle, Hughes Drummond, Fritz Trillmich, and an anonymous reviewer for helpful comments on previous versions of the text.

References

Aguirre, J. D. and Marshall, D. J. (2012). Does genetic diversity reduce sibling competition? *Evolution* 66, 94–102.

Alexander, R. D. (1974). The evolution of social behaviour. *Annual Review of Ecology and Systematics* 5, 325–83.

Amundsen, T. and Slagsvold, T. (1991). Hatching asynchrony: facilitating adaptive or maladaptative brood reduction? *Acta International Ornithological Congress, Christchurch, New Zealand* 20, 1707–19.

Anderson, D. J. and Ricklefs, R. E. (1995). Evidence of kin-selected tolerance by nestlings in a siblicidal bird. *Behavioral Ecology and Sociobiology* 37, 163–8.

Bell, M. B. V. (2007). Cooperative begging in banded mongoose pups. *Current Biology* 17, 717–21.

Blanco, G., Davila, J. A., Lopez Spetiem, J. A., Rodriguez, R., and Martinez, F. (2002). Sex-biased initial eggs favours sons in the slightly size-dimorphic scops owl (*Otus scops*). *Biological Journal of the Linnean Society* 76, 1–7.

Boncoraglio, G., Caprioli, M., and Saino, N. (2009). Fine-tuned modulation of competitive behaviour according to kinship in barn swallow nestlings. *Proceedings of the Royal Society of London, Series B* 276, 2117–23.

Bortolotti, G. R. (1986). Influence of sibling competition on nestling sex ratios of sexually dimorphic birds. *American Naturalist* 127, 495–507.

Botelho, E. S., Gennaro, A. L., and Arrowood, P. C. (1993). Parental care, nestling behaviors and nestling interactions in a Mississippi kite (*Ictinia mississippiensis*) nest. *Journal of Raptor Research* 27, 16–20.

Braun, B. M. and Hunt, G. L. (1983). Brood reduction in black-legged kittiwakes. *Auk* 100, 469–76.

Briskie, J. V., Naugler, C. T., and Leech, S. M. (1994). Begging intensity of nestling birds varies with sibling relatedness. *Proceedings of the Royal Society of London, Series B* 258, 73–8.

Brody, G. H. (1998). Sibling relationship quality: its causes and consequences. *Annual Review of Psychology* 49, 1–24.

Bulmer, E., Celis, P., and Gil, D. (2007). Parent-absent begging: evidence for sibling honesty and cooperation in the spotless starling (*Sturnus unicolor*). *Behavioural Ecology* 19, 279–84.

Cash, K. J. and Evans, R. M. (1986). Brood reduction in the American white pelican (*Pelecanus erythrorhynchos*). *Behavioral Ecology and Sociobiology* 18, 413–18.

Creighton, J. C. and Schnell, G. D. (1996). Proximate control of siblicide in cattle egrets: a test of the food-amount hypothesis. *Behavioral Ecology and Sociobiology* 38, 371–7.

Dobler, R. and Kölliker, M. (2010). Kin-selected siblicide and cannibalism in the European earwig. *Behavioral Ecology* 21, 257–63.

Dor, R., Kedar, H., Winkler, D. W., and Lotem, A. (2007). Begging in the absence of parents: a 'quick on the trigger' strategy to minimize costly misses. *Behavioral Ecology* 18, 97–102.

Dreiss, A., Lahlah, N., and Roulin, A. (2010). How siblings adjust sib-sib communication and begging signals to each other. *Animal Behaviour* 80, 1049–55.

Drummond, H. (1993). Have avian parents lost control of offspring aggression? *Etologìa* 3, 187–98.

Drummond, H. (2001). A revaluation of the role of food in broodmate aggression. *Animal Behaviour* 61, 517–26.

Drummond, H. (2006). Dominance in vertebrate broods and litters. *Quarterly Review of Biology* 81, 3–32.

Drummond, H. and Rodriguez, C. (2009). No reduction in aggression after loss of a broodmate: a test of the brood size hypothesis. *Behavioral Ecology and Sociobiology* 63, 321–7.

Drummond, H., Rodriguez, C., Vallarino, A., Valderrabano, C., Rogel, G., and Tobon, E. (2003). Desperado siblings: uncontrollably aggressive junior chicks. *Behavioral Ecology and Sociobiology* 53, 287–96.

Drummond, H., Vásquez, E., Sánchez-Colón, S., Martínez-Gómez, M., and Hudson, R. (2000). Competition for milk in the domestic rabbit: survivors benefit from littermate deaths. *Ethology* 106, 511–26.

Forbes, L. S. (1991). Insurance offspring and brood reduction in a variable environment: the costs and benefits of pessimism. *Oikos* 62, 325–32.

Forbes, L. S. (1993). Avian brood reduction and parent-offspring conflict. *American Naturalist* 142, 82–117.

Forbes, L. S. (2007). Sibling symbiosis in nestling birds. *Auk* 124, 1–10.

Forbes, L. S. and Wiebe, M. (2010). Egg size and asymmetric sibling rivalry in red-winged blackbirds. *Oecologia* 163, 361–72.

Fox, L. R. (1975). Cannibalism in natural populations. *Annual Review of Ecology and Systematics* 6, 87–106.

Fraser, D. and Thompson, B. K. (1991). Armed sibling rivalry among suckling piglets. *Behavioral Ecology and Sociobiology* 29, 9–15.

Fraser, O. N. and Bugnyar, T. (2011). Ravens reconcile after aggressive conflicts with valuable partners. *PLos ONE* 6, e18118.

Gilchrist, J. S. (2008). Aggressive monopolization of mobile carers by young of a cooperative breeder. *Proceedings of the Royal Society of London, Series B* 275, 2491–8.

Godfray, H. C. J. (1995). Signaling of need between parents and young: parent-offspring conflict and sibling rivalry. *American Naturalist* 146, 1–24.

Godfray, H. C. J. and Parker, G. A. (1992). Clutch size, fecundity and parent-offspring conflict. *Philosophical Transaction of the Royal Society of London, Series B* 332, 67–79.

Gonzalez-Voyer, A., Székely, T., and Drummond, H. (2007). Why do some siblings attack each other? Comparative analysis of aggression in avian broods. *Evolution* 61, 1946–55.

Hamilton, W. D. (1964). The genetical evolution of social behaviour. I. *Journal of Theoretical Biology* 7, 1–16.

Hodge, S. J., Thornton, A., Flower, T. P., and Clutton-Brock, T. H. (2009). Food limitation increases aggression in juvenile meerkats. *Behavioral Ecology* 20, 930–5.

Hudson, R. and Trillmich, F. (2008). Sibling competition and cooperation in mammals: challenges, developments and prospects. *Behavioral Ecology and Sociobiology* 62, 299–307.

Jeon, J. (2008). Evolution of parental favoritism among different-aged offspring. *Behavioral Ecology* 19, 344–52.

Johnstone, R. A. (2004). Begging and sibling competition: how should offspring respond to their rivals? *American Naturalist* 163, 388–406.

Johnstone, R. A. and Roulin, A. (2003). Sibling negotiation. *Behavioral Ecology* 14, 780–6.

Kilner, R. (1995). When do canary parents respond to nestling signals of need? *Proceedings of the Royal Society of London, Series B* 260, 343–8.

Kölliker, M., Richner, H., Werner, I., and Heeb, P. (1998). Begging signals and biparental care: nestling choice between parental feeding locations. *Animal Behaviour* 55, 215–22.

Krishnamurthy, K. S., Shaanker, R. U., and Ganeshaiah, K. N. (1997). Seed abortion in an animal dispersed species, *Syzygium cuminii* (L.) Skeels (Myrtaceae): The chemical basis. *Current Science* 73, 869–73.

Lack, D. (1947). The significance of clutch size. *Ibis* 89, 302–52.

Le Cam, S., Pechenik, J. A., Cagnon, M., and Viard, F. (2009). Fast versus slow larval growth in an invasive marine mollusc: Does paternity matter? *Journal of Heredity* 100, 455–64.

Lessells, C. M. (2002). Parentally biased favouritism: why should parents specialize in caring for different offspring? *Philosophical Transaction of the Royal Society of London, Series B* 357, 381–403.

Li, X. H., Li, D. M., Ma, Z. J., Zhai, T. Q., and Drummond, H. (2004). Ritualized aggression and unstable dominance in broods of crested ibis (*Nipponia nippon*). *Wilson Bulletin* 116, 172–6.

Lyon, B. E., Eadie, J. M., and Hamilton, L. D. (1994). Parental choice selects for ornamental plumage in American coot chicks. *Nature*, 371, 240–3.

Macnair, M. R. and Parker, G. A. (1979). Models of parent-offspring conflict. III. Intra-brood conflict. *Animal Behaviour* 27, 1202–9.

Madden, J. R., Kunc, H. P., English, S., Manser, M. B., and Clutton-Brock, T. H. (2009). Calling in the gap: competition or cooperation in littermates' begging behaviour? *Proceedings of the Royal Society of London, Series B* 276, 1255–62.

Marti, C. D. (1989). Food sharing by sibling common barn owls. *Wilson Bulletin* 101, 132–4.

Martyka, R., Rutkowska, J., Dybek-Karpiuk, A., Cichon, M., and Walasz, K. (2010). Sexual dimorphism of egg size in the European blackbird *Turdus merula*. *Journal of Ornithology* 151, 827–31.

Mathevon, N. and Charrier, I. (2004). Parent-offspring conflict and the coordination of siblings in gulls. *Proceedings of the Royal Society of London, Series B* 271, 145–7.

Miller, R. S. (1973). The brood size of cranes. *Wilson Bulletin* 85, 436–41.

Mock, D. W. (1987). Siblicide, parent-offspring conflict and unequal parental investment by egrets and herons. *Behavioral Ecology and Sociobiology* 20, 247–56.

Mock, D. W. and Lamey, T. C. (1991). The role of brood size in regulating egret sibling aggression. *American Naturalist* 138, 1015–26.

Mock, D. W., Lamey, T. C., Williams, C. F., and Ploger, B. J. (1987). Proximate and ultimate roles of food amount in regulating egret sibling aggression. *Ecology* 68, 1760–72.

Mock, D. W. and Parker, G. A. (1997). *The Evolution of Sibling Rivalry*. Oxford University Press, Oxford.

Mock, D. W. and Ploger, B. J. (1987). Parental manipulation of optimal hatch asynchrony in cattle egrets—an experimental study. *Animal Behaviour* 35, 150–60.

Nathan, A., Legge, S., and Cockburn, A. (2001). Nestling aggression in broods of a siblicidal kingfisher, the laughing kookaburra. *Behavioral Ecology* 12, 716–25.

O'Connor, R. J. (1978). Brood reduction in birds: selection for fratricide, infanticide and suicide? *Animal Behaviour* 26, 79–96.

Pfennig, D. W. (1997). Kinship and cannibalism. *Bioscience* 47, 667–67.

Ploger, B. J. (1997). Does brood reduction provide nestling survivors with a food bonus? *Animal Behaviour* 54, 1063–76.

Réale, D., Reader, S. M., Sol, D., McDougall, P. T., and Dingemanse, N. J. (2007). Integrating animal temperament within ecology and evolution. *Biological Review* 82, 291–318.

Roulin, A. (2002). The sibling negotiation hypothesis. In Wright J. and Leonard M., eds. *The Evolution of Begging: Competition, Cooperation and Communication*, pp. 107–27. Kluwer, Dordrecht.

Roulin, A., Dreiss, A., and Kölliker, M. (2010). Evolutionary perspective on the interplay between family life, and parent and offspring behavioural syndromes. *Ethology* 116, 787–96.

Roulin, A., Kölliker, M., and Richner, H. (2000). Barn owl (*Tyto alba*) siblings vocally negotiate resources. *Proceedings of the Royal Society of London, Series B* 267, 459–63.

Royle, N. J., Hartley, I. R., Owens, I. P. F., and Parker, G. A. (1999). Sibling competition and the evolution of growth rates in birds. *Proceedings of the Royal Society of London, Series B* 266, 923–32.

Royle, N. J., Hartley, I. R., and Parker, G. A. (2002). Begging for control: when are offspring solicitation behaviours honest? *Trends in Ecology and Evolution* 17, 434–40.

Royle, N. J., Wiebke, S., and Dall, S. R. X. (2010). Behavioral consistency and the resolution of sexual conflict over parental investment. *Behavioral Ecology* 21, 1125–30.

Sanchez-Macouzet, O. and Drummond, H. (2011). Sibling bullying during infancy does not make wimpy adults. *Biology Letters* 7, 869–87.

Simmons, R. (1988). Offspring quality and the evolution of canaism. *Ibis* 130, 339–57.

Slagsvold, T. (1986). Asynchronous versus synchronous hatching in birds: experiments with the pied flycatcher. *Journal of Animal Ecology* 55, 1115–34.

Slagsvold, T., Sandvik, J., Rofstad, G., Lorentsen, O., and Husby, M. (1984). On the adaptive value of intraclutch egg-size variation in birds. *Auk* 101, 685–97.

Smale, L., Holekamp, K. E., Weldele, M., Frank, L. G., and Glickman, S. E. (1995). Competition and cooperation between litter-mates in the spotted hyaena, *Crocuta crocuta*. *Animal Behaviour* 50, 671–82.

Smiseth, P. T., Lennox, L., and Moore, A. J. (2007). Interaction between parental care and sibling competition: parents enhance offspring growth and exacerbate sibling competition. *Evolution* 61, 2331–9.

Smiseth, P. T. and Moore, A. J. (2002). Does resource availability affect offspring begging and parental provisioning in a partially begging species? *Animal Behaviour* 63, 577–85.

Stamps, J. A. (2007). Growth-mortality tradeoffs and 'personality traits' in animals. *Ecology Letters* 10, 355–63.

Stearns, S. C. (1992). *The Evolution of Life Histories*. Oxford University Press, Oxford.

Stenning, M. J. (1995). Hatching asynchrony, brood reduction and other rapidly reproducing hypotheses. *Trends in Ecology and Evolution* 11, 243–6.

Sulloway, F. (2001). Birth order, sibling competition, and human behaviour. In P.S Davies and H.R Holcomb eds. *Conceptual Challenges in Evolutionary Psychology: Innovative Research Strategies*, pp. 39–83. Kluwer, Dordrecht.

Tanner, M., Kölliker, M., and Richner, H. (2008). Differential food allocation by male and female great tit, *Parus major*, parents: are parents or offspring in control? *Animal Behaviour* 75, 1563–9.

Trillmich, F. and Wolf, J. B. W. (2008). Parent-offspring and sibling conflict in Galapagos fur seals and sea lions. *Behavioral Ecology and Sociobiology* 62, 363–75.

Trivers, R. L. (1974). Parent-offspring conflict. *American Zoology* 14, 249–64.

Viñuela, J. (1999). Sibling aggression, hatching asynchrony, and nestling mortality in the black kite (*Milvus migrans*). *Behavioral Ecology and Sociobiology* 45, 33–45.

West, S. A., Pen, I., and Griffin, A. S. (2002). Conflict and cooperation—Cooperation and competition between relatives. *Science* 296, 72–5.

White, P. P. (2008). Maternal response to neonatal sibling conflict in the spotted hyena, *Crocuta crocuta*. *Behavioral Ecology and Sociobiology* 62, 353–61.

CHAPTER 9

Sexual conflict

C. M. Lessells

9.1 Introduction

Evolutionary conflicts of interest, including those over parental care, arise because interacting individuals are not perfectly related, so a trait that maximizes the fitness of one individual may not maximize the fitness of the others. For this reason, families are rife with evolutionary conflicts of interest (Parker et al. 2002). The previous two chapters dealt with parent–offspring conflict and sibling conflict—cases where the interacting individuals generally have a relatedness of one-quarter or one-half. This chapter concerns the third major evolutionary conflict within families over parental care—sexual conflict between the parents—where the interacting individuals are typically unrelated. Trivers (1972) was the first to point out the scope for this to create evolutionary conflicts of interest. His concept of parental investment (PI) encapsulates the trade-off faced by parents between current and future reproduction: by definition PI enhances the fitness of offspring at a cost to the fitness through future reproduction ('residual fitness') of the parent making the investment. As a result, both parents gain from PI made by either parent, but pay the cost of only their own PI, with the result that each parent would have higher fitness overall if the other parent did a larger share of the work.

Sexual conflict can occur over virtually any aspect of a reproductive attempt—from whether to mate in the first place, to the sex ratio of offspring (Chapter 10) or, in cooperatively breeding species, whether to accept helpers at the nest (Lessells 1999; Arnqvist and Rowe 2005; Chapter 12). However, sexual conflict over PI is unique in being an (almost) inevitable consequence of sexual reproduction. The only time that the evolutionary interests of the two sexes coincide entirely is when there is complete obligate lifelong monogamy in both sexes, but apart from experimental laboratory systems this seems to exist only as a theoretical reference point (Lessells 2006).

The term sexual conflict—'a conflict between the evolutionary interests of individuals of the two sexes' (Parker 1979)—refers to the way in which selection acts on the two sexes, with the optimal value of the trait over which there is conflict differing between them (Fig. 9.1). Between these two values there is sexually antagonistic selection. The extent of conflict can be measured by the difference between the two parental optima—the 'battleground' (Godfray 1995)—or by the amount by which the fitness of each sex is below that at its optimal value for the trait—the 'conflict load' (Lessells 2006). Because sexual conflict refers to the way that selection pressures act, it does not necessarily involve overt behavioural conflict, and is still present when there is no post-copulatory behavioural interaction between the parents.

The chief question posed by the existence of sexual conflict over PI is what the evolutionary outcome—the 'resolution' of sexual conflict—will be. This will occur when each sex has no further options to manipulate care by the other parent, or these options have fitness costs that balance the resultant reduction in conflict load. The extent of sexual conflict is not necessarily reduced at resolution. In particular, a resolution in which the two parents provide equal amounts of care does not imply that there is no sexual conflict. Conversely, there is not necessarily conflict over all aspects of parental care. For example, if care is divided among the offspring, and the fitness of each offspring depends on how much care it receives (but not family size *per* se), there is an optimal amount of care to give

The Evolution of Parental Care. First Edition. Edited by Nick J. Royle, Per T. Smiseth, and Mathias Kölliker.

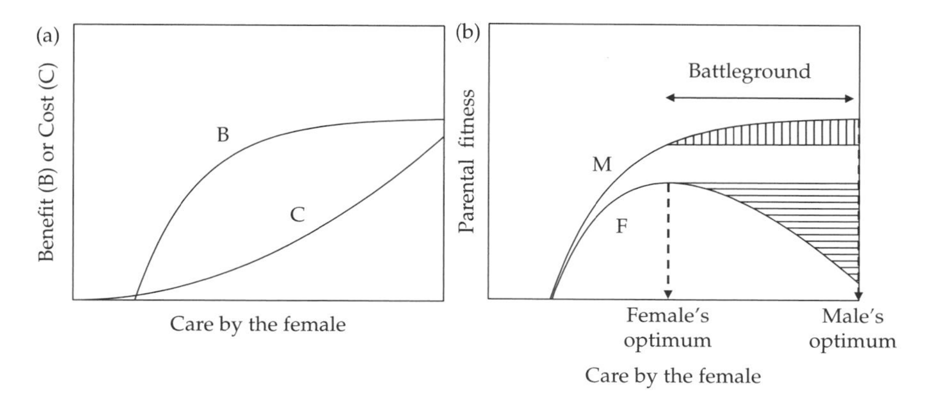

Figure 9.1 Sexual conflict over parental care when there is uniparental care by the female. (a) The fitness benefit (B) through the offspring, fitness cost (C) to the female, and (b) net fitness of the male and female, in relation to the amount of care by the female. In (b), only the female pays the cost of her care, so the male's fitness (M) is equal to B, and the female's fitness (F) is equal to B–C, and the male's and female's optimal amount of care by the female differ. Between the two optima, in the 'battleground', there is sexually antagonistic selection, and each parent's fitness is lower than at its optimum level of care by the female. This reduction in fitness—the 'conflict load'—is represented by the vertical extent of the horizontally (female's conflict load) or vertically (male's conflict load) shaded areas. (After Lessells 2006.)

to each offspring which does not depend on the total amount of care given to the family (Smith and Fretwell 1974). This amount would be the same for both parents, irrespective of whether they were a care-giver, so there would be no sexual conflict over care per offspring. In contrast, there would be sexual conflict over the total amount of parental effort (and hence the number of offspring).

This chapter is about how the existence of sexual conflict over PI influences the evolution of parental care. The main questions that it addresses are how the sexual conflict is resolved—in particular whether the evolutionary stable pattern of care is no care, uniparental care or biparental care, and in the case of uniparental care whether it is the male or female parent who gives care—and what the main selection pressures are which influence the outcome.

9.2 How is sexual conflict over parental care resolved?

There are essentially two ways in which sexual conflict can be resolved. First, new traits may evolve that directly or indirectly manipulate the values of the traits under conflict. Alternatively, PI by each sex coevolves to values where neither sex can gain fitness by changing its own behaviour.

9.2.1 Evolution of manipulation

Manipulative traits involved in the resolution of sexual conflict over PI fall into three categories. First, one sex may directly manipulate PI by the other, usually by exploiting a signalling system. For example, the seminal fluid of *Drosophila* males contains over 80 different accessory gland proteins (Acps), some of which modify female reproductive traits, including fecundity (Chapman et al. 2003). In some mammals, males manipulate maternal investment in offspring by genomic imprinting (Haig 2004): a modification of the DNA during gametogenesis reveals the parental origin of alleles, and paternally and maternally derived alleles are subsequently differentially expressed in the offspring. The loci involved affect embryonic and placental growth—and thus reflect offspring resource consumption at the expense of the mother (Chapter 17). An analogous process might occur in birds, with females manipulating paternal care via maternally derived yolk hormones that influence subsequent nestling begging behaviour (Müller et al. 2007), although there is currently little experimental evidence for this (e.g. Laaksonen et al. 2011).

Second, one sex may coerce the other into making more PI. In polygynous species, females may increase the amount of paternal PI in their own brood by preventing their male from remating

(Sandell and Smith 1996) or destroying his other broods (Hansson et al. 1997). Coercion over PI within mated pairs is extremely rare. In aquaria, parental individuals of a mouth-brooding cichlid, mango tilapia (also known as Galilee St Peter's fish) *Sarotherodon galilaeus*, that attempt to leave the spawning site without taking up eggs for brooding may be chased and butted by the other parent (Balshine-Earn and Earn 1997).

Third, one sex may deceive the other into making more PI. In Eurasian penduline tits *Remiz pendulinus*, either sex may desert leaving the other to raise the brood. Females increase the chance of being able to desert before the male by hiding the eggs in the nest material during laying, thereby concealing the state of advancement of the breeding attempt (Valera et al. 1997).

9.2.2 Coevolution of parental care in the two sexes without manipulation

In most cases, the resolution of sexual conflict over PI appears not to involve manipulation. In these cases, each parent controls its own PI, but not that of its mate. Nevertheless, parental care by the sexes coevolves, because the behaviour of each sex influences the fitness consequences of the other's behaviour. When there are evolutionary conflicts of interest between individuals with coevolving traits, intuition and verbal reasoning become poor guides to the evolutionary outcome. For this reason, progress in understanding the resolution of conflict over PI has relied heavily on theoretical modelling.

Game theory has been widely used because it considers situations where fitness consequences depend on what other individuals are doing. This is true for parental care because fitness pay-offs depend on the other parent's behaviour, and sometimes also on the behaviour of individuals in the population as a whole. Game theory models seek evolutionarily stable strategies (ESSs)—'a strategy such that, if all members of a population adopt it, then no mutant strategy could invade the population' (Maynard Smith 1982). These models are phenotypic (i.e. they ignore genetic mechanisms), and as such are only able to predict evolutionary endpoints, not trajectories. Nevertheless, they have the compensating advantage that it is easier to model selection pressures as the outcome of interactions between individuals.

The trade-off between current and future reproduction—offspring benefits and parental costs—is pivotal to the definition of PI and to understanding the resolution of sexual conflict. The mechanisms invoking costs and benefits are reviewed in Chapter 3, but the quantitative form of the cost and benefit functions is generally unknown. Instead, qualitative predictions are based on assumptions about the general form of these relationships. In general, models of parental care assume that the fitness benefit function through the offspring increases with diminishing returns to an asymptote, and the fitness cost function for the parent either arises from population feedback or increases at an accelerating rate, with the duration or amount of care (e.g. Fig. 9.1a).

In modelling parental care, there are two major issues. The first is whether the focal trait is the specific level of parental effort, or a behavioural rule that determines the level of effort taking into account the mate's behaviour. Models for the specific level of parental effort are referred to as 'sealed bid' models, because each parent makes a single decision that cannot be altered in response to its mate's effort. However, parents often make repeated bouts of investment in the same offspring, in which case parents may modify their care in relation to the effort of their mate, rather than making a single one-off decision (McNamara et al. 1999, 2003). Moreover, there is ample evidence that parents do respond to the behaviour of their mate: in species where either parent may desert leaving the other to care for the young, double desertions in which the offspring are orphaned are rarer than expected by chance (McNamara et al. 2002). Similarly, in species with biparental care, parents alter their care in relation to mate loss, and changes in their mate's work rate (see Section 9.4.2). Such responses on a *behavioural* time scale are referred to as negotiation, and the trait that evolves is the behavioural response rule to the mate's behaviour, rather than a specific level of parental effort. The ESS pattern of parental care will usually depend on whether the amount of care is a sealed bid, or the result of behavioural negotiation.

The second major issue, 'consistency' (Houston and McNamara 2005), occurs when there is feedback between the value of a trait in a population and selection on that trait. For example, when there is an opportunity cost to caring in terms of lost mating opportunities, the parental care behaviour of each sex determines the availability of those opportunities. Remating probabilities therefore cannot be treated as externally fixed values, but must be consistent with the population's behaviour, when the goal is to predict evolutionary endpoints. If, however, the goal is to empirically test whether the observed pattern of care in a population is the ESS (e.g. Balshine-Earn and Earn 1997), or to predict phenotypic plasticity in relation to temporal or spatial variation in remating (e.g. Carlisle 1982), models that ignore the feedback, and treat remating probability as a constant, will be adequate (Kokko and Jennions 2008).

The remainder of this chapter focuses on the resolution of sexual conflict over parental care in the absence of manipulation of the mate's behaviour by either parent.

9.3 How long to care? Offspring desertion

The first of the two major decisions regarding parental care is how long to care for the offspring—in other words, when to desert them—including whether to provide any care at all beyond gamete provisioning. The strategies adopted by males and females determine whether the offspring receive no care, uniparental care by the male or female, or biparental care, and hence the parental care patterns reviewed in Chapters 4 and 5. Sexual conflict results from a trade-off between current and future reproduction, and in the case of offspring desertion this may occur either because providing care for longer reduces the time available for other activities (an opportunity cost), or because caring is a riskier activity than the alternatives, for example by increasing vulnerability to predators. The opportunity costs have often been thought of as being invoked through missed mating opportunities—either through extra-pair copulations or remating. However, such costs may also occur in seasonal iteroparous breeders when there are no further breeding opportunities available in the current year, if spending more time on other activities, or carrying them out earlier, increases survival over the non-breeding season. For example, in shorebirds, the duration of parental care (at least by males) is shorter in species with longer migratory distances, and may allow the deserting parent to depart earlier on migration (Garcia-Pena et al. 2009). The benefits of a particular duration of parental care are dependent on other individuals (and hence need to be modelled using game theory) both because the survival of the offspring will depend on how long the other member of the brood cares, and because the benefits of desertion through other mating opportunities will depend on the availability of mates, which in turn depends on the desertion behaviour of other breeding males and females. When the benefit of desertion is gained through alternative mating opportunities, the trade-off encapsulated by PI is between parental care and competition for mates (in the broadest sense, including searching for mates) with members of the same sex. The question of how long to care is therefore intimately related to questions about mating systems and sex roles. The major questions addressed by evolutionary studies on the duration of parental care therefore include not only understanding how an evolutionary resolution to sexual conflict over the duration of parental care is reached, but also understanding the selection pressures involved in the existence of divergent sex roles, with one sex primarily caring and the other primarily competing for mates (see also Chapter 6).

Maynard Smith (1977) was the first to use game theory models to analyse how sexual conflict over PI would be resolved. In his first two models, each parent is allowed to choose between guarding (caring for the offspring) and deserting, and deserting males have the opportunity to search for a second mate. The survival of the brood depends on the number, but not sex, of the guarding parents. The cost of guarding is expressed in males as the loss of the opportunity to find a second mate, and (in model 2) in females as a reduction in fecundity.

Maynard Smith's (1977) model 2, while providing many of the elements of later models, has three obvious shortcomings (Lessells 1999): first, the amount of investment is fixed. Second, there is no opportunity for negotiation between the pair, and third, as Maynard Smith himself recognized, the model lacks consistency: the probability of remating after desertion should not be a fixed value, but must depend on the desertion strategies of all the other males and females in the population (Grafen and Sibly 1978; Webb et al. 1999; Houston et al. 2005; Kokko and Jennions 2008). These shortcomings acted as a powerful stimulus for further work and have been addressed to varying extents in later models of offspring desertion (Table 9.1). In terms of the parental care decision, these models fall into two main groups (see Table 9.1): in the first (as in Maynard Smith's first two models), the parents simply decide whether or not to care for the brood ('Care/no care' in Table 9.1), and in the second, how long to care for them ('Duration of care' in Table 9.1). Care/no care models either consider a breeding season with a fixed number (two) of breeding opportunities, with the cost of caring for the first brood usually being the loss of the second breeding opportunity, or continuous breeding (or a breeding season of fixed length), with the cost of caring being the time spent caring for the brood, which reduces the time available for mate searching and hence the overall number of broods produced. Duration of care models all consider continuous breeding. Models of continuous breeding seasons are sometimes referred to as 'time-in/time-out' models because time is divided between two activities assumed to be mutually exclusive: mate-searching (time in the mating pool) and parental care (plus a possible refractory period).

Of the shortcomings outlined above, the feedback between the desertion strategies and mating opportunities of the two sexes is critical to understanding the evolutionary resolution of sexual conflict over the duration of parental care. Desertion strategies and mating opportunities are inextricably linked because each offspring produced in the population has one father and one mother (the 'sex-ratio constraint'; Grafen and Sibly 1978), with the result that the total reproductive rate of males in the population must equal that of females (the 'Fisher condition'; Houston et al. 2005; Chapter 10). However, in Maynard Smith's models 1 and 2, females never remate and produce a second brood. Consequently, in a population with equal numbers of breeding males and females, there should be no female mates available in the second breeding attempt for males who desert their first brood, yet the model assumes that these males have some chance of remating (Houston and McNamara 2005). Instead, the probability of finding a new mate should depend on the relative numbers of males and females searching for mates, and hence on the desertion strategies of each of the two sexes. In a time-in/time-out model, if the sex ratio of breeding adults (the 'adult sex ratio', ASR) is, for example, 1:1, the total length of a breeding cycle (mate-searching plus parental care, including any refractory period) must be the same in both sexes. This causes a form of frequency-dependence: if one sex has a shorter duration of parental care, it must as a consequence have a longer mate-search duration. In other words, the sex with the shorter duration of parental care finds it harder to find a mate. The ASR also affects the duration of mate search: if one sex is rarer in the breeding population, it must have a shorter breeding cycle (the ratio of cycle lengths in male and females must be proportional to the ASR). In this case it is the more common sex in the breeding population that finds it harder to find a mate. This feedback between the parental care strategy and mating success must occur in any population where there is an opportunity cost to parental care due to lost mating opportunities. Such models should therefore meet the Fisher condition when the goal is to predict the ESS pattern of care. In contrast, if the goal is to predict phenotypic plasticity in offspring desertion in response to spatial or temporal variation in the availability of mates, it is only the current sex-ratio of individuals receptive to mating (the operational sex ratio, OSR) that is relevant to an individual deciding whether to desert (Kokko and Jennions 2008; McNamara et al. 2000; Chapter 6), and models predicting such decisions need not incorporate feedback between parental care and mating rate (e.g. Carlisle 1982).

Inclusion of the frequency-dependence generated by the Fisher condition has two important consequences for the predictions that desertion models make. The first is that it limits the divergence of sex roles: as individuals of one sex invest less in parental care, they become more common in the mating pool, their mating rate drops, and parental

Table 9.1 Game theory models of offspring desertion under sexual conflict

	Assumptions of model									
	Parental care decision	Offspring benefit as a function of duration of care	Breeding season	Currency of cost of parental care	Refractory period[1]	Search time of rarer searching sex[2]	Sex difference in parental ability considered[3]	Fisher condition met	Effect of adult sex ratio (ASR) considered	Negotiation
Maynard Smith 1977, Model 1	Care/no care	-	2 breeding opportunities	Lost breeding opportunity	-	-	No	No	-	No
Maynard Smith 1977, Model 2	Care/no care	-	2 breeding opportunities	♂Lost breeding opportunity, ♀reduced clutch size	-	-	No	No	-	No
Webb et al. 1999	Care/no care at each of breeding attempts	-	2 breeding opportunities	Lost breeding opportunity	-	-	No	No & Yes	No	No
Wade and Shuster 2002[4]	Care/no care	-	2 breeding opportunities	♂Lost breeding opportunity, ♀reduced clutch size	-	-	No	Yes	No	No
Fromhage et al. 2007	Care/no care	-	2 breeding opportunities	Lost EPC opportunities	-	-	No	Yes	No	No
Yamamura and Tsuji 1993	Care/no care	-	Continuous	Time	Yes	0	No	Yes	Yes	No
Balshine-Earn and Earn 1997	Care/no care	-	Continuous	Time	Yes, depends on care/no care decision	0	No	Yes	Yes	No
Ramsey 2010	Care/no care	-	Continuous	Time	Yes[5]	> 0	Yes	Yes	Yes	No
McNamara et al. 2000	Care/no care	-	Fixed-length breeding season	Time	No (but [6])	> 0	Yes	Yes	Yes	Yes (male decides first)

(continued)

Table 9.1 Continued

	Assumptions of model									
	Parental care decision	Offspring benefit as a function of duration of care	Breeding season	Currency of cost of parental care	Refractory period[1]	Search time of rarer searching sex[2]	Sex difference in parental ability considered[3]	Fisher condition met	Effect of adult sex ratio (ASR) considered	Negotiation
Barta et al. 2002	Care/no care	-	Fixed-length breeding season	Time and energy reserves (energetic costs can differ between uni- and bi-parental care)	Need to recoup reserves (and [6])	>0	Yes (and energetic search costs can differ)	Yes	Yes	Yes (male decides first)
Maynard Smith 1977, Model 3	Duration of care	Linear or decelerating	Continuous	Time	Yes	≥ 0	No	Yes	Yes	No
Grafen and Sibly 1978	Duration of care	Asymptotes	Continuous	Time	Yes	0	Yes	Yes	Yes	No
Kokko and Jennions 2008	Duration of care	Sigmoidal function of $\sum$parental care durations	Continuous	Time (+ mortality risk)	No	≥ 0	No	Yes	Yes (+ can be influenced by care duration)	No
Lazarus 1990	Duration of care	Accelerating (and decelerating costs to parent)	Unspecified	Unspecified	-	-	No (but parental costs can differ)	No	-	Yes

[1] Refractory period: the period from the termination of parental care until the adult is again receptive to mates.
[2] Search time of the rarer searching sex: models either assumed that the rarer searching sex found a mate immediately (search time of rarer sex = 0), or that search time was a function of the number of receptive mates etc. (search time of rarer sex > 0). Only in the latter case does the model predict the operational sex ratio (OSR; the sex ratio of receptive adults).
[3] Sex difference in parental ability: this was modelled as a difference in the benefit through offspring under uniparental care depending on the sex of the caring parent.
[4] See criticisms by Houston and McNamara 2005; Fromhage et al. 2007.
[5] Deserting individuals of both sexes have a refractory period represented by the rate at which they return to the mating pool. In addition, females that are not committed to a breeding attempt may switch between receptive and non-receptive states, creating additional 'refractory' time that is a constant multiple of the time spent mate searching, rather than being independent of this time.
[6] Both parents must make the same minimum period of care.

care becomes a relatively more profitable activity. Indeed, in populations in which there are no sex differences other than anisogamy, sexually egalitarian PI (gamete provisioning plus post-zygotic care) is the ESS (Kokko and Jennions 2008; see Table 9.1 for the assumptions of the model). Thus sexual conflict over parental care does not account by itself for the evolutionary divergence of sex roles. In Kokko and Jennions' (2008) model, some disparity in sex roles is selected by incomplete parentage (through multiple mating or communal breeding of females), by above-random variance in male mating success, or by a biased ASR. Moreover, the divergence can become self-reinforcing if adult mortality rates differ between mate-searching and caring phases of the breeding cycle (see Chapter 6 for further details).

The second general consequence of including frequency-dependence in mating rates resulting from the Fisher condition is that it may lead to mixed ESSs, in which caring behaviour varies within a parental sex (Webb et al. 1999). Without frequency-dependence, the relative pay-offs of caring and desertion remain the same, no matter the proportion of a sex that cares or deserts, so only pure ESSs occur. With the negative frequency-dependence caused by the Fisher condition, the profitability of deserting relative to caring decreases as the proportion of that sex deserting increases, so that some intermediate frequency of desertion may give equal pay-offs to deserters and carers. Mixed ESSs may occur in care/no care models (Webb et al. 1999; Yamamura and Tsuji 1993; Fromhage et al. 2007; Ramsey 2010), and also in models of the duration of care if the benefits through the offspring accelerate with increasing duration of care (Yamamura and Tsuji 1993). Such accelerating benefits would occur if some minimum period of care is required for the young to survive.

Although the models in Table 9.1 comprise varying combinations of assumptions, those without negotiation make the following general predictions: any of the four parental care patterns (no care, male uniparental care, female uniparental care, and biparental care) may be a pure ESS; biparental care is the ESS when two parents are more than twice as good at raising offspring as one (Maynard Smith 1977; Grafen and Sibly 1978; Yamamura and Tsuji 1993), and no care is the ESS when young survive well without parental care (Yamamura and Tsuji 1993). The first of these conditions will be met when care by the two parents is complementary—for example, if they take turns guarding, incubating, or brooding eggs or young—while desertion is more likely when one parent is nearly as effective as two (Maynard Smith 1977; Grafen and Sibly 1978), and is more likely by the sex who is less effective at providing parental care (Grafen and Sibly 1978; McNamara et al. 2000) or is rarer in the breeding population (Grafen and Sibly 1978; Yamamura and Tsuji 1993; Kokko and Jennions 2008).

Most time-in/time-out models consider a continuous breeding season because this enables predictions to be most easily drawn from the models about how various parameters will influence the pattern of parental care. However, the species to which time-in/time-out models are most likely to apply generally have finite breeding seasons. In this case, desertion decisions should depend on the time within the breeding season, because this determines the expected reproductive success over the whole of the season in relation to whether its mate has deserted. This requires a dynamic programming model incorporating the Fisher condition. Such a model (McNamara et al. 2000) predicts that the pattern of care changes during the season: towards the end of the season, a deserting adult is unlikely to have the time to find a mate and raise offspring, and young produced late in the season typically have low reproductive value, so biparental care is the ESS at this time (see also Webb et al. 1999). Consistent with this prediction, polyandrous female Eurasian dotterels *Eudromas morinellus* and spotted sandpipers *Tringa macularia* help to incubate eggs produced late in the season, but not in earlier clutches (Oring 1986). In snail kites *Rostrhamus sociabilis*, a species in which either sex parent may desert the brood, the duration of biparental care increases over the course of the breeding season (Beissinger 1990). Earlier in the season, the model predicts any of the four usual ESSs (no care, uniparental care by either sex, or biparental care), but in addition there may be stable oscillations between different patterns of care (McNamara et al. 2000). Examples of species showing such oscillations do not readily spring to mind, and this pattern demands synchronization of date-dependent desertion strategies of all breeding adults. Moreover, it is not clear whether oscillations would persist if the models considered mixed ESSs for desertion behaviour on a given day rather than only pure strategies.

In McNamara et al.'s (2000) model, the only cost of parental care is an opportunity cost based on time. Barta et al. (2002) considered a similar model in which there was an additional energetic cost of offspring production for the female, and of parental care for both sexes, and in which unmated individuals not caring for offspring must choose between foraging to replenish energetic reserves and searching for a mate. Parents can only survive the period of parental care if their reserves are above a critical level, and this offers the opportunity for one sex (in the model the female, because males decide first whether to desert; see Barta et al. 2002) to force the other parent to care by maintaining its reserves below that level. This is the ESS when uni- and biparental care are equally energetically expensive. When biparental care is substantially cheaper, the ESS is biparental care with both parents starting a reproductive attempt with intermediate reserves. Houston et al. (2005) have argued that 'self-handicapping' to extract more PI from a mate may occur more generally than females having low reserves (as in Barta et al.'s (2002) model). Self-handicapping must represent a 'credible threat' in order to be evolutionarily stable. For example, a mate's reserves must be directly observable (Barta et al. 2002).

Most of the offspring desertion models in Table 9.1 are sealed bid models in which neither parent's behaviour is conditional on the behaviour of its mate, but there is evidence that negotiation does occur over offspring desertion (McNamara et al. 2002). When one parent has the opportunity to desert first, and the remaining parent modifies its behaviour in relation to that of the first, the first parent may be able to leave the second in a 'cruel bind' (Trivers 1972; Dawkins and Carlisle 1976). Models confirm that the parent with the opportunity to desert tends to have the upper hand in sexual conflict (McNamara et al. 2002, 2003), although the individual that chooses second may be able to force the first to care by, for example, maintaining body reserves at too low a level to sustain parental care (Barta et al. 2002). If the order in which parents can choose to desert is not fixed (as it is, for example, in species with internal fertilization, where the male can leave first), the evolutionary outcome can be preemptive desertion, in which each parent is selected to desert progressively earlier in an attempt to avoid being deserted itself. This can drive the duration of parental care—and resultant fitness of the parents—to levels well below those which are evolutionarily stable when parents do not lay themselves open to being placed in a cruel bind by being responsive to their mate's behaviour (Lazarus 1990).

There is a second way that desertion decisions might be conditional on the behaviour of others which has received surprisingly little attention: if parental care and mate-searching are not mutually exclusive activities, parents might only desert their offspring when they have secured a new mate. Because of the stochastic nature of mate search, this could result in the parent of the sex that would not be predicted to desert on the basis of the probability of finding a new mate occasionally finding a new mate first and being the one to desert. Limited observations of four bird species in which either sex can desert (cited in Székely et al. 1996) suggest that three species desert before searching for a new mate, whereas the fourth (little egrets *Egretta garzetta*) develop courtship colouration while still caring for their young and may start searching for a new mate before deserting.

9.3.1 Experimental studies: sex differences in the benefits of care and desertion

Models of desertion identify offspring fitness in relation to the number and sex of caring parents, and the rate at which deserting parents of each sex can remate, as major determinants of evolutionarily stable desertion strategies. As always for fitness costs and benefits, measurement of these fitness consequences requires experimental studies, because individuals that choose to care or desert may differ with respect to traits that also affect fitness.

Experiments have been widely used to measure the benefits of biparental care over uniparental care (for reviews see Bart and Tornes 1989; Clutton-

Brock 1991; Székely et al. 1996). The majority of these studies have compared broods raised by experimentally-widowed females with control broods where both parents provided care. In some species, male removal had a detrimental effect on offspring growth and survival (e.g. Lyon et al. 1987), but in others it had little apparent impact on offspring fitness (e.g. Gowaty 1983). In willow ptarmigan *Lagopus lagopus*, male removal does not affect chick survival (Martin and Cooke 1987). In this species, males may remain with their mate and brood not because of the benefits of biparental care, but because the chance of remating is so low that the best opportunity for males to father further offspring is with the current mate should the breeding attempt fail. Studies removing females are rare, at least partly because it can often be assumed that this would result in complete failure (for example, if males lack the ability to incubate or lactate). Male and female European pied flycatchers *Ficedula hypoleuca* experimentally widowed when their chicks were 5 days-old did not differ in the number of chicks fledged, but were both less successful than control pairs (Alatalo et al. 1988). In contrast, male willow ptarmigan are less successful at raising chicks alone than females (Martin and Cooke 1987).

Clutton-Brock (1991) has pointed out that mate removal experiments in species with universal biparental care tell us little about the selective pressures giving rise to different mating patterns because once biparental care has evolved, subsequent coevolution of male and female parental behaviour may result in one or both parental sex becoming unable to care alone. In this respect experiments on species with variable mating patterns are more interesting, because both sexes have retained this ability. In Kentish plovers *Charadrius alexandrinus*, experimental removal of one parent at hatching reduced chick survival at one of two experimental sites in Turkey (Székely and Cuthill 1999), and chick survival did not depend on which sex was removed. However, at a study area in Portugal with high breeding density, broods cared for by the male were more successful, perhaps because males are more effective at defending chicks against conspecifics (Székely 1996). Greater success of broods tended by males would explain part of the bias towards female desertion in this species.

Species with variable mating patterns are also particularly suitable for measuring sex differences in the benefits of desertion through remating, because this is a normal part of the behavioural repertoire of both sexes. Female Kentish plovers remate about five times faster (median of 5 days versus 25 days) than males do when widowed and their clutch removed during incubation (Székely et al. 1999). This appears to be related to a male-biased ASR, which in turn may be generated by heavily female-biased chick mortality (Székely et al. 2004). The observed sex bias in remating time must create a substantial selection pressure in favour of the observed bias towards female desertion.

9.3.2 Experimental evidence for a trade-off between the benefits of parental care and desertion

The models in Table 9.1 are predicated on a trade-off between the benefits of parental care and desertion, with the latter usually assumed to be the opportunity to remate. Evidence for the existence of such a trade-off can be gained by manipulating the benefits of care or desertion. The benefit of care can be manipulated by increasing or decreasing brood size. In snail kites, the frequency of desertion by one parent decreased with increasing manipulated brood size (Beissinger 1990). In Kentish plovers, the duration of care by females increased with brood size in early hatched broods, when the female could remate after brood desertion, but not in late hatched broods, when she could not (Székely and Cuthill 2000). These results imply a trade-off for females between the experimentally-variable benefit of care and the seasonally-variable benefit of desertion.

An alternative to manipulating the benefits of care is to manipulate the benefit of desertion. Common starlings *Sturnus vulgaris* reduced their parental care when provided with increased mating opportunities by the addition of nest boxes early in incubation, but not late in incubation or during chick rearing (Smith 1995), a pattern which recalls the behaviour of female Kentish plovers in the previously described study. The benefits of desertion have also been manipulated via the adult sex ratio. When ASR was manipulated in groups of the cichlid fish *Herotilapia multispinos*, brood-guarding males frequently deserted their broods when there was a surplus of females, but females did not alter their parental care in relation to the experimental ASR (Keenleyside 1983). In similar experiments on Galilee St Peter's fish, both sexes increased their frequency of desertion when of the minority sex, but cared with the same frequency in groups with an unbiased sex ratio, or when their own sex was in surplus (Balshine-Earn and Earn 1998). Thus when the benefits of desertion are higher, parents of one or both sexes are more willing to desert.

9.4 How much to care?

The second of the two major decisions regarding parental care is how *much* parents should invest in offspring. The original, and still undoubtedly the most influential, model is that of Houston and Davies (1985) who were concerned with the question of how individuals could reach an agreement over how hard each would work in the face of sexual conflict. They assumed that the fitness of the brood depends on the combined amount of care provided by the two parents, but that each parent pays a fitness cost for only its own care. Thus the fitness of an individual depends on its mate's behaviour. However, the cost functions do not depend on the care strategies of other members of the population, which implies that there are no opportunity costs through missed mating opportunities (see Kokko 1999 for such a model). The model therefore applies most easily to species that do not change mates within a breeding season.

Given the assumptions about the costs and benefits of care, the model can be used to calculate the 'best response'—the amount of care that maximizes a parent's own fitness—given any level of care by its mate (Fig. 9.2a and b). When the best response curves cross in the direction shown in Fig. 9.2c—a situation in which, if one parent gives more or less care, the other compensates, but not by the full amount—biparental care is the ESS. Partial compensation for a change in PI by the other parent in species with biparental care is a key prediction of the Houston–Davies model.

In fitness terms, the Houston–Davies ESS is not the most efficient way for two parents to work together to raise offspring: both would have higher fitness at the ESS if their individual levels of parental care were tied in some way to match the other parent's effort (assuming the parents have the same cost functions) (McNamara et al. 2003). This 'cooperative solution' involves the parents making more effort, and the offspring having higher fitness, than in the Houston–Davies game. However, the cooperative solution is only evolutionarily stable when the efforts of the two parents are unbreakably tied. Nevertheless, it provides a theoretical refer-

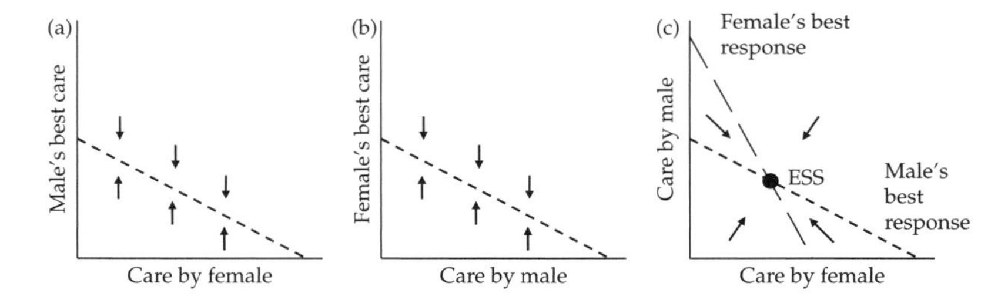

Figure 9.2 The Houston–Davies (1985) model for the evolution of parental effort. This model assumes that the fitness benefit through the offspring depends on the total amount of care given by the parents (increasing to an asymptote with increasing care), and that the fitness cost to each parent depends on the amount of care that it gives (increasing at an accelerating rate with increasing care). Given these fitness costs and benefits, it is possible to calculate (a) the male's optimal level of care for any given level of care by the female ('best response curve'). The vertical arrows indicate the direction in which the male's level of care is selected in pairs with levels of care by each sex represented by points that lie off the male's best response curve. Similarly, it is possible to calculate (b) the female's optimal level of care for any given level of care by the male. By reflecting (b) about the diagonal, both response curves can be plotted on the same axes (c), where the arrows show the combined selection on the male's and female's levels of care. If the best response curves cross in the direction shown in (c), the intersection shows the evolutionarily stable combination of levels of care by the male and female, at which neither can increase their fitness by changing their level of care. (After Houston and Davies 1985.)

ence against which the Houston–Davies ESS can be judged, and appears to suggest that sexual conflict results in parents losing fitness when raising offspring together rather than individually. However, this is not the case: under the Houston–Davies assumptions concerning costs and benefits, parents invest the same, and the resultant total fitness of al the offspring in the family is the same, whether the parents care together for the family or each cares separately for part of the family (provided that it is divided between the parents according to their relative amounts of effort under biparental care: for example, two parents of equal quality each care alone for half the brood) (Lessells 2002; Lessells and McNamara 2012).

9.4.1 Negotiation

The Houston–Davies model assumes that investment by each parent is a 'sealed bid' with neither parent having the opportunity to respond to the investment of their mate. The 'best response curves' summarize how selection acts on the level of care by each of the parents and represent the best *evolutionary* response to a fixed amount of investment made by parents of the other sex. However, for more than a decade, it was mistakenly assumed that the Houston–Davies best response curves were also the best negotiation rules (see references in McNamara et al. 1999). McNamara et al. (1999) pointed out that this was not the case: mutant individuals who always make a fixed effort a little less than the Houston–Davies ESS, rather than responding to the efforts of their mate, have slightly higher fitness. In other words, they are able to exploit their mate. This insight emphasizes the need for models in which parents negotiate over effort.

Negotiation models are complicated for several reasons. Parental care generally consists of many repeated bouts of investment, so negotiation can potentially continue throughout the period of parental care: parental effort is then simultaneously a negotiation bid and PI that carries costs to the parents and benefits to the offspring. Moreover, assumptions need to be made about how the parents interact behaviourally, and how costs and benefits of these bouts combine over the period of parental care.

McNamara et al. (1999, 2003) tackled these problems by assuming negotiation to be cost- and (implicitly) benefit-free. The parents make alternate bids, according to their negotiation rules, in response to their partner, until the bids settle down to stable values that are the outcome of the negotiation. These are the amounts of care that are invested in the offspring, and this investment carries the same costs and benefits as in the Houston–Davies model. Under these assumptions, the evolutionarily stable negotiation rule is less responsive than the Houston–Davies best response rules. The negotiated amounts of effort are 'honest' in the sense that, after negotiation, an individual can infer the quality of its mate—yet the negotiated amounts of effort are less than the parents would have made if they had been able to observe the quality of their mate directly. This means that if the negotiation and investment phases are entirely separate (one interpretation of the lack of negotiation costs; Lessells and McNamara 2012), with no possibility for either parent to modify its investment in relation to that of its mate, a mutant that invested *more* than its final negotiated bid would have higher fitness: without costs to the bidding process, there is nothing to bind a parent to investing the outcome of negotiation, and McNamara et al.'s (1999) solution would not be evolutionarily stable (Lessells and McNamara 2012). This emphasizes the need for models in which negotiation carries costs (Houston et al. 2005), as it would do if negotiation and investment are simultaneous, and parents respond behaviourally to the parental investment of their mates.

McNamara et al.'s (1999, 2003) models are predicated on variation in the quality of parents, and Houston et al. (2005) suggested that negotiation is only expected when there is something (mate quality) to learn from it. However, two recent models have considered negotiation in the absence of variation in quality within each parental sex. Thus, negotiation cannot involve signalling of parental quality, and any differences in the ESS pattern of care from the Houston–Davies model must reflect differential effects of sexual conflict when there are repeated bouts of investment. The first of these models (Lessells and McNamara 2012, see also Ewald et al. 2007 for a differential model) is simply a series of a fixed number of Houston–Davies sealed bid games, but in which the parents can monitor offspring growth before each bout of investment and adjust their amount of care accordingly. The fitness of offspring is assumed to depend only on the total amount of care they receive, while parental fitness costs are an accelerating function of care per bout, summed across bouts. When a single parent cares alone, and the cost function is the same in all bouts, the optimal pattern of care is a uniform spread across the period of parental care (Fig. 9.3a). This is because the temporal pattern of a given total amount of care does not affect the fitness benefit, but total costs are minimized by an even spread of care. In contrast, when two parents care together, ESS levels of care are initially low, and increase to a maximum in the last bout of investment (Fig. 9.3a). This is because investment in early bouts can be exploited by the mate, who would reduce its later investment in response. Investment in the final bout cannot be exploited in this way. As a result, the total amount of care, fitness benefit through the offspring, fitness cost to the parent, and net fitness gain of the parent are all reduced when the parents care together for the brood rather than each caring alone for half of it (Fig. 9.3b). This model emphasises that variation in parental quality is not a prerequisite for negotiation over parental investment to be evolutionarily stable.

9.4.2 Experimental changes in partner effort: mate removal and manipulation

The main prediction of the Houston–Davies (1985) model is that, in species with biparental care, a reduction or increase in parental care by one parent should be partially compensated by the other parent. This prediction has been a powerful stimulus for experimental studies manipulating the parental care by one member of the pair and observing the response in the amount of care given by the other parent. It has since been recognized that the Houston–Davies model refers to evolutionary selection pressures, not negotiation rules on a behavioural time scale (McNamara et al. 1999), but negotiation models also predict partial compensation (McNamara et al. 2003). These latter models also potentially predict over-compensation

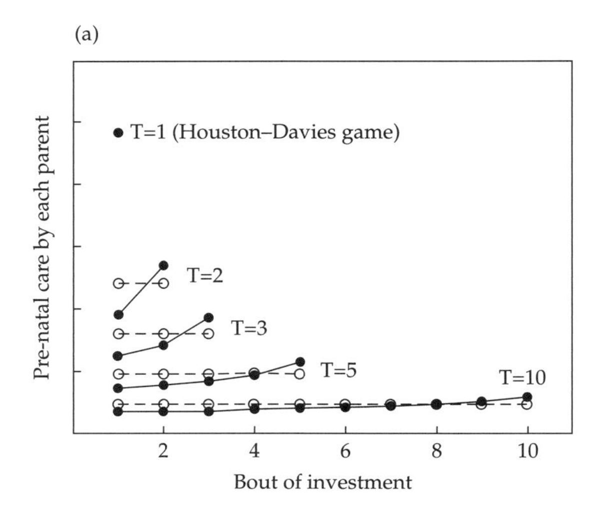

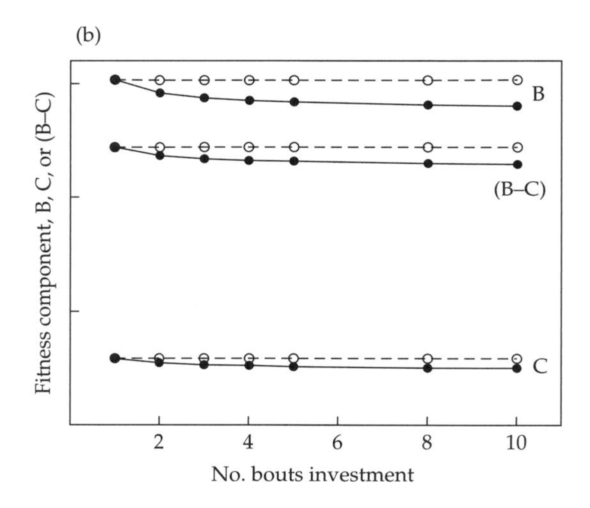

Figure 9.3 Predicted patterns of parental care when two parents make repeated bouts of investment in the offspring (Lessells and McNamara 2012): (a) The amount of care given in successive bouts of investment, and (b) the fitness benefit through the offspring (B), fitness cost to each parent (C), and net fitness gain to each parent (B–C) in relation to the number of bouts of investment. Parents either care together for the whole brood (filled symbols and solid lines), or each care alone for half the brood (unfilled symbols and broken lines). T is the total number of bouts of investment. The model assumes that offspring increase in mass in direct proportion to the amount of care that they receive, and that their fitness depends only on their final mass, increasing to an asymptote with increasing mass, and that the total cost to the parents is the sum of the cost of each bout of care, which increases at an increasing rate with increasing care provided. Parents decide the amount of care to be given during a bout on the basis of the current mass of the offspring, so that when two parents care together each bout consists of a sealed bid Houston–Davies game. (From Lessells and McNamara 2012.)

following mate removal, although this may result from the assumption that there are no costs during the negotiation phase, and would result in biparental care being evolutionarily unstable (Houston et al. 2005).

Two sorts of experiments have been used to change partner effort (but see also Smiseth and Moore 2004): in the first, one of the pair is experimentally widowed by removal of its mate, while the second involves manipulation of the parental effort of a mate left *in situ*. In this second case, partner effort has been experimentally reduced by handicapping through the addition of weights or clipping of flight feathers, or by diverting males to other activities through the use of testosterone implants, manipulation of sexually selected signals, or the provision of additional nesting sites (see Harrison et al. 2009 for references). Partner effort has been experimentally increased by selective playback of begging signals to one of the two parents (Ottosson et al. 1997; Hinde 2006; Hinde and Kilner 2007).

Experiments have been carried out on a taxonomic range of species including insects (e.g. Rauter and Moore 2004; Smiseth et al. 2005; Suzuki and Nagano 2009), fish (e.g. Mrowka 1982), and birds (Harrison et al. 2009). The results of these experiments have been disparate, with responses to changes in partner effort varying from matching, through no, partial, and full compensation, to over-compensation (see below in this section). The question of whether these results as a whole represent evidence for partial compensation has been addressed in a meta-analysis of the studies carried out on birds—the most strongly represented group among the experimental studies. The meta-analysis includes 54 studies in which partner effort was experimentally reduced by mate removal (25 male removals, 11 female removals) or manipulations (26 of male care, 11 of female care) (Harrison et al. 2009). Overall, these studies support partial compensation in both provisioning, and incubation or brooding, with the parent whose mate had been removed or manipulated increasing their effort (although this was marginally non-significant for incubation or brooding), while overall effort decreased (Fig. 9.4).

The results of the meta-analysis are therefore consistent with the predictions of both the Houston–Davies model (with the best evolutionary response curves being interpreted as behavioural negotiation rules) and McNamara et al.'s (1999, 2003) negotiation model. However, the negotiation model additionally predicts that mate removal and manipulation of the effort of a mate left *in situ* should give

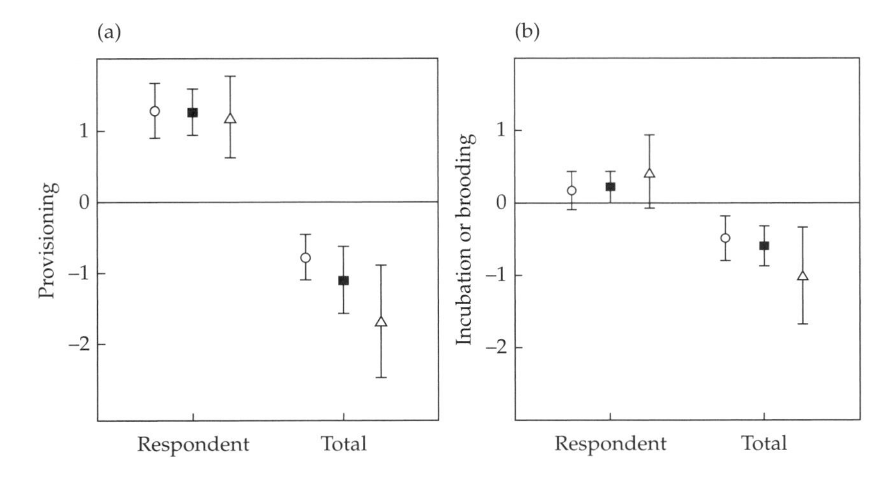

Figure 9.4 A meta-analysis of avian studies of the response in (a) offspring provisioning and (b) incubation or brooding by a parent when its mate's effort was experimentally reduced by manipulation *in situ* or removal of the mate. Plotted values are the weighted mean effect sizes ± 95% confidence intervals. 'respondent' values refer to the change in care provided by the parent whose mate was manipulated or removed, and 'total' the change in care by the two parents combined. Open circles refer to studies where the female was the respondent, open triangles to studies where the male was the respondent, and filled squares to these two sets of studies combined. (From Harrison et al. (2009), with permission from John Wiley & Sons.)

different results, with widowed birds showing a relatively larger degree of compensation because their increased effort is not open to exploitation by their mate. Two studies support this prediction: experimentally widowed female house sparrows *Passer domesticus* compensated for the decrease in their mate's effort to a proportionately larger degree than females whose mate was still present (Lendvai et al. 2009). Similarly, male burying beetles *Nicrophorus quadripunctatus* increased their effort when their mate was removed, but not when their effort was reduced by handicapping with weights (females did not respond at all) (Suzuki and Nagano 2009).

Although the meta-analysis suggests that partial compensation is the most common response to a reduction in parental care of the partner, responses in individual studies can vary widely and have acted as a stimulus for a range of additional explanations in terms of the selection pressures shaping the outcome of sexual conflict over parental care. Alternative responses can be grouped into four categories. The first of these is 'matching', which is the opposite of compensation, with parents adjusting their effort in the same direction as the experimental modification of their mate's effort. There are three broad explanations for this: reduction in care by one parent may mean that the other parent can no longer successfully raise the offspring, and ceases to care for the offspring. A possible example occurs in western sandpipers, *Calidris mauri*, where mate removal results in the immediate desertion of the other parent (Erckman 1983). Second, where there is some degree of task specialization in parental care by males and females, reduction in the care provided by one sex may lead the other sex to divide their effort more equally between different aspects of parental care, entraining a reduction in some care behaviours. In this case, there may still be partial compensation in terms of overall effort, but matching in specific parental care behaviours. For example, in unmanipulated pairs of the burying beetle *Nicrophorus vespilloides*, there is task specialization, with the antimicrobial activity of the anal exudates used to protect the carcass being higher in females than males. After mate removal, males compensate by increasing the antimicrobial activity of their anal exudates, whereas females do the opposite, resulting in widowed males and females having similar levels of antimicrobial activity (Cotter and Kilner 2010). Third, parents may respond directly to the work rate of their partner. For example, in short-term playback experiments in great tits *Parus major*, where chick begging calls were selectively broadcast to one of the parents, both parents increased the rate at which they provisioned the young (Hinde 2006; Hinde and Kilner 2007). Hinde and Johnstone (Hinde 2006; Johnstone and Hinde 2006) have shown using a negotiation model of the same type as McNamara et al.'s (1999, 2003) models that using the other parent's work rate as a measure of brood need could act as a selection pressure for such matching.

The second alternative to partial compensation is *no compensation*. For example, in three different species of *Nicrophorus* burying beetles, males, but not females, compensate for the removal of their mate (Rauter and Moore 2004; Smiseth et al. 2005, Suzuki and Nagano 2009). This may simply be because females are already working at close to their maximum capacity: females in pairs carry out more direct care than the males, and females of *N. orbicollis* do compensate for male removal when caring for small broods (Rauter and Moore 2004). In the meta-analysis of bird studies, experimentally widowed males increased their feeding effort more when females normally took a greater share in chick provisioning (Harrison et al. 2009), consistent with the idea that there may be some maximum capacity that limits the amount of parental care provided.

Third, a reduction of partner effort may be met with *complete compensation*. For example, in the cichlid *Aequidens paraguayensis*, an initial period in which eggs laid on the substrate are guarded and fanned is followed by a mouth-brooding phase in which the parents take turns at carrying the young in their mouth. Experimentally widowed fish of both sexes completely compensate during both phases of parental care (Mrowka 1982). Jones et al. (2002) have suggested that in some species, including mouth-brooders and birds whose eggs and chicks are very vulnerable to predation or cooling, the offspring benefit curve may be accelerating, rather than decelerating as usually assumed. They used a theoretical model to show that, in such cases, reduction in the care provided by one parent will be met with either complete compensation or desertion by the other. Biparental care may, nevertheless, be evolutionarily stable if two parents are needed to raise the young, and in these cases exploitation of one sex by the other is restrained by the risk of pushing the other into desertion.

Lastly, a reduction in partner effort may be met with *over-compensation*. This is the most surprising response, since it poses the questions both why individuals should more than compensate for reductions in effort by their mates, and (as is also the case for full compensation) how biparental care can be evolutionarily stable since their partner would appear to benefit by desertion in such circumstances. One possible example (albeit non-experimental) occurs in rock sparrows *Petronia petronia*, in which 20% of females are double-brooded (Pilastro et al. 2001) and either the male or female may desert the first brood when the chicks are about a week old to initiate a second brood with a different partner, while some broods are reared to fledging by both the parents. Compared to the feeding rates of pairs raised by both parents, deserted males partially compensate for the absence of their mate, while females over-compensate in terms of the rates of both feeding visits and food delivery (Griggio and Pilastro 2007). Griggio and Pilastro (2007) explain over-compensation by deserted females using McNamara et al.'s (1999, 2003) models. An additional possibility, which they did not consider, is that desertion by the male reduces the fitness cost of care to the female—specifically, the opportunity cost related to second broods—so that her optimal level of care is increased. This opportunity cost may be lower in deserted females because a quarter of double-brooded females produce their second brood with their first-brood mate (who is no longer available after desertion), or because the time and effort involved in raising the first brood alone prevents the female from attempting a second brood. Over-compensation by deserted females also begs the question of why males do not always desert their first-brood female; the answer may again lie in the opportunity that their first-brood mate presents as a partner for a second brood.

9.4.3 Brood division

The above studies consider what should happen when one parent is removed but the brood size is left unchanged. This is not a strongly discriminating test, because both sealed bid and negotiation models predict that the remaining parent should increase its effort, but that offspring may do worse, following mate removal. The only discriminating prediction is that offspring may also do better when effort is negotiated, rather than a sealed bid (McNamara et al. 2003), and this prediction is not made by a negotiation model with repeated bouts of investment (Lessells and McNamara 2012). An alternative manipulation is to remove one parent, but to keep the potential workload for the parent the same by simultaneously removing some of the young. The Houston–Davies model then predicts that neither parental effort nor the fitness of each of the offspring will change (provided that the parent is left with a proportion of the brood that is equal to the proportion of the care that it provides under biparental care) (Lessells 2002). In contrast, negotiation models predict that parental effort and offspring fitness should both increase (McNamara et al. 1999; Lessells and McNamara 2012).

These predictions have been tested on zebra finches *Taeniopygia guttata* in the laboratory (Royle et al. 2002). Broods were manipulated when the chicks were about five days old to give either broods of about four chicks subsequently reared by the pair, or two chicks subsequently reared by the female alone. Females rearing the brood alone fed larger meals to the chicks, and laid fewer and smaller eggs in their following clutch, indicating that they invested more in the brood. The offspring in the broods raised by the female alone had higher *per capita* fitness: although they were not larger or heavier at fledging, male chicks were more attractive to females as adults. These results could not be attributed to females carrying out more than half the work in biparental broods, and so imply that zebra finches negotiate over parental effort, and demonstrate that sexual conflict can reduce parental and offspring fitness in broods raised by both parents (Royle et al. 2002).

Both sealed bid and negotiation models predict that parents should divide broods and care alone for part of the brood: negotiation models predict this even when the brood is fairly divided by the parents (Lessells and McNamara 2012), and although parents are predicted to be indifferent to brood division in sealed bid models (Houston and Davies 1985; Lessells 2002), sexual conflict selects for preemptive desertion by one of the parents taking a little less than its fair share of the brood (Lessells 2002). A few species of birds do, indeed, divide their broods after leaving the nest (Lessells 2002, Table 3), but the vast majority of biparental species do not. There are three reasons why this might be the case. First, benefits through offspring may depend on brood size *per se* (e.g. several of the specific hypotheses listed by McLaughlin and Montgomerie 1985). Larger broods having higher fitness (for example, as a result of thermoregulatory benefits) would favour keeping the brood together. Second, all existing models of the amount of parental care assume that the care provided by the parents is divided out among the offspring such that only one offspring benefits from any unit of care ('shared' sensu Lazarus and Inglis 1986, 'depreciable' PI sensu Clutton-Brock 1991), rather than all young benefitting simultaneously ('unshared'/'depreciable' PI). Possible examples of unshared PI include nest building and certain forms of protection against predators. Although no formal models exist, it is clear that unshared PI will favour broods being kept together. Third, the care by each parent may have a more than additive effect on offspring fitness (see Section 9.4.4). Lastly, broods may be divided, but in a way that is not immediately obvious to an observer: for example individual great tit parents feed from consistent positions on the nest rim, which may result in parents feeding different subsets of the chicks (Tanner et al. 2007).

9.4.4 Complementarity and task specialization

Models of offspring desertion predict that biparental care will be stable when it more than doubles the fitness of offspring compared with broods cared for by one parent (Maynard Smith 1977; Grafen and Sibly 1978; Yamamura and Tsuji 1993). Similarly, the Houston–Davies model for the amount of parental care assumes that parental care is additive in its effect on offspring fitness, but a positive interaction between the care provided by each of the parents favours biparental care (Motro 1994). Such effects that are more than additive may come about either through complementarity in providing a single kind of care, for example because of turn-taking, or because of task specialization such that each parent can care less adequately for the offspring alone. Task specialization is likely to be favoured by some kind of physiological or morphological adaptation that improves the efficiency of providing a particular kind of care. In this case, biparental care and task specialization might be self-reinforcing, with biparental care allowing task specialization, and task specialization favouring biparental care. Task specialization might also impact on the outcome of sexual conflict by reducing the extent to which one sex can exploit the other when there is biparental care. Little research has been carried out on these issues, but it is clear that the way that the two parents work together, particularly in providing multidimensional care, is likely to have a big impact on the resolution of sexual conflict over parental care.

9.5 Interaction with other evolutionary conflicts within the family

This chapter has focused almost exclusively on sexual conflict, but it is worth remembering that patterns of PI in nature reflect the simultaneous resolution of all the conflicts between family members (Parker et al. 2002). In particular, the occurrence of one kind of conflict may impinge on the resolution of another. For example, sibling conflict that is resolved by a dominance hierarchy among offspring can result in an inequitable distribution of care, which reduces parental fitness. In a model by Lessells (2002), parents who divided the brood have higher fitness because this results in a more equitable distribution, even when the offspring still control the distribution of care. In an experiment on great tits, food was more equitably distributed between chicks when parents could feed from two locations, rather than one (Tanner et al. 2007). This suggests that sibling conflict acts as a selection pressure for the parents to cooperate in preventing the monopolization of food by dominant offspring.

9.6 Conclusions and prospects

Trivers' 1972 paper revolutionized our view of parental care, because it could no longer be regarded as an entirely cooperative undertaking between the parents. This insight prompted the development of theoretical models attempting to understand the selection pressures shaping patterns of parental care (Maynard Smith 1977; Grafen and Sibly 1978; Houston and Davies 1985). The initial models made the simplest of assumptions with each parent's care strategy characterized by a single number, and no variation between individual parents within a sex. Nevertheless, they provided insights into the selection pressures that might account for broad patterns of parental care and, perhaps more importantly, demonstrated how evolutionarily stable outcomes, including biparental care, could emerge despite sexual conflict favouring individuals who exploit their mates' parental efforts. The empirical studies that followed from these early models confirmed their broad conclusions and also revealed additional ways in which selection in general, and sexual conflict in particular, shapes patterns of parental care. Nevertheless, there are still areas that pose considerable challenges.

The first of these is negotiation. Negotiation is important because it alters the evolutionary outcome of sexual conflict (McNamara et al. 1999), and is particularly challenging when repeated bouts of parental effort are made in a brood because effort in early bouts is then both immediate investment in the offspring, and part of the negotiation process over later investment in the same offspring. Models have only recently been developed, and there is still a complete lack of models that integrate investment and signalling of parental quality as functions of early bouts of care. Such models also throw up subsidiary questions, such as how fitness costs and benefits combine over successive bouts of care, and what the mechanisms are which determine this. Negotiation also throws the spotlight onto behavioural mechanisms, because these are critically important in determining evolutionary outcomes (e.g. McNamara et al. 2003; Lessells and McNamara 2012). More studies of *how* negotiation takes place are needed.

The second area concerns complementarity (for example turn-taking in guarding) and task specialization in parental care. Theoretically, these are powerful selection pressures for biparental care, but many questions remain. What factors select for complementarity or task specialization that is reversible or irreversible (for example because of morphological adaptation) on a behavioural time scale, and do these in turn have different effects on the stability of biparental care? And, if task specialization stabilizes biparental care, and reduces the extent to which one parent can exploit the other's efforts, does the likelihood of an evolutionary transition from uniparental to biparental care depend on the kind of care provided by each sex? These questions raise related issues such as the currently-neglected distinction between shared and unshared PI (Lazarus and Inglis 1986) and its effect on the resolution of sexual conflict.

The third challenge is the interaction with other areas of reproductive biology. The feedback between desertion decisions and remating success creates a strong link between parental care and mating systems. Similar feedbacks exist with other areas such as sexual selection (Chapter 6, Kokko and Jennions 2008) and parentage (Chapter 11) and are important to include in frameworks for studying the resolution of sexual conflict (Székely et al. 2000; Alonzo 2010). Another area that parental care may interact with in the resolution of sexual conflict is the developmental strategy of the offspring. This must determine how the benefits of care combine over repeated bouts of investment, and hence the evolutionarily stable pattern of care. The pattern of care (particularly its predictability) will then in turn feed back on the optimal developmental strategy of offspring. Linking related areas allows an increasing number of traits to be explained as the outcome of selection, rather than being treated as fixed. This presents the challenge of maintaining consistency between areas, but will ultimately provide the most complete explanations of parental care.

Acknowledgements

I am grateful to John McNamara, Per Smiseth, Tamas Székely, and an anonymous referee for helpful comments.

References

Alatalo, R. V., Gottlander, K., and Lundberg, A. (1988). Conflict or cooperation between parents in feeding nestlings in the pied flycatcher *Ficedula hypoleuca*. *Ornis Scandinavica* 19, 31–4.

Alonzo, S. H. (2010). Social and coevolutionary feedbacks between mating and parental investment. *Trends in Ecology & Evolution* 25, 99–108.

Arnqvist, G. and Rowe, L. (2005). *Sexual Conflict*, Princeton University Press, Princeton.

Balshine-Earn, S. and Earn, D. J. D. (1997). An evolutionary model of parental care in St. Peter's fish. *Journal of Theoretical Biology* 184, 423–31.

Balshine-Earn, S. and Earn, D. J. D. (1998). On the evolutionary pathway of parental care in mouth-brooding cichlid fish. *Proceedings of the Royal Society of London, Series B* 265, 2217–22.

Bart, J. and Tornes, A. (1989). Importance of mongamous male birds in determining reproductive success. Evidence for house wrens and a review of male-removal studies. *Behavioral Ecology and Sociobiology* 24, 109–16.

Barta, Z. N., Houston, A. I., McNamara, J. M., and Székely, T. (2002). Sexual conflict about parental care: The role of reserves. *American Naturalist* 159, 687–705.

Beissinger, S. R. (1990). Experimental manipulations and the monoparental threshold in snail kites. *American Naturalist* 136, 20–38.

Carlisle, T. R. (1982). Brood success in variable environments: implications for parental care allocation. *Animal Behaviour* 30, 824–36.

Chapman, T., Bangham, J., Vinti, G., Seifried, B., Lung, O., Wolfner, M. F., Smith, H. K., and Partridge, L. (2003). The sex peptide of Drosophila melanogaster: Female post-mating responses analyzed by using RNA interference. *Proceedings of the National Academy of Sciences of the United States of America* 100, 9923–8.

Clutton-Brock, T. H. (1991). *The Evolution of Parental Care*. Princeton University Press, Princeton.

Cotter, S. C. and Kilner, R. M. (2010). Sexual division of antibacterial resource defence in breeding burying beetles, Nicrophorus vespilloides. *Journal of Animal Ecology* 79, 35–43.

Dawkins, R. and Carlisle, T. R. (1976). Parental investment, mate desertion and a fallacy. *Nature* 262, 131–3.

Erckman, W. J. (1983). The evolution of polyandry in shorebirds: an evaluation of hypotheses. In S. K. Wasser, ed. *Social Behavior of Female Vertebrates*. Academic Press, New York.

Ewald, C. O., McNamara, J., and Houston, A. (2007). Parental care as a differential game: A dynamic extension of the Houston-Davies game. *Applied Mathematics and Computation* 190, 1450–65.

Fromhage, L., McNamara, J. M., and Houston, A. I. (2007). Stability and value of male care for offspring: is it worth only half the trouble? *Biology Letters* 3, 234–6.

Garcia-Pena, G. E., Thomas, G. H., Reynolds, J. D., and Székely, T. (2009). Breeding systems, climate, and the evolution of migration in shorebirds. *Behavioral Ecology* 20, 1026–33.

Godfray, H. C. J. (1995). Evolutionary theory of parent-offspring conflict. *Nature* 376, 133–8.

Gowaty, P. A. (1983). Male parental care and apparent monogamy among eastern bluebirds (*Sialia sialis*). *American Naturalist* 121, 149–57.

Grafen, A. and Sibly, R. (1978). A model of mate desertion. *Animal Behaviour* 26, 645–52.

Griggio, M. and Pilastro, A. (2007). Sexual conflict over parental care in a speces with female and male brood desertion. *Animal Behaviour* 74, 779–85.

Haig, D. (2004). Genomic imprinting and kinship: How good is the evidence? *Annual Review of Genetics* 38, 553–85.

Hansson, B., Bensch, S., and Hasselquist, D. (1997). Infanticide in great reed warblers: secondary females destroy eggs of primary females. *Animal Behaviour* 54, 297–304.

Harrison, F., Barta, Z., Cuthill, I., and Székely, T. (2009). How is sexual conflict over parental care resolved? A meta-analysis. *Journal of Evolutionary Biology* 22, 1800–12.

Hinde, C. A. (2006). Negotiation over offspring care? a positive response to partner-provisioning rate in great tits. *Behavioral Ecology* 17, 6–12.

Hinde, C. A. and Kilner, R. M. (2007). Negotiations within the family over the supply of parental care. *Proceedings of the Royal Society, Series B* 274, 53–60.

Houston, A. I. and Davies, N. B. (1985). The evolution of cooperation and life-history in the dunnock. In R. M. Sibly and R. H. Smith, eds. *Behavioural Ecology*, pp. 471–87. Blackwell, Oxford.

Houston, A. I. and McNamara, J. M. (2005). John Maynard Smith and the importance of consistency in evolutionary game theory. *Biology and Philosophy* 20, 933–50.

Houston, A. I., Székely, T., and McNamara, J. M. (2005). Conflict between parents over care. *Trends in Ecology and Evolution* 20, 33–8.

Johnstone, R. A. and Hinde, C. A. (2006). Negotiation over offspring care—how should parents respond to each other's efforts? *Behavioral Ecology*, 17, 818–27.

Jones, K. M., Ruxton, G. D., and Monaghan, P. (2002). Model parents: is full compensation for reduced partner nest attendance compatible with stable biparental care? *Behavioral Ecology* 13, 838–43.

Keenleyside, M. H. A. (1983). Mate desertion in relation to adult sex ratio in the biparental cichlid fish *Herotilapia multispinosa*. *Animal Behaviour* 31, 683–8.

Kokko, H. (1999). Cuckoldry and the stability of biparental care. *Ecology Letters* 2, 247–55.

Kokko, H. and Jennions, M. D. (2008). Parental investment, sexual selection and sex ratios. *Journal of Evolutionary Biology* 21, 919–48.

Laaksonen, T., Adamczyk, F., Ahola, M., Möstl, E., and Lessells, C. M. (2011). Yolk hormones and sexual conflict over parental investment in the pied flycatcher. *Behavioral Ecology and Sociobiology* 65, 257–64.

Lazarus, J. (1990). The logic of mate desertion. *Animal Behaviour* 39, 672–84.

Lazarus, J. and Inglis, I. R. (1986). Shared and unshared parental investment, parent-offspring conflict and brood size. *Animal Behaviour* 34, 1791–804.

Lendvai, A. Z., Barta, Z., and Chastel, O. (2009). Conflict over parental care in house sparrows: do females use a negotiation rule? *Behavioral Ecology* 20, 651–6.

Lessells, C. M. (1999). Sexual conflict in animals. In L. Keller, ed. *Levels of Selection in Evolution*, pp. 75–99. Princeton University Press, Princeton.

Lessells, C. M. (2002). Parentally biased favouritism: why should parents specialize in caring for different offspring? *Philosophical Transactions of the Royal Society of London, Series B* 357, 381–403.

Lessells, C. M. (2006). The evolutionary outcome of sexual conflict. *Philosophical Transactions of the Royal Society, Series B* 361, 301–17.

Lessells, C. M. and McNamara, J. M. (2012). Sexual conflict over parental investment in repeated bouts: negotiation reduces overall care. *Proceedings of the Royal Society, Series B* 279, 1506–14.

Lyon, B. E., Montgomerie, R. D., and Hamilton, L. D. (1987). Male parental care and monogamy in snow buntings. *Behavioral Ecology and Sociobiology* 20, 377–82.

Martin, K. and Cooke, F. (1987). Bi-parental care in willow ptarmigan: a luxury? *Animal Behaviour* 35, 369–79.

Maynard Smith, J. (1977). Parental investment: a prospective analysis. *Animal Behaviou*, 25, 1–9.

Maynard Smith, J. (1982). *Evolution and the Theory of Games*. Cambridge University Press, Cambridge.

McLaughlin, R. L. and Montgomerie, R. D. (1985). Brood division by Lapland longspurs. *Auk* 102, 687–95.

McNamara, J. M., Gasson, C. E., and Houston, A. I. (1999). Incorporating rules for responding into evolutionary games. *Nature* 401, 368–71.

McNamara, J. M., Houston, A. I., Barta, Z., and Osorno, J. L. (2003). Should young ever be better off with one parent than with two? *Behavioral Ecology* 14, 301–10.

McNamara, J. M., Houston, A. I., Székely, T., and Webb, J. N. (2002). Do parents make independent decisions about desertion? *Animal Behaviour* 64, 147–9.

McNamara, J. M., Székely, T., Webb, J. N., and Houston, A. I. (2000). A dynamic game-theoretic model of parental care. *Journal of Theoretical Biology* 205, 605–23.

Motro, U. (1994). Evolutionary and contuous stability in asummetric games with continuous strategy sets: the parental investment conflict as an example. *American Naturalist* 144, 229–41.

Mrowka, W. (1982). Effect of removal of the mate on the parental care behaviour of the biparental cichlid *Aequidens paraguaryensis*. *Animal Behaviour* 30, 295–7.

Müller, W., Lessells, C. M., Korsten, P., and von Engelhardt, N. (2007). Manipulative signals in family conflict? On the function of maternal yolk hormones in birds. *American Naturalist*, 169 E84–96.

Oring, L. W. (1986). Avian polyandry. In R. F. Johnston, ed. *Current Ornithology*, pp. 309–51, Plenum Press, New York.

Ottosson, U., Backman, J., and Smith, H. G. (1997). Begging affects parental effort in the pied flycatcher, *Ficedula hypoleuca*. *Behavioral Ecology and Sociobiology* 41, 381–4.

Parker, G. A. (1979). Sexual selection and sexual conflict. In M. S. Blum and N. A. Blum, eds. *Sexual Selection and Reproductive Competition in Insects*, pp. 123–66. Academic Press, London.

Parker, G. A., Royle, N. J., and Hartley, I. R. (2002). Intrafamilial conflict and parental investment: a synthesis. *Philosophical Transactions of the Royal Society of London, Series B* 357, 295–307.

Pilastro, A., Biddau, L., Marin, G., and Mingozzi, T. (2001). Female brood desertion increases with number of available mates in the rock sparrow. *Journal of Avian Biology* 32, 68–72.

Ramsey, D. M. (2010). A large population parental care game: Polymorphisms and feedback between patterns of care and the operational sex ratio. *Journal of Theoretical Biology* 266, 675–90.

Rauter, C. M. and Moore, A. J. (2004). Time constraints and trade-offs among parental care behaviours: effects of brood size, sex and loss of mate. *Animal Behaviour* 68, 695–702.

Royle, N. J., Hartley, I. R., and Parker, G. A. (2002). Sexual conflict reduces offspring fitness in zebra finches. *Nature* 416, 733–6.

Sandell, M. I. and Smith, H. G. (1996). Already mated females constrain male mating success in the European starling. *Proceedings of the Royal Society of London, Series B* 263, 743–7.

Smiseth, P. T., Dawson, C., Varley, E., and Moore, A. J. (2005). How do caring parents respond to mate loss? Differential response by males and females. *Animal Behaviour* 69, 551–9.

Smiseth, P. T. and Moore, A. J. (2004). Behavioral dynamics between caring males and females in a beetle with facultative biparental care. *Behavioral Ecology* 15, 621–8.

Smith, C. C. and Fretwell, S. D. (1974). The optimal balance between size and number of offspring. *American Naturalist* 108, 499–506.

Smith, H. G. (1995). Experimental demonstration of a trade-off between mate attraction and paternal care. *Proceedings of the Royal Society of London, Series B* 260, 45–51.

Suzuki, S. and Nagano, M. (2009). To compensate or not? Caring parents respond differentially to mate removal and mate handicapping in the burying beetle, *Nicrophorus quadripunctatus*. *Ethology* 115, 1–6.

Székely, T. (1996). Brood desertion in Kentish Plover *Charadrius alexandrinus*: An experimental test of parental quality and remating opportunities. *Ibis* 138, 749–55.

Székely, T. and Cuthill, I. C. (1999). Brood desertion in Kentish plover: the value of parental care. *Behavioral Ecology* 10, 191–7.

Székely, T. and Cuthill, I. C. (2000). Trade-off between mating opportunities and parental care: brood desertion by female Kentish plovers. *Proceedings of the Royal Society, Series B* 267, 2087–92.

Székely, T., Cuthill, I. C. and Kis, J. (1999). Brood desertion in Kentish plover: sex differences in remating opportunities. *Behavioral Ecology* 10, 185–90.

Székely, T., Cuthill, I. C., Yezerinac, S., Griffiths, R., and Kis, J. (2004). Brood sex ratio in the Kentish plover. *Behavioral Ecology* 15, 58–62.

Székely, T., Webb, J. N., and Cuthill, I. C. (2000). Mating patterns, sexual selection and parental care: an integrative approach. In M. Apollonio, M. Festa-Bianchet, and D. Mainardi, eds. *Vertebrate Mating Systems*, pp. 194–223. World Science Press, London.

Székely, T., Webb, J. N., Houston, A. I., and McNamara, J. M. (1996). An evolutionary approach to offspring desertion in birds. In V. Nolan Jr and E. D. Ketterson, eds. *Current Ornithology*, pp. 271–330, Plenum Press, New York.

Tanner, M., Kölliker, M., and Richner, H. (2007). Parental influence on sibling rivalry in great tit, *Parus major*, nests. *Animal Behaviour* 74, 977–83.

Trivers, R. L. (1972). Parental investment and sexual selection. In B. Campbell, ed. *Sexual Selection and the Descent of Man 1871–1971*, pp. 139–79. Aldine, Chicago.

Valera, F., Hoi, H., and Schleicher, B. (1997). Egg burial in penduline tits, *Remiz pendulinus*: Its role in mate desertion and female polyandry. *Behavioral Ecology* 8, 20–7.

Wade, M. J. and Shuster, S. M. (2002). The evolution of parental care in the context of sexual selection: A critical reassessment of parental investment theory. *American Naturalist* 160, 285–292.

Webb, J. N., Houston, A. I., McNamara, J.M,. and Székely, T. (1999). Multiple patterns of parental care. *Animal Behaviour* 58, 983–93.

Yamamura, N. and Tsuji, N. (1993). Parental care as a game. *Journal of Evolutionary Biology* 6, 103–27.

CHAPTER 10

Sex allocation

Jan Komdeur

10.1 Introduction

Sex allocation is the apportioning of parental resources to male versus female offspring in a sexually reproducing species. Sex allocation theory assumes that natural selection favours parents that modify their investment into male and female offspring in such a way that it maximizes the parent's fitness (Fisher 1930; Charnov 1982). Fisher's equal allocation theory (1930) states that selection should favour an unbiased sex ratio at the population level. Because each offspring has a mother and a father, males and females will on average make equal genetic contributions to the next generation. Offspring of the minority sex will yield a greater *per capita* return on investment, putting a premium on the production of a larger number of offspring of this sex. Thus, frequency-dependent selection acts to favour an equal number of male and female offspring provided that males and females are equally costly to produce. In the case where the costs of producing males and females differ, the prediction must be phrased in terms of the investment ratio into each sex rather than the sex ratio. For example, parents may be selected to bias the offspring sex ratio towards the cheaper sex, such that the overall resource investment in the sexes is equal.

Hamilton (1967) was the first to identify conditions that violate the assumptions underlying Fisher's (1930) equal allocation theory. Hamilton pointed out that equal allocation is not expected in species where within-group interactions have a differential effect on the fitness of males and females. For example, if brothers compete over a limited number of mates (i.e. local mate competition), parental fitness will increase if more daughters are produced, as shown in studies on parasitic wasps (Hamilton 1967). Thus, selection should favour a sex ratio that is biased towards the sex experiencing the least amount of competition with close kin. In some cases, this is likely to be the dispersing sex (Clark 1978; Silk 1983). Conversely, selection may also favour a sex ratio that is biased towards the sex that improves within-group conditions for kin (i.e. local resource enhancement). This might for example be the helping sex in cooperatively breeding birds (Emlen et al. 1986; Lessells and Avery 1987). Finally, Trivers and Willard (1973) showed that individuals could be selected to adjust the sex of their offspring in response to environmental conditions. When environmental conditions have a differential effect on the fitness of male and female offspring, parents should bias the sex ratio of their offspring towards the sex that contributes more to parental fitness (Trivers and Willard 1973). Such facultative biasing by individual parents can occur despite strong selection for equal investment in daughters and sons in the population.

Theory predicts a number of situations in which individuals should adjust their relative allocation to male and female reproduction. From 1980 onwards there was a profusion of empirical studies testing the various forms of the Trivers–Willard hypothesis (Charnov 1982; Frank 1990; Hardy 2002; West 2009). Many of these studies were conducted on haplodiploid Hymenopterans (ants, bees, wasps), where females can control the sex of their offspring precisely because females develop from fertilized eggs and males develop from unfertilized egg (Trivers and Willard 1973; Charnov 1982; Frank 1990; Hardy 2002). For example, the clearest patterns of individual sex ratio adjustment come from studies on parasitic wasps where females lay a higher proportion of female eggs in large than in small hosts. The reason for this is that there is a strong positive relationship between the size of the host and the number of eggs

The Evolution of Parental Care. First Edition. Edited by Nick J. Royle, Per T. Smiseth, and Mathias Kölliker.

subsequently produced by daughters, whereas the relationship between host size and mating success is weaker in sons (Jones 1982).

In this chapter, I place the topic of sex allocation within the context of parental care by focusing on sex allocation in birds and mammals. Birds and mammals are unique in that several species have been studied in detail with respect to both parental care and sex allocation. First, I give an overview of mechanisms of sex ratio adjustment in birds and mammals. Second, I review a variety of specific social and ecological circumstances that could drive variation in adaptive sex allocation in birds and mammals at both the population and the individual level and discuss how well the observed sex ratio can be explained by traditional sex allocation models. Third, I outline some of the unresolved issues in studies of sex ratio adjustment, suggest alternative predictions for sex ratio adjustment, and discuss future research objectives.

10.2 Sex ratio adjustment in birds and mammals

Traditionally, it was thought that birds, mammals and other vertebrates with chromosomal (i.e. genetic) sex determination would have little scope for skewing the sex ratio at birth, because random meiosis would generate a mean sex ratio of 0.5 (Williams 1979; Charnov 1982; Maynard Smith 1978, 1980; Krackow 1995; Leimar 1996; Pen and Weissing 2002). Thus, the very possibility of adaptive sex ratio manipulation at conception was questioned, leaving selective mortality of male and female eggs or embryos as the only plausible mechanism for sex ratio adjustment (Maynard Smith 1978; Williams 1979; Charnov 1982; Clutton-Brock 1986; Krackow 1995; Palmer 2000). However, this view is challenged by more recent studies on a wide range of birds and mammals that have reported significant parental control of offspring sex ratios (Hardy 2002; West et al. 2002).

In species that exhibit parental care, manipulation of the proportion of male and female offspring is also possible through differential investment in males and females during the parental care period. For example, in species with sexually size dimorphic young, offspring of the larger sex often consume more parental resources than those of the smaller sex (Anderson et al. 1993). As a consequence, offspring of the larger sex are more susceptible to starvation than offspring of the smaller sex. Indeed, offspring of the larger sex have been reported to show slower growth (Velando 2002) and/or greater mortality (Wiebe and Bortolotti 1992; Torres and Drummond 1999) when food availability is low. Parents may adjust the relative food provisioning to male and female offspring depending on environmental conditions. For example, when food availability is low, parents may provision more resources to offspring of the less costly sex to enhance their quality and survival. However, the offspring may also be able to influence the allocation of parental investment by demanding more care from the parents. For example, in elephants, male calves receive more milk from their mothers because they demand more from their mothers than the smaller female calves (Lee and Moss 1986). Furthermore, offspring may indirectly influence parental sex allocation through sibling competition. When food is limited and brood reduction becomes advantageous, early sibling competition may result in biased mortality towards the smaller sex if offspring of the larger sex outcompete offspring of the smaller sex (Chapter 8). This results in brood reduction and an offspring sex ratio that is skewed towards the larger sex. For example, golden eagles (*Aquila chrysaetos*) do not adjust their hatching sex ratio, but in years of poor food availability, there is a post-natal sex bias in favour of the larger and stronger female nestlings that is caused by brood reduction (Bortolotti 1986).

10.3 Difficulties applying the theory

The observed sex ratios in birds and mammals have often been interpreted within the framework of classic sex allocation theory. However, there is poor concordance between observed and expected sex ratios in birds and mammals, which might reflect that both the mechanisms of sex determination and the general life-histories differ between birds and mammals and the haplo-diploid insects upon which the standard models of sex allocation are built (Komdeur and Pen 2002; Pen and Weissing 2002; Sheldon and West 2004). Thus, in order to test sex allocation theory in a given species, it is essential to determine whether the species meets the assumptions of the models and also identify specific conditions that may affect the relative cost of producing each sex or the reproductive potential of the sexes. The following key assumptions are made in sex allocation theory.

First, sex ratio manipulation is assumed to incur no cost to the individual in control (Maynard Smith 1980). However, cost-free sex ratio manipulation seems unlikely in species with chromosomal sex determination because there is no obvious mechanism by which parents can bias sex ratios at fertilization. If sex ratio manipulation requires selective killing of offspring at some point during development, this may be costly as it results in a loss of invested resources (Myers 1978). In the Seychelles warbler (*Acrocephalus sechellensis*), it has recently been demonstrated that sex ratio control arises with virtually no costs (Komdeur et al. 2002; see Box 10.1). Thus, the assumption that sex ratio manipulation incurs no cost may be valid for birds where females may have more control over sex allocation because they are the heterogametic sex. It is less likely that the assumption holds for mammals, where males may have more control over sex determination because they are the heterogametic sex. There is experimental evidence for male contribution to sex ratio adjustments at birth from a study on red deer (*Cervus elaphus*), in which females were artificially inseminated with ejaculates from high and low quality males. The study found that high-quality males produced more sons than low-quality males (Gomendio et al. 2006). There was also a positive correlation between the percentage of morphologically normal spermatozoa and the proportion of male offspring, suggesting that males with a higher proportion of normal spermatozoa may benefit from producing sons that inherit this trait (Gomendio et al. 2006). A possible mechanism by which males may adjust the sex ratio is that ejaculates may differ in the proportion of Y-bearing spermatozoa depending on the male's quality (Chandler et al. 2002), thus resulting in biases in sex ratio at birth. For example, high-quality males could have a higher proportion of Y-bearing spermatozoa in the ejaculate than low-quality males.

Second, sex allocation models assume that generations are non-overlapping and that parents have a fixed total amount of resources for reproduction. However, birds and mammals usually have multiple breeding attempts and the resulting complex interactions between overlapping generations provide challenges when attempting to apply the theory to these species (Cockburn et al. 2002). Birds and mammals face a fundamental decision about how much to invest in a particular reproductive episode, which further complicates predictions for the adaptive sex ratio. For example, it is unclear whether ungulate, marsupial, and primate sex ratios are shaped by the overlapping generations of these species, or by other factors such as polygynous mating systems, competition among related females, cooperation among related females and inheritance of maternal rank by daughters (e.g. Trivers and Willard 1973; Clark 1978; Silk 1983; Cockburn et al. 2002). In this context, parental resources are synonymous with parental investment as defined by Trivers (1972). Sex allocation theory generally assumes that parents have a fixed amount of resources for reproduction. However, this assumption seems unlikely to be met for income breeders, such as animals that bring in food for their young from their environment, in which reproduction is financed using current energetic income. Furthermore, empirical tests of sex allocation theory based on measures of parental investment face the challenge that parental investment is likely to include more than one resource (such as time and energy), and that any single resource can often be invested in different ways (Chapter 1).

Third, sex allocation models assume that environmental conditions are predictable. Females that live in predictable environments will have access to reliable cues to assess factors that influence the optimum sex ratio, such as the amount of competition over reproduction with other females or the amount of resources available for reproduction. However, environments are often variable in both space and time, which would hamper an organism's ability to assess the relevant environmental factors that determine the optimal sex ratio. For example, the longer the period of parental investment, the more likely it is that environmental unpredictability may constrain any offspring sex ratio bias, because it would be more difficult for parents to predict the amount of resources they would have available for rearing offspring. There is some evidence that environmental predictability plays such a role from studies showing larger sex ratio biases in ungulates with shorter gestation periods (West and Sheldon 2002) and primates with shorter maturation times (Schino 2004).

Fourth, sex allocation models assume a single short period of investment and that there are no family conflicts. However, birds and mammals tend to have extended parental care (Chapter 4), which makes it difficult to estimate the relative investment in sons and daughters, especially if differential mortality takes place during the period of parental care (Komdeur and Pen 2002). Furthermore, both parents cooperate to provide care in most birds (Chapter 9), and the two parents may not necessarily agree on the optimal sex allocation (Charnov 1982). If so, the outcome of sexual conflict over the sex ratio may depend on which parent controls what aspect of allocation (Pen and Weissing 2002). For example, male parents may have no control over, or information about, the sex ratio among embryos, but they may have the opportunity to modify sex allocation after hatching or birth. In species with biparental care, selection may also favour specialization whereby each parent invests more in offspring of a particular sex (Lessells 1998). Finally, in mixed-sexed broods or litters, there may be scope for sex-specific sibling competition that may alter the relative reproductive value of each sex due to differences in competitive ability. For example, in birds where nestlings hatch asynchronously and interact aggressively, sibling competition may alter the relative value of each sex, particularly if it leads to siblicide (Chapter 8). If certain combinations of offspring sexes exacerbate the intensity of sibling competition, there should be selection against these combinations because they may increase the probability of brood reduction (Bortolotti 1986; Bednarz and Hayden 1991). However, there may be selection for these same combinations under other circumstances, such as when food is limited and brood reduction becomes advantageous. Given that there may be conflicting selection pressures acting on offspring and parents, and on male and female parents, it is now essential to incorporate the effects of within-family conflict into sex allocation theory (West et al. 2001).

Finally, sex allocation models use a simple measure of fitness. In order to test sex allocation in the field it is necessary to have an adequate measure of fitness. The fitness measure of choice is reproductive value, which is a measure of the long-term contribution to the gene pool (Fisher 1930). Fitness is often estimated in terms of offspring recruitment, but such a simple fitness measure may not always suffice. For example, Leimar (1996) showed that even if high-quality males have higher reproductive success than high-quality females, a sufficiently strong correlation between maternal quality and offspring quality may increase the reproductive value of high-quality daughters above that of high-quality males. Under these circumstances, it may pay for high-quality mothers to produce daughters, as occurs in some primates and ungulates (Silk 1983; Leimar 1996; Sheldon and West 2004), thus reversing Trivers and Willard's (1973) prediction.

10.4 Sex ratio bias at the population level

Many previous studies have examining broad-scale patterns of sex ratio variation by focusing on the mean and variance in population sex ratios (e.g. Williams 1979; Clutton-Brock 1986; Clutton-Brock and Iason 1986; Palmer 2000). The finding that most birds and mammals show primary sex ratios close to equality (e.g. Clutton-Brock 1986; Ewen 2001; Ewen et al. 2004; Donald 2007) appears to support Fisher's equal allocation theory. There are a number of potential explanations for the lack of a bias in the population sex ratio. First, selection on the sex ratio may be very weak, as would be expected if the fitness costs and benefits of producing sons and daughters are similar. Second, there may be several opposing selective forces acting on the sex ratio, resulting in weak net selection (Cockburn et al. 2002). Thus, examining population sex ratios may be a poor test of adaptive sex allocation at the individual level because it is often hard to predict the expected pattern of sex ratio adjustment for a given species, or between species, without detailed data on for example the fitness returns of raising sons and daughters (see Section 10.5).

The problem discussed above is illustrated by two meta-analyses of the literature on sex ratio variation across avian species that came to opposing conclusions with respect to facultative primary sex ratio adjustment in wild bird populations. West and Sheldon (2002) restricted their analysis to studies with clear *a priori* predictions, and concluded that birds can show strong sex ratio biases. In contrast, when Ewen et al. (2004) also included studies with weaker *a priori* predictions, they found no evidence for the general occurrence of avian primary sex ratio adjustment (see also Cassey et al. 2006). The majority of studies make no clear prediction about which examined factor should influence the sex ratio, and in what direction this effect should be. This situation makes the assignment of positive or negative signs to effect sizes an *ad hoc* process. Nevertheless, Ewen et al. (2004) identified a few influential case studies that showed particularly large effect sizes, although it is unclear if these studies represent rare biological exceptions in which the study species indeed exhibits sex ratio control, or whether these studies represent false positives (i.e. statistical type-I errors). It is now crucial to replicate key studies to evaluate the robustness and generality of the reported patterns (Palmer 2000; Griffith et al. 2003). Currently, replication of studies on the same species is scarce (Palmer 2000), which may partly reflect a publication bias towards significant results (Palmer 2000; West 2009). When considering the effects of single traits, it is not necessarily the case that the same pattern should be expected across different populations or species (West and Sheldon 2002). For example, opposite patterns of sex ratio adjustment were found in two populations of the house finch (*Carpodacus mexicanus*) In this species, selection differed between the two populations, with sons and daughters doing better in different laying order positions and mothers adjusting their offspring sex ratio accordingly. In one population, the first-laid eggs were mostly female, while in the other population they were mostly male (Badyaev et al. 2002). The finding that the direction of sex ratio adjustment may differ between populations of the same species clearly illustrates the difficulty of making *a priori* predictions in cases where the fitness benefits of sex ratio adjustment have not been demonstrated.

10.5 Tests of population sex ratio models

One way to improve tests of population sex ratio models is to employ a cost–benefit analysis to make testable predictions about sex ratios under the assumption that facultative sex ratio adjustment will only be favoured when the fitness benefits outweigh the fitness cost. The most extreme and precise sex ratio adjustments are expected in species where the fitness benefits of facultative sex ratio adjustment are high and the costs are low. The benefit of facultative sex ratio variation will depend heavily on the fitness benefits gained from shifting the offspring's sex ratios. Good examples of tests of population sex ratio models involve species where it is possible to estimate the strength of selection for such sex ratio adjustment, for example species with sexual size dimorphism and species that breed cooperatively (e.g. West and Sheldon 2002; Griffin et al. 2005; Benito and Gonzális-Solís 2007).

In sexually size dimorphic species, it is predicted that the sex ratio will be biased towards the smaller and cheaper sex at the end of parental care period in such a way that the total investment of parental resources in the two sexes is equal. Parents could balance the total investment of parental resources in the two sexes by adjusting sex ratios between birth and the end of parental care. Under the assumption that species that are more sexually size dimorphic will have a greater sex difference in demand for parental resources (Trivers and Hare 1976; Clutton-Brock 1985; Magrath et al. 2007), it is expected that the change in offspring sex ratio from birth toward the end of the parental care period should correlate with the investment ratio in male and female offspring. In support of this prediction, a comparative analysis on birds with sexual size dimorphism showed that population-level sex ratios are biased toward the smaller sex at fledging, but that this is not the case at hatching (Pen et al. 2000). However, another analysis on birds where females are the larger sex showed that there was a tendency for parents to produce a higher proportion of sons at both fledging and hatching, although this trend was not statistically significant (Benito and Gonzális-Solís 2007).

There is little evidence for consistent primary sex ratio biases at the population level in sexually size-dimorphic species (Hartley et al. 1999; Radford and Blakey 2000; Magrath et al. 2007), with the exception of a recent study on the highly size-dimorphic blue-footed booby (*Sula nebouxii*; Torres and Drummond 1999). A potential factor that may constrain adaptive sex ratio biases at the population level is increased mortality of offspring of the larger offspring sex (Clutton-Brock et al. 1985; Benito and Gonzális-Solís 2007). If such differences in mortality are not due to parental decisions, they may moderate the strength of selection for sex ratio biases because a higher baseline mortality of the larger sex would reduce the differences in the costs of producing males and females.

For cooperatively breeding species with sex-specific helping behaviour, it is possible to estimate the strength of selection on sex ratio adjustment. Many studies have reported that helpers increase the reproductive success or survival of their breeding parents (Griffin and West 2003; Chapter 12). In cooperative breeders, selection may favour parents that produce more offspring of the helping sex (Pen and Weissing 2000). The strength of selection for such sex ratio adjustment is determined by the benefit provided by helpers, which varies across species (Griffin and West 2003). Helping by philopatric individuals can be thought of as a means for reducing the overall cost of parental investment because such offspring repay their parents. In some cases, helpers have been shown to have large positive effects on the fitness of their parents (Griffin and West 2003), in which case there is strong selection for sex ratio adjustment towards the philopatric and helping sex (Emlen et al. 1986; Lessells and Avery 1987). As expected, a meta-analysis across cooperatively breeding species based on nine species of bird and two species of mammal found that there was a significant sex ratio bias at the population level towards the helping sex in species where the presence of helpers leads to greater fitness benefits to the parents (Griffin et al. 2005; Fig. 10.1).

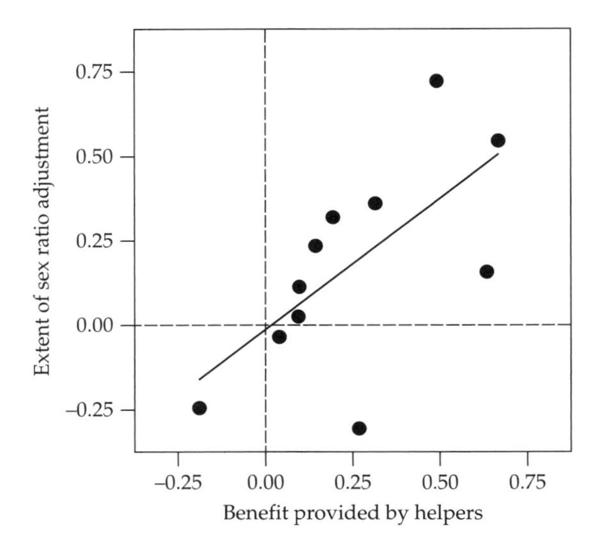

Figure 10.1 Correlation between the extent of sex ratio adjustment and the benefit that helpers provide across cooperatively breeding species. The effect size of the relationship between offspring sex ratio and the number of helpers is plotted against the effect size of the relationship between the benefit of helping (e.g. number of offspring fledged) and the level of helping (e.g. number of helpers). The data points represent nine bird and two mammal species (adapted from Griffin et al. 2005, with permission from The University of Chicago Press).

10.6 Facultative sex ratio variation

In contrast to the general pattern that there is often no sex ratio bias at the population level, there is mounting evidence for facultative variation in sex ratios in species with chromosomal sex determination. Here I review some relevant case studies classified according to the specific factors that were found to correlate with sex ratios.

10.6.1 Food availability

In many animal taxa, the amount of food resources available to the parent determines the amount of care they provide, which in turn influences offspring growth and development (Clutton-Brock 1991; Chapter 1). If the abundance of food has a differential effect on the fitness of male and female offspring, then selection may favour conditional sex allocation. The first to report evidence for conditional sex allocation was Howe (1977), who showed that the sex ratio was more female-biased later in the season in great-tailed grackles

(*Quiscalus mexicanus*). This effect was attributed to increasing differential mortality biased towards offspring of the larger sex (in this case males) due to decreasing food availability. In tawny owls (*Strix aluco*), parents adjust the primary sex ratio within broods in response to the density of prey (voles) on breeding territories (Appleby et al. 1997). In this species, a higher density of prey was associated with a primary sex ratio that was biased toward the larger sex (in this case females), because female, but not male, offspring had enhanced breeding success when the prey density was high. In contrast, a more recent study on tawny owls that included a much larger sample size failed to confirm that more females were produced when food conditions were favourable (Desfor et al. 2007). The inconsistency between the sex ratio allocation patterns within the same species suggests that adaptive sex allocation strategies could differ across populations.

10.6.2 Maternal condition or quality

The first strong evidence for an effect of maternal condition on sex ratio adjustment, as suggested by Trivers and Willard (1973), came from a study on red deer. In this species, high-ranking females were found to consistently bias their sex ratio toward male calves, while low-ranking females produced an excess of daughters (Clutton-Brock et al. 1984). Because high-ranking females were in better condition, they could afford to invest more parental resources in their offspring than low-ranking females. Male offspring produced by high-ranking females stand to gain more from their mother's quality than daughters (Clutton-Brock et al. 1982). The reason for this is that stronger males are better able to defend harems of females during the rutting, and that such males therefore are likely to have a higher reproductive success than males produced by low-ranking females. Since this groundbreaking study, the relationship between maternal condition and offspring sex ratio has been examined in numerous other ungulates with mixed results. Some studies have found support that females in better condition were more likely to produce sons, while others found no relationship or even the opposite pattern with better condition females producing more daughters (Sheldon and West 2004; West 2009).

One possible reason for these apparently inconsistent results is that sex allocation theory can predict sex ratio adjustment in the opposite direction, such as when social rank is inherited from mother to daughter. For example, in the Cape mountain zebra (*Equus zebra zebra*), only daughters appear to benefit from having a high-ranking mother. As expected, this species shows a sex ratio shift in the opposite direction with high-ranking females being more likely to produce daughters (Lloyd and Rasa 1989). Red deer and Cape mountain zebras are therefore at the opposite ends of a continuum ranging from male offspring benefitting from the additional investment provided by high-quality females to female offspring benefitting from such additional investment. Another explanation for the observed differences between studies is that the sex ratio may be affected by maternal age, which in turn may covary with maternal condition (Martin and Festa-Bianchet 2011). In bighorn sheep (*Ovis canadensis*), the effect of maternal condition on sex allocation reverses as mothers age. In this species, young females generally produced a sex ratio biased toward sons while old ewes generally produced a sex ratio biased toward daughters. Old ewes could maximize their survival by either skipping reproduction or producing the cheaper sex (in this case females). One of the few studies where maternal condition was manipulated experimentally was conducted on the highly endangered kakapo (*Strigops habroptilus*), a flightless nocturnal parrot endemic to New Zealand (Clout et al. 2002). In this species, supplementary feeding led to the offspring sex ratio changing from female-biased to male-biased, thus providing evidence for a causal link between female nutritional condition and offspring sex ratio.

It is important to note that, in order to demonstrate adaptive sex allocation, it is necessary to demonstrate that maternal quality has a differential effect on the reproductive value of sons and daughters. At present, there are some studies where both the pattern of sex ratio adjustment and the fitness consequences have been investigated. For example, in the sexually size-dimorphic lesser black-backed gull (*Larus fuscus*), the survival of the larger sex (in this case males) was substantially reduced when they hatched from less well-provisioned eggs (Bolton et al. 1992). Nager et al. (1999) simultaneously reduced maternal condition by removing eggs, thereby inducing females to lay costly replacement eggs that become progressively smaller, and enhanced maternal condition by supplementary feeding. As maternal condition declined, females progressively skewed the sex ratio toward the cheaper sex (in this case females). However, if maternal condition was subsequently enhanced through supplementary feeding, and the quality of replacement eggs was rescued, more males were produced to the extent that there no longer was a sex ratio bias toward females (Nager et al. 1999).

10.6.3 Attractiveness or quality of males

Weatherhead and Robertson (1979) were the first to suggest that females should adjust the sex ratio in response to the attractiveness or quality of their mate. If male attractiveness is causally linked with the quality of sons, either through higher quality care or the inheritance of good genes, sons of attractive males might have a higher reproductive value than daughters. It has therefore been argued that it would be adaptive for females to modify the sex ratio in response to male attractiveness. This idea is similar to the classic Trivers and Willard (1973) argument; the only difference being that it is mate quality rather than maternal condition that influences offspring fitness. Analytical and simulation models showed that irrespective of whether male ornamental traits evolve in response to Fisher's runaway process or female preferences for traits that indicate good genes, females mated to more attractive males should be under selection to produce a higher proportion of sons (Fawcett et al. 2007). This prediction was confirmed by Burley (1981) who found that female zebra finches (*Taeniopygia guttata*) produced a male-biased sex ratio when they were paired to attractive males with brighter bills.

Most studies investigating whether there is a relationship between male attractiveness and the sex ratio have been conducted on birds. These studies have shown that some males enhance their reproductive success by mating with females other than their social partner (Birkhead and Møller 1992). Such behaviour can cause substantial variance in male reproductive success, suggesting that males are more likely to benefit from an increase in mating opportunities than females. Despite considerable research effort, there is mixed evidence for a link between sex ratios and paternal attractiveness with some studies reporting a relationship between male attractiveness and production of sons and others reporting no such relationship (West 2009). There are also indications that the relationship between male attractiveness and sex-ratio variation is inconsistent across years or populations of the same species. For example, a study on blue tits (*Parus caeruleus*) conducted by Sheldon et al. (1999) found that females adjusted the brood sex ratios in response to manipulations of the ultraviolet reflectance of the male's crown feathers, which is a sexually selected trait in this species. Griffith et al. (2003) found correlational evidence confirming this pattern from the same population, but Dreiss et al. (2005) found no association between male plumage colour and offspring sex ratios in a different population of blue tits. Finally, Korsten et al. (2006) replicated the experimental treatment in the study of Sheldon et al. (1999), and found an association between the ultraviolet reflectance of the male's crown feathers and the sex ratio in only one out of two years. This example illustrates inconsistencies in patterns of sex allocation between studies on the same species. Even the findings of Burley (1981) in zebra finches, which initiated this research area, have not been replicated in later studies (e.g. Zann and Runciman 2003; Rutstein et al. 2005); this includes one study using the same experimental design as Burley, but birds from a different population (Von Engelhardt et al. 2004).

There are several reasons why only some studies may report a relationship between male attractiveness and offspring sex ratio. First, the advantage of sex ratio adjustment may vary across populations or species due to variation in factors that influence the benefit of sex ratio adjustment, such as influence of mate attractiveness on offspring fitness. Second, several studies have analysed brood sex ratios in relation to the attractiveness of the female's social mate rather than the actual sire, which may have been an extra-pair mate. If sons benefit primarily by inheriting attractive ornaments from their genetic fathers (as would be the case if females gain indirect benefits from mate preferences), then females should adjust the sex ratio in relation to the ornaments of the sires regardless of whether they are within-pair or extra-pair partners. It is unclear whether females are able to bias the sex of young sired by extra-pair males without also biasing the sex of young sired by the social partner in the same brood. Moreover, theoretical models suggest that the strength of selection for sex-ratio adjustment in relation to male attractiveness is weak, and any costs and constraints on sex ratio adjustment would therefore make it less likely that such sex ratio adjustment would evolve (Fawcett et al. 2007). Indeed, only four studies of birds have reported evidence of a male-biased sex ratio among offspring sired by extra-pair males, which are presumably more attractive (Du and Lu 2010; Kempenaers et al. 1997; Schwarzova et al. 2008; Johnson et al. 2009).

10.6.4 Social environment

The social environment in which a breeding pair lives may also affect sex allocation, for example through local resource competition or through local resource enhancement. Sex allocation in response to local resource competition was initially suggested to explain the male-biased sex ratios in the prosiminian primate *Otolemur crassicaudatus* (Clark 1978). Clark (1978) suggested that parents could limit competition among offspring of the philopatric sex (in this case females) by producing more offspring of the dispersing sex (in this case males). Conversely, parents should allocate more resources towards the philopatric sex when competition over resources is low as in many insects (West et al. 2005). In the common brushtail possum (*Trichosurus vulpecula*), where females are the philopatric sex and males the dispersing sex, females spend the day in hollows in large trees sheltering from predators. Females use a number of such dens within their territory and do not share them with other females, including their daughters or sisters. When the number of den sites on the territory are limited, females produce an excess of males to reduce competition for dens (Johnson et al. 2001). Another example comes from the collared flycatcher *Ficedula albicollis*, where males are the philopatric sex. In this species, males defend breeding territories, which is a limiting resource. If parents bred in areas where potential breeding territories were abundant, the primary sex ratio was biased toward males (Hernquist et al. 2009).

In cooperatively breeding species, where helping may be sex-specific, theoretical models were originally used to predict a population-wide bias in the primary sex ratio (see Section 10.5). However, such models may also be applied at the family or brood level (Trivers and Hare 1976; Pen and Weissing 2000) because the effects of repayment through helping, local resource competition, and local mate competition may operate to varying extent among different families within the population. First, each breeding pair may have different optima depending upon whether they already have some helping offspring, and the fitness effects of additional helpers is likely to show diminishing returns. Second, the territory of the breeding pair may vary in quality, and may not be able to support extra helpers. Helpers may even experience a net loss of inclusive fitness if they consume scarce resources on a territory that otherwise could have been spent on the production of offspring. In both situations, it is predicted that the dominant breeders that lack helpers should bias their offspring sex ratio towards the production of the helping sex, but that the presence of helpers in a social group may be sufficient to cause females to increase the production of the non-helping and dispersing sex. Overall, as shown above, the data show consistent support for this prediction (West and Sheldon 2002; West et al. 2005). For example, offspring sex ratios vary with helper number in the predicted direction in a number of cooperatively breeding birds and mammals (reviewed in West 2009; Box 10.1). However, there are also several species where sex ratios are not adjusted in response to the number of helpers. An explanation for the variation in extent of offspring sex ratio adjustment across species is that species may differ in the extent of benefit and/or costs of sex ratio adjustment (West and Sheldon 2002; Komdeur 2004).

Box 10.1 Sex allocation in the Seychelles warbler

The Seychelles warbler is a rare island endemic that occurred only on Cousin Island in the Seychelles until 1988. Breeding pairs remain together on the same territory for up to nine years. Offspring can delay dispersal from the natal territory and remain to help rear their parents' offspring (Fig. 10.2). Helpers are usually daughters from previous broods (Komdeur 1996; Richardson et al. 2002). Helping includes territory defence, nest building, incubation, nest guarding, and feeding of young. The advantage of having a helper depends on territory quality (measured as insect food abundance; Komdeur 1994; Richardson et al. 2002). Helpers are costly for parents inhabiting poor territories with fewer insects (i.e. less food), because helpers deplete insect prey. On high-quality territories, the presence of one or two helpers increases the reproductive success of breeding pairs, but the presence of more helpers is detrimental. As predicted by the local resource competition hypothesis, pairs breeding on poor territories maximize their fitness by biasing the sex ratio towards males (i.e. the dispersing sex). Pairs breeding on medium-quality territories produced sex ratios around parity. Finally, pairs breeding on high-quality territories without helpers or with one helper biased the sex ratio toward daughters, whereas such pairs that already had two helpers present produced mainly sons (Fig. 10.3).

Helper removal experiments confirm that sex ratio bias is causally linked to helpers. When pairs breeding on high-quality territories had one of their two helpers removed, they switched from producing all sons to producing 83% females (Komdeur et al. 1997). Furthermore, when an additional population was established on nearby Aride Island in 1988 in an effort to conserve the species, data on the same parents transferred between islands confirmed that the sex ratio differences were related to territory quality. Breeding pairs that were transferred from low- to high-quality territories, switched from producing 90% sons to producing 85% daughters. Pairs switched between high-quality territories showed no change in sex ratios, producing 80% daughters before and after the switch (Komdeur et al. 1997). Sex ratio control in the Seychelles warbler appears to be virtually cost-free. Sex-specific embryo mortality between egg laying and hatching can be ruled out because the sex of dead embryos as revealed by molecular techniques was not biased towards the less adaptive sex (Komdeur et al. 1997, 2002). Thus, in these cases, the sex ratio must have been biased inside the mother.

Observational analysis on the fitness consequences of producing sons and daughters by unassisted breeding pairs in different quality territories confirms the advantage of producing sons in low-quality territories and daughters on high-quality territories. This arises through effects due to the offspring helping and through their reproduction. On low-quality territories, neither helping sons nor helping daughters significantly influence the raising of offspring. In contrast, daughters do provide significant help in raising

Figure 10.2 (a) Cousin Island (photograph by L. Brouwer). (b) Seychelles warblers feeding nestling (photograph by D. Ellinger). (c) Seychelles warbler feeding fledgling.

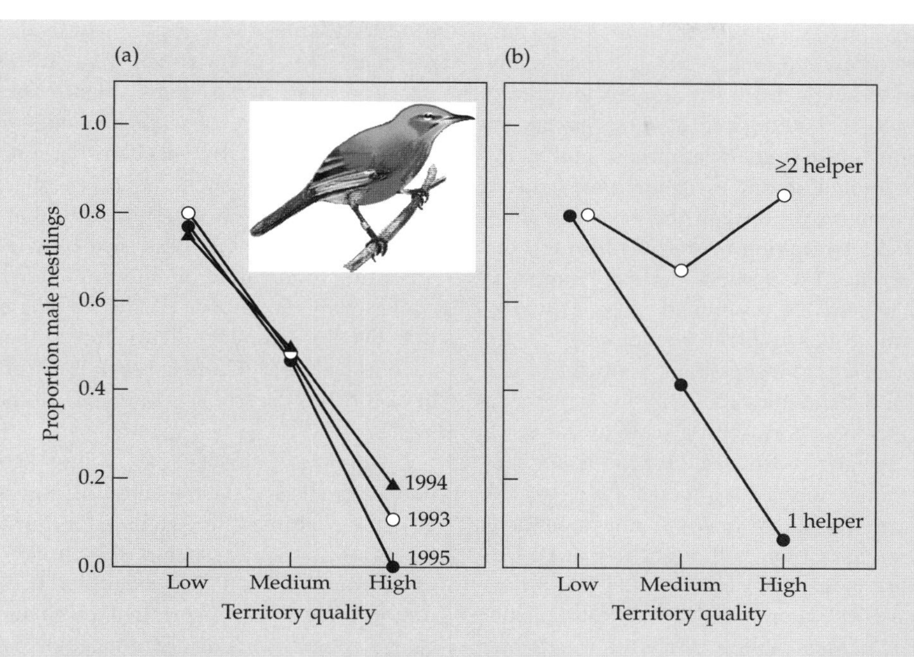

Figure 10.3 (a) Sex ratio of nestlings produced by Seychelles warbler pairs in relation to quality class of breeding territory (territory quality classes: low-quality territory, medium-quality territory; and high-quality territory; 1993–1995). Young were hatched from one-egg clutches only in different years. No additional young were present on the territory. (b) Sex ratio of nestlings produced in relation to quality class of breeding territory when one or two or more helpers were present on the territory (1995) (adapted from Komdeur et al. 1997).

offspring on high-quality territories. Furthermore, territory quality has a greater impact on the breeding success of daughters. This observational evidence was confirmed by a cross-fostering experiment of nestlings. In this experiment, nestlings were swapped between unassisted breeding pairs on low- and high-quality territories that were feeding a nestling of the putatively adaptive sex. By swapping nestlings immediately after hatching, some breeding pairs were forced to raise either a foster son or a foster daughter, allowing comparison of the subsequent inclusive fitness gains for pairs raising male and female offspring. Pairs breeding on low-quality territories that were allocated foster sons gained significantly higher inclusive fitness benefits than those raising foster daughters, while the reverse was true for pairs breeding on high-quality territories (Komdeur 1998). This finding provides strong evidence that sex allocation in the Seychelles warbler is adaptive for the breeding pair.

10.6.5 Sibling competition

Competition among siblings for access to parental resources that are required for the offspring's development and survival (Chapter 8) may have an important effect on sex allocation because males and females often differ in morphology, physiology, and behaviour. Thus, male and female offspring may differ in their requirements for parental resources, the way in which their survival is influenced by the environmental conditions they experience during early development, and/or their competitive abilities. If certain sex ratio combinations among the offspring exacerbate sibling competition, there could be selection against such sex ratios because they could lead to sub-optimal development and wasteful brood reduction. Recent work on a variety of vertebrates suggests that sex-specific sibling interactions may be common, and asymmetrical interactions between males and females in mixed-sexed litters or broods can also have long-term consequences for survival and reproduction (Uller 2006). Sibling competition among males and females may start as early as *in utero*. For example, in polytocous mammals that produce many offspring in a single birth, androgens produced by male foetuses could have adverse effects on the development of female embryos, and competition for limited resources *in utero* may therefore be asymmetrical between the sexes. In mice and other rodents, leaking of steroids through the foetal membranes leads to an increased exposure to steroids in females positioned close to males, resulting in more masculinized phenotypes at birth, reduced sexual attractiveness, and reduced fecundity (Ryan and Vandenbergh 2002). In Soay sheep (*Ovis aries*), females born with a male twin have a lower birth weight and lifetime reproductive success than females born with a female twin. In contrast, the birth weight of males is not dependent on the sex of their twin, although it is unknown whether there is an effect on their lifetime reproductive success (Korsten et al. 2009). Similar results showing that *in utero* sibling competition between males and female can have consequences for birth weight have been reported for humans (James 2002) and other mammals (Kühl et al. 2007).

Sibling competition between male and female offspring may also occur after birth, for example due to sex differences in food requirements. A sex ratio biased toward the larger and more demanding sex can lead to reduced growth and survival for individual offspring of that sex (reviewed in Uller 2006). For example, in the lesser black-backed gull, the pre-fledgling survival of offspring of the larger sex (in this case males) was strongly reduced in an experimental brood comprising male offspring only, while the pre-fledgling survival of the smaller sex (in this case females) was unaffected by the sex composition of the brood (Nager et al. 2000).

When the offspring's fitness is dependent on the sex composition of the brood before and/or after birth, selection may favour maternal adjustments of within-brood sex ratios. However, there is currently no clear evidence for sex ratio adjustment among broods in response to sex-specific sibling interactions (e.g. Cockburn et al. 2002; Krackow 2002). For example, there is no evidence for a bias against mixed-sex litters in Soay sheep even though selection is expected to favour females that produced same-sex litters. In Soay sheep, twins are relatively rare (15%), which may further weaken selection against mixed-sex litters (Korsten et al. 2009). The absence of such an adjustment could also be explained by mechanistic constraints or some as yet unknown opposing selective pressures (Uller 2003). It is noteworthy that the clearest evidence for adaptive sex ratio adjustment according to maternal or environmental conditions comes from species that produce a single offspring that develops independently from other siblings, such as red deer (Clutton-Brock et al. 1984; Kruuk et al. 1999) or Seychelles warblers (Komdeur et al. 1997).

Deleterious effects of sibling competition between male and female offspring may be reduced if parents adjust the hatching or birth order of males and females within the brood (reviewed in Carranza 2004). For example, in some size dimorphic species, offspring of the smaller sex may be at a considerable disadvantage when competing for food against offspring of the larger sex (Stamps 1990). Mothers may seek to offset this disadvantage by ensuring that the members of this sex are produced earlier in the hatching order. For example, in the Harris's hawk (*Parabuteo unicinctus*), males are more likely to hatch first in the clutch than their larger sisters, and this has been interpreted as a mechanism to avoid maladaptive brood-reduction (Bednarz and Hayden 1991). Furthermore, parents may mitigate the adverse consequences of sibling competition by preferentially provisioning food to offspring of the smaller sex to counteract the effects of sex-biased sibling competition.

10.6.6 Sexual conflict

Parents may rarely have an equal degree of control over the sex ratio of the offspring they produce. In general, the heterogametic sex might be expected to have some control through influences on the segregation of the sex chromosomes at meiosis, while females might be expected to have some control through sex-specific fertilization and/or embryo resorption or abortion. Female birds are expected to have greater control over the sex ratio at hatching than female mammals. The reason is that females are the heterogametic sex in birds, and therefore they might exercise control both through influences on the segregation of the sex chromosomes and through sex-specific embryo resorption or abortion. In contrast, males are the heterogametic sex in mammals, leaving females with control through sex-specific fertilization and/or embryo resorption or abortion. A study on red deer suggests that male mammals may influence the sex ratio of their offspring (Gomendio et al. 2006), thus creating an unforeseen evolutionary scenario that includes conflicts of interest between males and females. For instance, a fertile male may benefit from producing sons, but the costs of raising a male may be high for a female in poor physical condition (e.g. Gomendio et al. 1990). This level of sexual conflict between parents may improve our ability to explain biases in sex ratio at birth.

With sufficient scope for sexual conflict over the sex ratio of offspring, mathematical models suggest that the optimal sex ratio bias may depend on which parent is in control (Mesterton-Gibbons and Hardy 2001). Especially in species with distinct sex roles, it is straightforward to envisage scenarios in which the fitness of the parents is not affected equally by the offspring sex ratio. For example, in the Eurasian sparrow hawk (*Accipiter nisus*) where the parent of the smaller sex (in this case males) does most of the food provisioning and the parent of the larger sex (in this case females) guards the nest, the greater energy consumption by female offspring in the nestling phase (Vedder et al. 2005) may have a greater effect on the residual reproductive value of the male parent.

In the context of sexual conflict in cooperative breeding systems, the value of sons and daughters may depend on the costs and benefits for parents of having helpers (Pen and Weissing 2000), which may not necessarily be equal for male and female parents. For example, in the superb fairy wren (*Malurus cyaneus*), the presence of male helpers enhances the survival probability of the breeding female to the next year (Russell et al. 2007), while the dominant male loses more paternity to extra-group males (Mulder et al. 1994). In this species, male helpers may release the female from dependency on care from the dominant male, giving her more freedom to engage in extra-pair copulations. Hence, while females obviously benefit from producing the helping sex, males may suffer through a greater loss of paternity.

10.7 Concluding remarks and future directions

In this chapter I have reviewed studies on sex allocation linked to parental care in birds and mammals. The review highlights that evidence for adaptive sex-ratio adjustment is equivocal. Sex ratio biases may occur at birth, but may also occur during the period after birth when parents provide care. Whether parents adjust the sex ratio before or after birth may depend on the degree to which environmental conditions are predictable. When conditions are unpredictable, parents may do better by adjusting the sex ratio after birth when the parents have access to better information on the actual food abundance. Differential parental care towards sons and daughters is evident in many sexually dimorphic species, and may reflect the fact that parents have been selected to respond to the physiological needs of their male and female offspring (Chapter 7). However, other studies report no evidence for a sex ratio bias, either at birth or at the end of parental care period, despite the fact that the conditions predicted to favour adaptive sex ratio adjustments are satisfied. This discrepancy between the results of studies conducted under what appear to be similar conditions emphasizes that we need a more sophisticated theory for an understanding of when offspring sex ratio manipulation should be expected, as well as more empirical data on when sex ratio manipulation occurs. Furthermore, we need to recognize that the predictions from sex allocation theory ultimately are about the total investment of resources into male and female offspring rather than the sex ratio.

In addition, there is a need for more experimental and long-term studies of the fitness effects of sex ratio adjustments. Experimental studies are necessary to get a better understanding of the causal and functional significance of sex allocation, including studies seeking to disentangle the extent to which sex ratio variation is generated through mechanisms operating before and after birth. Although experimental studies on brood sex ratio variation in birds and mammals are scarce, the first promising results are now emerging (Komdeur and Pen 2002; Robert and Schwanz 2011). Such experimental studies should be designed to target the factor that is assumed to favour adjustments to the sex ratio, and the resulting changes in sex ratio adjustment can subsequently be observed. In cases where a significant effect is reported, such studies provide good evidence for a causal relationship between the target factor and the sex ratio. However, it should be noted that non-significant results do not necessarily justify the conclusion that there is no causal relationship between the target factor and the sex ratio, because concurrent selection pressures other than the manipulated factor may potentially mask the effect of the manipulated factor on sex allocation (Grindstaff et al. 2001; Alonzo and Sheldon 2011). For example, in cooperatively species, it is possible that local resource enhancement and local resource competition operate simultaneously because different groups may be exposed differently to factors such as local competition and number of helpers present in the group. To address this issue, research should focus on quantifying and separating the effects of the factor under investigation with other factors on sex allocation strategies.

Many studies of sex ratio variation draw strong inferences about the adaptive value of sex ratio strategies without having access to sufficient information on the effect of sex allocation on individual fitness. Therefore, it is also important to obtain more detailed information on the fitness functions of both parents and male and female offspring (Leimar 1996; Pen and Weissing 2000). The lack of experimental work examining the fitness effects of sex allocation does not seem to be due to difficulties associated with conducting experiments but may reflect the difficulties associated with how to measure fitness effects (e.g. Komdeur 2004; Alonzo and Sheldon 2011). Until now, most studies have focused on short-term fitness effects of sex allocation, which may not accurately reflect the long-term fitness consequences. The reason for this is that the environment in which an individual lives is likely to vary both spatially and temporally and, as such, the short-term fitness benefits for offspring and parents of both sexes can be offset later by, for example, increased local competition between individuals of the same sex (West et al. 2001). For the majority of populations, information on the long-term fitness effects of sex allocation for parents and male and female offspring is not available and may prove difficult to obtain (Lessells et al. 1996). One should keep in mind that estimates of fitness should ideally include the lifetime reproductive success of all sons and all daughters produced over the female parent's and the male parent's lifetime.

A major task for both theoretical and empirical work on sex allocation is to consider the influence of variable selection pressures and constraints when applying the theory to particular species. There is a need for theoretical models that can predict the observed variation in the amount and precision of sex ratio manipulation in response to the specific factors that are thought to influence the selection pressure. In addition, there is also a need for empirical work that tests these revised theoretical models. Another exciting area of research on sex allocation concerns the intersection between proximate and ultimate explanations of sex allocation. Such research may help explain how and why animals deviate from equal sex allocation, as well as the degree to which the underlying mechanisms evolve both within and between species. Once more studies on sex allocation are available, a meta-analysis should examine the association between specific factors and sex ratio adjustment. Such an analysis would benefit from clear predictions about the relationship between a factor of interest and offspring sex ratios. Given the variable conditions encountered by different species, it might be unreasonable to assume that there would be a common effect size across all studies. Hence, caution is required when pooling studies in a meta-analysis. A first step in this direction would be to conduct a meta-analysis on the relationship between male attractiveness and the sex ratio of offspring, because there are now a substantial number of published studies on this issue. A non-formal summary of the published work suggests a trend towards the production of sons when mating more attractive males (reviewed in West 2009), but a meta-analysis would be needed to confirm this pattern.

Acknowledgements

I am grateful to Per Smiseth, Nick Royle, and Mathias Kölliker for the invitation to write this chapter. I thank many colleagues who have helped shape my thoughts for this chapter. I thank Hannah Dugdale, Tim Fawcett, Cor Dijkstra, and Reinder Radersma for their helpful and constructive comments. Finally, I acknowledge Ian Hardy, Kate Arnold, Per Smiseth and an anonymous reviewer for constructive comments on an earlier draft of this chapter.

References

Alonzo, S. H. and Sheldon, B. C. (2011). Population density, social behaviour and sex allocation. In T. Székely, A. J. Moore, and J. Komdeur, eds. *Social Behaviour: Genes, Ecology and Evolution*, pp. 474–88. Cambridge University Press, Cambridge.

Anderson, D. J., Budde, C., Apanius, V., Martinez Gomez, J. E., Bird, D., and Weathers, W. W. (1993). Prey size influences female competitive dominance in nestling American kestrels *Falca sparverius*. *Ecology* 74, 367–76.

Appleby, B. M., Petty, S. J., Blakey, J. K., Rainey, P., and Macdonald, D. W. (1997). Does variation of sex ratio enhance reproductive success of offspring in tawny owls (*Strix aluco*)? *Proceedings of the Royal Society of London, Series B* 264, 1111–16.

Badyaev, A. V., Hill, G. E., Beck, M. L., Dervan, A. A., Duckworth, R. A., McGraw, K. J., Nolan, P. M., and Whittingham, L. A. (2002). Sex-biased hatching order and adaptive population divergence in a passerine bird. *Science* 295, 316–18.

Bednarz, J. C. and Hayden, T. J. (1991). Skewed brood sex ratio and sex-biased hatching sequence in Harris's hawks. *American Naturalist* 137, 116–32.

Benito, M. M. and Gonzális-Solís, J. (2007). Sex ratio, sex-specific chick mortality, and sexual size dimorphism in birds. *Journal of Evolutionary Biology* 20, 1333–8.

Birkhead, T. R. and Möller, A. P. (1992). *Sperm Competition in Birds: Evolutionary Causes and Consequences*. Academic Press, London.

Bolton, M., Houston, C., and Monaghan, P. (1992). Nutritional constraints on egg formation in the lesser black-backed gull: an experimental study. *Journal of Animal Ecology* 61, 521–32.

Bortolotti, G. R. (1986). Influence of sibling competition on nestling sex ratios of sexually dimorphic birds. *American Naturalist* 127, 495–507.

Burley, N. (1981). Sex ratio manipulation and selection for attractiveness. *Science* 211, 721–2.

Carranza, J. (2004). Sex allocation within broods: the intra-brood sharing-out hypothesis. *Behavioural Ecology* 15, 223–32.

Cassey, P., Ewen, J. G., and Møller, A. P. (2006). Revised evidence for facultative sex ratio adjustment in birds: a correction. *Proceedings of the Royal Society of London, Series B* 273, 3129–30.

Chandler, J. E., Canal, A. M., Paul, J. B., and Moser, E. B. (2002). Collection frequency affects percent Y-chromosome bearing sperm, sperm head area and quality of bovine ejaculates. *Theriogenology* 57, 1327–46.

Charnov, E. L. (1982). *The Theory of Sex Allocation*. Princeton University Press, Princeton.

Clark, A. B. (1978). Sex ratio and local resource competition in a prosimian primate. *Science* 201, 163–5.

Clout, M. N., Elliott, G. P., and Robertson, B. C. (2002). Effects of supplementary feeding on the offspring sex ratio of kakapo: a dilemma for the conservation of a polygynous parrot. *Biological Conservation* 107, 13–18.

Clutton-Brock, T. H. (1985). Size, sexual dimorphism and polygamy in primates. In W. L. Jungers, ed. *Size and Scaling in Primate Biology*, pp. 211–37. Plenum Press, New York.

Clutton-Brock, T. H. (1986). Sex ratio variation in birds. *Ibis* 128, 317–29.

Clutton-Brock, T. H. (1991). *The Evolution of Parental Care*. Princeton University Press, Princeton.

Clutton-Brock, T. H., Albon, S. D., and Guinness, F. E. (1984). Maternal dominance, breeding success, and birth sex ratios in red deer. *Nature* 308, 358–60.

Clutton-Brock, T. H., Albon, S. D., and Guinness, F. E. (1985). Parental investment and sex differences in juvenile mortality in birds and mammals. *Nature* 313, 131–3.

Clutton-Brock, T. H., Guiness, F. E., and Albon, S. D. (1982). *Red Deer: Behavior and Ecology of Two Sexes*. University of Chicago Press, Chicago.

Clutton-Brock, T. H. and Iason, G. R. (1986). Sex ratio variation in mammals. *Quarterly Review of Biology* 61, 339–74.

Cockburn, A., Legge, S., and Double, M. C. (2002). Sex ratios in birds and mammals: can the hypotheses be disentangled? In I. C. W. Hardy, ed. *Sex Ratios: Concepts and Research Methods*, pp. 266–86. Cambridge University Press, Cambridge.

Desfor, K. B., Boomsma, J. J., and Sunde, P. (2007). Tawny owls *Strix aluco* with reliable food supply produce male-biased broods. *Ibis* 149, 98–105.

Donald, P. F. (2007). Adult sex ratios in wild bird populations. *Ibis* 149, 671–92.

Dreiss, A., Richard, M., Moyen, F., White, J., Møller, A. P., and Danchin, E. (2005). Sex ratio and male sexual characters in a population of blue tits, *Parus caeruleus*. *Behavioural Ecology* 17, 13–19.

Du, B. and Lu, X. (2010). Sex allocation and paternity in a cooperatively breeding passerine: evidence for the male attractiveness hypothesis? *Behavioural Ecology and Sociobiology* 64, 1631–9.

Emlen, S. T., Emlen, M., and Levin, S. A. (1986). Sex ratio selection in species with helpers-at-the-nest. *American Naturalist* 127, 1–8.

Ewen, J. G. (2001). *Primary sex ratio variation in the Meliphagidae (honeyeaters)*. Ph. D. dissertation, La Trobe University, Melbourne, Australia.

Ewen, J. G., Cassey, P., and Møller, A. P. (2004). Facultative primary sex ratio variation: a lack of evidence in birds? *Proceedings of the Royal Society of London, Series B* 271, 1277–82.

Fawcett, T. W., Kuijper, B., Pen, I., and Weissing, F. J. (2007). Should attractive males have more sons? *Behavioural Ecology* 18, 71–80.

Fisher, R. A. (1930). *The Genetical Theory of Natural Selection*. Clarendon, Oxford.

Frank, S. A. (1990). Sex allocation theory for birds and mammals. *Annual Review of Ecology, Evolution, and Systematics* 21, 13–56.

Gomendio, M, Clutton-Brock, T. H., Albon, S. D., Guinness, F. E., and Simpson, M. J. A. (1990). Mammalian sex ratios and variation in costs of rearing sons and daughters. *Nature* 343, 261–3.

Gomendio, M., Malo, A. F., Soler, A. J., Fernández-Santos, M. R., Esteso, M. C., García, A. J., Roldan, E. R. S., and Garde, J. (2006). Male fertility and sex ratio at birth in red deer. *Science*, 314, 1445–7.

Griffin, A. S., Sheldon, B. C., and West, S. A. (2005). Cooperative breeders adjust offspring sex ratios to produce helpful helpers. *American Naturalist* 166, 628–32.

Griffin, A. S. and West, S. A. (2003). Kin discrimination and the benefit of helping in cooperatively breeding vertebrates. *Science* 302, 634–6.

Griffith, S. C., Örnborg, J., Russell, A. F., Andersson, S., and Sheldon, B. C. (2003). Correlations between ultraviolet coloration, overwinter survival and offspring sex ratio in the blue tit. *Journal of Evolutionary Biology* 16, 1045–54.

Grindstaff, J. L., Buerkle, C. A., Casto, J. M., Nolan Jr V., and Ketterson, D. (2001). Offspring sex ratio is unrelated to male attractiveness in dark-eyed juncos (*Junco hyemalis*). *Behavioural Ecology and Sociobiology* 50, 312–16.

Hamilton, W. D. (1967). Extraordinary sex ratios. *Science* 156, 477–88.

Hardy, I. C. W. (2002). *Sex Ratios: Concepts and Research Methods*. Cambridge University Press, Cambridge.

Hartley, L. R., Griffith, S. C., Wilson, K., Shepherd, M., and Burke, T. (1999). Nestling sex ratios in the polygynously breeding corn bunting, *Miliaria calancra*. *Journal of Avian Biology* 30, 7–14.

Hernquist, M. B., Hjernquist, K. A. T., Forsman, J. T., and Gustafsson, L. (2009). Sex allocation in response to local resource competition over breeding territories. *Behavioral Ecology* 20, 335–9.

Howe, H. F. (1977). Sex-ratio adjustment in the common grackle. *Science* 198, 744–6.

James, W. H. (2002). Birth weight in dizygotic twins. *Twin Research* 5, 309.

Johnson, C. N., Clinchy, M., Taylor, A. C., Krebs, C. J., Jarman, P. J., Payne, A., and Ritchie, E. G. (2001). Adjustment of offspring sex ratios in relation to the availability of resources for philopatric offspring in the common brushtail possum. *Proceedings of the Royal Society of London, Series B* 268, 2001–6.

Johnson, L. S., Thompson, C. F., Sakaluk, S. K., Neuhäuser, M., Johnson, B. G. P., Soukup, S. S., Forsythe, S. J. and Masters, B. S. (2009). Extra-pair young in house wren broods are more likely to be male than female. *Proceedings of the Royal Society of London, Series B* 276, 2285–9.

Jones, W. T. (1982). Sex ratio and host size in a parasitic wasp. *Behavioural Ecology and Sociobiology* 10, 207–10.

Kempenaers, B., Congdon, B., Boag, P. and Robertson, R. J. (1997). Extrapair paternity and egg hatchability in tree swallows: evidence for the genetic compatibility hypothesis. *Behavioural Ecology* 10, 304–11.

Komdeur, J. (1994). Experimental evidence for helping and hindering by previous offspring in the cooperative breeding Seychelles warbler (*Acrocephalus sechellensis*). *Behavioural Ecology and Sociobiology* 34, 31–42.

Komdeur, J. (1996). Facultative sex ratio bias in the offspring of Seychelles warblers. *Proceedings of the Royal Society of London, Series B* 263, 661–6.

Komdeur, J. (1998). Long-term fitness benefits of egg sex modification by the Seychelles warbler. *Ecology Letters* 1, 56–62.

Komdeur, J. (2004). Sex-ratio manipulation. In W. D. Koenig and J. L. Dickinson, eds. *Ecology and Evolution of Cooperative Breeding in Birds*, pp. 102–16. Cambridge University Press, Cambridge.

Komdeur, J., Daan, S., Tinbergen, J. M., and Mateman, C. (1997). Extreme adaptive modification in sex ratio of the Seychelles warbler's eggs. *Nature* 385, 522–5.

Komdeur, J., Magrath, M. J. L. and Krackow, S. (2002). Preovulation control of hatchling sex ratio in the Seychelles warbler. *Proceedings of the Royal Society of London, Series B* 357, 373–86.

Komdeur, J. and Pen, I. (2002). Adaptive sex allocation in birds: the complexities of linking theory and practice. *Philosophical Transactions of the Royal Society B* 357, 373–80.

Korsten, P., Clutton-Brock, T., Pilkington, J. G., Pemberton, J. M., and Kruuk, L. E. B. (2009). Sexual conflict in twins: male co-twins reduce fitness of female Soay sheep. *Biology Letters* 5, 663–6.

Korsten, P., Lessells, C. M., Mateman, A. C., van der Velde, M., and Komdeur, J. (2006). Primary sex-ratio adjustment to experimentally reduced male UV attractiveness in blue tits. *Behavioural Ecology* 17, 539–46.

Krackow, S. (1995). Potential mechanisms for sex ratio adjustment in mammals and birds. *Biological Reviews* 70, 225–41.

Krackow, S. (2002). Why parental sex ratio manipulation is rare in higher vertebrates. *Ethology* 108, 1041–56.

Kruuk, L. E. B., Clutton-Brock, T. H., Albon, S. D., Pemberton, J. M., and Guinness, F. E. (1999). Population density affects sex ratio variation in red deer. *Nature* 399, 459–61.

Kühl, A., Mysterud, A., Erdnenov, G. I., Lushchekina, A. A., Grachev, I. A., Bekenov, A. B., and Milner-Gulland, E. J. (2007). The 'big spenders' of the steppe: sex-specific maternal allocation and twinning in the Saiga antelope. *Proceedings of the Royal Society of London, Series B* 274, 1293–9.

Lee, P. C. and Moss, C. J. (1986). Early maternal investment in male and female African elephant calves. *Behavioural Ecology and Sociobiology* 18, 353–61.

Leimar, O. (1996). Life history analysis of the Trivers and Willard sex ratio problem. *Behavioural Ecology* 7, 316–25.

Lessells, C. M. (1998). A theoretical framework for sex-biased parental care. *Animal Behaviour* 56, 395–407.

Lessells, C. M. and Avery, M. I. (1987). Sex ratio selection in species with helpers at the nest: some extensions of the repayment model. *American Naturalist* 129, 610–20.

Lessells, C. M., Mateman, A. C., and Visser, J. (1996). Great tit hatchling sex ratios. *Journal of Avian Biology* 27, 135–42.

Lloyd, P. H. and Rasa, O. A. E. (1989). Status, reproductive success and fitness in Cape mountain zebras (*Equus zebra zebra*). *Behavioural Ecology and Sociobiology* 25, 411–20.

Magrath, M. J. L., Lieshout, E. V., Pen, I., Visser, G. H., and Komdeur, J. (2007). Estimating expenditure on male and female offspring in a sexually size-dimorphic bird: a comparison of different methods. *Journal of Animal Ecology* 76, 1169–80.

Martin, J. G. A. and Festa-Bianchet, M. (2011). Sex ratio bias and reproductive strategies: What sex to produce when? *Ecology* 92, 441–9.

Maynard Smith, J. (1978). *The Evolution of Sex*. Cambridge University Press, Cambridge.

Maynard Smith, J. (1980). A new theory of sexual investment. *Behavioural Ecology and Sociobiology* 7, 247–51.

Mesterton-Gibbons, M. and Hardy, I. C. W. (2001). A polymorphic effect of sexually differential production costs when one parent controls the sex ratio. *Proceedings of the Royal Society of London, Series B* 68, 1429–34.

Mulder, R. A., Dunn, P. O., Cockburn, A., Lazenby-Cohen, K. A., and Howell, M. J. (1994). Helpers liberate female fairy-wrens from constraints on extra-pair mate choice. *Proceedings of the Royal Society of London, Series B* 255, 233–229.

Myers, J. H. (1978). Sex ratio adjustment under food stress: maximization of quality or numbers of offspring. *American Naturalist* 112, 381–8.

Nager, R. G., Managhan, P., Griffiths, R., Houston, D. C., and Dawson, R. (1999). Experimental demonstration that offspring sex ratio varies with maternal condition. *Proceedings of the National Academy of Sciences of the United States of America* 96, 570–3.

Nager, R. G., Monaghan, P., Houston, D. C., and Genovart, M. (2000). Parental condition, brood sex ratio and differential young survival: an experimental study in gulls (*Larus fuscus*). *Behavioural Ecology and Sociobiology* 48, 452–7.

Palmer, A. R. (2000). Quasireplication and the contract of error: lessons from sex ratios, heritabilities and fluctuating asymmetry. *Annual Review of Ecology, Evolution, and Systematics* 31, 441–80.

Pen, I. and Weissing, F. J. (2000). Sex ratio optimizations with helpers at the nest. *Proceedings of the Royal Society of London, Series B* 267, 539–44.

Pen, I. and Weissing, F. J. (2002). Optimal sex allocation: steps towards a mechanistic theory. In I. C. W. Hardy, ed. *Sex Ratios: Concepts and Research Methods*, pp. 26–47. Cambridge University Press, Cambridge.

Pen, I., Weissing, F. J., Dijkstra, C., and Daan, S. (2000). Sex ratios and sex-biased mortality in birds. In I. Pen *Sex allocation in a life history context*. PhD thesis, University of Groningen, The Netherlands.

Radford, A. N. and Blakey, J. K. (2000). Is variation in brood sex ratios adaptive in the great tit (*Parus major*)? *Behavioural Ecology* 11, 294–8.

Richardson, D. S., Burke, T., and Komdeur, J. (2002). Direct benefits and the evolution of female-biased cooperative breeding in Seychelles warblers. *Evolution* 56, 2313–21.

Robert, K. A. and Schwanz, L. E. (2011). Emerging sex allocation research in mammals: marsupials and the pouch advantage. *Mammalian Review* 41, 1–22.

Russell, A. F., Langmore, N. E., Cockburn, A., Astheimer, L. B., and Kilner, R. M. (2007). Reduced egg investment can conceal helper effects in cooperatively breeding birds. *Science* 317, 941–4.

Rutstein, A. N., Gorman, H. E., Arnold, K. E., Gilbert, L., Orr, K. J., Adam, A., Nager, R., and Graves, J. A. (2005).

Sex allocation in response to paternal attractiveness in the zebra finch. *Behavioural Ecology* 16, 763–9.

Ryan, B. C. and Vandenbergh, J. G. (2002). Intrauterine position effects. *Neuroscience and Biobehavioral Reviews* 26, 665–78.

Schino, G. (2004). Birth sex ratio and social rank: consistency and variability within and between primate groups. *Behavioural Ecology* 15, 850–6.

Schwarzova, L., Simek, J., Coppack, T., and Tryjanowski, P. (2008). Male-biased sex of extra pair young in the socially monogamous redbacked shrike *Lanius collurio*. *Acta Ornithologica* 43, 235–40.

Sheldon, B. C., Anderson, S., Griffith, S. C., Ornborg, J., and Sendecka, J. (1999). Ultraviolet colour variation influences blue tit sex ratios. *Nature* 402, 874–7.

Sheldon, B. C. and West, S. A. (2004). Maternal dominance, maternal condition, and offspring sex ratio in ungulate mammals. *American Naturalist* 163, 40–54.

Silk, J. B. (1983). Local resource competition and facultative adjustment of the sex-ratios in relation to competitive abilities. *American Naturalist* 121, 56–66.

Stamps, J. A. (1990). When should avian parents differentially provision sons and daughters? *American Naturalist* 135, 671–85.

Torres, R. and Drummond, H. (1999). Variably male-biased sex ratio in a marine bird with females larger than males. *Oecologia* 118, 16–22.

Trivers, R. L. (1972). Parental investment and sexual selection. In B. Campbell, ed. *Sexual Selection and the Descent of Man*, pp. 136–79. Aldine, Chicago.

Trivers, R. L. and Hare, H. (1976). Haplodiploidy and the evolution of the social insects. *Science* 191, 249–63.

Trivers, R. L. and Willard, D. E. (1973). Natural selection of parental ability to vary the sex ratio of offspring. *Science* 179, 90–2.

Uller, T. (2003). Viviparity as a constraint on sex ratio evolution. *Evolution* 57, 927–31.

Uller, T. (2006). Sex-specific sibling interactions and offspring fitness in vertebrates: patterns and implications for maternal sex ratios. *Biological Reviews* 81, 207–17.

Vedder, O, Dekker, A. L., Visser, H. G., and Dijkstra, C. (2005). Sex-specific energy requirements in nestlings of an extremely sexually size dimorphic bird, the European sparrowhawk (*Accipiter nisus*). *Behavioural Ecology and Sociobiology* 58, 429–36.

Velando, A. (2002). Experimental manipulation of maternal effort produces differential effects in sons and daughters: implications for adaptive sex ratios in the blue-footed booby. *Behavioural Ecology* 13, 443–449.

Von Engelhardt, N., Witte, K., Zann, R., Groothuis, T. G. G., Weissing, F. J., Daan, S., Dijkstra, C., and Fawcett, T. W. (2004). Sex ratio manipulation in colour-banded populations of zebra finches. In N. von Engelhard *Proximate control of avian sex allocation: a study on zebra finches*, pp. 13–29. PhD thesis, University of Groningen.

Weatherhead, P. J. and Robertson, R. J. (1979). Offspring quality and the polygyny threshold: 'the sexy son hypothesis'. *American Naturalist* 113, 201–8.

West, S. A. (2009). *Sex Allocation*. Princeton University Press, Princeton.

West, S. A., Murray, M.G, Machado, C., Griffin, A. S., and Herre, E. A. (2001). Testing Hamilton's rule with competition between relatives. *Nature* 409, 510–13.

West, S. A., Reece, S. E., and Sheldon, B. C. (2002). Sex ratios. *Heredity* 88, 117–24.

West, S. A. and Sheldon, B. C. (2002). Constraints in the evolution of sex ratio adjustment. *Science* 295, 1685–8.

West, S. A., Shuker, D. M., and Sheldon, B. C. (2005). Sex-ratio adjustment when relatives interact: a test of constraints on adaptation. *Evolution* 59, 1211–28.

Wiebe, K. L. and Bortolotti, G. R. (1992). Facultative sex ratio manipulation in American kestrels. *Behavioural Ecology and Sociobiology* 30, 379–86.

Williams, G. C. (1979). On the question of adaptive sex ratio in outcrossed vertebrates. *Proceedings of the Royal Society of London, Series B* 205, 567–80.

Zann, R. and Runciman, D. (2003). Primary sex ratios in zebra finches: no evidence for adaptive manipulation in wild and semi-domesticated populations. *Behavioural Ecology and Sociobiology* 54, 294–302.

CHAPTER 11

Paternity, maternity, and parental care

Suzanne H. Alonzo and Hope Klug

11.1 An overview of parentage and parental effort

Extensive research has focused on asking whether and how much care parents should provide to their offspring (Chapters 1 and 2). Multiple maternity or paternity (i.e. more than one female or male having sired offspring within a brood) complicates expected patterns of parental care because individuals may provide care for unrelated young (Westneat and Sherman 1993; Houston and McNamara 2002; Sheldon 2002; Alonzo 2010). Selection on parental effort (see Chapter 1 for a definition) is made even more complex when both sexes provide care because conflict between males and females over individual effort can arise (Houston et al. 2005; Alonzo 2010, Chapter 9). In general, it is argued that individuals should invest more in the care of young to which they are more likely to be related. Empirical results have been mixed, however, implying that the relationship between parentage and care is not so simple (Kempenaers and Sheldon 1997; Sheldon 2002; Alonzo 2010). Following a general overview, we discuss theoretical predictions and empirical patterns relating paternity and maternity to parental effort. As we focus on this question, though, it will become clear that to understand how parentage affects parental care one must consider how patterns of relatedness and the costs and benefits of care arise from interactions within and between the sexes. We finish, therefore, by asking what is missing from our understanding of the relationship between parentage and parental effort and where further research could improve our knowledge of this complex but fundamental evolutionary question.

11.1.1 The effect of relatedness on parental effort

As described above, selection should favour higher parental effort, all else being equal, when the parent is more likely to be related to the offspring in its care (Trivers 1972). This argument is a general extension of Hamilton's rule, which predicts that helping another individual will be favoured by selection when the fitness costs of helping to the helper (c) are outweighed by the fitness benefits to the recipient (b), weighted by the relatedness (r) between the helper and the recipient (i.e. selection will favour individuals that provide help if $rb>c$, Hamilton 1963; West et al. 2007). Despite this relatively intuitive argument, the abundant theoretical and empirical research on this topic has revealed a more complex relationship between parentage and parental care than that captured by this simple prediction (Kempenaers and Sheldon 1997; Houston and McNamara 2002; Sheldon 2002; Alonzo 2010; Chapter 6). Given this complexity, it is important to ask whether the underlying positive relationship is always expected, whether it may be masked by other biological factors, such as variation among individuals, or whether there are conditions in which no relationship or even a negative relationship between parentage and parental effort is expected.

11.1.2 Variation in maternity and paternity

We define parentage as the proportion of offspring in the brood that are the genetic offspring of the male or female under consideration (Westneat and Sherman 1993). For males, uncertainty of paternity

The Evolution of Parental Care. First Edition. Edited by Nick J. Royle, Per T. Smiseth, and Mathias Kölliker.

can arise if females mate with multiple males within a single reproductive bout. For females, maternity can become uncertain when individuals leave their offspring in another female's care. In birds, for example, intra-specific brood parasitism can cause females to have both related and unrelated young in their nests, while in other species communal breeding causes complex patterns of egg laying that lead to mixed and uncertain maternity within a communal nest (Lyon and Eadie 2008). There are two distinct components of this variation in parentage. First, there may be variation in the expected parentage because offspring have been sired or produced by others. Here, the uncertainty is in the parentage of any particular offspring, even if the proportion of offspring produced or sired is known with certainty. Second, there may be variation in the certainty of information about expected parentage. For example, males may be uncertain whether females have mated with and sired offspring with other males. Both components of uncertainty in parentage can vary among species, between individuals of the same species, and among different reproductive bouts for the same individual.

Although there has been little quantitative comparison among species (but see Gowaty 1996), maternity is likely to be generally more certain than paternity simply because conspecific brood parasitism and communal breeding occur in relatively few species (Gowaty 1996; Lyon and Eadie 2008), while multiple mating by females is found in most species (e.g. Jennions and Petrie 2000; Avise et al. 2002; Griffith et al. 2002; Clutton-Brock and Isvaran 2006). The vast majority of studies have focused on the effect of variation in paternity on paternal effort. However, empirical studies examining maternity and paternity simultaneously have yielded important insights into our understanding of differences between the sexes in patterns of mating and parental effort (e.g. Gowaty and Karlin 1984; Gowaty 1991), and we argue that further consideration of the potential effects of maternity on interactions within and between the sexes is warranted.

11.1.3 Parentage is important, but it is not fitness

The expected positive relationship between parentage and parental effort is based on the assumption that higher parentage is associated with higher expected fitness. Yet, fitness depends more directly on the total number of surviving offspring produced, which may not always be tightly coupled with parentage. Imagine, for example, a species of bird in which a female cares for her own offspring but also suffers reduced maternity due to other females laying eggs in her nest. If these extra young have no net negative (or even a positive) effect on expected maternal or offspring survival and future reproductive success, then the female's maternity will have decreased though her fitness was unchanged or even increased (Lyon and Eadie 2008). Similarly, when males care for the young of multiple females, it is possible for a successful male to have higher total reproductive success but lower paternity if males with high mating rates also experience greater sperm competition (Alonzo 2008; Alonzo and Heckman 2010). It is therefore important to keep in mind that, although we often focus on parentage and parental effort, as explained above, fitness and parentage may not always be tightly coupled. Instead the evolution of parental care depends on fitness variation arising from heritable variation in parental care patterns rather than variation in parentage per se. For example, male adjustment of parental effort in response to variation in paternity is likely to be heritable and under direct selection in many species.

11.1.4 Interactions within and between the sexes drive parentage and parental effort

Parentage is not a trait that evolves but is instead an outcome that arises from interactions within and between the sexes. For example, multiple paternity and maternity can arise from competition within a sex, and interactions between males and females further influence patterns of parentage. Hence, an essential component of understanding and predicting the relationship between parentage and parental effort is first considering how interactions within and between the sexes simultaneously affect the evolution and expression of traits related to mating, parentage, and parental effort (Fig. 11.1).

Although reproductive cooperation is possible, interactions between the sexes often involve sexual conflict. For example, there can be conflict with respect to whether and how often to mate (Arn-

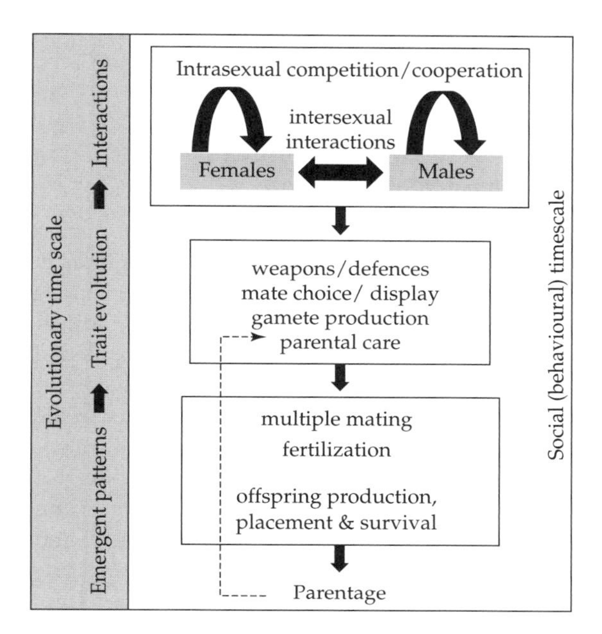

Figure 11.1 The complex and indirect connection between parentage and parental effort: on a social (or behavioural) time scale (right side), interactions within and between the sexes (top) affect the expression of multiple male and female traits (middle) whose expression affect emergent patterns such as mating, survival, and parentage (bottom). Although these patterns are not traits, they can influence the evolution of heritable components of male and female traits, whose expression further affect interactions between individuals on evolutionary time scales (left side). The dashed arrow highlights the very indirect nature of the relationship between parentage (an emergent property) and parental effort (a socially influenced trait), both of which are influenced by interactions within and between the sexes on both social and evolutionary time scales.

qvist and Rowe 2005). In addition, conflict between the sexes will commonly arise with respect to how much parental care each sex provides as both typically benefit from the opposite sex providing more of the care (Chapter 9). How these conflicts are resolved both evolutionarily and behaviourally will influence patterns of mating and parental effort and thus also the strength and direction of the relationship between parentage and care (Fig. 11.1).

Extensive research, a review of which is beyond the scope of this chapter, has considered why and when females choose specific mates or mate with multiple males (Jennions and Petrie 2000; Griffith 2007). Multiple mating by females may be favoured if females gain direct or genetic benefits (e.g. Foerster et al. 2003; Fisher et al. 2006; Griffith and Immler 2009). Competition among males, even when at a cost to female fitness, can also lead to multiple mating by females (Arnqvist and Rowe 2005). Regardless of why females sire offspring with multiple males, the typical effect of multiple paternity is to reduce the relatedness between a male and the offspring for which he might provide care. Despite extensive research on individual components of these interactions, the connection between mating patterns and the evolution of parental care remains poorly understood (Alonzo 2010). We first highlight theoretical and empirical examples that illustrate the importance of these interactions within and between the sexes for our understanding of the relationship between parentage and care. We finish by discussing what we may be missing in our understanding of these relationships and suggest some areas for future study.

11.2 Theoretical predictions: does higher parentage always favour greater parental effort?

11.2.1 Paternity is expected to affect parental effort

In his seminal paper on parental investment, Trivers (1972) argued that paternity should affect male parental investment. He defined parental investment as 'any investment by the parent in an individual offspring that increases the offspring's chance of surviving (and hence reproductive success) at the cost of the parent's ability to invest in other offspring' (Trivers 1972). In general, Trivers argued that differences between the sexes in parental investment can be explained by their initial differences in investment in gametes and the consequences these differences have for the number of gametes produced and competition among individuals of the same sex (Trivers 1972). In short, males typically produce more sperm than females produce eggs, which leads to competition among males to fertilize eggs. This is argued to drive sex differences in reproductive investment: females are selected to invest more in their offspring while males are selected to invest more in mate attraction and success in fertilization (Trivers 1972). Dawkins and Carlisle (1976) argued that this commits the 'Concorde fallacy' by assuming, incorrectly, that past investment should directly influence future investment. The core of Trivers' argument remains, however, that sex differences in parental investment may arise from sex differences in the way that other aspects of reproduction such as investment in mate attraction and gamete production affect lifetime expected fitness (Kokko and Jennions 2008; Chapter 6).

11.2.2 The biological importance of model self-consistency

For diploid sexual organisms, each offspring will have exactly one mother and one father (or for hermaphrodites each zygote results from one sperm and one egg). This biological fact leads to the mathematical requirement that the number of offspring produced at the population level by all females must equal the number of offspring sired by all males in the population. While intuitive, this condition was often not met in early theory relating paternity to parental effort (reviewed in Queller 1997; Houston and McNamara 2002; Houston and McNamara 2005; Houston et al. 2005). A biologically reasonable model must also ensure consistency of paternity; that is, any paternity lost by one male must be gained by another male (Houston and McNamara 2002; Houston and McNamara 2005; Houston et al. 2005). As an example, Maynard Smith (1977) did not ensure self-consistency in his influential model of male and female parental care behaviour. This theory assumed that males providing care would suffer a cost in terms of lost future mating opportunities. However, this assumption violates self-consistency as it is not possible for males that do not provide care to find additional females to mate with if all the females in the population remain to care for their offspring and are thus unavailable (Maynard Smith 1977). While this theory has been immensely useful heuristically, it is important to remember that predictions from these models may not hold when altered to ensure self-consistency (reviewed in more detail in Queller 1997; Houston and McNamara 2002; Houston and McNamara 2005; Houston et al. 2005; Kokko and Jennions 2008). In this chapter, we therefore focus our review on the predictions of more recent self-consistent theory predicting the expected relationship between parental effort and parentage.

11.2.3 A specific self-consistent model relating parentage and care

In an especially clear and elegant paper, Queller (1997) considered the question of sex differences in parental care and the effect of paternity on male parental effort. His arguments not only clarified why males and females might differ in parental care, but also clearly illustrated the effect of parentage on these differences and the importance of the above-mentioned Fisher condition when developing theory on the co-evolution of male and female traits (Alonzo 2010). In this paper, Queller examines the situation where the fitness benefits (b) and the fitness costs (c) of providing care are the same for both males and females. To address the 'Fisher Condition' described above, let x represent the total number of offspring produced in the population, m the number of males, and f the number of females in the population such that average female reproduction will be x/f and average male reproductive success will be x/m. If males and females pay the same proportional cost of care, then the costs for males will be cx/m and the cost of care for females if they provide care will be cx/f. Now imagine that an individual that provides care gains a fitness benefit b. In general, for selection to favour the evolution of care or an increase in parental effort, the benefits of providing care must exceed the costs of providing care. Imagine that maternity is certain (females only care for their own young) but that males sire a proportion p of the offspring to which they might provide care. Then the conditions that will favour an increase in maternal care are

$$b > \frac{cx}{f} \tag{11.1}$$

and the conditions required for selection to favour the evolution of increased male care are

$$pb > \frac{cx}{m} \tag{11.2}$$

illustrating that all else being equal, multiple paternity reduces the benefit to males of providing care. This implies that the minimum benefit required for selection to favour parental care will be higher for

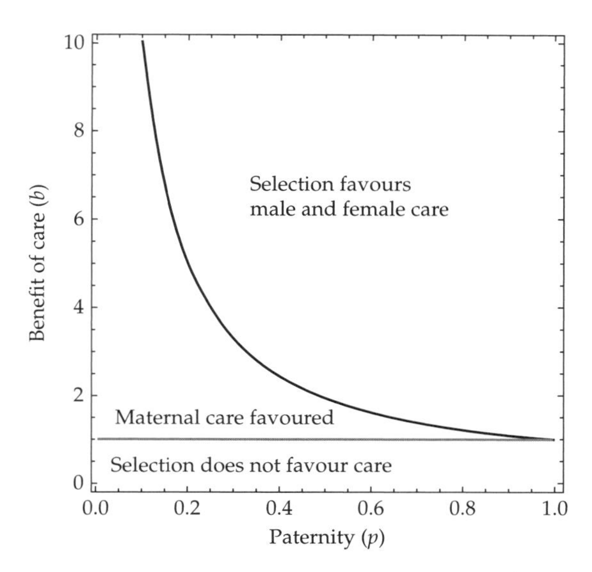

Figure 11.2 The effect of parentage on whether selection favours parental care. This figure shows a graphical representation of Equations 1 and 2 from Queller (1997) where each line gives the boundary at which selection switches from favouring to disfavouring the evolution of parental care as a function of parentage (p) and the relative benefit of care (b). Results are shown for $m = f$ and $c = 1$.

males than for females (Fig. 11.2). Remember, however, that if expected maternity is low relative to paternity, the reverse would be true. One can again see the connection between the equations above and Hamilton's rule ($rb>c$). The arguments confirm the logic of parentage affecting selection for (or against) care behaviour while also demonstrating the need to consider selection on both sexes and the connection between male and female fitness. A limitation of this model, however, is that it does not consider how parentage arises from interactions within and between the sexes and how this potentially also affects the costs and benefits of maternal and paternal care (Fig. 11.1).

11.2.4 Interactions between the sexes affect parentage and parental care

Parentage is not a trait that can evolve. Instead, traits affecting interactions within and between the sexes evolve and lead to patterns of mating and parentage that, in turn, influence selection on parental effort. Recent theory is beginning to explore how mating patterns arise and affect both paternity and parental effort. These analyses suggest that interactions within and between the sexes must be considered if we wish to understand the complex relationship between parentage and care.

For example, Kokko and Jennions (2008) examined the evolution of male and female parental effort in circumstances where individuals spend time either as available to mate or as unavailable to mate when producing a new brood or providing parental care. This self-consistent model predicts that increased sexual selection and decreased parentage in a brood leads to decreased parental effort for both sexes. However, these analyses assumed that the mating group size is fixed and the number of same-sex competitors in the mating group determines parentage. This model therefore looked at the consequences of multiple mating and sexual selection but did not allow mating behaviour to co-evolve with parental effort (but see Kokko 1999). In contrast, theory examining the co-evolution of female mating preferences with male and female parental effort predicts that intersexual selection can favour the evolution of paternal care and weaken the effect of relatedness on paternal care evolution (Alonzo 2012). Hence, it is important to consider not only the consequences of various mating patterns for parentage and parental care, but also to consider how these mating patterns themselves arise and evolve.

Variation among individuals that affects expected parentage or the costs and benefits of care has the potential to make it difficult to detect an underlying positive relationship between parentage and parental effort (Kempenaers and Sheldon 1997; Sheldon 2002). If individuals vary in the total amount of energy available for current reproduction, for example, a positive relationship between parentage and parental effort may arise simply because some individuals have invested more in both mating and parental effort. Recent theory has also shown that individual variation can alter the predicted relationship between parentage and care. For example, a model examining variation in male quality found that the population-level relationship between paternity and paternal effort may not always be positive and even predicts that the relationship could be negative if low quality males have low paternity but provide more care when they actually manage to attract a mate (high quality males with high paternity provide less care because they are more likely to attract additional mates; Houston and McNamara 2002). This model not only considered variation among males but also allowed parentage to arise from interactions between the sexes and competition within a sex. Additional theory examining extra-pair young that are competitively superior to within-pair nestmates also suggests that paternity and parental effort could be negatively correlated if parents are compensating for the negative effects of competition between extra-pair and within-pair young (Holen and Johnstone 2007). In addition, when males vary in the quality of the territories they defend, female mating behaviour may depend on territory quality in ways that confound the relationship between the direct benefits of paternal care and paternity, (Eliassen and Kokko 2008). For example, a negative relationship between parentage and paternal effort can arise if males of lower genetic quality provide more parental care and but also experience more extra-pair paternity due to female mating behaviour.

Together, the theory reviewed in this section demonstrates that assumptions made about interactions between the sexes and variation within a sex influence the predictions of the model. Furthermore, attempting to relate paternity to paternal effort without thinking about how paternity arises from these interactions can lead to false inferences about whether and how paternity affects the evolution of parental care. Our ability to make testable *a priori* predictions about the relationship between parentage and parental effort will improve if we increase our understanding of how variation among individuals affects interactions within and between the sexes, and the patterns of parentage and care that these interactions generate (Alonzo 2010).

11.2.5 Information about expected parentage affects predictions

Variation among and interactions between individuals strongly influences the relationship between parentage and parental effort, but information about expected paternity or maternity also affects these predicted patterns. For example, many models assume that all individuals experience the same average expected paternity and that this average paternity determines selection on patterns of care (Queller 1997; Kokko and Jennions 2008). In contrast, other theories found that whether males gain information about female mating patterns can influence male parental care behaviour and allow the coexistence of extra-pair mating and high male parental effort (Kokko 1999). Similarly, the presence or absence of kin recognition strongly influences behaviour patterns (Kempenaers and Sheldon 1996; Richardson et al. 2003), as also illustrated by research on cooperation (Keller 1997). The question of why organisms do or do not evolve kin recognition is beyond the scope of this chapter. However, this basic question is in many ways at the heart of how relatedness affects parental care (Johnstone 1997; Gardner and West 2007; Kilner and Hinde 2008). Imagine, for example, that males can reliably detect which offspring are or are not their own. If this is the case, the question of how individual males should respond to average paternity becomes irrelevant because males may direct parental effort preferentially to related young. Empirical evidence suggests that in many species individuals cannot reliably recognize which offspring are their own. However, cues that indicate multiple mating or brood parasitism are commonplace but vary among species depending on reproductive physiology and mating system. For example, in species with multiple mating by females, the presence of additional males or the absence of the female may reliably predict expected parentage in the brood. It is therefore important to know what potential cues of parentage and relatedness exist for a particular species when comparing model predictions to empirical data. Hence, our ability to make predictions about the relationship between parentage and care could be greatly improved by improving our proximate knowledge of whether and when individuals across a range of taxa can detect cues of relatedness or expected parentage.

11.3 Empirical patterns: what determines the relationship between parentage and parental effort?

A large number of studies have asked whether variation in paternity or a male's certainty of paternity covaries with variation in male parental effort. In contrast, relatively few studies have examined the relationship between maternity and maternal care. Attempts to summarize studies examining the effect of paternity (or certainty of paternity) on measures of male parental effort yield no clear consensus (Kempenaers and Sheldon 1997; Sheldon 2002; Alonzo 2010). A recent review found that only half of the studies examining the relationship reported the expected positive relationship between paternity and parental effort (Alonzo 2010). While a meta-analysis is needed to synthesize across studies, the absence of a clear pattern demonstrates that while an underlying effect of paternity and paternal care can exist, it is also complicated or masked by other factors. As described above, theoretical research has attempted to provide possible explanations for these mixed results. The most likely candidates to explain these apparently equivocal patterns are: the effects of among-individual variation in quality, availability, and reliability of parentage cues; the effect of within-individual variation in paternity and reproductive success across broods; and the potential for interactions within and between individuals with respect to mating and parental effort to weaken and even alter the relationship between paternity and paternal effort (Kempenaers and Sheldon 1997; Sheldon 2002; Alonzo 2010). There have not, however, been any rigorous empirical studies testing *a priori* predictions regarding whether these factors can explain the presence or absence of a detectable effect of paternity on parental effort (Alonzo 2010). This seems surprising given that the need for rigorous experiments has long been recognized (Kempenaers and Sheldon 1997; Sheldon 2002).

What can we infer about the effect of parentage on parental care evolution from the available empirical information? Given the absence of a clear consensus pattern, it is helpful to consider the effect of paternity on parental care in a few well-studied systems and ask what we understand and what remains unclear about how patterns of relatedness generally affect the evolution of parental care (Table 11.1). Below, we outline several such examples. These specific examples were chosen to illustrate that: 1) the relationship between parentage and parental effort is highly variable across species; 2) understanding this relationship in a given population often requires detailed knowledge of patterns of mating, kin recognition, and future reproductive opportunities; and 3) it has often necessary to rely on *post hoc* explanations because observed patterns are not predicted *a priori* by current theory.

11.3.1 Paternity affects parental care in the bluegill sunfish

In the bluegill sunfish (*Lepomis macrochirus*), territorial males defend nests and provide parental care but lose paternity to sneaker males and female mimics that release sperm during a spawning event between a female and a territorial male (Neff 2001). Consistent with the predictions of classic theory, parental males have been found to decrease care of developing eggs and fry in response to increased sneaker presence in the nest and decreased parental male paternity (Neff and Gross 2001; Neff 2003a). Furthermore, males can use olfactory cues to discriminate their own fry from those of other males (Neff and Sherman 2003), and thus, after hatching they reassess their paternity and adjust further parental effort in response to actual paternity (Neff and Gross 2001). In a separate study, parental males were found to increase cannibalism of young after hatching (but not before) if they had been cuckolded (Neff 2003b). While it is not yet known exactly why this difference in the ability to detect relatedness exists between developing egg and fry stages, it is hypothesized that olfactory cues are available from newly hatched eggs that developing eggs do not possess (Neff and Sherman 2003). When direct cues of relatedness were not available (e.g. during the developing egg phase of parental care), parental males responded to indirect cues of paternity, decreasing effort when many sneaker males had been present at the nest (Neff and Gross 2001; Neff 2003a). This work illustrates how perceived and realized paternity can influence parental care decisions in a dynamic way. This is also one of

Table 11.1 The relationship between parentage and parental effort varies across species. For several species with mixed parentage we describe the source of mixed parentage, parental response, and possible cues used by parents. Additionally, we note whether the patterns are consistent with classic theory that predicts a decrease in parental effort when parentage is uncertain. These specific examples have been selected to highlight how the relationship between parentage and parental effort is highly variable. References for all examples can be found in the text

Example	Source of mixed parentage	Effect of mixed parentage on parental effort	Cues of mixed parentage used by parents	Consistent with classic predictions? If not, possible explanation for observed pattern
Bluegill sunfish	Sneakers and female mimics fertilize eggs in the nests of parental males.	Males decrease parental effort and exhibit more cannibalism of fry when paternity is mixed.	The presence of sneakers in the nest and direct recognition of offspring through olfactory cues.	YES
Dunnocks	Multiple mating leads to broods with mixed paternity.	Males that experienced polyandrous or polygynandrous matings were more likely to feed young if they had some share of paternity, although they did not preferentially feed their own young.	Access with females during the mating period.	YES
Canada geese	Adoption of unrelated offspring leads to mixed paternity and maternity among young.	Parents care for unrelated young, but their own offspring tend to be kept closer to the adult pair.	New, unfamiliar young in the brood.	NO Presence of adopted offspring increases survival of a parent's own young, i.e. there are direct benefits of mixed parentage.
Ocellated wrasse	Sneakers fertilize eggs in the nests of parental males.	The presence of sneakers at a nest is positively correlated with male care.	Unclear whether males use clues of paternity to adjust levels of care.	NO Successful males have more sneakers at their nests but, despite lost paternity, they also have greater numbers of their own offspring than less successful males.

only a few cases where careful experiments have been able to clearly demonstrate the presence of kin recognition in parental males.

11.3.2 Paternal care is more likely with high sperm competition in the ocellated wrasse

In contrast, very different patterns are observed in the ocellated wrasse (*Symphodus ocellatus*), another fish species with male alternative strategies that include both sneakers and territorial males that provide care. In this species, the presence of sneakers at the nest (a likely cue of paternity) is positively associated with the probability that a territorial male will provide parental care such that males that experience high sperm competition from sneaker males are more likely to provide care (Fig. 11.3; Alonzo and Heckman 2010). While this pattern seems counterintuitive at first, it becomes intuitive when one examines how the total number of larvae sired covaries with the number of sneakers at a nest. Successful males lose more paternity to sneakers but also have more of their own offspring in their nest (Alonzo and Heckman 2010), thus illustrating the fact that paternity and male reproductive success are not the same thing and may not even be positively associated in all species. These empirical patterns are likely to be explained by the fact that females prefer successful nesting males because they are more likely to provide parental care (Alonzo 2008). Sneaker males are also found in larger numbers at these nests because these nests provide both greater mating opportunities and higher chances of any young they sire being cared for by the nesting male. This unexpected pattern demonstrates the importance of considering both female mating patterns and competition with other males when attempting to relate certainty of paternity with paternal care (Alonzo and Heckman 2010).

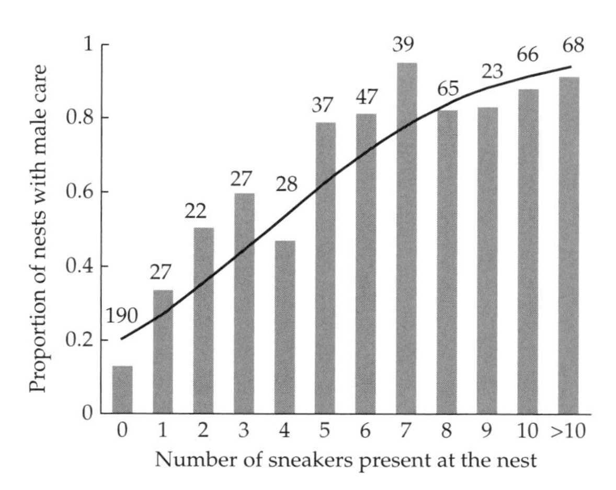

Figure 11.3 The presence of sneakers at the nest (a cue correlated with sperm competition and likely associated with certainty of paternity) is positively correlated with parental care in the ocellated wrasse. The maximum number of sneakers observed at the nest is presented here. The average number of sneakers at the nest yields similar results. The solid line gives the predicted relationship between sneakers and the probability of male care (logistic regression, $\beta = 0.38$, $\chi^2 = 170.3$, $p<0.001$, intercept $= -1.36$, $\chi^2 = 82.0$ $p<0.001$), bars give the observed proportion of males caring for the nest and the numbers above the bars are the sample sizes (number of nests observed per sneaker number). (Figure redrawn from Alonzo and Heckman 2010.)

11.3.3 Paternity does not affect paternal effort in western bluebirds

Western bluebirds (*Sialia mexicana*) breed primarily in socially monogamous pairs Approximately 45% of females have at least one chick in their nest that is not sired by their social partner and extra-pair copulations with neighbouring males are relatively common (Dickinson 2003). To evaluate the effect of lost paternity on paternal effort in western bluebirds, Dickinson (2003) conducted an experiment in which males were either un-manipulated, removed from the nest during the second day of laying in order to reduce certainty of paternity, or removed during incubation. Males removed during the second day of laying were denied copulatory access and the ability to mate-guard during a period when two to three eggs remained unfertilized. These males showed no significant reduction in provisioning relative to the un-manipulated males or males that were removed during incubation. A separate experiment also revealed that resident males that visually observed an extra-pair male visit their mate did not provide significantly less provisioning than resident males whose mate was not visited by an extra-pair intruder (Dickinson 2003). Likewise, the proportion of extra-pair chicks in a male's nest did not significantly affect provisioning. These findings suggest that neither apparent nor realized loss of paternity affect provisioning rate in the western bluebirds studied. Previous work (Leonard et al. 1995) suggests that males cannot recognize their own young, and as a result reduced paternal effort in response to extra-pair young might reduce the survival of a male's own offspring (Dickinson 2003). Additionally, breeding males have a relatively low chance of surviving to breed the following year, and thus it is potentially adaptive for them to not reduce provisioning in response to lost paternity if it ensures survival of their own offspring (Dickinson 2003).

11.3.4 Interactions between the sexes affect parentage and parental care in the dunnock

In the dunnock (*Prunella modularis*), a small passerine bird with a highly variable mating system, Burke et al. (1989) found that paternity affects male parental care. When a mating pair was monogamous, chicks were fed only by the monogamous male and female. In contrast, when mating was polyandrous or polygynandrous, females fed the young in their nest in all cases. At least one male typically helped feed the chicks, and DNA fingerprinting revealed that males were more likely to help feed if they had some share of paternity. However, males did not preferentially feed their own young. As an explanation for this pattern, Burke et al. (1989) hypothesized that males increase their reproductive success by feeding offspring in relation to their access with females during the mating period. In a later study, Davies et al. (1992) experimentally removed males from the nest at varying stages of the mating cycle to create variation in mating success and paternity. When mating was polyandrous, removed males fed chicks only if they had mated during the egg laying period, suggesting that mating access, which is correlated with paternity, affects parental effort. In contrast to the polyandrous case, when mating was monogamous, experimental manipulation of mating access did not affect parental effort, despite paternity loss. The findings of Davies et al. (1992) raise the question of why males are not better at discriminating their own offspring from those of other males. Davies et al. (1992) discuss two possibilities. First, unlike in the bluegill example above, they suggest that dunnocks are simply unable to distinctively label their offspring in a way that would make them recognizable. Second, they hypothesized that conflict between males and females and chicks might have prevented a gene for direct paternity markers from spreading. If males recognized their own offspring, chicks would likely only be fed by their own father, and this would decrease overall provisioning of a female's young. This hypothesis highlights the complex interaction between sexual conflict, paternity, kin recognition, and parental effort.

11.3.5 Alloparental care in Canada geese can increase offspring survival

Another mechanism that gives rise to mixed parentage is communal breeding or the adoption of unrelated offspring. For example, adoption has been recorded in more than 120 mammals and 150 bird species (Riedman 1982). Adopting offspring is expected to be costly to parents because it reduces resources that can be invested in current parental care or future reproduction. To examine the fitness implications of adopting offspring, Nastase and Sherry (1997) quantified the relationship between gosling status (adopted or natural offspring), distance to the parent or adult, and survival to five years in Canada geese, *Branta canadensis*. They found that adopted goslings tended to be further from adults than the offspring of the adult pair, and natural offspring that were reared in mixed broods had higher survival rates than adopted goslings or offspring that were raised in unmixed broods (Nastase and Sherry 1997). While it is typically assumed that caring for unrelated young comes at a cost, adopting young, which leads to mixed maternity and paternity in a brood, actually improves the survival of their own offspring in Canada geese through direct benefits associated with large broods. While this care could still come at a cost to the parent's future reproductive success, in species where offspring survival increases with density (for example due to the dilution of predation risk with density), decreased parentage resulting from adoption may actually represent a form of increased parental effort in the sense that it increases individual offspring survival.

11.3.6 Alloparental care in eiders might increase survival of a caring mother's chicks

In eiders (*Somateria mollissima*) communal brood care is the norm and broods tended by a single female often contain unrelated young (Öst 1999; Kilpi et al. 2001). Initially, females form large coalitions with several females, but typically some mothers leave or groups split, leaving two to four mothers (Öst 1999). Öst and Bäck (2003) found that in eider groups, a female's own young tended to be closer to her than other females. Chicks on the edge of a group experience the highest predation rates in eiders (Swennen 1989) and females are more aggressive to unrelated chicks than their own (Öst and Bäck 2003). Because females typically remain in the centre of the brood, a female's own young are expected to benefit most from parental guarding in comparison to unrelated offspring in single female groups (Öst and Bäck 2003). In groups of more than one female, the position of females is determined by a hierarchy system in which subordinates are usually near the edges (Öst 1999). Offspring survival is therefore thought to be closely linked to the status of the mother (Öst and Bäck 2003). Given that eider females preferentially protect their own offspring, we are faced with two additional questions: 1) why do some females leave their young with an unrelated female, and 2) why do mothers accept unrelated chicks? With regard to the first question, females who leave young with an unrelated female are often in poor condition and potentially abandon young to increase their own likelihood of surviving and reproducing in the future (Kilpi et al. 2001). Because chicks on the edges suffer more predation, and because females keep their own young closest to them in the centre, accepting unrelated offspring potentially increases the survival of a female's own young (Öst and Bäck 2003). Female variation in condition therefore may explain whether females provide brood care because it affects the potential costs and benefits for caring for both related and unrelated young.

11.3.7 Lessons from comparative studies

As the examples above illustrate, studies of individual species provide a wealth of information on factors that may explain variation in the relationship between parentage and parental effort. Yet the relationship between parentage and parental effort is mixed and no single variable seems to predict whether or not the expected positive relationship exists (Alonzo 2010). Another approach is to ask whether greater clarity arises if we synthesize information across studies on different species. Comparative phylogenetic studies of paternity and parental care have mainly focused on birds and fishes, due to the availability of studies on these taxa. Comparative analysis found a significant positive relationship between average paternity and male food provisioning across species of birds, but did not find a relationship between interspecific variation in paternity and other forms of paternal care (Møller and Birkhead 1993; Møller and Cuervo 2000). Similar analyses report a negative relationship between extra-pair paternity and the importance of male care to female fitness across species (Møller 2000). A study estimating the effect of within-species variation in paternity on paternal effort also found that females often experience a net loss of fitness as a result of male reduction in care in response to extra-pair paternity (Arnqvist and Kirkpatrick 2005), though it remains unclear whether females gain other direct or indirect benefits from extra-pair mating (Griffith 2007). While comparative studies of birds do generally support the idea that paternity is positively associated with paternal care, these studies have not identified factors that could explain the presence or absence of this predicted relationship among species. In fishes, external fertilization was found to be positively associated with the evolution of male-only care (Mank et al. 2005). Another study found no effect of the presence or absence of sperm competition from non-parental males on the evolution of male care in fishes (Ah-King et al. 2005). We are aware of no studies examining the relationship between multiple maternity and patterns of care. To date, comparative studies thus find the same basic result as individual studies: higher parentage is sometimes but not always associated with greater parental effort. Detailed meta-analytic and comparative phylogenetic studies will be needed to reveal the degree to which various factors (such as female choice, sperm competition among males, male harassment, cues of relatedness, and variation among individuals) explain the strength and direction of the relationship between parentage and parental effort.

11.4 Parentage and parental care: what are we missing?

As the theoretical and empirical patterns reviewed above illustrate, the relationship between parentage and parental effort is not a simple one: a positive relationship between parentage and effort is neither universally predicted nor found. Whether and how much parents adjust their parental effort in response to uncertain parentage is likely to depend on numerous environmental, physiological, genetic, and social factors. These factors affect the costs of providing care to males and females and benefits of care to both related and unrelated young, mating dynamics and the patterns of relatedness they create, and the reliability and availability of relatedness and parentage cues. While the question may seem hopelessly complex, we argue that the patterns reviewed above pinpoint a few key issues, detailed below, where further theoretical and empirical attention is likely to provide new insights.

11.4.1 *A priori* predictions should replace *post hoc* explanations

Numerous studies have illustrated the mixed effects that parentage can have on parental effort. For the most part, however, these studies have explained the observed patterns using *post hoc* explanations related to ecology, life-history, mating system, conflict, and/or kin recognition. Further studies correlating paternity or certainty of paternity with measures of parental effort are unlikely to yield new insights. Progress could be made with further theory that makes concrete and testable predictions regarding which variables are expected to cause a change in the strength and direction of adjustment of parental effort in response to changes in actual or perceived paternity. To be useful, however, these *a priori* predictions must then be tested using a combination of experimental manipulation of key variables identified by this theory as well as rigorous comparative studies of interspecific variation.

11.4.2 The importance of kin recognition and cues of extra-pair paternity

The studies reviewed above demonstrate that both the theoretical predictions and our understanding of empirical patterns depend on the existence and reliability of parentage cues and kin recognition mechanisms. The evolutionary dynamics of kin recognition mechanism are complex, as strong selection for genetic markers of relatedness may make the maintenance of variation in genetic kin markers unlikely (Rousset and Roze 2007; Gardner and West 2007). Although generally rare, kin recognition has been found in some species (Tibbetts and Dale 2007; Widdig 2007). For example, in bluegill sunfish (Neff and Sherman 2003), dunnocks (Davies et al. 1992), and eiders (Öst and Bäck 2003) males or females adjust parental effort in a way that is consistent with the existence of either direct or indirect cues of parentage. A challenge, however, is that when individuals fail to adjust parental effort in response to paternity or maternity, it is difficult to know whether cues of parentage are simply absent or undetectable, or whether such cues exist but an adjustment of parental effort has not been favoured by selection. For instance, monogamous dunnocks did not decrease parental effort when male mating access was limited and paternity was lost to other males (Davies et al. 1992). It is possible that males could not detect the loss in paternity; alternatively the costs of reducing care when paternity is relatively low might outweigh any future benefits to male fitness. Additionally, in some cases caring for unrelated young actually increases the current or future reproductive success of a parent. For instance in geese, a parent's own offspring have higher survival rates if they are raised in mixed broods with unrelated young (Nastase and Sherry 1997). One way that research can circumvent this confound is by understanding and experimentally manipulating the sensory and cognitive processes known to underlie kin recognition and parentage cues. For example, imagine that one could manipulate pharmacologically a male's ability to detect paternity cues and examine the consequences for male effort and male and female fitness. Further research on the mechanistic underpinnings and evolutionary dynamics of the cognitive and sensory processes used to assess relatedness and parentage could therefore greatly enhance our ability to predict theoretically, and study empirically, the effect of parentage on parental effort.

11.4.3 Co-evolutionary and social feedbacks between the sexes

One key factor affecting the relationship between parentage and parental effort that we have highlighted above is how interactions within and between the sexes affect mating patterns, information about parentage, and the costs and benefits of care. Our incomplete understanding of the relationship between paternity and paternal effort may be blamed on the tendency to treat paternity as a property of a species or an individual male rather than an emergent property of interactions within and between the sexes (Fig. 11.1; Alonzo 2010). For example, whether males respond to paternity (assuming reliable cues exist) will depend on their future expected mating success and paternity as well as how females respond to male adjustment in paternal effort. Furthermore, we cannot understand male parental effort without knowing how female choice for direct and genetic benefits affects female mating strategies and thus patterns of paternity and male expected reproductive success. It is therefore counter-productive to study parentage and parental effort in isolation from other aspects of the mating system that affect the evolution and expression of mate choice, fertilization, and parental care. The examples above illustrate that we must also keep in mind how individual variation (e.g. in condition, genotype, or experience) affects both observed patterns and evolutionary dynamics.

Theoretical and empirical studies are just beginning to explore how these interactions within and between the sexes arise from social interactions among individuals and the co-evolutionary dynamics of multiple male and female traits (reviewed in Stiver and Alonzo 2009; Alonzo 2010). Further theory is needed that considers how multiple socially influenced traits evolve simultaneously in both sexes. For example, theory that examines how traits affecting whether and how males adjust paternal effort in response to paternity and mating opportunities co-evolve with traits affecting female choice among males that differ in both genetic benefits and paternal adjustment of effort. It is important to not only consider the evolution of multiple interrelated traits simultaneously but also to consider how the expression of these traits is influenced by social interactions.

11.4.4 Rigorous experiments will be needed to test how multiple traits interact

Carefully controlled experiments will be critical for understanding patterns that are initially perplexing. For instance, the ocellated wrasse is a nice example of how observational patterns can be misleading. In this species, males with more sneakers (and likely lower expected paternity) exhibit higher levels of parental care (Alonzo and Heckman 2010). If one were to infer causation from this unexpected pattern, the conclusion might be that lower expected paternity leads to increased parental effort. However, the mating dynamics of this fish are complex, and if such complexity is taken into account more plausible explanations emerge. Specifically, more successful males are preferred by females because they are more likely to care for eggs, and this higher number of females likely leads to the increased numbers of sneakers at the nests of males who provide more care (Alonzo and Heckman 2010). Understanding the effect of sneakers on parental care in this species would require carefully controlled experimental manipulation of sneaker presence. However, examining the effect of multiple traits on social interactions and fitness is even more challenging. Either experimental evolution or phenotypic manipulation of multiple traits or behaviours followed by observation of social interactions and individual fitness, for example, could allow one to examine how multiple traits interact to shape social interactions and fitness.

11.4.5 Connecting relatedness, sexual selection, social interactions, and parental care: cooperative breeding as a case study

The empirical examples above illustrate both the underlying importance of relatedness between parents and their young for the evolution of parental behaviours and that, while important, parentage alone cannot fully explain observed patterns of care. We have argued above that the evolution of parental care can be informed by a generalization of Hamilton's rule that considers how interactions within and between the sexes simultaneously affect patterns of relatedness and the costs and benefits of care (i.e. the variables r, b, and c actually emerge from interactions within and between the sexes, see Fig. 11.1; Chapter 1). The challenge is to understand not only how various factors affect r, b, and c, but also how these important variables are interconnected with one another. We argue that more information is needed on how relatedness and the costs and benefits of care arise from social interactions and evolution and expression of multiple male and female traits.

A particularly challenging but informative set of case studies for understanding the evolution of parental care in this context are cooperative and other social breeders (Chapter 12). The specific case of cooperative breeding also illustrates a more general point about parental care evolution: The only way to understand these patterns is to consider how competition, conflict between the sexes, and cooperation among related and unrelated individuals jointly influence selection on multiple aspects of reproduction.

For example, while a potential cost of multiple mating for females is the associated reduction in male care resulting from decreased expected paternity, in species with cooperative breeding females are sometimes argued to be freed from needing their social mates to provide parental care because helpers provide care (Hughes et al. 2003; Rubenstein 2007; Chapter 12). Australian magpies (*Gymnorhina tibicen*, Hughes et al. 2003), superb fairy wrens (*Malurus cyaneus*, Mulder et al. 1994; Dunn and Cockburn 1996, Cockburn et al. 2009), and a few other cooperative breeders exhibit very high levels of extra-pair paternity and extra-group paternity. These high levels of extra-group paternity have been hypothesized to arise to decrease inbreeding within the group and because of the presence of maternally related helpers at the nest to provide care. However, a recent meta-analysis among birds did not find general support for these hypotheses, instead finding support for low extra-group paternity being associated with the evolution of cooperative breeding (Cornwallis et al. 2010). A comparative study of insects also found low promiscuity to be associated with the origin of eusociality (Hughes et al. 2008), thus supporting recent predictions that eusociality and promiscuity may be unlikely to exist simultaneously (Boomsma 2007; Rankin 2010).

In cooperative breeders, it is generally acknowledged that paternity and parental care may be uncoupled, but that relatedness and interactions between individuals are essential for understanding observed patterns (Mulder et al. 1994; Cockburn 1998; Hughes et al. 2003; Richardson et al. 2003; Boomsma 2007; Cornwallis et al. 2010). Here we argue that social interactions and patterns of relatedness are also essential to consider when we attempt to understand the effect of parentage on care in species that do not exhibit cooperative breeding. Furthermore, understanding the evolution and expression of social behaviours, such as parental care, requires thinking about how interactions within and between groups of individuals and patterns of relatedness influence the co-evolution of multiple male and female traits and thus the observed relationship between observed variables such as parentage and parental care.

11.4.6 Do we need to change the question?

The patterns above clearly demonstrate that parentage and parental effort can be but are not always positively correlated. As discussed above, one way forward is to tease apart the factors that influence the strength and direction of the relationship between parentage and parental effort. An alternative conclusion that could be drawn is that future research should not focus on the relationship between parentage and parental effort. Since parentage is neither a trait that evolves nor a fixed characteristic of a species, it may not make sense to search for a functional relationship between these two variables (see Fig. 11.1). The fundamental evolutionary relationship does not exist between parentage and parental effort, but instead between heritable male and female reproductive traits that determine patterns of mating, fertilization, and parental effort. More insight could be gained by asking new questions that reflect this distinction. For example, how is the co-evolution of mating preferences and adjustment in parental effort influenced by the availability of kin recognition and cues of parentage? How does individual variation in condition and experience simultaneously affect the plastic adjustment of male and female parental effort and multiple mating? Answers to these questions will of course inform our understanding of the relationship between parentage and care. These questions, however, also have the potential to address bigger and more general questions about how multiple traits co-evolve and how trait evolution is influenced by the social environment the traits themselves create.

11.5 Conclusions

The relationship between parentage and parental effort is complex and much remains to be discovered about what factors affecting patterns of mating, fertilization, and parental care. Despite this complexity, generalities emerge that inform our current understanding of parental care and set the stage for future work on the topic. First, it is clear that decreased parentage often, but not always, leads to decreased parental effort. We have argued that understanding the relationship between parentage and parental care requires thinking in a more refined way about how this relationship is affected by kin recognition and cues of parentage, individual variation, interactions within and between the sexes, and parental effort in the context of the mating and social system of a species.

Research on parentage and care, however, demonstrates some more general issues relevant to our understanding of mating and social systems. First, attempts to simply correlate one variable with another often lead to equivocal results and potentially false inference about underlying processes. Second, a close connection between theoretical predictions and empirical tests is needed to make sense of complex patterns. Third, patterns of reproduction (such as the relationship between parentage and parental care) can be more readily understood if we think carefully about the co-evolution and expression of multiple male and female reproductive traits (Fig. 11.1; Alonzo 2010). While mating and social systems are complex, our message is not simply that 'it's complicated'. Instead, we argue that making sense of this apparent complexity is tractable. However, it will require that we let go of our focus on these simple predictions (such as looking for a positive correlation between parentage and care) and *post hoc* explanations of patterns that deviate from such simple expectations. We argue that the best way to move forward in understanding the relationship between parentage and parental effort is to ask new questions about how the evolution and expression of multiple male and female traits affect patterns of mating, fertilization, and care within and among species.

References

Ah-King, M., Kvarnemo, C., and Tullberg, B. S. (2005). The influence of territoriality and mating system on the evolution of male care: A phylogenetic study on fish. *Journal of Evolutionary Biology* 18, 371–82.

Alonzo, S. H. (2008). Female mate choice copying affects sexual selection in wild populations of the ocellated wrasse. *Animal Behaviour* 75, 1715–23.

Alonzo, S. H. (2010). Social and coevolutionary feedbacks between mating and parental investment. *Trends in Ecology and Evolution* 25, 99–108.

Alonzo, S. H. (2012). Sexual selection favours male parental care, when females can choose. *Proceedings of the Royal Society of London, Series B* 279, 1784–90.

Alonzo, S. H. and Heckman, K. (2010). The unexpected but understandable dynamics of mating, paternity and paternal care in the ocellated wrasse. *Proceedings of the Royal Society of London, Series B* 277, 115–22.

Arnqvist, G. and Kirkpatrick, M. (2005). The evolution of infidelity in socially monogamous passerines: The strength of direct and indirect selection on extrapair copulation behavior in females. *American Naturalist* 165, S26–37.

Arnqvist, G. and Rowe, L. (2005). *Sexual Conflict.* Princeton University Press, Princeton.

Avise, J. C., Jones, A. G., Walker, D., Dewoody, J. A., and Collaborators (2002). Genetic mating systems and reproductive natural histories of fishes: Lessons for ecology and evolution. *Annual Review of Genetics* 36, 19–45.

Boomsma, J. J. (2007). Kin selection versus sexual selection: Why the ends do not meet. *Current Biology* 17, R673–83.

Burke, T., Davies, N. B., Bruford, M. W., and Hatchwell, B. J. (1989). Parental care and mating behavior of polyandrous dunnocks, *Prunella modularis,* related to paternity by DNA fingerprinting. *Nature* 338, 249–51.

Clutton-Brock, T. H. and Isvaran, K. (2006). Paternity loss in contrasting mammalian societies. *Biology Letters* 2, 513–16.

Cockburn, A. (1998). Evolution of helping behavior in cooperatively breeding birds. *Annual Review of Ecology and Systematics* 29, 141–77.

Cockburn, A., Dalziell, A. H., Blackmore, C. J., Double, M. C., Kokko, H., Osmond, H. L., Beck, N. R., Head, M. L., and Wells, K. (2009). Superb fairy-wren males aggregate into hidden leks to solicit extragroup fertilizations before dawn. *Behavioral Ecology* 20, 501–10.

Cornwallis, C. K., West, S. A., Davis, K. E., and Griffin, A. S. (2010). Promiscuity and the evolutionary transition to complex societies. *Nature* 466, 969–U91.

Davies, N. B., Hatchwell, B. J., Robson, T., and Burke, T. (1992). Paternity and parental effort in dunnocks, *Prunella modularis*. How good are male chick feeding rules? *Animal Behaviour* 43, 729–45.

Dawkins, R. and Carlisle, T. R. (1976). Parental investment, mate desertion and a fallacy. *Nature* 262, 131–3.

Dickinson, J. L. (2003). Male share of provisioning is not influenced by actual or apparent loss of paternity in western bluebirds. *Behavioral Ecology* 14, 360–6.

Dunn, P. O. and Cockburn, A. (1996). Evolution of male parental care in a bird with almost complete cuckoldry. *Evolution* 50, 2542–8.

Eliassen, S. and Kokko, H. (2008). Current analyses do not resolve whether extra-pair paternity is male or female driven. *Behavioral Ecology and Sociobiology* 62, 1795–804.

Fisher, D. O., Double, M. C., Blomberg, S. P., Jennions, M. D., and Cockburn, A. (2006). Post-mating sexual selection increases lifetime fitness of polyandrous females in the wild. *Nature* 444, 89–92.

Foerster, K., Delhey, K., Johnsen, A., Lifjeld, J. T., and Kempenaers, B. (2003). Females increase offspring heterozygosity and fitness through extra-pair matings. *Nature* 425, 714–17.

Gardner, A. and West, S. A. (2007). Social evolution: The decline and fall of genetic kin recognition. *Current Biology* 17, R810–12.

Gowaty, P. A. (1991). Nestbox availability affects extra-pair fertilizations and conspecific nest parasitism in Eastern bluebirds, *Sialia sialis*. *Anim Behaviour* 41, 661–75.

Gowaty, P. A. (1996). Field studies of parental care in birds: New data focus questions on variation in females. I. *Advances in the Study of Behaviour* 25, 476–531.

Gowaty, P. A. and Karlin, A. A. (1984). Multiple maternity and paternity in single broods of apparently monogamous Eastern bluebirds (*Sialia sialis*). *Behavioral Ecology and Sociobiology* 15, 91–5.

Griffith, S. C. (2007). The evolution of infidelity in socially monogamous passerines: Neglected components of direct and indirect selection. *American Naturalist* 169, 274–81.

Griffith, S. C. and Immler, S. (2009). Female infidelity and genetic compatibility in birds: The role of the genetically loaded raffle in understanding the function of extrapair paternity. *Journal of Avian Biology* 40, 97–101.

Griffith, S. C., Owens, I. P. F., and Thuman, K. A. (2002). Extra pair paternity in birds: a review of interspecific variation and adaptive function. *Molecular Ecology* 11, 2195–212.

Hamilton, W. D. (1963). The evolution of altruistic behavior. *American Naturalist* 97, 354–6.

Holen, O. H. and Johnstone, R. A. (2007). Parental investment with a superior alien in the brood. *Journal of Evolutionary Biology* 20, 2165–72.

Houston, A. I. and Mcnamara, J. M. (2002). A self-consistent approach to paternity and parental effort. *Philosophical Transactions of the Royal Society B* 357, 351–62.

Houston, A. I. and Mcnamara, J. M. (2005). John Maynard Smith and the importance of consistency in evolutionary game theory. *Biology and Philosophy* 20, 933–50.

Houston, A. I., Szekely, T., and Mcnamara, J. M. (2005). Conflict between parents over care. *Trends in Ecology and Evolution* 20, 33–8.

Hughes, J. M., Mather, P. B., Toon, A., Ma, J., Rowley, I., and Russell, E. (2003). High levels of extra-group paternity in a population of Australian magpies *Gymnorhina tibicen*: Evidence from microsatellite analysis. *Molecular Ecology* 12, 3441–50.

Hughes, W. O. H., Oldroyd, B. P., Beekman, M., and Ratnieks, F. L. W. (2008). Ancestral monogamy shows kin selection is key to the evolution of eusociality. *Science* 320, 1213–16.

Jennions, M. D. and Petrie, M. (2000). Why do females mate multiply? A review of the genetic benefits. *Biological Reviews* 75, 21–64.

Johnstone, R. A. (1997). Recognition and the evolution of distinctive signatures: when does it pay to reveal identity? *Proceedings of the Royal Society B* 264, 1547–53.

Keller, L. (1997). Indiscriminate altruism: Unduly nice parents and siblings. *Trends in Ecology and Evolution* 12, 99–103.

Kempenaers, B. and Sheldon, B. C. (1996). Why do male birds not discriminate between their own and extra-pair offspring? *Animal Behaviour* 51, 1165–73.

Kempenaers, B. and Sheldon, B. C. (1997). Studying paternity and paternal care: Pitfalls and problems. *Animal Behaviour* 53, 423–7.

Kilner, R. M. and Hinde, C. A. (2008). Information warfare and parent-offspring conflict. *Advances in the Study of Behavior* 38, 283–336.

Kilpi, M., Öst, M., Lindstrom, K., and Rita, H. (2001). Female characteristics and parental care mode in the creching system of eiders, *Somateria mollissima*. *Animal Behaviour* 62, 527–34.

Kokko, H. (1999). Cuckoldry and the stability of biparental care. *Ecology Letters* 2, 247–55.

Kokko, H. and Jennions, M. D. (2008). Parental investment, sexual selection and sex ratios. *Journal of Evolutionary Biology* 21, 919–48.

Leonard, M. L., Dickinson, J. L., Horn, A. G., and Koenig, W. (1995). An experimental test of offspring recognition in Western Bluebirds. *Auk* 112, 1062–4.

Lyon, B. E. and Eadie, J. M. (2008). Conspecific brood parasitism in birds: A life-history perspective. *Annual Review of Ecology Evolution and Systematics* 39, 343–63.

Mank, J. E., Promislow, D. E. L., and Avise, J. C. (2005). Phylogenetic perspectives in the evolution of parental care in ray-finned fishes. *Evolution* 59, 1570–8.

Maynard Smith, J. (1977). Parental investment: A prospective analysis. *Animal Behaviour* 25, 1–9.

Møller, A. P. (2000). Male parental care, female reproductive success, and extrapair paternity. *Behavioral Ecology* 11, 161–8.

Møller, A. P. and Birkhead, T. R. (1993). Certainty of paternity covaries with paternal care in birds. *Behavioral Ecology and Sociobiology* 33, 261–8.

Møller, A. P. and Cuervo, J. J. (2000). The evolution of paternity and paternal care in birds. *Behavioral Ecology* 11, 472–85.

Mulder, R. A., Dunn, P. O., Cockburn, A., Lazenby-Cohen, K. A., and Howell, M. J. (1994). Helpers liberate female fairy-wrens from constraints on extra-pair mate choice. *Proceedings of the Royal Society B* 255, 223–9.

Nastase, A. J. and Sherry, D. A. (1997). Effect of brood mixing on location and survivorship of juvenile Canada geese. *Animal Behaviour* 54, 503–7.

Neff, B. D. (2001). Genetic paternity analysis and breeding success in bluegill sunfish (Lepomis macrochiros). *Journal of Heredity* 92, 111–19.

Neff, B. D. (2003a). Decisions about parental care in response to perceived paternity. *Nature* 422, 716–19.

Neff, B. D. (2003b). Paternity and condition affect cannibalistic behavior in nest-tending bluegill sunfish. *Behavioral Ecology and Sociobiology* 54, 377–84.

Neff, B. D. and Gross, M. R. (2001). Dynamic adjustment of parental care in response to perceived paternity. *Proceedings of the Royal Society B* 268, 1559–65.

Neff, B. D. and Sherman, P. W. (2003). Nestling recognition via direct cues by parental male bluegill sunfish (*Lepomis macrochirus*). *Animal Cognition* 6, 87–92.

Öst, M. (1999). Within-season and between-year variation in the structure of Common eider broods. *Condor* 101, 598–606.

Öst, M. and Bäck, A. (2003). Spatial structure and parental aggression in eider broods. *Animal Behaviour* 66, 1069–75.

Queller, D. C. (1997). Why do females care more than males? *Proceedings of the Royal Society B* 264, 1555–7.

Rankin, D. J. (2010). Kin selection and the evolution of sexual conflict. *Journal of Evolutionary Biology* 24, 71–81.

Richardson, D. S., Burke, T., and Komdeur, J. (2003). Sex-specific associative learning cues and inclusive fitness benefits in the Seychelles warbler. *Journal of Evolutionary Biology* 16, 854–61.

Riedman, L. M. (1982). The evolution of alloparental care and adoption in mammals and birds. *Quarterly Review of Biology* 57, 405–35.

Rousset, F. and Roze, D. (2007). Constraints on the origin and maintenance of genetic kin recognition. *Evolution* 61, 2320–30.

Rubenstein, D. R. (2007). Female extrapair mate choice in a cooperative breeder: Trading sex for help and increasing offspring heterozygosity. *Proceedings of the Royal Society B* 274, 1895–903.

Sheldon, B. C. (2002). Relating paternity to paternal care. *Philosophical Transactions of the Royal Society of London B Biological Sciences* 357, 341–50.

Stiver, K. A. and Alonzo, S. H. (2009). Parental and mating effort: Is there necessarily a trade-off? *Ethology* 115, 1101–26.

Swennen, C. (1989). Gull predation upon eider, *Somateria mollissima*, ducklings: Destruction or elimination of the unfit. *Ardea* 77, 21–45.

Tibbetts, E. A. and Dale, J. (2007). Individual recognition: it is good to be different. *Trends in Ecology and Evolution* 22, 529–37.

Trivers, R. L. (1972). Parental investment and sexual selection. In B. Campbell (ed.) *Sexual Selection and the Descent of Man*. Aldine-Atherton, Chicago.

West, S. A., Griffin, A. S., and Gardner, A. (2007). Evolutionary explanations for cooperation. *Current Biology* 17, R661–72.

Westneat, D. F. and Sherman, P. W. (1993). Parentage and the evolution of parental behavior. *Behavioral Ecology* 4, 66–77.

Widdig, A. (2007). Paternal kin discrimination: The evidence and likely mechanisms. *Biological Reviews* 82, 319–34.

CHAPTER 12

Cooperative breeding systems

Michael A. Cant

12.1 Introduction

Cooperative breeding is a type of social system in which some group members (referred to as 'helpers') routinely provide care for offspring that are not their own, but retain the potential to reproduce themselves either currently or in the future. This broad definition (which derives from those suggested by Cockburn 1998; Crespi and Yanega 1995; Emlen 1991) includes a range of species, from primitively eusocial insects such as paper wasps, hover wasps, halictid bees, and ambrosia beetles; to avian, mammalian, and fish 'helper-at-the-nest' systems in which offspring delay dispersal and help dominant breeders with subsequent breeding attempts; and also larger animal societies with multiple male and female breeders and helpers per group (Fig. 12.1). From current information, 9% of birds (852 species; Cockburn 2006) around 2% of mammals (Lukas and Clutton-Brock in press; Riedman 1982), <0.5% of fishes (20–38 species; Taborsky 1994; Taborsky 2009), and hundreds of species of insect can be classed as cooperative breeders. There are also examples from arachnids (Salomon and Lubin 2007) and crustaceans (Duffy and Macdonald 2010). These societies, while very diverse in terms of social structure and basic biology, share some common features. Populations are usually subdivided into groups of kin (although non-kin individuals may also be present) with strong ecological constraints on dispersal or independent breeding (Hatchwell 2009). Within groups, there is usually (but not always) a reproductive division of labour in which high ranked or socially dominant individuals breed, and lower ranked individuals help (Field and Cant 2009b). Because helpers retain the ability to reproduce themselves, their behaviour reflects a trade-off between current and future fitness, and between direct and indirect components of their inclusive fitness. In this way cooperative breeders differ from eusocial species which have distinct reproductive and worker castes and helpers remain functionally or morphologically sterile throughout their lives (Bourke 2011).

Cooperative breeders have been the focus of intense research in behavioural ecology for two main reasons. First, they embody a major puzzle of evolutionary theory: how can altruistic behaviour be favoured by natural selection? Helpers pay a fitness cost to boost the reproductive output of other group members. For example, subordinate foundresses of the paper wasp *Polistes dominulus* risk their lives foraging to feed larvae to which they are often unrelated (Leadbeater et al. 2010; Queller et al. 2000). Using the classification of social behaviours introduced by Hamilton (1964), helping is a form of altruism when it involves a lifetime direct fitness cost to the helper, and results in a lifetime direct fitness benefit to the recipient of help. In the case of paper wasps, foraging involves clear fitness costs because foundresses that do more foraging suffer higher mortality (Cant and Field 2001). Cooperative breeding systems provide concrete examples of altruism together with the possibility of measuring the fitness consequences of helping, and hence an opportunity to test evolutionary theories of cooperation.

Second, cooperative breeders have proved to be excellent models for the study of evolutionary conflict and its consequences for behaviour and group dynamics. Evolutionary conflict arises whenever the optimum fitness outcomes for the participants in an interaction cannot all be achieved simultaneously. In the case of cooperative breeders, the role of breeder is usually more profitable (in terms of fitness) than the role of helper, which generates

The Evolution of Parental Care. First Edition. Edited by Nick J. Royle, Per T. Smiseth, and Mathias Kölliker.

evolutionary conflict over reproductive roles and shares of reproduction. Conflict arises over helping effort because investment in helping usually trades off against a helper's own residual reproductive value, so each helper would prefer other group members to invest more. Studies of within-group conflicts in cooperative breeders have provided insights into how groups remain stable despite selection for selfishness, and the ways in which evolutionary conflicts of interest within groups can be resolved on an evolutionary time scale, for example, by the evolution of morphological specialization (Bourke 1999), or fertility schedules which eliminate reproductive overlap within groups (Cant and Johnstone 2008); and also on a behavioural time scale, through mechanisms of 'negotiation' (Cant 2011; McNamara et al. 1999). The general principles arising from these studies can help us to understand how conflict is resolved in a range of contexts, such as sexual conflict over mating and conflict between parents (Chapter 9), and between parents and offspring over parental investment (Chapter 7).

The topic of cooperative breeding has been well-reviewed in birds (Cockburn 1998; Hatchwell 2009; Koenig and Dickinson 2004; Stacey and Koenig 1990) and mammals (Russell 2004; Solomon and French 1997). Insect cooperative breeders are usually described as 'primitively eusocial' (because they lack sterile castes) and are not usually considered alongside vertebrates, although there are clear similarities between insect and vertebrate systems (Field and Cant 2009a). Here I focus on four questions that are of interest to researchers working on both taxa. First I discuss current understanding of evolutionary routes to cooperative breeding, and the constraints on dispersal that can lead to the formation of cooperative groups. Second, I outline the main hypotheses for the evolution of helping

Figure 12.1 Examples of cooperative breeding systems. (a) Tropical hover wasps (Stenogastrinae; such as this group of *Parischnogaster alternata*) breed year round in South East Asia in semi-permanent mud nests. Females mate and either attempt to breed independently or form a strict age-based queue to inherit the position of breeder (Field et al. 2006; Photo by Adam Cronin). (b) A group of cooperatively breeding spiders *Stegodyphus dumicola* preying upon a cricket. This species forms colonies of tens to hundreds of individuals in which a large proportion of females are non-breeders. These females help by regurgitating food for the offspring of other females, and are eventually consumed by them (Salomon and Lubin 2007; photo by Mor Salomon-Botner). (c) The African cichlid *Neolamprologus pulcher* forms cooperative groups in which reproduction is monopolized by a single breeding pair. Subordinates delay dispersal and help if there is a shortage of suitable breeding habitat, and prefer to settle with non-kin over kin (Heg et al. 2008; photo by Michael Taborsky). (d) Pied babblers *Turdoides squamiceps* form cooperative groups of 2–10 individuals in the Kalahari Desert. Helpers engage in sophisticated sentinel behaviour, vocal negotiation over cooperation, and active teaching of fledglings (Raihani and Ridley 2008; photo by Alex Thornton). (e) Banded mongooses *Mungos mungo* in Uganda live in groups of 8–60 individuals in which multiple females give birth together in each breeding attempt. After pups emerge from the den they form one-to-one relationships with adult 'escorts' who guard and provision them (Bell et al. 2010; Cant et al. 2010; photo by the author). (f) Human reproductive life-history is characterized by a short-interbirth interval and long period of offspring dependency (Mace and Sear 2005). Offspring are reliant upon the investment of their parents, grandparents, and older siblings for many years, and are cooperative breeders *par excellence* (photo of members of the forager-horticulturalist Phari Korwa tribe, India, by Shakti Lamba).

behaviour based on direct and indirect fitness benefits, and assess the evidence for these in insects and vertebrates. I focus in particular on recent developments in kin selection theory which examine the impact of demography on the evolution of social behaviour and life-history in cooperative species. Third, I discuss recent attempts to incorporate behavioural negotiation into evolutionary models of parental care and cooperation. Fourth, I consider reproductive conflicts that arise within groups over helping effort and reproduction, and theory and empirical tests of how evolutionary conflict over reproduction is resolved. Much of this conflict theory applies equally well to non-cooperative species and can be used to derive insights into how within-family conflict over parental care is resolved on evolutionary and behavioural time scales (see also Chapters 7, 8, and 9).

12.2 Routes to cooperative breeding

In most cooperatively breeding fish, birds, and mammals, groups form when offspring delay dispersal and remain on their natal territory to help their parents rear subsequent broods. In birds, 852 out of 9268 bird species for which parental care systems are known or can be inferred are cooperative breeders (Cockburn 2006). Avian cooperative breeders almost always evolved from socially monogamous biparental ancestors, the mating system exhibited by around 80% of extant birds (Cockburn 2006; Cornwallis et al. 2010). The ancestral mating system of mammals is polygyny rather than social monogamy (Clutton-Brock 1989), but recent phylogenetic analysis suggests that most cooperatively breeding mammals evolved from monogamous ancestors (Lukas and Clutton-Brock in press). In both birds and mammals, therefore, monogamy appears to set the stage for the evolution of cooperative breeding.

In many cooperative insect lineages an important precursor to cooperative breeding is the evolution of progressive provisioning, where mothers remain to guard and provision their offspring during development (Field and Brace 2004). Progressive provisioning and extended parental care facilitate the evolution of cooperation because additional helpers can provide insurance against the death of the mother and defend offspring against predators and parasites while mothers forage (Field and Brace 2004; Gadagkar 1990).

In insects two main evolutionary routes to cooperative breeding and eusociality have been proposed. The *subsocial* route (Wheeler 1928) is similar to that proposed for cooperative vertebrates, namely, that transitions to cooperation occurred through offspring remaining in their natal nest to help their mother (or mother and father, in the case of termites; Korb 2008). The second, *semisocial* route (Lin and Michener 1972; Michener 1958), suggests that cooperation arose among same-generation females who could gain mutualistic and kin-selected benefits from breeding together. Same-generation associations are commonly seen in some bees, ants, and polistine wasps (Lin and Michener 1972). Phylogenetic analysis suggests that high relatedness as a consequence of monogamy is associated with evolutionary transitions from solitary breeding to cooperative breeding and eusociality in Hymenoptera (Hughes et al. 2008) and in termites (Boomsma 2009). Ancestral monogamy in Hymenoptera has been taken as evidence in support of the subsocial route to cooperative breeding, because in subsocial associations helpers can expect to raise full rather than half siblings, whereas monogamy would appear to offer no clear advantage to sociality via the semisocial route (Boomsma 2009).

These comparative analyses emphasize the importance of kin structure for cooperative transitions, but it is clear that ecological conditions also play a major role in the origin and maintenance of cooperative breeding. In birds, offspring delay dispersal when there is a shortage of suitable breeding habitat (e.g. Komdeur et al. 1995), where there are high indirect or direct fitness benefits of philopatry, for example through inheritance of breeding positions (Dickinson and Hatchwell 2004); and, across species, where ecological environments are temporally variable, since this allows groups to breed in both harsh and benign years (Rubenstein and Lovett 2007). Similar factors may also promote group formation in cooperatively breeding cichlids (Wong and Balshine 2011). In mammals, delayed dispersal of offspring does not appear to be the result of habitat saturation, since vacant habitat often remains unutilized (Russell 2004). Rather, dispersal is often costly in mammals because males aggressively defend access to mates and because dispersing individuals are susceptible to predation and attack by conspecifics. By delaying dispersal, young adults can remain in a 'safe haven' and may inherit breeding status if same-sex dominants die. Such are the benefits of philopatry for young adults that in many species of mammal dominant individuals often go to considerable lengths to forcibly evict subordinates, while the subordinates themselves appear highly reluctant to leave (Clutton-Brock 2002; Johnstone and Cant 1999).

In insects harsh ecological conditions and high adult mortality are suggested to promote the formation of groups via the subsocial or semisocial routes. In primitively eusocial wasps, for example, dispersal is typically not constrained by a lack of available breeding habitat: nests can be constructed on a range of vegetation types or substrates. There are nevertheless severe constraints on independent breeding because mothers have a high probability of dying in the extended period for which offspring are dependent upon their care. Offspring that stay to help their mother or join the nesting associations of same-generation females can provide insurance against the failure of the nest due to the death of their mother or the dominant female on the nest (Gadagkar 1990; Queller 1994). In addition, insect cooperative breeders frequently form social queues and can often gain greater direct fitness from queuing than they can from independent nesting (Field and Cant 2009b; Leadbeater et al. 2011). In some socially polymorphic species (e.g. some halictine bees and *Polistes* species) individuals adopt a cooperative or non-cooperative life history depending on latitude and length of the summer breeding season. In the bee *Halictus rubicundus*, for example, overwintering females transplanted from northern to more southern latitudes switched from a solitary to a cooperative life-history; bees transplanted in the reverse direction switched to solitary breeding (Field et al. 2010; Fig. 12.2). The key ecological variable in this case is the length of the summer season: at the more southern latitudes foundresses can raise two broods in a season rather than one, so that first brood offspring have the option to become helpers.

From this brief survey it is clear that multiple factors may contribute to delayed dispersal of offspring and subsequent evolution of cooperative breeding. Monogamy makes the transition to cooperative breeding easier for both insects and vertebrates, and in insects progressive provisioning sets the stage for cooperative breeding because helpers then become particularly useful. In vertebrates, delayed dispersal arises where there is a shortage of suitable habitat or mates or strong barriers to group entry. In insects helpers may do best to remain in their natal group because independent nesting entails high mortality, and in Hymenoptera from a life-history in which solitary breeders have a high chance of dying before their offspring are fully developed. In each case the consequence of constraints on dispersal is a genetically structured population, in which relatedness to local group members is on average greater than relatedness to the breeding population at large (Hatchwell 2010). In the next section I explore how population structure influences selection for any traits that have social effects, such as helping or breeding.

12.3 Selection for helping behaviour

Helpers contribute to the rearing of offspring in a variety of ways: assisting with nest construction, provisioning of offspring, babysitting young, and defence of a nest or territory against conspecifics and predators. These behaviours involve measurable survival costs or costs to attributes which are likely to correlate with direct fitness, such as mating success or condition. A central challenge has been to explain how these behaviours can evolve and persist in populations despite these costs. Selection will favour alleles for helping behaviour only if the fitness costs are offset by benefits either to the helper themselves or to other individuals in which copies of the 'helping' alleles reside. This has led to four main hypotheses to explain helping behaviour, the first two of which are usually lumped together as forms of 'kin selection': 1) indiscriminate helping may be favoured if dispersal is limited so that the recipients of help are on average more closely related than the population at large (Hamilton 1964); 2) individuals may recognize kin and preferentially direct care towards them (Hamil-

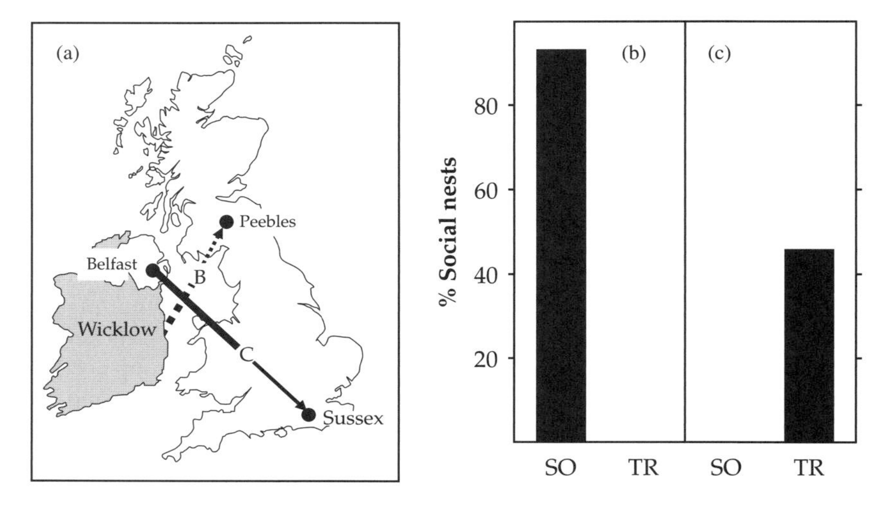

Figure 12.2 Sweat bees (Halictinae) include both solitary and cooperatively breeding species, and in some species cooperative behaviour varies between populations. Females of *Halictus rubicundus* dig a nest burrow in spring. In southern Ireland, *Halictus rubicundus* (Wicklow, Eire) breeds cooperatively. Mated overwintering females dig a burrow nest in spring and produce a first brood of 5–7 offspring which remain to help their mother raise a second, larger brood of reproductives. In Northern Ireland populations (Belfast), nests are started later in spring and only one brood is produced per season, with no helping among offspring. Transplantation of overwintered females to southern and northern latitudes of Great Britain (a) resulted in females switching from cooperative to non-cooperative life-histories (b) and (c). Panel (b) shows the percentage of nests from the cooperative Wicklow population that became cooperative at the source site (SO) and at the site to which they were transplanted (TR). Panel (c) shows the results for the transplant experiment on the Belfast population. Reproduced from Field et al. (2010) with permission from Elsevier.

ton 1964); 3) there may be immediate or delayed direct fitness benefits which outweigh the immediate fitness costs (Kokko et al. 2001); and 4) helping may be enforced by social punishment, so that the alternative, not helping, results in even greater fitness costs (Clutton-Brock and Parker 1995; Gaston 1978). It is important to recognize that these hypotheses are not mutually exclusive: for example, helping may benefit relatives while at the same time result in direct fitness benefits to the actor.

Given that the great majority of cooperatively breeding species are composed of groups of genetic relatives, mechanisms based on kin selection seem to offer a plausible and general explanation for the evolution of helping behaviour. However, the discovery of hardworking, unrelated helpers in both insect (Leadbeater et al. 2010; Queller et al. 2000) and vertebrate (Clutton-Brock et al. 2000; Reyer 1984) systems indicates that direct fitness benefits or coercion must also be important in maintaining helping behaviour, at least in some systems. Moreover, it is important to remember that the factors that could have initially promoted cooperative breeding may be very different from those that select for cooperative behaviour in extant systems.

12.3.1 Demography and indiscriminate altruism

Altruistic traits can be favoured by natural selection if they satisfy Hamilton's rule (Hamilton 1964)

$$r_{xy}B - C > 0 \qquad (12.1)$$

In which C is the lifetime direct fitness cost to the focal bearer of a trait; B is the lifetime direct fitness benefit to social partners resulting from the trait; and r_{xy} is the coefficient of relatedness of the focal individual x to its social partner y, that is, their genetic similarity relative to the population mean (Hamilton 1964). Since costs and benefits are defined in terms of lifetime fitness effects, acts which involve immediate costs need not be altruistic. For example, an actor who pays an immediate cost to help another is not behaving altruistically if this cost is repaid later, for example, through reciprocation by the recipient. Failure to consistently employ Hamilton's definition of altruism in terms of lifetime fitness effects has caused much misunderstanding in the past, and is one of the sources of confusion underlying recent attacks on Hamilton's theory (e.g. Nowak et al. 2010; but see Rousset and Lion 2011 for a counter-critique). Note also that a focus on Hamilton's rule doesn't mean ignoring the role of ecology in the evolution of cooperation, such as whether animals build a safe nest or live in saturated habitats. Factors such as these can raise the benefits, B, and lower the costs, C, of helping, and thereby favour the evolution of altruism (Alexander et al. 1991).

From the outset Hamilton recognized two distinct ways in which altruism may evolve. First, actors might recognize and preferentially direct help toward their genetic relatives. Second, and more controversially, constraints on dispersal may ensure that actors interact primarily with close kin. The problem with this second mechanism is that limited dispersal (or population 'viscosity') increases both average relatedness between social partners and the intensity of local competition for resources or reproductive opportunities. To illustrate, suppose that an actor engages in some behaviour (such as guarding the nest, or provisioning young) which results in b extra offspring for a recipient, at a cost of c fewer offspring for the actor. As a consequence of the act, the overall change in the number of offspring produced locally is $(b - c)$. Will this act be favoured by natural selection? If all these extra offspring disperse away from the patch and compete with unrelated individuals, there are no further fitness consequences of helping and the bs and cs (which measure numbers of offspring) can be used as reasonable proxies for the Bs and Cs in Hamilton's rule (which measure lifetime fitness effects) to determine the direction of selection on the helping act. If, however, there are constraints on dispersal then extra offspring produced may remain locally and compete with offspring to which the actor may be related (through local density dependent regulation). Selection will then favour helping if the following extended version of Hamilton's rule is satisfied (Queller 1992):

$$r_{xy}b - c - r_{xe}(b - c) > 0 \tag{12.2}$$

where r_{xe} is mean relatedness of the actor to the offspring that are displaced by competition as a result of the helping act. Note that this inequality can equally be used to predict selection for 'harming' behaviour or traits, which are defined as traits that *reduce* the fecundity of local breeders, such that b is negative (Johnstone and Cant 2008; West et al. 2002). Where competition is global $r_{xe} = 0$ because displaced individuals are a genetically random sample of the population. When competition is local $r_{xe} > 0$, and hence inequality (12.2) is harder to satisfy for a given (positive) b and c. In fact the first model to examine explicitly the consequences of local kin competition for the evolution of altruism (Taylor 1992, building on the 'infinite island' population genetic model of Wright 1943) found that the positive effect of limited dispersal on selection for altruism was exactly cancelled by the negative effects of increased competition. In other words, according to Taylor's model, Hamilton's second mechanism for the evolution of altruism didn't work.

Subsequent theoretical work by Taylor and others has shown that this 'cancelling' result arises because of the simplifying assumptions of the model, and that incorporating plausible demographic and life-history features (e.g. overlapping generations, sex-biased dispersal, budding dispersal, individual variation in fecundity) recovers the prediction that dispersal constraints select for helping (Lehmann and Rousset 2010). Empiricists know that experimental manipulation of ecological constraints can affect individual dispersal and helping decisions on a behavioural time scale (see Section 12.2 above)—what the new models suggest is that severe ecological constraints on dispersal over many generations can favour the evolutionary origin of local helping. Indeed, not only helping behaviour but any social trait which increases the fecundity of local group members is more likely to evolve when there are strong dispersal constraints.

Two factors which promote the evolution of indiscriminate altruism are particularly relevant to both insect and vertebrate systems: sex-biased dispersal and budding dispersal. These are considered below.

12.3.1.1 *Sex biased dispersal*

In most cooperative breeders one sex disperses more frequently, or further, from their natal patch than the other (Lawson Handley and Perrin 2007). Sex biased dispersal of this kind likely evolved as a strategy to reduce inbreeding and its associated deleterious effects, and because the advantages of philopatry are often different for males and females (Lawson Handley and Perrin 2007). But it also has profound consequences for the evolution of social behaviour in males and females, and hence social structure and mating system.

Johnstone and Cant (2008) extended Taylor's (1992) approach to explore how sex differences in dispersal influence selection for 'helping' and 'harming' behaviour. They showed that where there is a strong sex bias in dispersal, selection favours harming behaviour among adults of the dispersing sex; and in general favours helping among adults of the philopatric sex. This prediction agrees with observations of dispersal and helping in cooperatively breeding birds and mammals, where in general it is the philopatric sex which provides most help (Cockburn 1998; Russell 2004). In cooperatively breeding cichlids, both sexes disperse and both sexes provide help (Stiver et al. 2004). Extending the model to haplodiploid organisms (Johnstone et al. 2012) show that male-biased dispersal, which is widespread in social Hymenoptera, strongly selects for helping among females. Thus, incorporating sex biased dispersal and local competition into Hamilton's rule suggests that Hamilton (1964) was in fact correct in his original claim that 'family relationships in Hymenoptera are potentially very favourable to the evolution of reproductive altruism'.

An important message of these models is that not only behaviour but also life-history traits are shaped by kin selection in the same way as are acts of helping or harming. Decisions such as whether to breed or not and how many offspring to produce, and life-history traits such as the rate of senescence, will affect the fitness of other local group members and thus constitute forms of indiscriminate help or harm. Consider, for example, the timing of reproduction. Where selection favours helping early in life and harming later, females may gain from delaying reproduction and instead helping early in life; where the reverse is true, females may gain from early reproductive cessation and late-life helping, in other words, menopause (Cant and Johnstone 2008). It turns out that sex biased dispersal and patterns of mating interact to determine the strength of selection for helping and harming across the lifespan (Johnstone and Cant 2010). Specifically, local mating and male-biased dispersal, the demographic pattern exhibited by most cooperative mammals, results in females becoming less closely related to local group members as they get older, and hence stronger selection for helping early in life compared to later. However, two unusual (and different) demographic patterns result in the opposite pattern. Specifically, a pattern of female-biased dispersal and local mating (thought to characterize ancestral humans), and non-local mating with low dispersal by both sexes (characteristic of pilot whales *Globicephala spp* and killer whales *Orcinus orca*, the two other species which exhibit menopause), results in selection for early reproductive cessation and late life helping.

12.3.1.2 Budding dispersal

The usual assumption of infinite island models is that individual offspring disperse alone and join groups in which there are no other relatives present. Gardner and West (2006), however, explored selection for helping under 'budding dispersal', that is, when juveniles disperse in groups. Budding dispersal is conducive to the evolution of helping behaviour because it decouples the positive and negative consequences of dispersal: extra offspring produced as a result of help can disperse away from their parents (so avoiding competition with kin) but still form groups in which relatedness is high.

This model helps to explain the evolution of group dispersal in cooperative insects and vertebrates. In many species, young adults form dispersal coalitions to seek out or compete for vacant territories, or to take over existing groups. In acorn woodpeckers (*Melanerpes formicivorus*), for example, helpers form sibling groups which compete intensely for any reproductive vacancies that appear (Koenig et al. 1998; similar dispersal coalitions are found in other cooperatively breeding birds and mammals, e.g. Port et al. 2010; Sharp et al. 2008). Dispersal coalitions may also arise because multiple young adults are evicted by older dominant breeders (e.g. Cant et al. 2010). A form of budding dispersal occurs in 'independent-founding' social vespid wasps of the genera *Polistes*,

Belonogaster, *Mischocyttarus*, *Parapolybia*, and *Ropalidia*, in which colonies are founded by multiple inseminated foundresses, independently of any workers (Gadagkar 1991; Reeve 1991). The benefits of joining a dispersal coalition will depend on social structure and the distribution of reproduction within groups, for example, whether breeding is monopolized by a single individual (as is the case in most independent founding wasps) or shared more evenly among group members (as occurs in banded mongooses). The evolution of group dispersal may also involve an element of positive feedback: once group dispersal is common, dispersers may be selected to join larger coalitions which can compete more effectively; or evictors selected to expel larger numbers of their offspring, but these benefits will eventually be offset by the costs of elevated within-group competition. Group dispersal in cooperative breeders is a promising area for further theoretical and empirical work, and may have hitherto unexplored impacts on individual behaviour, group structure, and population dynamics.

12.3.2 Discriminate altruism: kin directed care

Kin discrimination should promote the evolution of helping because Hamilton's rule is easier to satisfy if helpers can direct care towards more closely related group members. In this case relatedness r_{xy} is by definition higher than the average relatedness to all potential recipients. Since the ability to preferentially aid kin increases the inclusive fitness pay-off of costly helping, we might expect selection for mechanisms which enable helpers to single out close relatives among members of their social group. Helpers could learn to recognize kin if other factors ensured that individuals with whom helpers interact most frequently, or those in closest physical proximity, were reliably genetically related. In cooperatively breeding birds, kin recognition is typically based on cues that are learnt during development in the nest (Komdeur et al. 2008). There is less evidence that genetic similarity can be recognized directly and there are some theoretical difficulties with this idea. Selection could in principle favour the spread of 'marker' alleles which allow relatives to recognize and direct care toward each other, but the more common such markers became the less useful they would be in distinguishing kin from non-kin (Crozier 1986). For genetic kin recognition systems to work and to promote the evolution of altruism requires that genetic 'rarity' is associated with some other fitness advantage, such as resistance to parasites or pathogens (Rousset and Roze 2007). This may explain why known examples of genetic kin recognition in mice and humans involve detection of similarity at major histocompatibility (MHC) genes which are highly polymorphic and involved in immune function (Rousset and Roze 2007).

Helpers may potentially be able to discriminate kin from non-kin using learned or genetic cues, but do they use these cues when allocating help? In cooperative birds and mammals, some studies find evidence of kin-directed care while others do not: across 9 species, variation in relatedness explains 10% of the variation between helpers in the probability of helping (Cornwallis et al. 2009; Griffin and West 2003). One result to emerge from these comparative analyses is that there is typically no relationship between relatedness and helping effort (as opposed to the probability of helping; Cornwallis et al. 2009; Griffin and West 2003). This suggests that the costs and benefits of helping are more important determinants of individual helping effort than is variation in relatedness (Cornwallis et al. 2009).

In cooperatively breeding insects, there is little evidence that helpers discriminate between kin and non-kin within groups. Typically helpers distinguish between nestmates and non-nestmates, but do not distinguish degrees of relatedness among nestmates (Keller 1997). In *Polistes dominulus*, for example, 20–30% of helpers are non-relatives but there is no difference between related and unrelated helpers in foraging effort, nest defence, aggression or inheritance rank (Leadbeater et al. 2010; Queller et al. 2000). As with vertebrates, kin-biased helping may not be favoured because of the costs of recognition errors. In insects that share a nest, there may be few environmental cues to distinguish kin from non-kin. In paper wasps, for example, it has been argued that cues based on cuticular hydrocarbons (implicated in kin discrimination among larvae in solitary insects) may be unreliable since these can be acquired from contact with the nest (Gamboa 2004). However, recent studies have shown that unrelated helpers in *Polistes dominulus* have measurably different hydrocarbon profiles (E. Leadbeater and J. Field, personal communication). While cues to discriminate kin exist, therefore, they are not used by wasps in helping decisions.

12.3.3 Direct fitness benefits

Examples of species with hardworking, unrelated helpers (for example, *Polistes dominulus*, Leadbeater et al. 2010; Queller et al. 2000) suggest that helping can also yield direct fitness benefits, and that in some cases these direct benefits alone may be sufficient to outweigh the fitness costs of helping. When considering direct fitness benefits of helping it is useful to distinguish those benefits that are non-enforced or enforced (Gardner and Foster 2008). In both cases helping is more profitable than non-helping, but in the case of enforced benefits it is the threat or action of social partners that reduces the pay-off of the non-helping option.

12.3.3.1 Non-enforced benefits

Helping can be readily explained if it results in some form of immediate or delayed direct fitness benefit which offsets the initial cost of the helpful act. In this case helping is a form mutualism (Gardner and Foster 2008; West et al. 2007). Several behavioural mechanisms which result in delayed benefits of helping have been proposed to operate in cooperative breeders, including the acquisition of parenting skills; the recruitment of offspring into the group which later become helpers themselves; and elevated social status or dominance (reviewed by Dickinson and Hatchwell 2004; Koenig and Walters 2011). The last two mechanisms have received some theoretical attention under the terms 'group augmentation' (Kokko et al. 2001) and 'prestige' (Zahavi 1995) respectively.

An actor can gain group augmentation benefits if the extra offspring produced as a result of helping remain in their natal group and boost the actor's future survival or reproductive success (Kokko et al. 2001). The delayed direct fitness of helping will be especially important where helpers have a good chance of inheriting breeding status, and where large group size is associated with elevated survival or reproductive output—two conditions which commonly hold in cooperatively breeding insects and vertebrates. The key assumptions to test are 1) that helping leads to increased recruitment and larger future group size, and 2) that a larger group size is beneficial to the direct fitness of helpers (Wong and Balshine 2011). Studies of insect and vertebrate cooperative breeders often report an association between group size and breeder productivity, but do not show that helping per se leads to elevated future fitness for helpers. In paper wasps and hover wasps group augmentation benefits do not appear to be a major determinant of helper effort: helpers reduce their helping effort as they get closer to inheriting, a pattern which is opposite to that predicted by the group augmentation hypothesis (Field and Cant 2007). In birds there is also scant evidence that variation in group augmentation benefits underlies variation in helping effort (but see Kingma et al. 2011 for one such case).

The prestige hypothesis suggests that helping evolves as a costly signal of quality, and that the costs of helping are offset by the fitness benefits of improved mating access or dominance status that result from this honest advertisement of quality. Initial evidence for the hypothesis came from observations of Arabian babbler *Turdoides squamiceps* helpers competing with each other to help (Carlisle and Zahavia 1986); although a subsequent study on the same species did not replicate this finding (Wright 1999). The key predictions of the prestige hypothesis are 1) that helpers should help more in the presence of an audience; and 2) that elevated helping effort should causally increase social status or mating success. McDonald et al. (2008) tested the first of these predictions in cooperatively breeding bell miners, but found that helpers did not adjust their helping effort to the presence or absence of an audience (the breeding male or female). In other species, helpers that invest most have a higher probability of obtaining breeding status, but this may simply reflect variation in helper quality rather a causal link between helping and future mating success (Cant and Field 2005). Overall, evidence for the prestige hypothesis is scarce.

12.3.3.2 Enforced benefits

Much theoretical interest in evolutionary biology has focused on the use of punishment and threats to induce cooperation and helping (Cant 2011; Ratnieks and Wenseleers 2008). In the context of cooperative breeding, the pay to stay hypothesis (Gaston 1978) suggests that dominants can exploit the gains that subordinates derive from group membership to charge 'rent' in the form of help. This mechanism is based on the use of a threat or 'last move' in the interaction: helpers work to rear the offspring of dominants to avoid expulsion from the group. When the threat is clear and credible, no evictions will be observed, so an effective threat is a highly cost-effective means of social control (Cant 2011). Alternatively, dominants might coerce subordinates into helping via the use of punishment. Punishment differs from threat in that it involves a repeated interaction rather than a last move: dominants might pay an immediate cost to punish a lazy helper if this act induces the helper to reciprocate by working harder in the future.

The key prediction of the pay to stay hypothesis is that experimental reduction of helper effort should lead to eviction from the group. In splendid fairy wrens and cooperative cichlids, experiments to temporarily remove subordinates helpers or reduce their helping effort typically led to increased aggression from dominants and higher rates of subordinate helping thereafter (Balshine-Earn et al. 1998; Bergmüller and Taborsky 2005; Mulder and Langmore 1993). However, none of these experimental manipulations led to helpers being expelled from the group, as would be expected if helpers were induced to help because of the hidden threat of eviction by dominants.

In other species, there is evidence that dominants use aggression to activate lazy workers. In naked mole rats (*Heterocephalus glaber*), for example, queens direct aggression (in the form of 'shoves') toward lazy workers to increase their activity level (Reeve 1992). In meerkats (*Suricata suricatta*), hardworking male helpers are subject to less aggression from dominant breeders than lazy male helpers, and males that 'false feed' (i.e. bring food over to a pup but then eat the item themselves) are subject to more aggression than those that do not engage in this behaviour (Clutton-Brock et al. 2005). In the paper wasp *Polistes fuscatus*, the removal or inactivation of dominant foundresses (by cooling them) leads to reduced helper effort (Reeve and Gamboa 1987); and wing-clipping of subordinate helpers leads to increased aggression from dominants (Reeve and Nonacs 1997), as expected if aggression is used to enforce help. A difficulty with these studies is that the function of behaviours classed as aggressive is often unclear. In *Polistes* for example, 'dart' behaviour is usually classed as aggression but may instead serve as a cooperative signal to coordinate worker activity (Nonacs et al. 2004). In *P. dominulus*, dominants are often aggressive to subordinates, but this appears to be linked to conflict over social rank rather than conflict over help (Cant et al. 2006).

In summary, patterns of helping in cooperative breeders provide general evidence for Hamilton's first mechanism based on indiscriminate altruism, namely that increasing constraints on dispersal and the presence of relatives should promote helping and inhibit harming behaviour. Inclusive fitness models which incorporate demography and population structure are also an important step toward an evolutionary theory of 'cooperative life history'. Variation in relatedness does not correlate well with helping effort within groups, which parallels findings on the relationship between parentage and parental care (Chapter 11). There is little evidence that threats of eviction induce helping, although dominants in some species do use punishment to enforce helping. In some cooperatively breeding insects, helpers adjust their effort according to their expected future fitness (Field and Cant 2009b). However, more research is required to understand individual variation in helping and, importantly, variation in the consistency of individual contributions to helping (their 'cooperative personality' Bergmuller et al. 2010; English et al. 2010).

12.4 Negotiation over help

The preceding discussion highlights the evolutionary conflicts that exist within cooperatively breeding groups over levels of investment. Coercion—via punishment and threats—is one behavioural manifestation of this conflict. In many species, however, coercion may be impractical or inefficient: in birds, for example, the targets of punishment can fly away. Nevertheless, individuals may be able to induce others to help more by adjusting their own helping effort contingent on the helping effort of their social partners, in a process of bargaining or 'negotiation'. Negotiation is by definition a behavioural interaction: a process of bid and counter-bid. By contrast, classic models of parental care exclude bargaining because they solve for evolutionarily stable combinations of genetically specified fixed 'sealed bid' efforts, so called because it is assumed that players can't change their effort after observing that of their partner (Houston and Davies 1985). Sealed bid models have the advantage of mathematical tractability, but their assumptions are at odds with a wealth of evidence that animals typically observe and adjust their helping effort to that of their social partners (Johnstone 2011; McNamara et al. 1999). Allowing players to observe and respond to each other's helping effort on a behavioural time scale can render sealed bid equilibria evolutionarily unstable, at least in cases where there is some variation in individual efforts due to noise or variation in quality (McNamara et al. 1999; Johnstone 2011).

To address this issue, McNamara et al (1999) developed one type of 'negotiation' model (again in the context of biparental care) in which they solved for evolutionarily stable levels of behavioural responsiveness or 'rules for responding', rather than evolutionarily stable fixed efforts. Where increasing investment brings diminishing productivity returns, and each party has perfect information about offspring need, the ESS rule for responding is to partially compensate for changes in each other's effort levels. However, other response rules (for example, effort 'matching', where an increased effort by one parent results in increased effort by its partner) can be evolutionarily stable where parents have incomplete information about the level of offspring need, and use each other's effort levels to estimate this (Johnstone and Hinde 2006). Variation in how parents obtain information may help to explain the range of responses to experimental manipulation of parental effort that have been observed in biparental birds, including no change in partner effort level, partial or full compensation, and effort matching (Hinde 2006; Chapter 9). In the context of cooperative breeding, Johnstone (2011) adapted McNamara's negotiation approach to show that helping may often benefit mothers and fathers (via 'load lightening') as much as offspring, so that relatedness of helpers to each parent and the responsiveness of parents may be a more important determinant of helping effort than average relatedness to the brood. The predictions of this model have not yet been tested. Moreover, there have been no tests of the assumptions or predictions of any negotiation model in cooperatively breeding insects.

McNamara et al.'s (1999) model is an important first step towards an evolutionary theory of behavioural negotiation, but there are many other forms that bargaining or negotiation might take. Some biological interactions, for example, can be thought of as consisting of a sequence of 'moves', in which one player commits to a level of investment which is observed and responded to by their social partners. For example, in birds and mammals mothers effectively make a 'first move' by allocating resources to the egg, while their mate (or a helper) is placed in the role of a 'second mover' who must choose how much effort to invest after birth or hatching. Selection should favour mothers that adjust their investment in eggs according to the amount of help their offspring will receive later (Russell et al. 2007). Empirical tests of this idea present an interestingly mixed picture. In splendid fairy wrens (*Malurus splendens*; Russell et al. 2007), carrion crows (*Corvus corone*; Canestrari et al. 2011), and cooperative cichlids (*Neolamprologus pucher*; Taborsky et al. 2007), mothers reduce their investment in eggs when helpers are present, but in the wrens and crows this reduction is offset by the investment received from helpers post-hatching. By contrast, in acorn woodpeckers (*Melanerpes formicivorus*), mothers do not decrease investment per egg when helpers are present, but instead lay a greater number of eggs, so each offspring is worse off overall (Koenig et al. 2009). A recent model by Savage et al. (in press) suggests that these contrasting empirical patterns might be explained by variation in the personal fitness costs of increasing clutch size. Where clutch size is constrained by high costs of egg production, mothers are predicted to reduce their investment per egg, and helpers to compensate for this reduction so that each offspring receives greater resources overall. Where clutch size is more flexible, females should produce more eggs in the presence of helpers, to the detriment of each individual offspring (a situation similar to the results found in acorn woodpeckers). More work is needed to test the specific predictions of the model, and to determine the general conditions for which mothers might gain from 1) producing fragile offspring to attract compensatory care from helpers; versus 2) producing hardy clutches, or perhaps a mixture of hardy and fragile young, when helpers are present.

12.5 Reproductive conflict

Within cooperatively breeding groups there are usually strong asymmetries in fitness between breeders and helpers which leads to evolutionary conflict over reproduction, and often to intense competition among group members to monopolize reproduction (Cant and Johnstone 2009). Much research over the last 30 years has focused on understanding how this conflict is resolved, and the evolutionary causes of variation in the distribution of reproduction, or the degree of reproductive skew, between groups and between species. Many researchers were drawn to working on reproductive skew because there existed a simple candidate model which potentially applied very widely (the 'concession' model; Reeve 1991; Vehrencamp 1983). This model assumed that a single dominant individual controlled reproduction in the group, but that subordinates could use the threat of departure from the group to extract a reproductive concession or 'staying incentive' from dominant individuals, at least in cases where the presence of subordinates boosted the reproductive success of dominants. Thus in this model the level of skew in groups was determined by the inclusive fitness value of 'outside options' to subordinates, that is, their fitness pay-off should they choose to disperse to breed elsewhere. The model suggested that variation in skew both within and between species was explicable by variation in three parameters: relatedness, ecological constraints, and the productivity benefit of retaining subordinates.

Starting in the 1990s a number of other models appeared which relaxed the assumptions of the concession model, and produced quite different predictions. For example, 'incomplete control' models (Cant 1998; Reeve et al. 1998) assumed that no single individual had cost-free control over the allocation of reproduction, but rather that both dominants and subordinates could invest costly effort to increase their share of reproduction; while the 'restraint' model (Johnstone and Cant 1999) assumed that subordinate reproduction was limited only by the threat of eviction from the group. These and other models led to a confusing array of predictions and a tangled theoretical picture, at least initially. As Gardner and Foster (2008) put it:

> From this simple beginning skew theory diversified into a comedy of additional models, each differing in their specific assumptions about the power of individuals, the information available, and whether and how individuals negotiate their reproductive share.

The problem with skew research, however, does not lie in the diversification of models: as a rule the development of different models represents a progression in the understanding of a natural system, since models can only be rejected by comparison with other models (Hilborn and Mangel 1997; Lakatos 1978). In the case of skew theory, the various models can be classified into those that assume the resolution of conflict is influenced by threats to exercise 'outside options' (such as leaving the group, or evicting a competitor), versus those that assume outside options are irrelevant. This distinction provides an opportunity to distinguish between and eliminate models, and to clear away some of the theoretical tangle.

The main problem with the research on reproductive skew has been a dearth of empirical tests of the models' assumptions, particularly experimental tests. Two notable exceptions are the experimental studies of Langer et al. (2004) (on a social bee *Exoneura nigrescens*) and Heg et al. (2006) (on a cooperative cichlid *Neolamprologus pulcher*), both of which manipulated the value of outside options (i.e. breeding opportunities outside the group) to subordinates to test whether this influenced the outcome of reproductive conflict. In both cases varying outside options had no effect on skew, suggesting that threats of departure do not determine the pattern of reproduction in natural groups. In cooperatively breeding birds, experimental suppression of paternity share (by removing males during the female's fertile period) has not been shown to lead to the departure of subordinates, as expected if the pattern of reproductive sharing among males reflects a subordinate's threat of departure (Cant 2006). In banded mongooses (*Mungos mungo*), experimental suppression of dominant or subordinate breeders (using short-acting contraceptives) never leads to their departure from the group, or to the eviction of subordinates. Suppression does, however, trigger mass infanticide shortly after birth. In banded mongooses, therefore, it appears that low reproductive skew arises from a 'Mexican standoff'—any female that attempts to monopolize reproduction is very likely to have her litter killed (see also Hodge et al. 2011; Fig. 12.3).

Current evidence, therefore, suggests that the value or availability of outside options does not determine the level of reproductive skew within groups, although more experiments are needed. This fits with recent theory which suggests that outside options will be least relevant in groups of relatives and where the productivity benefits of association are high—exactly the conditions that apply to most cooperative breeders (Cant and Johnstone 2009). Rather, the outcome of reproductive conflict within groups appears to depend on the ability of one party to suppress the reproductive attempts of other group members, and the costs of these attempts at suppression. Consequently, the hope that the simple framework offered by early skew models could provide a universal explanation for variation in skew both within and between species seems to have faded. The spotlight has shifted to understanding the evolution of conflict strategies within cooperative groups: how animals suppress each other's breeding attempts, how conflicts are settled on a behavioural time scale, and why the outcome of reproductive conflict is so variable. The study of 'reproductive skew' has therefore given way to the study of 'reproductive conflict', which highlights conflict mechanisms that operate on a behavioural time scale as well as evolutionary outcomes, and the value of experiments over correlations (see also Chapters 7, 8, and 9).

In many cooperative species, reproductive conflict is resolved in a costly and wasteful manner, for example through egg destruction, nest destruction, infanticide, and aggression (Cant 2012). An important question is whether evolutionarily stable resolution of conflict always requires individuals to invest costly effort in conflict, or whether selection can favour less costly and more efficient resolutions mechanisms. Theory suggests two ways in which evolutionary conflicts need not be manifested in actual conflict: 1) through the use of an effective deterrent threat (Cant 2011; Cant et al. 2010); and 2) through mechanisms which make conflict investment unprofitable (Cant 2012). Threats can lead to efficient conflict resolution because they need only be carried out when the social rules they enforce are broken. For a threat (e.g. of departure, eviction, or attack) to be effective in this way requires effective communication and the ability to discriminate transgressors. Data from social hymenoptera, social fish, and mammals indicates that such effective communication exists in some systems, because threats of attack or infanticide do successfully deter subordinates from breeding or challenging the position of dominant (reviewed in Cant 2011). In fish size hierarchies, for example, dominant fish use the threat of eviction from the group to limit the growth and competitive ability of subordinates, but conflict is resolved without them having to carry out this threat (Cant 2011).

Second, evolutionary conflict (i.e. a disparity between the fitness optima of social partners; Chapter 9) may exist, but need not be manifested in costly or destructive acts. Outcomes featuring zero actual conflict are possible when biological conflict takes the form of 'suppression competition' (for example, infanticide, policing, mate guarding) in which success in competition depends on eliminating or nullifying the competitive acts of others. By contrast overt conflict is always expected in 'production competition' where success depends on maximizing proportional effort or competitive acts (as is the case, for example, in biological 'scrambles'; Chapter 8). To illustrate the biological distinction between production and suppression competition, consider two female birds laying eggs in a shared nest. Where competition takes the form of a scramble between offspring after hatching, each female's fitness pay-off will depend on her proportional representation in the communal clutch. A female who

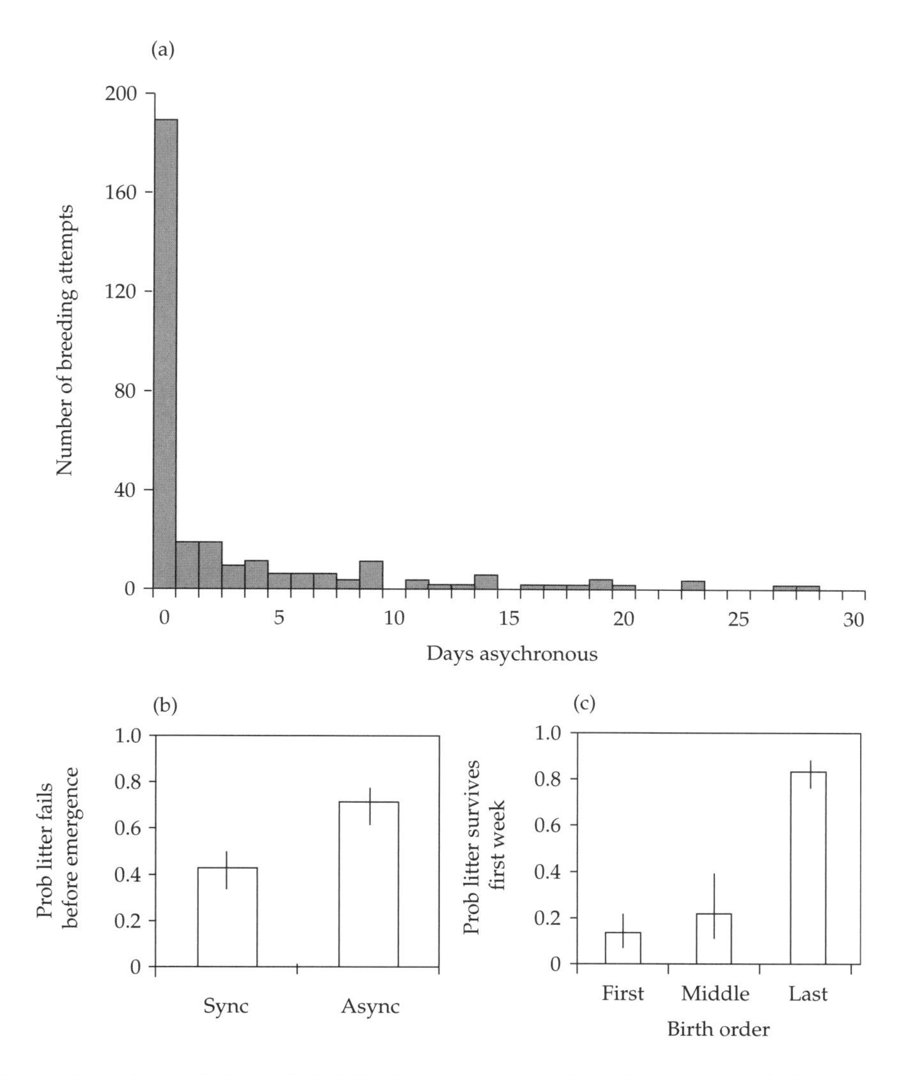

Figure 12.3 Extreme birth synchrony and evidence of infanticide in banded mongooses. In this species most adult females breed in each breeding attempt and typically give birth on exactly the same day. (a) Frequency histogram of synchronous and asynchronous breeding attempts. Females give birth on the same day in 63% of breeding attempts. (b) Synchronous communal litters are less likely to fail in the den. (c) In those litters that are asynchronous, females that give birth first are very likely to fail in the first week compared to those that give birth last. This dependency of early pup survival on the pregnancy status of co-breeders is a signature of infanticide. Extreme birth synchrony in this species appears to be an adaptation to avoid infanticide and minimize competitive disparities between young. Reproduced with permission from Hodge et al. (2011).

invests nothing in competition (i.e. lays no eggs) is certain to get zero fitness pay-off. If, however, competition takes the form of infanticide after eggs are laid, females who invest nothing in infanticide may still achieve some reproductive success, particularly if egg discrimination is not perfect. A recent model explores this type of suppression competition and shows that asymmetry in strength and uncertainty about strength or conflict effort can promote peaceful resolution of evolutionary conflict, even among unrelated individuals (Cant 2012). The level of information in the contest has a strong effect on the costliness of behavioural conflict resolution, and can be expected to shape signalling strategies and dominance interactions (e.g. Chapters 7 and 8). Experiments to manipulate the status quo distribution of reproduction to reveal hidden threats, or to alter the level of uncertainty about relative strength, would help to test these models in the context of reproductive conflict and conflict over parental care.

12.6 Conclusion and future research

Cooperative breeding species are excellent subjects for research on the evolution of parental and alloparental care; how cooperation evolves; and the behavioural mechanisms by which animals resolve conflicts and exercise control over each other's behaviour. A recurring theme of this chapter is the need for more experimental tests of models of helping and reproductive conflict. Any proposed behavioural mechanism of social control (such as punishments or threats) can only be investigated rigorously through the use of manipulation experiments. Manipulations of this type are often challenging logistically, but without them there is a danger of developing a biased and inaccurate picture of the forces sustaining cooperation and resolving reproductive conflict in natural systems. For example, where threats of attack or infanticide are effective, observed acts of aggression may be just the tip of the iceberg of forces influencing behaviour in cooperative groups. Detecting hidden threats requires experiments to disturb the status quo.

There are many promising areas for future research; I mention four here. First, detailed longitudinal studies of cooperative animals in their natural environment provide an opportunity to investigate selection on aging in highly viscous populations. Field studies have an enormous advantage over laboratory studies to address these questions because key life-history trade-offs might only be manifested in an environment where individuals are exposed to their natural predators, parasites, and pathogens.

Longitudinal studies of cooperative breeders also offer unparalleled opportunities to examine transgenerational effects and epigenetic inheritance under natural conditions (see also Chapters 14 and 17). Studies of laboratory rodents have revealed mechanisms by which pre- and post-natal care can result in heritable changes in patterns of gene expression in adulthood, but much less is known about these mechanisms in wild populations and their ecological and evolutionary significance (Bossdorf et al. 2008). For example, the level of investment received in early life offers a strong candidate explanation for observed differences in later life-history and the consistency of contributions to helping in some cooperatively breeding species (e.g. English et al. 2010). Moreover, individuals that receive greater investment when young may be more or less likely to help themselves as adults, or more or less likely to disperse from the group, leading to positive or negative feedback in the quality of early versus late cooperative environments and potentially dramatic impacts on group stability and population dynamics over time.

Third, there is much scope for research on the behavioural processes by which animals resolve conflict over reproduction and helping. For example, low-level social aggression is a conspicuous feature of animal societies, but the function of aggressive interactions is often unclear. Are aggressive dominants advertising their strength to deter subordinates from challenging their status? Does submissive behaviour serve to conceal a subordinate's true strength or motivation to challenge dominants? How frequently should dominants interact with subordinates to maintain their social status, and how often should subordinates probe dominant strength? Research on these questions would help to understand why some societies are peaceful while others are overtly fractious, and how social conflict influences the evolution of cognitive and neural processes such as individual recognition and social memory.

Fourth, current research is starting to reveal the neural and hormonal mechanisms controlling cooperative behaviour, and how these mechanisms are themselves shaped by natural selection. This is helping to break down the traditional barriers between studies of proximate mechanism and ultimate function. This barrier can be traced to an influential paper by Tinbergen (1963) which divided research questions into four categories (sometimes called the 'four whys'): physiological causation, development, evolutionary history, and adaptive value. However, in that paper Tinbergen was at pains to point out that these four research foci should be viewed as complementary and their protagonists united by a common aim, that is, to understand why animals behave in the way that they do. 'Cooperation between all these workers is within reach', he wrote, 'and the main obstacle seems to be a lack of appreciation of the fact that there is a common aim'. In cooperative breeding species, research on physiological mechanisms would help to understand the proximate control of helping, punishment, winner–loser effects, and reproductive suppression (see also Chapter 3). All of these social responses have large impacts on inclusive fitness and hence the mechanisms controlling their development and expression are subject to selection and will be shaped by the social and ecological environment. Research programmes which blend evolutionary theory, life-history analysis, and investigation of physiological mechanism offer exciting possibilities to advance knowledge about cooperation and parental care. Much of this information may be relevant to ourselves: after all, our morphology, physiology, fertility, rate of senescence, and long period of offspring dependency all reflect an evolutionary history of cooperative breeding, not more recent technological developments.

Acknowledgements

Thanks to Jeremy Field, Mathias Kölliker, Nick Royle, Andy Russell, Michael Taborsky, and an anonymous referee for helpful comments. Funding was provided by a Royal Society University Research Fellowship.

References

Alexander, R. D., Noonan, K. M., and Crespi, B. (1991). The evolution of eusociality. In P. W. Sherman, J. Jarvis, and R. D. Alexander, eds. *The Biology of the Naked Mole Rat*. Princeton University Press, Princeton.

Balshine-Earn, S., Neat, F. C., Reid, H., and Taborsky, M. (1998). Paying to stay or paying to breed? Field evidence for direct benefits of helping behavior in a cooperatively breeding fish. *Behavioral Ecology* 9, 432–8.

Bell, M. B. V., Radford, A. N., Smith, R. A., Thompson, A. M., and Ridley, A. R. (2010). Bargaining babblers: vocal negotiation of cooperative behaviour in a social bird. *Proceedings of the Royal Society B-Biological Sciences* 277, 3233–28.

Bergmuller, R., Schurch, R., and Hamilton, I. M. (2010). Evolutionary causes and consequences of individual variation in cooperative behaviour. *Philosophical Transactions of the Royal Society B* 365, 2751–64.

Bergmüller, R. and Taborsky, M. (2005). Experimental manipulation of helping in a cooperative breeder: helpers 'pay-to-stay' by pre-emptive appeasement. *Animal Behaviour* 69, 19–28.

Boomsma, J. J. (2009). Lifetime monogamy and the evolution of eusociality. *Philosophical Transactions of the Royal Society B-Biological Sciences* 364, 3191–207.

Bossdorf, O., Richards, C. L., and Pigliucci, M. (2008). Epigenetics for ecologists. *Ecology Letters* 11, 106–15.

Bourke, A. F. G. (1999). Colony size, social complexity and reproductive conflict in social insects. *Journal of Evolutionary Biology* 12, 245–57.

Bourke, A. F. G. (2011). *Principles of Social Evolution*. Oxford University Press, Oxford.

Canestrari, D., Marcos, J. M., and Baglione, V. (2011). Helpers at the nest compensate for reduced maternal investment in egg size in carrion crows. *Journal of Evolutionary Biology* 24, 1870–8.

Cant, M. A. (1998). A model for the evolution of reproductive skew without reproductive suppression. *Animal Behaviour* 55, 163–9.

Cant, M. A. (2006). A tale of two theories: parent-offspring conflict and reproductive skew. *Animal Behaviour* 71, 255–63.

Cant, M. A. (2011). The role of threats in animal cooperation. *Proceedings of the Royal Society B-Biological Sciences* 278, 170–8.

Cant, M. A. (2012). Suppression of social conflict and evolutionary transitions to cooperation. *American Naturalist* 179, 293–301.

Cant, M. A. and Field, J. (2001). Helping effort and future fitness in cooperative animal societies. *Proceedings of the Royal Society of London Series B-Biological Sciences* 268, 1959–64.

Cant, M. A. and Field, J. (2005). Helping effort in a dominance hierarchy. *Behavioral Ecology* 16, 708–15.

Cant, M. A., Hodge, S. J., Gilchrist, J. S., Bell, M. B. V., and Nichols, H. J. (2010). Reproductive control via eviction (but not the threat of eviction) in banded mongooses. *Proceedings of the Royal Society of London B* 277, 2219–26.

Cant, M. A. and Johnstone, R. A. (2008). Reproductive conflict and the separation of reproductive generations in humans. *Proceedings of the National Academy of Sciences of the United States of America* 105, 5332–6.

Cant, M. A. and Johnstone, R. A. (2009). How threats influence the evolutionary resolution of within-group conflict. *American Naturalist* 173, 759–71.

Cant, M. A., Llop, J. B., and Field, J. (2006). Individual variation in social aggression and the probability of inheritance: Theory and a field test. *American Naturalist* 167, 837–52.

Carlisle, T. R. and Zahavia, A. (1986). Helping at the nest, allofeeding and social status in immature Arabian babbler. *Behavioural Ecology and Sociobiology* 18, 339–51.

Clutton-Brock, T. (2002). Breeding together: kin selection and mutualism in cooperative vertebrates. *Science* 296, 69–72.

Clutton-Brock, T. H. (1989). Mammalian mating systems. *Proceedings of the Royal Society B-Biological Sciences* 236, 339–72.

Clutton-Brock, T. H., Brotherton, P. N. M., O'Riain, M. J., Griffin, A. S., Gaynor, D., Sharpe, L., Kansky, R., Manser, M. B., and McIlrath, G. M. (2000). Individual contributions to babysitting in a cooperative mongoose, Suricata suricatta. *Proceedings of the Royal Society of London Series B-Biological Sciences* 267, 301–5.

Clutton-Brock, T. H. and Parker, G. A. (1995). Punishment in animal societies. *Nature* 373, 209–16.

Clutton-Brock, T. H., Russell, A. F., Sharpe, L. L., and Jordan, N. R. (2005). 'False feeding' and aggression in meerkat societies. *Animal Behaviour* 69, 1273–84.

Cockburn, A. (1998). Evolution of helping behavior in cooperatively breeding birds. *Annual Review of Ecology and Systematics* 29, 141–77.

Cockburn, A. (2006). Prevalence of different modes of parental care in birds. *Proceedings of the Royal Society B-Biological Sciences* 273, 1375–83.

Cornwallis, C. K., West, S. A., and Griffin, A. S. (2009). Routes to indirect fitness in cooperatively breeding vertebrates: kin discrimination and limited dispersal. *J. Evol. Biol.* 22, 2445–57.

Cornwallis, C. K., West, S. A., Davis, K. E., and Griffin, A. S. (2010). Promiscuity and the evolutionary transition to complex societies. *Nature* 466, 969–U91.

Crespi, B. J. and Yanega, D. (1995). The definition of eusociality. *Behavioral Ecology* 6, 109–15.

Crozier, R. H. (1986). Genetic clonal recognition abilities in marine invertebrates must be maintained by selection for something else. *Evolution* 40, 1100–1.

Dickinson, J. and Hatchwell, B. J. (2004). Fitness consequences of helping. In W. Koenig and J. Dickinson, ed. *Ecology and Evolution of Cooperative Breeding in Birds*, pp. 48–66. Cambridge University Press, Cambridge.

Duffy, J. E. and Macdonald, K. S. (2010). Kin structure, ecology and the evolution of social organization in shrimp: a comparative analysis. *Proceedings of the Royal Society B-Biological Sciences* 277, 575–84.

Emlen, S. T. (1991). The evolution of cooperative breeding in birds and mammals. In J. R. Krebs and N. B. Davies, eds. *Behavioural Ecology*, 3rd edn, pp. 301–37. Blackwell Scientific Publications, Oxford.

English, S., Nakagawa, S., and Clutton-Brock, T. H. (2010). Consistent individual differences in cooperative behaviour in meerkats (Suricata suricatta). *Journal of Evolutionary Biology* 23, 1597–604.

Field, J. and Brace, S. (2004). Pre-social benefits of extended parental care. *Nature* 428, 650–2.

Field, J. and Cant, M. A. (2007). Direct fitness, reciprocity, and helping: a perspective from primitively eusocial wasps. *Behavioural Processes* 76, 160–2.

Field, J. and Cant, M. A. (2009a). Reproductive skew in primitively eusocial wasps: how useful are current models? In R. Hager and C. B. Jones, eds. *Reproductive Skew in Vertebrates: Proximate and Ultimate Causes*. Cambridge University Press, Cambridge.

Field, J. and Cant, M. A. (2009b). Social stability and helping in small animal societies. *Philosophical Transactions of the Royal Society B* 364, 3181–9.

Field, J., Cronin, A., and Bridge, C. (2006). Future fitness and helping in social queues. *Nature* 441, 214–17.

Field, J., Paxton, R. J., Soro, A., and Bridge, C. (2010). Cryptic plasticity underlies a major evolutionary transition. *Current Biology* 20, 2028–31.

Gadagkar, R. (1990). Evolution of eusociality—the advantage of assured fitness returns. *Philosophical Transactions of the Royal Society of London Series B-Biological Sciences* 329, 17–25.

Gadagkar, R. (1991). *Belonogaster*, *Mischocuttarus*, *Parapolybia*, and independent-founding *Ropalidia*. In K. G. Ross and R. W. Matthews, eds. *The Social Biology of Wasps*, pp. 149–87. Cornell University Press, Ithaca NY.

Gamboa, G. J. (2004). Kin recognition in eusocial wasps. *Annales Zoologici Fennici* 41, 789–808.

Gardner, A. and Foster, K. (2008). The evolution and ecology of cooperation—history and concepts. In J. Korb and J. Heinze, eds. *Ecology of Social Evolution*, pp. 1–36. Springer-Verlag, Berlin.

Gardner, A. and West, S. A. (2006). Demography, altruism, and the benefits of budding. *J Evol Biol* 19, 1707–16.

Gaston, A. J. (1978). The evolution of group territorial behavior and cooperative breeding. *American Naturalist* 112, 1091–100.

Griffin, A. S. and West, S. A. (2003). Kin discrimination and the benefit of helping in cooperatively breeding vertebrates. *Science* 302, 634–6.

Hamilton, W. D. (1964). The genetical evolution of social behaviour. *Journal of Theoretical Biology* 7, 1–16.

Hatchwell, B. J. (2009). The evolution of cooperative breeding in birds: kinship, dispersal and life history. *Philosophical Transactions of the Royal Society B-Biological Sciences* 364, 3217–27.

Hatchwell, B. J. (2010). Cryptic kin selection: kin structure in vertebrate populations and opportunities for kin-directed cooperation. *Ethology* 116, 203–16.

Heg, D., Bergmuller, R., Bonfils, D., Otti, O., Bachar, Z., Burri, R., Heckel, G., and Taborsky, M. (2006). Cichlids do not adjust reproductive skew to the availability of independent breeding options. *Behavioral Ecology* 17, 419–29.

Heg, D., Heg-Bachar, Z., Brouwer, L., and Taborsky, M. (2008). Experimentally induced helper dispersal in colonially breeding cooperative cichlids. *Environmental Biology of Fishes* 83, 191–206.

Hilborn, R. and Mangel, M. (1997). *The Ecological Detective: Confronting Models with Data*. Princeton University Press, Princeton.

Hinde, C. A. (2006). Negotiation over offspring care?—a positive response to partner-provisioning rate in great tits. *Behavioral Ecology* 17, 6–12.

Hodge, S. J., Bell, M. B. V., Gilchrist, J. S., and Cant, M. A. (2011). Reproductive competition and the evolution of extreme birth synchrony in a cooperative mammal. *Biology Letters* 7, 54–6.

Houston, A. I. and Davies, N. B. (1985). The evolution of cooperation and life history in the dunnock *Prunella modularis*. In R. M. Sibly and R. H. Smith, eds. *Behavioural Ecology: Ecological Consequences of Adaptive Behaviour*. Blackwell Scientific Publications, Oxford.

Hughes, W. O. H., Oldroyd, B. P., Beekman, M., and Ratnieks, F. L. W. (2008). Ancestral monogamy shows kin selection is key to the evolution of eusociality. *Science* 320, 1213–16.

Johnstone, R. A. (2011). Load lightening and negotiation over offpsring care in cooperative breeders. *Behavioral Ecology* 22, 436–44.

Johnstone, R. A. and Cant, M. A. (1999). Reproductive skew and the threat of eviction: a new perspective. *Proceedings of the Royal Society of London Series B-Biological Sciences* 266, 275–9.

Johnstone, R. A. and Cant, M. A. (2008). Sex differences in dispersal and the evolution of helping and harming. *American Naturalist* 172, 318–30.

Johnstone, R. A. and Cant, M. A. (2010). The evolution of menopause in cetaceans and humans: the role of demography. *Proceedings of the Royal Society B-Biological Sciences* 277, 3765–71.

Johnstone, R. A., Cant, M. A., and Field, J. (2012). Sex-biased dispersal, haplodiploidy, and the evolution of helping in social insects. *Proceedings of the Royal Society B-Biological Sciences* 279, 787–93.

Johnstone, R. A. and Hinde, C. (2006). Negotiation over offspring care—how should parents respond to each other's efforts? *Behavioral Ecology* 17, 818–27.

Keller, L. (1997). Indiscriminate altruism: Unduly nice parents and siblings. *Trends in Ecology and Evolution* 12, 99–103.

Kingma, S. A., Hall, M. L., and Peters, A. (2011). Multiple benefits drive helping behavior in a cooperatively breeding bird: an integrated analysis. *American Naturalist* 177, 486–95.

Koenig, W. D. and Dickinson, J. L. (ed.). (2004). *Ecology and Evolution of Cooperative Breeding in Birds*. Cambridge University Press, Cambridge.

Koenig, W. D., Haydock, J., and Stanback, M. T. (1998). Reproductive roles in the cooperatively breeding acorn woodpecker: Incest avoidance versus reproductive competition. *American Naturalist* 151, 243–55.

Koenig, W. D. and Walters, E. L. (2011). Age-related provisioning behaviour in the cooperatively breeding acorn woodpecker: testing the skills and pay-to-stay hypotheses. *Animal Behaviour* 82, 437–44.

Koenig, W. D., Walters, E. L., and Haydock, J. (2009). Helpers and egg investment in the cooperatively breeding acorn woodpecker: testing the concealed helper effects hypothesis. *Behavioral Ecology and Sociobiology* 63, 1659–65.

Kokko, H., Johnstone, R. A., and Clutton-Brock, T. H. (2001). The evolution of cooperative breeding through group augmentation. *Proceedings of the Royal Society of London Series B-Biological Sciences* 268, 187–96.

Komdeur, J., Huffstadt, A., Prast, W., Castle, G., Mileto, R., and Wattel, J. (1995). Transfer experiments of Seychelles Warblers to New Islands—changes in dispersal and helping-behavior. *Animal Behaviour* 49, 695–708.

Komdeur, J., Richardson, D. S., and Hatchwell, B. J. (2008). Kin recognition mechanisms in cooperative breeding systems: ecological causes and behavioural consequences of variation. In J. Korb and J. Heinze, eds. *Ecology of Social Evolution*, pp. 175–94. Springer-Verlag, Berlin.

Korb, J. (2008). *The Ecology of Social Evolution in Termites*. Ecology of Social Evolution. Springer-Verlag, Berlin.

Lakatos, I. (1978). *The Methodology of Scientific Research Programmes: Philosophical Papers Vol 1*. Cambridge University Press, Cambridge.

Langer, P., Hogendoorn, K., and Keller, L. (2004). Tug-of-war over reproduction in a social bee. *Nature* 428, 844–7.

Lawson Handley, L. J. and Perrin, N. (2007). Advances in our understanding of mammalian sex-biased dispersal. *Molecular Ecology* 16, 1559–78.

Leadbeater, E., Carruthers, J. M., Green, J. P., Rosser, N. S., and Field, J. (2011). Nest inheritance is the missing source of direct fitness in a primitively eusocial insect. *Science* 333, 874–6.

Leadbeater, E., Carruthers, J. M., Green, J. P., van Heusden, J., and Field, J. (2010). Unrelated helpers in a primitively eusocial wasp: is helping tailored toward direct fitness? *PLoS ONE* 5, e11997.

Lehmann, L. and Rousset, F. (2010). How life history and demography promote or inhibit the evolution of helping behaviours. *Philosophical Transactions of the Royal Society B-Biological Sciences* 365, 2599–617.

Lin, N. and Michener, C. D. (1972). Evolution of sociality in insects. *The Quarterly Review of Biology* 47, 131–59.

Lukas, D. and Clutton-Brock, T. H. (in press). Cooperative breeding and monogamy in mammalian societies. *Proceeding of the Royal Society of London B: Biological Sciences*.

Mace, R. and Sear, R. (2005). Are humans cooperative breeders? In E. Voland, A. Chasiotis, and W. Schiefenhoevel, eds. *Grandmotherhood: The Evolutionary Significance of the Second Half of Female Life*, pp. 143–59. Rutgers University Press, Piscataway, N.J.

McNamara, J. M., Gasson, C. E., and Houston, A. I. (1999). Incorporating rules for responding into evolutionary games. *Nature* 401, 368–71.

McDonald, P. G., te Marvelde, L., Kazem, A. J. N., Wright, J. (2008). Helping as a signal and the effect of a potential audience during provisioning visits in a cooperative bird. *Animal Behaviour* 75, 1318–30.

Michener, C. D. (1958). Evolution of social behaviour in bees. *Proceedings of the 10th International Congress of Entomology* 2, 441–8.

Mulder, R. A. and Langmore, N. E. (1993). Dominant males punish helpers for temporary defection in the superb fairy wren. *Animal Behaviour* 45, 830–3.

Nonacs, P., Reeve, H. K., and Starks, P. T. (2004). Optimal reproductive-skew models fail to predict aggression in wasps. *Proceedings of the Royal Society of London Series B-Biological Sciences* 271, 811–17.

Nowak, M. A., Tarnita, C. E., and Wilson, E. O. (2010). The evolution of eusociality. *Nature* 466, 1057–62.

Port, M., Johnstone, R. A., and Kappeler, P. M. (2010). Costs and benefits of multi-male associations in red-fronted lemurs (Eulemur fulvus rufus). *Biology Letters* 6, 620–2.

Queller, D. C. (1992). Does population viscosity promote kin selection? *Trends in Ecology and Evolution* 7, 322–4.

Queller, D. C. (1994). Extended parental care and the origin of eusociality. *Proceedings of the Royal Society of London Series B-Biological Sciences* 256, 105–11.

Queller, D. C., Zacchi, F., Cervo, R., Turillazzi, S., Henshaw, M. T., Santorelli, L. A., and Strassmann, J. E. (2000). Unrelated helpers in a social insect. *Nature* 405, 784–7.

Raihani, N. J. and Ridley, A. R. (2008). Experimental evidence for teaching in wild pied babblers. *Animal Behaviour* 75, 3–11.

Ratnieks, F. L. W. and Wenseleers, T. (2008). Altruism in insect societies and beyond: voluntary or enforced? *Trends in Ecology and Evolution* 23, 45–52.

Reeve, H. K. (1991). Polistes. In K. Matthews, ed. *The Social Biology of Wasps*, pp. 99–148. Cornell University Press, Ithaca, NY.

Reeve, H. K. (1992). Queen activation of lazy workers in colonies of the eusocial naked mole-rat. *Nature* 358, 147–9.

Reeve, H. K., Emlen, S. T., and Keller, L. (1998). Reproductive sharing in animal societies: reproductive incentives or incomplete control by dominant breeders? *Behavioral Ecology* 9, 267–78.

Reeve, H. K. and Gamboa, G. J. (1987). Queen regulation of worker foraging in paper wasps: a social feedback control system. *Behaviour* 102, 147–67.

Reeve, H. K. and Nonacs, P. (1997). Within-group aggression and the value of group members: Theory and a field test with social wasps. *Behavioral Ecology* 8, 75–82.

Reyer, H.-U. (1984). Investment and relatedness: a cost/benefit analysis of breeding and helping in the pied kingfisher. *Animal Behaviour* 32, 1163–78.

Riedman, M. L. (1982). The evolution of alloparental care and adoption in mammals and birds. *Quarterly Review of Biology* 57, 405–35.

Rousset, F. and Lion, S. (2011). Much ado about nothing: Nowak et al's charge against inclusive fitness theory. *Journal of Evolutionary Biology* 24, 1386–92.

Rousset, F. and Roze, D. (2007). Constraints on the origin and maintenance of genetic kin recognition. *Evolution* 61, 2320–30.

Rubenstein, D. R. and Lovett, I. J. (2007). Temporal environmental variability drives the evolution of cooperative breeding in birds. *Current Biology* 17, 1414–19.

Russell, A. F. (2004). Mammals: comparisons and contrasts. In W. D. D. Koenig, ed. *Ecology and Evolution of Cooperative Breeding in Birds*, pp. 210–27. Cambridge University Press, Cambridge.

Russell, A. R., Langmore, N. E., Cockburn, A., Astheimer, L. B., and Kilner, R. M. (2007). Reduced egg investment can conceal helper effects in cooperatively breeding birds. *Science* 317, 941–4.

Salomon, M. and Lubin, Y. (2007). Cooperative breeding increases reproductive success in the social spider Stegodyphus dumicola (Araneae, Eresidae). *Behavioral Ecology and Sociobiology* 61, 1743–50.

Savage, J. L., Russell, A. R., and Johnstone, R. A. (in press). Reproductive investment and maternal tactics with variable numbers of carers. *Behavioral Ecology*.

Sharp, S. P., Simeoni, M., and Hatchwell, B. J. (2008). Dispersal of sibling coalitions promotes helping among immigrants in a cooperatively breeding bird. *Proceedings of the Royal Society B-Biological Sciences* 275, 2125–30.

Solomon, N. G. and French, J. A. (1997). *Cooperative Breeding in Mammals*. Cambridge University Press, Cambridge.

Stacey, P. B. K. and Koenig, W. D. (1990). *Cooperative Breeding in Birds: Long-term Studies of Ecology and Behavior*. Cambridge University Press, Cambridge.

Stiver, K. A., Dierkes, P., Taborsky, M., and Balshine, S. (2004). Dispersal patterns and status change in a cooperatively breeding cichlid Neolamprologus pulcher: evidence from microsatellite analyses and behavioural observations. *Journal of Fish Biology* 65, 91–105.

Taborsky, B., Skubic, E., and Bruintjes, R. (2007). Mothers adjust egg size to helper number in a cooperatively breeding cichlid. *Behavioral Ecology* 18, 652–7.

Taborsky, M. (1994). Sneakers, satellites and helpers: parasitic and cooperative behaviour in fish reproduction. *Advances in the Study of Behavior* 23, 1–100.

Taborsky, M. (2009). Reproductive skew in cooperative fish groups: virtue andlimitations of alternative modeling approaches. In R. Hager and C. B. Jones, eds. *Reproductive Skew in Vertebrates: Proximate and Ultimate Causes*. Cambridge University Press, Cambridge.

Taylor, P. D. (1992). Altruism in viscous populations: an inclusive fitness model. *Evolutionary Ecology* 6, 352–6.

Tinbergen, N. (1963). On aims and methods in ethology. *Zeitschrift fur Tierpsychologie* 20, 410–33.

Vehrencamp, S. L. (1983). A model for the evolution of despotic versus egalitarian societies. *Animal Behaviour* 31, 667–82.

West, S. A., Griffin, A. S., and Gardner, A. (2007). Social semantics: altruism, cooperation, mutualism, strong reciprocity and group selection. *Journal of Evolutionary Biology* 20, 415–32.

West, S. A., Pen, I., and Griffin, A. S. (2002). Cooperation and competition between relatives. *Science* 296, 72–5.

Wheeler, W. M. (1928). *The Social Insects*. Kegan Paul, London.

Wong, M. and Balshine, S. (2011). The evolution of cooperative breeding in the African cichlid fish, Neolamprologus pulcher. *Biological Reviews* 86, 511–30.

Wright, J. (1999). Altruism as a signal—Zahavi's alternative to kin selection and reciprocity. *Journal of Avian Biology* 30, 108–15.

Wright, S. (1943). Isolation by distance. *Genetics* 28, 114–38.

Zahavi, A. (1995). Altruism as a handicap—the limitations of kin selection and reciprocity. *Journal of Avian Biology* 26, 1–3.

CHAPTER 13

Brood parasitism

Claire N. Spottiswoode, Rebecca M. Kilner, and Nicholas B. Davies

13.1 Introduction

Whenever parents provide care they are vulnerable to exploitation by brood parasites (Fig. 13.1). Brood parasitic offspring have no evolutionary interest in their foster siblings, or in their foster parents' residual reproductive value. These unconventional families provide startling images: a cliff swallow *Petrochelidon pyrrhonota* carrying in its bill a partially incubated egg of its own to another cliff swallow's nest (Brown and Brown 1988); a three day old greater honeyguide *Indicator indicator*, naked, blind, and heavily armed, stabbing and shaking to death a newly-hatched bee-eater chick in the darkness of a burrow (Spottiswoode and Koorevaar 2012); a large blue butterfly *Maculinea rebeli* caterpillar in an ant's nest, mimicking the stridulations of the ant queen to assure that it receives royal care from the workers it has previously fooled, with mimetic hydrocarbons, into taking it for an ant (Barbero et al. 2009); or a reed warbler *Acrocephalus scirpaceus* perching on the shoulder of a young common cuckoo *Cuculus canorus* nine times its size, stuffing food into its bright orange gape (Kilner et al. 1999). How are host parents duped into tending for an imposter, and how might interactions between hosts and parasitic offspring differ from those among genetic family members?

Figure 13.1 A red-chested cuckoo *Cuculus solitarius* being fed by Cape wagtail *Motacilla capensis* in South Africa (photo: Alan Weaving).

In this chapter, we suggest that the key to predicting the host's co-evolutionary response to brood parasitism, and to explaining how selection influences the behaviour of the young parasite, lies in the virulence of parasitic offspring. We define this as the fitness costs that the parasite imposes on its host. The costs of parasitism influence the strength of selection on hosts to defend themselves against parasitism and this, we argue, explains some of the vast diversity both in host defences and in subsequent parasite counter-adaptations (Sections 13.3 to 13.6). Furthermore, the virulence of the young parasite dictates the social environment in which parasitic offspring extract parental care from their hosts. This in turn explains some of the variation in brood parasitic tactics to secure care from foster parents (Sections 13.5.2 to 13.5.4). Finally, since variation in virulence explains so much about the interactions between brood parasites and their hosts, we consider the factors that cause variation in virulence in the first place (Section 13.7). Throughout, we focus primarily on the well-studied avian brood parasites (which include examples of both interspecific and conspecific parasitism), because interactions between brood parasitic offspring and their hosts have been relatively little studied in other taxa.

The Evolution of Parental Care. First Edition. Edited by Nick J. Royle, Per T. Smiseth, and Mathias Kölliker.

13.2 Who are the brood parasites, how virulent are they?

Brood parasitism has repeatedly arisen in taxa exhibiting parental care: it is well documented in birds, insects, and fish, and has recently been confirmed to occur in frogs (Brown et al. 2009). To date it is completely unknown from mammals, perhaps owing to live birth and early learning of offspring through olfactory cues. Brood parasitism takes two main forms: obligate (where the parasite is completely dependent on the parental care of another species), and facultative (where parasitism is an alternative tactic that supplements the parasite's own reproduction, or helps compensate for reproductive failure). In birds, obligate brood parasitism occurs in about 100 species (1% of bird species) spanning four orders (Davies 2000), and has evolved independently seven times (Fig. 13.2; Sorenson and Payne 2005): three times within the cuckoo family and once each in ducks, honeyguides, finches, and New World blackbirds (cowbirds). Facultative brood parasitism occurs much more widely across the avian phylogeny and typically occurs within individuals of the same species, or sometimes related species (e.g. Sorenson 1997). It is especially frequent among species that breed colonially (Yom-Tov and Geffen 2006) or have precocial young (Sorenson 1992). Among insects, interspecific brood parasitism is most common in the social Hymenoptera (often referred to as 'social parasites', including the 'slave-maker' ants): ants, bees, wasps, and bumblebees (reviewed by Kilner and Langmore 2011), as well as certain Coleoptera such as tenebrionid and dung beetles (e.g. Chapman 1869, Rasa 1996). Many of these groups also have conspecific brood parasitism. In fish, egg-guarding is the predominant form of parental care exploited, and parasitism is typically facultative and occurs among conspecifics. One exception is the obligately parasitic cuckoo catfish *Synodontis multipunctatus*, which feeds off all of its foster siblings while they are brooded within the cichlid host parent's mouth (Sato 1986). Brood parasitism also appears to be facultative in amphibians (Brown et al. 2009).

The main source of variation in virulence in brood parasites arises from the behaviour of parasitic offspring: some species kill their foster siblings, while others are raised alongside host young, and these alternative parasitic tactics have arisen independently among birds and other animals (Fig. 13.2). In birds, virulence varies from relatively low (e.g. in many conspecific brood parasites, reviewed by Lyon and Eadie 2008) to extremely severe, where parasitic hatchlings obligately kill host young and there is no scope for host re-nesting within the season (Fig. 13.3 and Table 13.1 which lists different forms of avian parasitic systems in ascending order of virulence; see also Brandt et al. 2005 for equivalent discussion on insect social parasites). There are two key points that can be emphasized from this diversity. First, virulence in the same parasitic species can vary depending on the host species, ranging from low (young of relatively large host species often survive alongside the parasite) to very high (young of relatively small host species rarely survive). Second, the highly virulent chick-killing brood parasites are not all equally virulent to all hosts: variation in parasite developmental rates and host breeding seasons means that re-nesting after successfully raising a parasitic chick is feasible for some host species but not for others (Brooker and Brooker 1998; Langmore et al. 2003).

We now review successive lines of parasite attack and host defence in the light of this variation in virulence.

13.3 The egg-laying stage

The first hurdle faced by brood parasites is gaining entry to host nests. Many avian and insect host parents recognize their parasite and fiercely mob or attack it; hosts are even able to identify the intentions of conspecific brood parasites, and repel them (reviewed by Lyon and Eadie 2008). To evade these defences, both insect and bird brood parasites use brute force, stealth, or deception (reviewed by Kilner and Langmore 2011). Mobbing by avian hosts deters parasitism (Welbergen and Davies 2009), alerts defences in neighbouring hosts (Davies and Welbergen 2009), increases the chances that the host will reject a parasitic egg (Davies and Brooke 1988, Lotem et al. 1995), and in extreme cases may lead to injury or even death

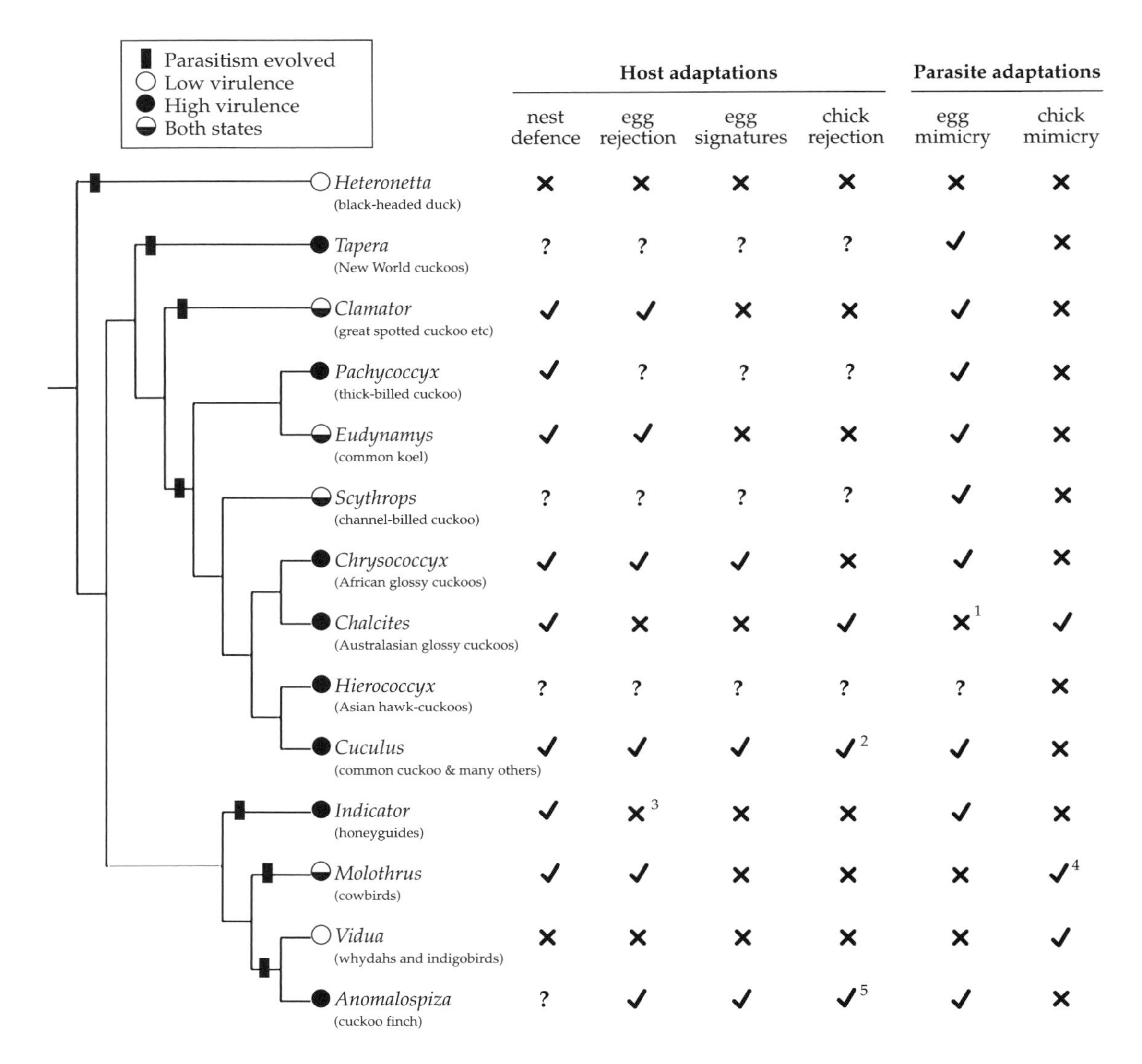

Figure 13.2 Parasitic genera mentioned in the text in a phylogenetic framework (topology from Payne 2005a; Hackett et al. 2008), showing the seven times brood parasitism has independently evolved, the variation in parasitic virulence among parasitic genera, and the presence or absence of host and parasite adaptations (so far as is known). In species showing both virulence states, whether host young die seems typically to depend on their relative size. *Pachycoccyx* is not discussed in the text but is shown simply to emphasize that occasionally low virulence in *Eudynamys* has arisen from a more virulent ancestor (Section 13.7); however, note that while *Eudynamys* and *Scythrops* are allocated both states of virulence because they sometimes fail to kill host young, it is unknown whether they are sometimes simply prevented from doing so owing to large host size. Footnotes: [1]except Horsfield's bronze-cuckoo eggs which resemble those of superb fairy wrens, even though egg rejection by fairy wrens is rare (Langmore and Kilner 2010); [2]to date detected only in one population of one host; [3]but only one host tested to date (C.N.S. unpubl. data); [4]only for screaming cowbirds *Molothrus rufoaxillaris*; [5]C.N.S. unpubl. data.

for the parasite (e.g. an adult lesser honeyguide can be killed by its larger barbet host, Moyer 1980). As a counterdefence against mobbing, many *Cuculus* cuckoos have evolved rapid and secretive laying, and body shapes and plumage patterns that closely resemble those of predatory hawks, which inhibits close approach (Davies and Welbergen 2008). There is also a striking resemblance between

Table 13.1 Variation in virulence in avian brood parasites, in ascending order of virulence. See also Fig. 13.3

	Examples of cost of parasitism to host		Examples of parasitic species
	Current brood	Future broods	
Interspecific brood parasite with negligible cost	Incubation of one extra egg (parasitic chick runs off soon after hatching)	None	black-headed duck *Heteronetta atricapilla*
Conspecific brood parasite (young feed themselves: parental care shareable among brood)	Loss of none to several eggs	Probably none	canvasback *Aythya valisineria*, common goldeneye *Bucephala clangula*
Conspecific brood parasite (young need to be fed; parental care unshareable among brood)	Sometimes loss of one egg; sometimes reduced viability of own young	Probably minor	American coot *Fulica americana*, cliff swallow *Petrochelidon pyrrhonota*
Interspecific, non-siblicidal brood parasite	Loss of at least one egg; host eggs deliberately damaged; reduced growth or viability of host young	Probably minor	great spotted cuckoo *Clamator glandarius*, shiny cowbird *Molothrus bonariensis*, pin-tailed whydah *Vidua macroura*
Interspecific brood parasite which kills entire brood (high cost)	Loss of entire brood	Potential for renesting within same season	Horsfield's bronze-cuckoo *Chalcites basalis* (ejects host young), cuckoo finch *Anomalospiza imberbis* (usually outcompetes host young)
Interspecific brood parasite which kills entire brood (extremely high cost)	Loss of entire brood	No potential for renesting within same season (most host species)	common cuckoo *Cuculus canorus* (ejects host young), greater honeyguide *Indicator indicator* and striped cuckoo *Tapera naevia* (each has independently evolved stabbing host young to death)

cuckoo finch females and those of the harmless bishopbirds *Euplectes* spp. As an additional first line of defence, many *Ploceus* weaver species have long woven tubes dangling below their nests, impeding or at least slowing down the entrance of diederik cuckoos *Chrysoccocyx caprius* (Freeman 1988; Davies 2000).

Do the hosts of more benign parasites show weaker defences? Hosts of the *Vidua* finches seem to ignore their parasites, who may even push the incubating host female aside to insert their egg into the clutch (Skead 1975). Defences may only evolve when the costs of parasitism are sufficiently severe to outweigh the costs of defence. In relatively non-virulent parasitic ducks, for example, the ferocity of host nest defence may be tempered by collateral damage to the host's own clutch (Sorenson 1997). Overall, Fig. 13.2 shows a pattern broadly consistent with the idea that weak defences are associated with benign parasites, but many gaps in our natural history knowledge remain.

13.4 The incubation stage

Typically, the incubation of a parasitic egg does not impose severe costs to hosts, although there are certain exceptions among smaller cowbird hosts whose eggs suffer reduced hatchability alongside the much larger cowbird egg (Rothstein 1975). However, detecting parasitism at the incubation stage may prevent potentially high costs at the chick stage, and it is hence at the egg stage when some of the most sophisticated co-evolutionary interactions between host and parasite occur. Host parents can eject a parasitic egg, selectively withhold incubation from it, or abandon the nesting attempt altogether and start again, but all of these defences depend crucially on prior egg recognition. When recognition evolves, it can unleash a cycle of adaptation and counter-adaptation in parasitic egg mimicry and host egg markings that act as 'signatures' to aid egg discrimination, resulting in interclutch polymorphisms (recent reviews: Davies 2011; Kilner and Langmore 2011; Langmore and Spottiswoode 2012).

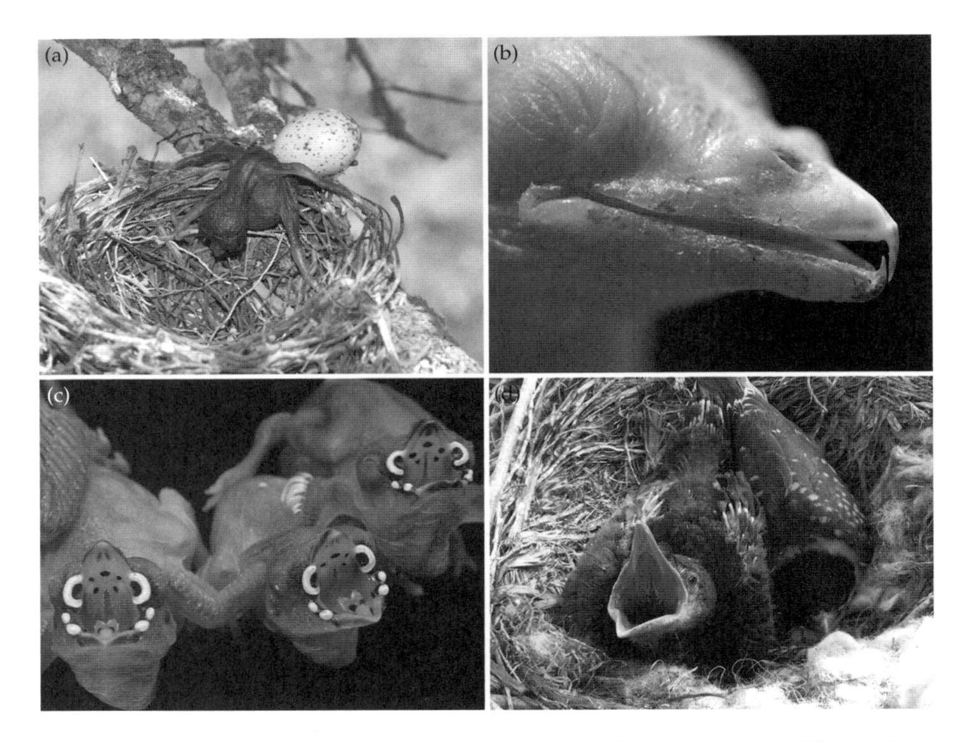

Figure 13.3 Variation in brood parasitic virulence. The top row shows two independently evolved highly virulent chick-killing brood parasites and their respective weaponry: (a) an African cuckoo *Cuculus gularis* hatchling evicts a fork-tailed drongo *Dicrurus adsimilis* egg in Zambia, and (b) a young greater honeyguide *Indicator indicator* shows its lethal bill hooks, also in Zambia (photos: Claire Spottiswoode). The bottom row shows two independently evolved relatively benign brood parasites: (c) a highly mimetic pin-tailed whydah *Vidua macroura* (chick at right) is raised alongside its common waxbill *Estrilda astrild* foster-siblings in South Africa (photo: Justin Schuetz), and (d) a great spotted cuckoo *Clamator glandarius* (chick at right) probably profits from the begging efforts of its carrion crow *Corvus corone* (chick at left) foster sibling in Spain (photo: Vittorio Baglione).

Discriminating against suspicious eggs can entail costs to hosts, either from mistakenly rejecting the host's own egg, or a result of damage to their own clutch in the process of ejecting what is often a large and thick-shelled foreign egg (Antonov et al. 2009; Davies and Brooke 1988). Nonetheless, selection has repeatedly favoured discriminating hosts, resulting in egg mimicry having repeatedly evolved among both avian and insect brood parasites: just as cuckoos and cuckoo finches mimic host eggs in colour and pattern in response to visual egg recognition by host parents, cuckoo bumblebees and socially parasitic ants mimic host egg hydrocarbon profiles in response to olfactory egg recognition by hosts. Likewise, in each taxon there is evidence that visual (birds) and olfactory (insects) host egg 'signatures' have diversified in escalated defence against parasitic mimicry (reviewed by Kilner and Langmore 2011).

However, egg mimicry is not the only way that parasites can escape host detection. The *Chalcites* cuckoos of Australasia lay dark-coloured eggs that do not mimic the eggs of their hosts but are cryptic within the dark interior of the domed nests of their hosts (Langmore et al. 2009b), and certain non-mimetic egg traits may even be attractive to hosts and thereby increase acceptance (Alvarez 2000). Hosts fooled by cryptic or attractive eggs may be forced to depend on subsequent lines of defence to combat parasitism (Section 13.5.5).

In the following survey, we will assume that parasitic mimicry is indicative of host defences (with the caveat that other sources of selection can generate mimicry: reviewed by Langmore and Spottiswoode 2012), and that the evolution of host egg 'signatures' (interclutch polymorphisms) is indicative of even stronger host defences. At first sight, broad patterns seem to be generally consistent with the idea that egg rejection is related to parasite virulence (Fig. 13.2): eggs of the benign black-headed duck and *Vidua* finches show no visual resemblance above that expected from common ancestry with their hosts. Among moderately virulent parasites, *Clamator* and *Eudynamys* cuckoos show egg mimicry but their hosts have not evolved signatures in response, while *Molothrus* cowbirds seem not to show widespread egg mimicry. By contrast, the highly virulent cuckoo genera *Cuculus* and *Chrysococcyx* and the cuckoo finch genus *Anomalospiza* have all evolved egg mimicry, and many of their hosts have in turn evolved egg signatures in defence. The highly virulent honeyguides, family Indicatoridae, may show host egg mimicry with respect to size and shape (Spottiswoode et al. 2011) as well as colour (Vernon 1987). This is a crude overview and many gaps in our knowledge still remain (e.g. concerning the highly virulent New World cuckoos), but it suggests that strength of defence at the incubation stage is related to the costs hosts face if they fail to identify an alien egg.

13.5 The chick-rearing stage

An exhausted songbird feeding a giant, solitary, cuckoo chick many times larger than itself (Fig. 13.1) is an arresting image that has captured the imagination of birdwatchers and biologists for hundreds of years, but this may obscure the fact that many other species of brood parasites have taken quite different and more subtle routes to achieving high levels of care during post-natal development. In this section, we consider how strategies of high and low virulence can each be highly successful for brood parasites.

13.5.1 How parasitic parents can improve the nestling environment

Parasitic parents show adaptations to maximize their offspring's ability to exploit host care, even prior to laying their egg. Parasites select host species with appropriate diets (Schulze-Hagen et al. 2009) and those individual nests that are likely to provide the best rearing conditions. Among the insects, for example, the digger wasp *Cerceris arenaria* preferentially chooses host nests containing greater food stocks (Field 1994). Among birds, we might speculate that parents of parasites that do not kill the foster siblings have the most to gain from targeting host pairs that provide superior parental care, if it is more energetically demanding to rear a brood where the parasitic chick is raised alongside host chicks rather than a parasitic chick on its own (Section 13.5.5). In broad accordance with this expectation, most examples of parasitic selectivity to date come from the non-killing *Clamator* cuckoos and cowbirds, although empirical studies are admittedly few (reviewed by Parejo and Avilés 2007). For example, the non-chick-killing great spotted cuckoo *Clamator glandarius* chooses individual magpie *Pica pica* hosts that enable better fledging success for their offspring (Soler et al. 1995a), whereas the chick-killing Horsfield's bronze-cuckoo *Chalcites basalis* apparently does not (Langmore and Kilner 2007). What cues might parasites use to assess parental quality? One possibility is that parasites eavesdrop on correlates of host parental quality such as sexual display (Parejo and Avilés 2007), as is the case with great spotted cuckoos and magpies (Soler et al. 1995a). Alternatively, parasites may assess parental quality directly, especially in the case of some conspecific brood parasites: cliff swallows also select superior host nests, and it is possible that their transfer of semi-incubated eggs described at the beginning of the chapter may allow them more time to assess the relative parental quality of prospective hosts (Brown and Brown 1991). Similarly, northern masked weavers *Ploceus taeniopterus* also transfer eggs that have developing embryos, which may have a similar function to that suggested for cliff swallows (Jackson 1993).

Prior to egg-laying, parasitic parents can manipulate the rearing environment of their offspring by removing host eggs (e.g. Massoni and Reboreda 1999; Soler and Martínez 2000), and by giving their own egg a head start in embryonic development. The latter facilitates early hatching relative to host chicks, thereby producing a corresponding size disparity in the parasite's favour, and is an adaptation shared by both chick-killing and non-chick-killing species of parasite. In chick-killing species, early hatching is advantageous not only because an egg or small host hatchling might be easier or cheaper to kill than a larger chick, but also because of the potential risk of being killed by any other parasite laid in the same host nest. In the case of non-chick-killing species, early hatching is expected to improve the parasite's ability to compete with host chicks. In both cases rapid embryonic development is an adaptation for dealing with rivals in the nest.

The mechanisms contributing to early hatching of parasitic eggs are only partially understood. A mechanism known to be shared by both cuckoos and honeyguides is internal incubation, whereby eggs are laid at 48 hour intervals, allowing 24 additional hours of embryonic development before laying (Birkhead et al. 2011). Subsequently, eggs laid by parasites generally have rapid development, which studies of cuckoos and cowbirds suggest may also to some degree be accounted for by their often relatively small overall size, but large yolk size and elevated yolk carotenoid (but apparently not androgen) content (Hauber and Pilz 2003; Török et al. 2004; Hargitai et al. 2010). Evidence to date for the adaptive role of these maternal factors in brood parasites is not yet wholly clear, and we have much still to learn about the physiological mechanisms contributing to brood parasites' rapid development and hatchling vigour.

13.5.2 Costs of chick-killing to parasites

In brood parasites, just as in pathogens, there are costs associated with being highly virulent (reviewed by Kilner 2005). In avian brood parasites, these costs may be threefold: first, the act of removing host chicks might be energetically expensive and potentially incur longer term costs. For example, cuckoo chicks certainly look exhausted when collapsing into the nest bowl after ejecting an egg, and honeyguides pant heavily after a bout of stabbing (Spottiswoode and Koorevaar 2012). However, the evidence to date suggests that at least in common cuckoos, the act of eviction imposes only short term costs to growth that are quickly regained (Anderson et al. 2009; Grim et al. 2009). Second, by killing host chicks, parasites might lose their assistance in stimulating the host parents to provision them with food. The clearest evidence comes from the brown-headed cowbird *Molothrus ater*. There is considerable variation among host species in the number of chicks that die as a result of brown-headed cowbird parasitism. Cowbird growth and viability was greatest in host species in which an intermediate number of host young typically survive alongside the cowbird, implying that host chick mortality conferred a cost to the cowbird (Kilner 2003; Kilner et al. 2004). Moreover, experiments in eastern phoebes *Sayornis phoebe* nests clearly showed that cowbirds acquire the most food when host young are present alongside them (Kilner et al. 2004). A third cost of virulence can be incurred if a parasite's sole occupation of the nest signals its alien identity. To date only one host species has been shown to use this as a cue of parasitism: the superb fairy wren *Malurus cyaneus* deserts about 40% of Horsfield's bronze-cuckoo chicks within days of hatching, and experiments confirmed that the number of nestlings in the nest contributed to triggering this parental defence—although this was not the only such cue—particularly in inexperienced fairy wren females unfamiliar with the appearance of their own young (Langmore et al. 2003; Langmore et al. 2009a).

13.5.3 Virulent chicks: how to solicit a foster-parent

The previous section has shown that monopolizing parental care comes at a cost to parasitic chicks, but so too does sharing parental provisioning with host young. In each case, parasitic adaptations have arisen to minimize the costs incurred by that strategy. In the case of the highly virulent chick-killing brood parasites that lose the begging assistance of their foster siblings, the challenge lies in compensating for the reduced visual and vocal begging signal that can be produced by a lone chick. The two most-studied chick-killing parasites, the Cuculinae cuckoos and the honeyguides, have evolved slightly different solutions to overcome this problem.

Most cuckoo species and host races are raised in relatively open nests with a sufficiently bright light environment that allows both visual and vocal signals to be involved in chick begging. Common cuckoos elicit parental provisioning using both types of signal: cuckoo gapes are large and very brightly coloured but this stimulus alone is insufficient to elicit the rates of care they require, as it cannot match the gape area of a brood of young (Kilner et al. 1999). This is especially so at the later stages of nestling development when the area of a single cuckoo gape is disproportionately smaller compared to a host brood. To compensate for this inferior visual stimulus, cuckoos supplement it with unusually rapid begging calls that sound like many hungry host chicks, and together these stimuli elicit the same degree of care provided to a brood of host young (Davies et al. 1998; Kilner et al. 1999). However, despite its lack of genetic interest in the host parents' residual reproductive value, the cuckoo is fed at the same rate as a brood of host chicks rather than at a supernormally high rate (Kilner et al. 1999), despite the fact that hosts are physically capable of higher short-term provisioning rates (Brooke and Davies 1989). It is possible that there are physical constraints upon the vocal signal produced by a lone chick (Kilner et al. 1999), in addition to a lack of host responsiveness to the bright colouration of the cuckoo's gape, which fails to compensate for its small size compared to the gape area of a brood of host chicks (Noble et al. 1999).

The Horsfield's hawk-cuckoo *Cuculus fugax* of eastern Asia has evolved a remarkable alternative solution to supplement its bright yellow gape's inadequate visual signal. Host nests experience high levels of predation and hence neither host nor parasite begs loudly. Instead, the cuckoo has evolved a silent accompaniment to increase its gape's apparent area: during begging it additionally flashes false gapes, in the form of bright yellow patches of naked skin beneath its wings, which so effectively stimulate the hosts that they sometimes attempt to feed the young cuckoo's wing rather than its mouth (Tanaka and Ueda 2005).

Most honeyguide species, by contrast, parasitize hosts that breed in deep holes, either within tree branches or in terrestrial burrows. In this unpromising visual environment, vocal cues might be expected to predominate, and indeed neither host nor parasitic young have brightly coloured gapes or are in any other way visually adorned. However, both host and parasite have loud and vigorous begging calls in the security of their holes and, in the nests of two species of bee-eaters *Merops* spp. that are parasitized by the greater honeyguide, the begging call of a single honeyguide strikingly resembles that of many host young calling at once (Fry 1974; CNS unpubl. data). It is as yet unknown whether either hawk-cuckoos or honeyguides are able to achieve a truly supernormal stimulus using their exaggerated visual and vocal cues respectively, and thus fully exploit their parents' provisioning ability; this would require showing that parasites are fed at a higher rate than a brood of host young.

13.5.4 Benign chicks: how to compete with foster-siblings

In the case of less virulent brood parasites that share the nest with host young, the challenge in successfully exploiting parental care depends on competition with host chicks. Chick mimicry can be at its most sophisticated in such mixed broods, suggesting that sibling competition can be at least as potent a selective force for mimicry as host rejection. Mimicry is only one weapon in the benign parasite's arsenal, however: non-chick-killing brood parasites have evolved several times, and in each instance a slightly different combination of traits ensures parasitic success. The black-headed duck becomes independent soon after hatching and so requires no special adaptations for extracting post-hatching care (Table 13.1). The remaining three groups of non-chick-killing parasites are the *Molothrus* cowbirds, the *Clamator* cuckoos, and the *Vidua* finches; we will discuss each of these in turn.

Most cowbird species share two advantages over their hosts that are common to many other brood parasitic systems. First, in many host species the brood parasite has an automatic upper hand in chick competition owing to its larger size (e.g. Soler et al. 1995b; Hauber 2003a). Second, release from kin selection allows it to beg more vigorously than hosts for a given level of hunger (Lichtenstein 2001; but see Box 13.1). This in turn opens up the possibility of parasites co-opting host begging to their own advantage, since they are able to take a disproportionate share of the food that is brought to the nest. Evidence for this idea comes from the brown-headed cowbird, which when parasitizing most host species receives more food than host chicks thanks to its more vigorous begging behaviour and its larger size (reviewed by Hauber 2003a). Owing to their advantage in competition for food provided by host parents and the flexibility of parental provisioning in relation to offspring demand, cowbirds can accrue a net benefit by sharing the nest with host young: hosts assist the cowbird to signal demand, but when food is provided it is usually monopolized by the cowbird (Kilner et al. 2004). Comparative analyses suggest that this advantage is greatest when an intermediate number of host young typically survive alongside the cowbird; perhaps too few hosts provide negligible assistance in begging, whereas too many hosts provide excessively stiff competition for resources (Kilner et al. 2004). Under these specific conditions (flexible parental provisioning, parasitic advantage in competition with host young), selection will favour an element of restraint in parasitic selfishness such that parasites do not cause host chick mortality. Thus in order maximally to exploit its unrelated foster parent, the parasite must restrain its impact on unrelated foster siblings (Kilner et al. 2004). This may be contrasted with conditions when parental resources are fixed (e.g. parasitoid eggs on a carcass), when it would pay a parasite always to outcompete non-relatives (Kilner 2003). Despite this restraint, cowbird parasitism may have a severe impact on the survival of host daughters: in at least one host species, the cowbird chick's superior competitive ability has a disproportionately heavy impact on less competitive female host young, resulting in higher female mortality and hence male-biased brood sex ratios at fledging (Zanette et al. 2005).

Cowbird behaviour can, in turn, have consequences for dynamics of interactions among host offspring even in unparasitized broods. From a host chick's perspective, a cowbird nestmate is of course unrelated, so when parasitism is frequent, it should favour an increase in selfishness also among host chicks, while the presence of other related individuals in the brood should favour a decrease in selfishness. This predicts that any cowbird-induced host selfishness should be most pronounced in hosts with smaller broods, when the introduction of a cowbird has a disproportionately large effect on the brood's average relatedness. Precisely such a pattern was found across a range of North American birds: species that endure high rates of brown-headed cowbird parasitism beg more loudly than species less severely affected, and this effect was strongest in species with small clutches (Boncoraglio et al. 2008). This pattern is consistent with the general pattern of begging in relation to brood relatedness across all bird species, parasitized or not, since those species experiencing high rates of extra-pair paternity—and thus lower average within-brood relatedness—also beg more loudly than their less promiscuous relatives (Briskie et al. 1994).

Among the host-tolerant *Clamator* cuckoos, great spotted cuckoos are the most studied to date, and as for cowbirds, their size and vigorous begging relative to their hosts contributes to their success (Soler et al. 1995b). However, in this species these advantageous traits are compounded by a third: great spotted cuckoos have conspicuously pale papillae on their palates, which despite not being present on host young have a strong stimulatory effect on parental provisioning (Soler et al. 1995b). Similar papillae are found in related non-parasitic cuckoos (Payne 1977) and may therefore persist in the great spotted cuckoo via common descent. It is unknown why they stimulate host parents so effectively.

Vidua finches are even more benign parasites than either of the previous examples. These brood parasites are closely related to their estrildid finch hosts, and resemble them in size. All estrildid finches have complicated patterns of spots, stripes, and bizarrely coloured reflective nodules inside their mouths and on their gape flanges (Fig. 13.3; reviewed by Payne 2005b). Most *Vidua* species specialize on a single host species, and different *Vidua* species show precisely the same markings as their specialist host (Neunzig 1929). It was long assumed that this close matching was a product of host rejection selecting for mimicry, just as cuckoo eggs match their hosts'. At first sight, host mouth markings do look like signatures that have been forged by the parasite. In other words, parasites are selected to mimic hosts. However, experiments by Schuetz have shown that hosts readily accept chicks with mis-matched mouths, although they feed them less (Schuetz 2005). The hosts' mouth patterns thus seem not to be defensive signatures, but rather stimuli attractive to parents (the chick's equivalent of a peacock's tail). The parasite will therefore be selected to innovate new mouth markings that host parents find highly attractive, because its needs are not tempered by any genetic interest in its brood mates. Once this happens, host chicks might in turn be selected to exaggerate their signals simply to compete effectively with the parasite, leading to a co-evolutionary race in which the hosts mimic the parasite (Hauber and Kilner 2007). According to this speculative scenario, elaborate gapes are still the product of co-evolution, but result not from a race to signal identity, but rather from competition to stimulate parental provisioning.

The need to compete effectively in within-brood competition might also account for other cases of parasitic chick mimicry in groups that are not known to encounter regular chick rejection, such as screaming cowbirds *Molothrus rufoaxillaris* parasitizing (distantly related) bay-winged cowbirds *M. badius* (Fig. 13.2; Fraga 1998), and (with the exception of the papillae described above), great spotted cuckoos parasitizing magpies (Soler et al. 1995b). In these cases, there is less reason to suspect that parasites have driven visual elaboration of hosts.

13.5.5 How can hosts defend themselves at the chick stage?

Given the costs of chick provisioning (e.g. Brooke and Davies 1989; Kilpatrick 2002) and the lengthy incubation period that preceded them, the nestling period is a late and expensive stage for hosts to detect and respond to parasitism (Section 13.6). Yet host parents can still escape further costly investment by detecting parasitic chicks after they have hatched, and we might expect selection for such defences to be strongest in the hosts of highly virulent brood parasites. In recent years, multiple examples have been uncovered (Fig. 13.2), particularly among the hosts of the (highly virulent) bronze-cuckoos *Chalcites* spp. of Australasia. Importantly, bronze-cuckoos have evaded earlier lines of host defence with their cryptic (or mimetic, Langmore and Kilner 2009) eggs (Section 13.4), which should exacerbate selection for chick rejection since this is the only stage at which hosts can still defend themselves. We have already noted in Section 13.5.2 that superb fairy wrens identify Horsfield's bronze-cuckoo chicks by virtue of their sole occupancy of the nest (Langmore et al. 2003), supplemented with visual cues that reduce mistaken rejection of single chicks of their own (Langmore et al. 2009a). Similarly, two gerygone *Gerygone* spp. hosts of the little bronze-cuckoo *C. minutillus* are able to reject cuckoo chicks (in this case by physically tossing them out of the nest with their bills) even when host chicks remain in the nest, suggesting that other cues must act in these cases (Sato et al. 2010, Tokue and Ueda 2010). Powerful evidence that *Chalcites* hosts use visual cues to prompt such rejection comes from the remarkably sophisticated phenotypic matching of several species of *Chalcites* cuckoos to their respective specialist host, mimicking skin and mouth colour, and the presence and structure of white downy feathers (Langmore et al. 2011).

In a population of reed warblers, host parents sometimes desert common cuckoo nestlings in response purely to the longer duration of parental care (Grim 2007). While host parents have by this stage already paid a considerable cost, by abandoning a large cuckoo chick they still avoid squandering additional weeks of parental effort, increasing the possibility of renesting. However, a behavioural rule to desert a chick after a lengthy period of care need not have evolved in response to parasitism but could, for example, have evolved to protect them from investing in a brood of their own which is likely to fail. A broad analogue in timing of defence in the insects has recently been discovered in the *Temnothorax* spp. hosts of the slave-making ant *Protomognathus americanus*: by the time of enslavement hosts have lost all chances of reproducing themselves, but by killing parasitic pupae they can reduce the future impact of parasitism on neighbouring colonies that are typically closely related to themselves (Achenbach and Foitzik 2009).

The previous examples all involve highly virulent parasites, but recent evidence suggests that even the relatively low virulence of conspecific parasitism can select for host defences at the chick stage. Such defences are made all the more remarkable by the sensory challenge of identifying a conspecific chick as a parasite. For example, American coots *Fulica americana* rely on the tendency for parasitic females to lay their eggs later in the laying sequence to imprint on the appearance of the first few chicks that hatch, and use this information to kill any subsequently hatched young that deviate from this template (Shizuka and Lyon 2010). Moreover, the accuracy of this learning process is enhanced by selectively moving suspected parasitic eggs to the outside of the clutch at the incubation stage, thus delaying their hatch date and minimizing the risk of learning the wrong template (Shizuka and Lyon 2011).

Finally, if hosts cannot or do not reject parasitic eggs or chicks, they can lessen the costs of parasitism by tailoring their life-histories towards parasitic tolerance. Tolerance is an adaptation that we might expect to see in the hosts of less virulent parasites where tolerance is a plausible compromise. Specifically, hosts might be expected to decrease their investment in each clutch, thus allowing them a greater number of reproductive attempts. Consistent with this idea, cowbird host species with a long history of co-evolution with the brood parasite lay smaller and more clutches than species that have only recently begun to be parasitized by cowbirds (Hauber 2003b). Conversely, larger clutches may be favourable under parasitism if they offset the costs of egg loss and if multiple clutches are impossible, as in magpies parasitized by great spotted cuckoos. Magpies sympatric with the cuckoo correspondingly laid larger clutches of smaller eggs than populations allopatric with the cuckoo (Soler et al. 2001). Both of these examples do indeed come from less virulent parasites, but equivalent analyses of the hosts of virulent parasites are clearly needed before concluding that only relatively benign parasites can select for adaptations that facilitate tolerance.

13.5.6 How are chick adaptations evolutionarily maintained?

Chick mimicry and other parasitic adaptations to different host species manifest at the nestling stage of parental care pose a special evolutionary problem because they are expressed in both sexes, whereas host-specific adaptations at the egg stage are expressed only by adult females. Thus, while female-specific traits may potentially be maternally inherited within parasite species via the female-specific W chromosome, allowing multiple parasitic host-races ('gentes') to specialize on different hosts at the egg stage (Punnett 1933; Gibbs et al. 2000; Spottiswoode et al. 2011), in the absence of genomic imprinting a maternal mode of inheritance cannot explain chick specialization. Three alternative solutions to the problem of chick adaptations to multiple host species might be envisaged: parasites could mate assortatively according to host use, ultimately leading to parasite speciation; parasites could produce intermediate signals that adequately fool several hosts; or parasites could develop specializations through phenotypic plasticity. The first of these solutions is beautifully illustrated by the parasitic indigobirds *Vidua* spp. in which each host species has a corresponding parasite species (Fig. 13.3, Sorenson et al. 2003; Langmore et al. 2011). The second solution, an intermediate phenotype, is seen in the Horsfield's bronze-cuckoo which parasitizes numerous host species and which shows an intermediate visual phenotype that matches no host perfectly (Langmore et al. 2011), despite selection from hosts that reject visually mis-matched chicks (Langmore et al. 2009a). It may be no coincidence that this species also shows the third solution, phenotypic plasticity in host specialization: Horsfield's bronze-cuckoo chicks give different, mimetic begging calls in the nests of different host species, and experiments have demonstrated that this develops through social shaping, whereby cuckoo chicks gradually hone their calls towards those that are most successful in eliciting parental care (Langmore et al. 2008). It is possible that this facultative adjustment compensates for this generalist species' jack-of-all-trades visual signal. Several other cuckoo species exploiting multiple hosts have also been found to show similar host-specific differentiation in begging calls (reviewed by Langmore et al. 2008), although the mechanism responsible for their maintenance remains to be identified.

13.6 Why are host parents often so gullible?

Why are host parents often susceptible to being parasitized? This is particularly perplexing at the chick stage when, with a very few exceptions (Section 13.5.5), host parents assiduously feed a chick that bears no resemblance to their own. At the

Box 13.1 Kin selection and the evolution of brood parasitism

We have assumed that a parasitic chick is always unrelated to its nestmates, and hence that within-brood relatedness should not temper its selfishness. Is this a reasonable assumption? It should be plausible for the chick-killing brood parasites, although there are rare instances of two parasitic chicks raised together (e.g. Skead 1970); however, since selection should not favour females that lay multiple eggs in the same nest, these are highly unlikely to be relatives. In the less virulent cuckoo finch, more than half of parasitized nests contain two parasitic chicks, which based on egg phenotype can be inferred to have been laid by the same parasitic female (CNS unpubl. data). Occasionally, a great spotted cuckoo female will lay more than one egg in the same host nest (Martinez et al. 1998). In such species sibling conflict might be tempered by kin selection, more similarly to conventional families but with the exceptions that: 1) the nestlings have no genetic interest in the hosts' future reproduction, and 2) brood parasites are commonly (but not exclusively) promiscuous owing to their release from parental care (Barnard 1998), suggesting that sibs are likely to be half-sibs and correspondingly more selfish (Briskie et al. 1994). For example, in only a third of nests multiply parasitized by great spotted cuckoos did the two cuckoo chicks share at least one parent (Martinez et al. 1998). We currently have no information on degrees of selfishness in parasites that are commonly raised alongside potential siblings. Conspecific brood parasitism, of course, allows fascinating possibilities to arise since host and parasite can be related; see Lyon and Eadie (2008) for an excellent recent review.

outset, parasites may be exploiting simple parental behavioural rules that serve host parents well in the absence of parasitism. For example, an incubating snow goose female will roll into her own nest an egg deposited near the nest rim by another female, most likely because conspicuous eggs attract predators (Lank et al. 1991). In this case, the costs of raising an extra precocial chick are sufficiently low that it still pays the host female to obey her rule of thumb and not risk losing her entire clutch. Similarly, a begging gape is a powerful parental stimulus: cuckoo chicks are sometimes fed by other birds passing by (McBride 1984; B.G. Stokke pers. comm.). Given the potential costs associated with feeding a gaping cuckoo chick, it might be expected that selection would favour host parents who *do* change their behavioural rules with respect to parental care. Yet, this is curiously rare, at least among the well-studied hosts of common cuckoos and brown-headed cowbirds.

Why might a changed rule of thumb fail to evolve despite its evident benefits? The traditional alternatives are that either a lack of defence is non-adaptive and there has simply not been sufficient time or genetic variation for defences to evolve ('evolutionary lag'), or that the costs of defence outweigh its benefits and hence it is adaptive for defences not to evolve ('evolutionary equilibrium'). While this is a long-running debate (Davies 2000), new theory and recent fieldwork on previously little-known model systems have shed light on some of the intriguing nuances these alternatives might take; these are discussed next.

Perhaps the strongest evidence to date for evolutionary lag comes from the finding that cowbird hosts that indiscriminately incubate parasitized clutches rather than deserting them are often species that have only recently been exposed to parasitism (Hosoi and Rothstein 2000). However, this hypothesis is difficult to falsify (Kilner and Langmore 2011). Evolutionary equilibrium arguments are instead based on the premise that host defences are costly in relation to the costs of parasitism. Overall, it might be predicted that evolutionary equilibrium should be more easily reached in the case of relatively benign parasites, since it is in these circumstances that the benefits of rejection are small and only relatively low costs of defence are sufficient to prevent rejection evolving. By contrast, only overwhelmingly high costs of defence would be sufficient to offset the costs of highly virulent parasites, and it may therefore be less easy to attain equilibrium (Stokke et al. 2007). This balance of costs can be further decomposed to lead to two interesting explanations for the puzzling mix of refined adaptation and the lack of it.

First, in the case of chick-killing parasites, learnt recognition of a parasitic chick may itself impose prohibitively high costs. An elegant hypothesis is based on the premise that chick recognition would by necessity involve the host learning the appearance of its own young. If so, costs arise because if the host parent is parasitized by a chick-killing brood parasite on its first nesting attempt, it would imprint on the wrong species, with disastrous consequences for future, unparasitized nesting attempts (Lotem 1993). Such high costs should then prevent chick recognition and rejection from evolving. These costs should not be so severe at the egg stage or in benign brood parasites, where host and parasite share the nest and the risk of misimprinting should be lower, thus allowing defences to evolve in these cases. Experiments on American coots show that misimprinting can indeed be costly (Shizuka and Lyon 2010), and this hypothesis may well account for many cases of puzzling absence of adaptation. Clearly, a genetically inherited recognition template would be advantageous compared to learnt recognition, if it removed the risk of misimprinting. The only evidence to date for any form of such innate recognition system comes from fairy-wren hosts that frequently do reject Horsfield's bronze-cuckoo chicks (Langmore et al. 2003). Fairy-wrens innately use single occupancy of the nest as a cue to prompt chick desertion (Langmore et al. 2009a), but this recognition template is error-prone because hosts occasionally encounter single chicks of their own. The error rate decreases in more experienced females, suggesting that the innate template is additionally refined through learning of their own offspring's appearance (Langmore et al. 2009a); presumably these errors do not outweigh the advantages of avoiding a prolongation of parasitism. An innately transmitted but highly accurate chick recognition template has apparently never evolved in any host species; some possible explanations are discussed by Lotem (1993).

Second, if early lines of defence such as nest defence and egg rejection are effective, they may result in diminishing returns from subsequent costly lines of defence, as parasitism becomes an increasingly weak selective force later in the reproductive cycle. This variant on the rare-enemy effect has been called 'strategy-blocking', since theoretical models show that an effective early line of defence can plausibly block the evolution of a subsequent one (Britton et al. 2007). Correspondingly, as successive lines of defence are breached by the parasite, successive strategies of defence should evolve (Welbergen and Davies 2009; Langmore and Kilner 2010), and the best host defences should be the earliest possible ones, since these allow the parasite to avoid further costly investment.

What is suggested by the empirical evidence? Most of the recently-discovered examples of chick rejection all occur in species whose earlier defences have indeed been breached (see example on *Chalcites* cuckoos; Section 13.5.5). However, in addition to chick rejection, American coots also show effective egg discrimination and the two defences appear even to act synergistically, suggesting that the earlier strategy has not blocked the adaptive value of chick rejection (Shizuka and Lyon 2011). Strategy blocking may sometimes help to explain surprising levels of parental gullibility at the egg stage as well as the chick stage. For example, Afrotropical jacobin cuckoos *Clamator jacobinus* lay huge, round, white eggs that could not contrast more with the smaller, tapered, mottled eggs of their Cape bulbul *Pycnonotus capensis* hosts, who nonetheless ignore cuckoo eggs. Bulbuls do however fiercely attack laying cuckoos, which might restrict their access to host nests, leading to many cuckoo eggs being laid too late and hence failing to hatch. In combination with high costs of rejecting a large, strong-shelled egg and of attempting a second breeding attempt, poor parasitic egg hatchability means that on average it pays the bulbuls to accept the evident imposter and rely on a high probability that it fails to hatch (Krüger 2011). In this instance, evolutionary equilibrium appears to explain the bulbul's apparent naïvety, and if vigorous nest defence does indeed substantially lower the hatching rate of cuckoo eggs, then this may be interpreted as an example of strategy blocking.

Both evolutionary lag and such a sequential accumulation of adaptive defences predict that systems of host defence should extend further into the reproductive cycle with increasing evolutionary age of the parasite–host association. For example, cowbirds are a relatively young parasitic lineage, and moreover many cowbird hosts have only been parasitized for a few centuries owing to anthropogenic change (Hosoi and Rothstein 2000). While many cowbird host species show vigorous nest defence, only some reject foreign eggs and none are known to reject foreign chicks. The interactions between common cuckoos and many hosts also appear to be relatively recent, with the common ancestor of the extant cuckoo gentes in the UK estimated using mitochondrial DNA to date back to around 80,000 years ago (Gibbs et al. 2000). Nest defence against common cuckoos is common, nearly all host species reject foreign eggs, and with one exception, all hosts accept foreign chicks (the one exception need not have evolved as a response to parasitism; Section 13.5.5). By comparison, the Australasian bronze-cuckoos (*Chalcites*) are more ancient species (Payne 2005a) likely to have experienced longer term interactions with their hosts, perhaps owing to the relative climatic stability of tropical systems. Bronze-cuckoo hosts have had their defences entirely breached at the egg stage, and three species are now known to reject foreign chicks (Section 13.5.5). Thus these three systems seem broadly consistent with the idea that variation in depth of defences through the reproductive cycle can to some degree be attributed to evolutionary age. However, counter-examples also exist: genetic evidence suggests that interactions between greater honeyguides and their hosts are very ancient (Spottiswoode et al. 2011), yet a common host shows only extremely vigorous nest defence and appears to have no defences thereafter (CNS unpubl. data).

It is interesting to speculate how host–parasite co-evolution might escalate with increasing parasitic trickery. For example, American coots have a sophisticated three-tiered system of defence against intraspecific brood parasitism involving egg rejection, selective incubation, and chick discrimination (Shizuka and Lyon 2011). Yet if parasitic females were to evolve internal incubation and early hatching like cuckoos (Birkhead et al. 2011), or transfer of partially incubated eggs like cliff swallows (Brown and Brown 1988), they could quickly destabilize the latter two defences.

13.7 Conclusions and speculations

Variation in virulence may set off a cascade of host and parasitic adaptations at different stages of the reproductive cycle. In hosts, variation in virulence seems likely to impact on the evolution of nest defence, egg rejection, egg signatures, and chick rejection (Kilner and Langmore 2011). In parasites, variation in virulence may account for variation in the evolution of egg-laying trickery, egg mimicry, killing or co-opting of foster siblings, chick mimicry, and perhaps even parasitic speciation.

What then dictates a parasite's level of virulence? The relative size of host and parasite may play a role: the benefits of low virulence are most likely to be accrued when hosts are similar sized or larger than parasites, and can thus provide effective assistance in soliciting parental care. High virulence may conversely be favoured when parasitizing smaller hosts whose parental care is easily stimulated and harder to share (Kilner 2005). But once highly virulent traits such as chick-killing have evolved in a parasitic lineage, they may be difficult to lose, even in a particular host environment when it might become advantageous to do so. Interestingly, the cuckoo genera *Scythrops* and *Eudynamys* that do not always kill their hosts are nested within clades of brood parasites that are obligate chick-killers (Fig. 13.2; Sorenson and Payne 2005), suggesting that chick-killing can be evolutionarily reversed. In this instance the loss of chick-killing might have been forced by the use of relatively large hosts that might be physically impossible to eject: *Eudynamys* populations in India parasitize relatively large hosts and do not eject host chicks, whereas elsewhere they do (Payne 2005a). Moreover, a parasite's virulence might be influenced by whether it is a generalist or a specialist. If the former, they may not have the ability to tailor their level of virulence for different hosts, and if hosts vary substantially in size then a generalist strategy may work better in some hosts than in others.

Perhaps parasitic chicks could facultatively reduce their virulence (e.g. by begging less to avoid starving host chicks through competition) when parasitizing hosts that it would be advantageous to keep alive, to assist them in soliciting parental care? Perhaps parasitic mothers could strategically influence their offspring's virulence by manipulating offspring phenotype via maternal effects such as yolk androgens or antioxidants (Royle et al. 2011 but see Hauber and Pilz 2003; Török et al. 2004)? Or if host death is inevitable, perhaps parasitic mothers could strategically compensate by laying two parasitic eggs in the same nest (Kilner 2005) (as cuckoo finches commonly do)? These intriguing possibilities all await experimental tests.

The studies reviewed in this chapter illustrate that the life of a full-time cheat is not quite as straightforward as we might think. Cheats have to evolve successive lines of trickery to dupe their hosts into providing them with parental care. They also have to evolve adaptations that allow them best to exploit another species' life-history, such as their begging behaviour, to maximize the gains they obtain from the host's care. Moreover, mounting host defences is also not as easy as we might think. A host needs hardwired mechanisms to ensure that it cares for its own young, and these lay it open to exploitation by parasites. While a loudly begging chick could be a parasite, it could also be a hungry and valuable large chick of its own. This chapter has reviewed multiple examples that show that host susceptibility to parasites is generally the rule, and examples of host counterdefences at the chick stage are still few. At the outset, we described a reed warbler perched on the shoulder of a young common cuckoo nine times its size, stuffing food into its bright orange gape. Are we any closer to understanding why it didn't behave like an Australian gerygone and toss out the cuckoo long before it reached such giant and costly proportions? To some extent, we are: we can speculate that the gerygone can probably afford to take chances with chick recognition since it has a longer lifetime than the reed warbler in which to perfect the ability. We can also speculate that the interactions between the gerygone and its cuckoo have probably been going on for substantially longer than the reed warbler's.

Satisfactory answers to these questions might well depend as much on advances in old-fashioned natural history as on advances in theory, and in many respects the latter could be argued to have outstripped the former. The natural history of the vast majority of brood parasites is unknown. Decades of research on classic north-temperate model species such as common and great spotted cuckoos and brown-headed cowbirds have been tremendously revealing, but in recent years fascinating advances have been made thanks to new study systems, often in concert with significant technical advances such as visual modelling (allowing signals to be quantified through a bird's eye) and digital filming. For example, given the lack of chick rejection by hosts of the common cuckoo, no one would have anticipated that rejection of parasitic chicks would prove to be widespread in the Australasian *Chalcites* cuckoos (Langmore et al. 2003; Sato et al. 2010; Tokue and Ueda 2010), precipitating a major rethink of theory. Further advances may turn out to be especially forthcoming from tropical systems where they may have been more time for complex species interactions to evolve. Surely there are many gems awaiting discovery among the invertebrates and the large array of avian parasitic systems that remain barely studied, especially the Neotropical cuckoos and the little-known Asian and African cuckoos and honeyguides, that may well transform the conclusions of future reviews.

Acknowledgements

We thank Scott Forbes, Mathias Kölliker, Nick Royle, Per Smiseth, and Bård Stokke for helpful comments, and Vittorio Baglione, Justin Schuetz, and Alan Weaving for kindly providing photographs. CNS was supported by a Royal Society Dorothy Hodgkin Research Fellowship.

References

Achenbach, A. and Foitzik, S. (2009). First evidence for slave rebellion: enslaved ant workers systematically kill the brood of their social parasite *Protomognathus americanus*. *Evolution* 63, 1068–75.

Alvarez, F. (2000). Response to Common Cuckoo *Cuculus canorus* model egg size by a parasitized population of Rufous Bush Chat *Cercotrichas galactotes*. *Ibis* 142, 683–6.

Anderson, M. G., Moskát, C., Bán, M., Grim, T., Cassey, P., and Hauber, M. E. (2009). Egg eviction imposes a recoverable cost of virulence in chicks of a brood parasite. *PLoS ONE* 4, e7725.

Antonov, A., Stokke, B. G., Moksnes, A., and Røskaft, E. (2009). Evidence for egg discrimination preceding failed rejection attempts in a small cuckoo host. *Biol. Lett.* 5, 169–71.

Barbero, F., Thomas, J. A., Bonelli, S., Balletto, E., and Schonrogge, K. (2009). Queen ants make distinctive sounds that are mimicked by a butterfly social parasite. *Science* 373, 782–5.

Barnard, P. (1998). Variability in the mating systems of parasitic birds. In S. I. Rothstein and S. K. Robinson, eds. *Parasitic Birds and their Hosts: Studies in Coevolution.* Oxford University Press, New York.

Birkhead, T. R., Hemmings, N., Spottiswoode, C. N., Mikulica, O., Moskát, C., Bán, M., and Schulze-Hagen, K. (2011). Internal incubation and early hatching in brood parasitic birds. *Proc. R. Soc. Lond B* 278, 1019–24.

Boncoraglio, G., Saino, N., and Garamszegi, L. Z. (2008). Begging and cowbirds: brood parasites make hosts scream louder. *Behav. Ecol.* 20, 215–21.

Brandt, M., Foitzik, S., Fischer-Blass, B., and Heinze, J. (2005). The coevolutionary dynamics of obligate social parasite systems—between prudence and antagonism. *Biol. Rev.* 80, 251–67.

Briskie, J. V., Naughler, C. T., and Leech, S. M. (1994). Begging intensity of nestling birds varies with sibling relatedness. *Proc. R. Soc. Lond. B* 258, 73–8.

Britton, N. F., Planqué, R., and Franks, N. R. (2007). Evolution of defence portfolios in exploiter-victim systems. *Bull. Math Biol.* 69, 957–88.

Brooke, M. de L. and Davies, N. B. (1989). Provisioning of nestling cuckoos *Cuculus canorus* by reed warbler *Acrocephalus scirpaceus* hosts. *Anim. Behav.* 131, 250–6.

Brooker, L. and Brooker, M. (1998). Why do splendid fairy-wrens always accept cuckoo eggs? *Behav. Ecol.* 9, 420–4.

Brown, C. R. and Brown, M. B. (1988). A new form of reproductive parasitism in cliff swallows. *Nature* 331, 66–8.

Brown, C. R. and Brown, M. B. (1991). Selection of high-quality host nests by parasitic cliff swallows. *Anim. Behav.* 41, 457–65.

Brown, J. L., Morales, V., and Summers, K. (2009). An experimental demonstration of tactical parasitism of conspecific reproduction via larval cannibalism in Peruvian poison frogs. *Biol. Lett.* 5, 148–51.

Chapman, T. A. (1869). *Aphodius porcus*, a cuckoo parasite on *Geotrupes stercorarius*. *Entomol. Month. Mag.* 5, 273–6.

Davies, N. B. (2000). *Cuckoos, Cowbirds and Other Cheats*, T & A D Poyser, London.

Davies, N. B. (2011). Cuckoo adaptations: trickery and tuning. *J. Zool.* 284, 1–14.

Davies, N. B. and Brooke, M. de L. (1988). Cuckoos versus reed warblers: adaptations and counteradaptations. *Anim. Behav.* 36, 262–84.

Davies, N. B., Kilner, R. M., and Noble, D. G. (1998). Nestling cuckoos, *Cuculus canorus*, exploit hosts with begging calls that mimic a brood. *Proc. R. Soc. Lond. B* 265, 673–8.

Davies, N. B. and Welbergen, J. A. (2008). Cuckoo-hawk mimicry? An experimental test. *Proc. R. Soc. Lond B* 275, 1817–22.

Davies, N. B. and Welbergen, J. A. (2009). Social transmission of a host defense against cuckoo parasitism. *Science* 324, 1318–20.

Field, J. (1994). Selection of host nests by intraspecific nest-parasitic digger wasps. *Anim. Behav.* 48, 113–18.

Fraga, R. M. (1998). Interactions of the parasitic screaming and shiny cowbirds (*Molothrus rufoaxillaris* and *M. bonariensis*) with a shared host, the bay-winged cowbird (*M. badius*). In S. I. Rothstein and S. K. Robinson, eds. *Parasitic Birds and Their Hosts.* Oxford University Press, New York.

Freeman, S. (1988). Egg variability and conspecific nest parasitism in the *Ploceus* weaverbirds. *Ostrich* 59, 49–53.

Fry, C. H. (1974). Vocal mimesis in nestling Greater Honey-guides. *Bull. Brit. Orn. Club* 94, 58–9.

Gibbs, H. L., Sorenson, M. D., Marchetti, K., Brooke, M. de L., Davies, N. B., and Nakamura, H. (2000). Genetic evidence for female host-specific races of the common cuckoo. *Nature* 407, 183–6.

Grim, T. (2007). Experimental evidence for chick discrimination without recognition in a brood parasite host. *Proc. R. Soc. Lond B* 274, 373–81.

Grim, T., Rutila, J., Cassey, P., and Hauber, M. E. (2009). The cost of virulence: an experimental study of egg eviction by brood parasitic chicks. *Behav. Ecol.* 20, 1138–46.

Hackett, S. J., Kimball, R. T., Reddy, S., Bowie, R. C. K., Braun, E. L., Braun, M. J., Chojnowski, J. L., Cox, W. A., Han, K.-L., Harshman, J., Huddleston, C. J., Marks, B. D., Miglia, K. J., Moore, W. S., Sheldon, F. H., Steadman, D. W., Witt, C. C., and Yuri, T. (2008). A phylogenomic study of birds reveals their evolutionary history. *Science* 320, 1763–8.

Hargitai, R., Moskát, C., Bán, M., Gil, D., Lopéz-Rull, I., and Solymos, E. (2010). Eggshell characteristics and yolk composition in the common cuckoo *Cuculus canorus*: are they adapted to brood parasitism? *J. Avian Biol.* 41, 177–85.

Hauber, M. E. (2003a). Hatching asynchrony, nestling competition, and the cost of interspecific brood parasitism. *Behav. Ecol.* 14, 227–35.

Hauber, M. E. (2003b). Interspecific brood parasitism and the evolution of host clutch sizes. *Evol. Ecol. Res.* 5, 559–70.

Hauber, M. E. and Kilner, R. M. (2007). Co-evolution, communication and host chick mimicry in parasitic finches: who mimics whom? *Behav. Ecol. Sociobiol.* 61, 497–504.

Hauber, M. E. and Pilz, K. M. (2003). Yolk testosterone levels are not consistently higher in the eggs of obligate brood parasites than their hosts. *Am. Midl. Nat.* 149, 354–62.

Hosoi, S. A. and Rothstein, S. I. (2000). Nest desertion and cowbird parasitism: evidence for evolved responses and evolutionary lag. *Anim. Behav.* 59, 823–40.

Jackson, W. M. (1993). Causes of conspecific nest parasitism in the northern masked weaver. *Behav. Ecol. Sociobiol.* 32, 119–26.

Kilner, R. M. (2003). How selfish is a cowbird nestling? *Anim. Behav.* 66, 569–76.

Kilner, R. M. (2005). The evolution of virulence in brood parasites. *Ornithol. Sci.* 4, 55–64.

Kilner, R. M. and Langmore, N. E. (2011). Cuckoos versus hosts in insects and birds: adaptations, counter-adaptations and outcomes. *Biol. Rev.* 86, 836–52.

Kilner, R. M., Madden, J. R., and Hauber, M. E. (2004). Brood parasitic cowbirds use host young to procure food. *Science* 305, 877–9.

Kilner, R. M., Noble, D. G., and Davies, N. B. (1999). Signals of need in parent-offspring communication and their exploitation by the common cuckoo. *Nature* 397, 667–72.

Kilpatrick, A. M. (2002). Variation in growth of Brown-headed Cowbird (*Molothrus ater*) nestlings and energetic impacts on their host parents. *Can. J. Zool.* 80, 145–53.

Krüger, O. (2011). Brood parasitism selects for no defence in a cuckoo host. *Proc. R. Soc. Lond B* 278, 2777–83.

Langmore, N. E., Cockburn, A., Russell, A. F., and Kilner, R. M. (2009a). Flexible cuckoo chick-rejection rules in the superb fairy-wren. *Behav. Ecol.* 20, 978–84.

Langmore, N. E., Hunt, S., and Kilner, R. M. (2003). Escalation of a coevolutionary arms race through host rejection of brood parasitic young. *Nature* 422, 157–60.

Langmore, N. E. and Kilner, R. M. (2007). Breeding site and host selection by Horsfield's bronze-cuckoos *Chalcites basalis*. *Anim. Behav.* 74, 995–1004.

Langmore, N. E. and Kilner, R. M. (2009). Why do Horsfield's bronze-cuckoo *Chalcites basalis* eggs mimic those of their hosts? *Behav. Ecol. Sociobiol.* 63, 1127–31.

Langmore, N. E. and Kilner, R. M. (2010). The coevolutionary arms race between Horsfield's bronze-cuckoos and superb fairy-wrens. *Emu* 110, 32–8.

Langmore, N. E., Maurer, G., Adcock, G. J., and Kilner, R. M. (2008). Socially acquired host-specific mimicry and the evolution of host races in Horsfield's Bronze-Cuckoo *Chalcites basalis*. *Evolution* 62, 1689–99.

Langmore, N. E. and Spottiswoode, C. N. (2012). Visual trickery in avian brood parasites. In D. P. Hughes, J. Brodeur, and F. Thomas, eds. *Host Manipulation by Parasites.* Oxford University Press, Oxford.

Langmore, N. E., Stevens, M., Maurer, G., Heinsohn, R., Hall, M. L., Peters, A., and Kilner, R. M. (2011). Visual mimicry of host nestlings by cuckoos. *Proc. R. Soc. Lond B* 278, 2455–63.

Langmore, N. E., Stevens, M., Maurer, G., and Kilner, R. M. (2009b). Are dark cuckoo eggs cryptic in host nests? *Anim. Behav.* 78, 461–8.

Lank, D. B., Bousfield, M. A., Cooke, F., and Rockwell, R. F. (1991). Why do snow geese adopt eggs? *Behav. Ecol.* 2, 181–7.

Lichtenstein, G. (2001). Selfish begging by screaming cowbirds, a mimetic brood parasite of the bay-winged cowbird. *Anim. Behav.* 61, 1151–8.

Lotem, A. (1993). Learning to recognize nestlings is maladaptive for cuckoo *Cuculus canorus* hosts. *Nature* 362, 743–5.

Lotem, A., Nakamura, H., and Zahavi, A. (1995). Constraints on egg discrimination and cuckoo-host co-evolution. *Anim. Behav.* 49, 1185–209.

Lyon, B. E. and Eadie, J. M. (2008). Conspecific brood parasitism in birds: a life-history perspective. *Ann. Rev. Ecol. Evol. Syst.* 39, 343–63.

Martinez, J. G., Burke, T., Dawson, D., Soler, J. J., Soler, M., and Møller, A. P. (1998). Microsatellite typing reveals mating patterns in the brood parasitic great spotted cuckoo (*Clamator glandarius*). *Mol. Ecol.* 7, 289–97.

Massoni, V. and Reboreda, J. C. (1999). Egg puncture allows shiny cowbirds to assess host egg development and suitability for parasitism. *Proc. R. Soc. Lond. B* 266, 1871–4.

McBride, H. C. A. (1984). Multiple feeding of a juvenile cuckoo. *Brit. Birds* 77, 422–3.

Moyer, D. C. (1980). On Lesser Honeyguide and Black-collared Barbet. *Zambian Ornithological Society Newsletter* 10, 159.

Neunzig, R. (1929). Zum Brutparasitismus der Viduinen. *J. Ornithol.* 77, 1–21.

Noble, D. G., Davies, N. B., Hartley, I. R., and McRae, S. B. (1999). The red gape of the nestling cuckoo (*Cuculus canorus*) is not a supernormal stimulus for three common hosts. *Behaviour* 136, 759–77.

Parejo, D. and Avilés, J. M. (2007). Do avian brood parasites eavesdrop on heterospecific sexual signals revealing host quality? A review of the evidence. *Anim. Cogn.* 10, 81–88.

Payne, R. B. (1977). The ecology of brood parasitism in birds. *Annu. Rev. Ecol. Syst.* 8, 1–28.

Payne, R. B. (2005a). *The Cuckoos,* Oxford University Press, Oxford.

Payne, R. B. (2005b). Nestling mouth markings and colors of Old World finches Estrildidae: mimicry and coevolution of nesting finches and their *Vidua* brood parasites.

Miscellaneous Publications, Museum of Zoology, University of Michigan, 194.

Punnett, R. C. (1933). Inheritance of egg-colour in the parasitic cuckoos. *Nature* 132, 892.

Rasa, O. A. E. (1996). Interspecific association in desert tenebrionid beetles: a cleptoparasite does not affect the host's reproductive success, but that of its offspring. *Naturwiss.* 83, 575–7.

Rothstein, S. I. (1975). An experimental and teleonomic investigation of avian brood parasitism. *Condor* 77, 250–71.

Royle, N. J., Hall, M. E., Blount, J. D., and Forbes, S. (2011). Patterns of egg yolk antioxidant co-variation in an avian brood parasite–host system. *Behav. Ecol. Sociobiol.* 65, 313–23.

Sato, N. J., Tokue, K., Noske, R. A., Mikami, O. K., and Ueda, K. (2010). Evicting cuckoo nestlings from the nest: a new anti-parasitism behaviour. *Biol. Lett.* 6, 67–9.

Sato, T. (1986). A brood parasitic catfish of mouthbrooding cichlid fishes in Lake Tanganyika. *Nature* 323, 58–9.

Schuetz, J. G. (2005). Low survival of parasite chicks may result from their imperfect adaptation to hosts rather than expression of defenses against parasitism. *Evolution* 59, 2017–24.

Schulze-Hagen, K., Stokke, B. G., and Birkhead, T. R. (2009). Reproductive biology of the European Cuckoo *Cuculus canorus*: early insights, persistent errors and the acquisition of knowledge. *J. Ornithol.* 150, 1–16.

Shizuka, D. and Lyon, B. E. (2010). Coots learn to recognize and reject brood parasitic chicks. *Nature* 463, 223–6.

Shizuka, D. and Lyon, B. E. (2011). Hosts improve the reliability of chick recognition by delaying the hatching of brood parasitic eggs. *Curr. Biol.* 21, 515–19.

Skead, C. J. (1970). Two Greater Honeyguide *Indicator indicator* chicks in the nest of the Greater Striped Swallow *Cecropis cucullata*. *Ostrich* 41, 217–18.

Skead, D. M. (1975). Ecological studies of four estrildines in the central Transvaal. *Ostrich* Suppl. 11, 1–55.

Soler, J. J., Martínez, J. G., Soler, M., and Møller, A. P. (2001). Life history of magpie populations sympatric or allopatric with the brood parasitic great spotted cuckoo. *Ecology* 82, 1621–31.

Soler, J. J., Soler, M., Møller, A. P., and Martínez, J. G. (1995a). Does the great spotted cuckoo choose magpie hosts according to their parenting ability? *Behav. Ecol. Sociobiol.* 36, 201–6.

Soler, M. and Martínez, J. G. (2000). Is egg-damaging behavior by great spotted cuckoos an accident or an adaptation? *Behav. Ecol.* 11, 495–501.

Soler, M., Martínez, J. G., Soler, J. J., and Møller, A. P. (1995b). Preferential allocation of food by magpies *Pica pica* to great spotted cuckoo *Clamator glandarius* chicks. *Behav. Ecol. Sociobiol.* 37, 7–13.

Sorenson, M. D. (1992). Comment: Why is conspecific nest parasitism more common in waterfowl than in other birds? *Can. J. Zool.* 70, 1856–8.

Sorenson, M. D. (1997). Effects of intra- and interspecific brood parasitism on a precocial host, the canvasback, *Aythya valisineria*. *Behav. Ecol.* 8, 155–61.

Sorenson, M. D. and Payne, R. B. (2005). A molecular genetic analysis of cuckoo phylogeny. In R. B. Payne, ed. *The Cuckoos*. Oxford University Press, Oxford.

Sorenson, M. D., Sefc, K. M. and Payne, R. B. (2003). Speciation by host switch in brood parasitic indigobirds. *Nature* 424, 928–31.

Spottiswoode, C. N. and Koorevaar, J. (2012). A stab in the dark: chick killing by brood parasitic honeyguides. *Biol. Lett.* 8, 241–4.

Spottiswoode, C. N., Stryjewski, K. F., Quader, S., Colebrook-Robjent, J. F. R., and Sorenson, M. D. (2011). Ancient host-specificity within a single species of brood parasitic bird. *Proc. Natl. Acad. Sci. USA* 108, 17738–42.

Stokke, B. G., Takasu, F., Moksnes, A., and Røskaft, E. (2007). The importance of clutch characteristics and learning for antiparasite adaptations in hosts of avian brood parasites. *Evolution* 61, 2212–28.

Tanaka, K. and Ueda, K. (2005). Horsfield's hawk-cuckoo nestlings simulate multiple gapes for begging. *Science* 308, 653.

Tokue, K. and Ueda, K. (2010). Mangrove gerygones reject and eject little bronze-cuckoo hatchlings from parasitized nests. *Ibis* 152, 835–9.

Török, J., Moskát, C., Michl, G., and Péczely, P. (2004). Common cuckoos (*Cuculus canorus*) lay eggs with larger yolk but not more testosterone than their great reed warbler (*Acrocephalus arundinaceus*) hosts. *Ethol., Ecol., Evol.* 16, 271–7.

Vernon, C. J. (1987). On the Eastern Green-backed Honeyguide. *Honeyguide* 33, 6–12.

Welbergen, J. A. and Davies, N. B. (2009). Strategic variation in mobbing as a front line of defence against brood parasitism. *Curr. Biol.* 19, 235–40.

Yom-Tov, Y. and Geffen, E. (2006). On the origin of brood parasitism in altricial birds. *Behav. Ecol.* 17, 196–205.

Zanette, L., Macdougall-Shakleton, E., Clinchy, M., and Smith, J. N. M. (2005). Brown-headed cowbirds skew host offspring sex ratios. *Ecology* 86, 815–20.

SECTION III

Evolutionary Genetics of Parental Care

CHAPTER 14

Parental effects in development and evolution

Tobias Uller

14.1 Introduction

In contrast to the traditional conception that the genome provides a blueprint for development (e.g. Mayr 1982; Williams 1992), parents transfer a variety of non-genetic resources and templates that are as necessary for development as the transfer of genes (Fig. 14.1). Even in organisms with minimal parental care, maternally derived mRNA and proteins accumulated in the egg during oogenesis regulate early development, such as embryonic axis and pattern formation (Pelegri 2003; Li et al. 2010; Gilbert 2010). Maternal influences on offspring development continue during embryonic life, with the egg usually providing all or most of the macro- and micronutrients that are necessary for normal development. Furthermore, viviparity has evolved repeatedly in several of the major phyla (Hogarth 1976; Reynolds et al. 2002) and is associated with a further range of dependencies of offspring development on maternal transfer of resources and signalling molecules (e.g. Fowden and Forhead 2009). Finally, mothers and fathers have a significant impact on their offspring well beyond birth in many species through resource provisioning and behavioural interactions, which play important roles in the ontogeny of species-typical phenotypes and individual phenotypic variation (Gottlieb 1997; Avital and Jablonka 2000; Maestripieri 2009; Michel 2011).

Under the definition employed in this book, all these mechanisms that contribute to the 'continuity of the phenotype' across generations (West-Eberhard 2003; Fig. 14.1) are parental effects—causal effects of the parental phenotype on offspring phenotype (Wolf and Wade 2009; Chapter 1). The observation that maternal transfer of macro- and micronutrients, behavioural interactions, and reconstruction of the ecological conditions in which development takes place are necessary for expression of functional phenotypes show that many parental effects are developmentally entrenched, that is the transfer of developmental templates and resources from parents to offspring form an integral part of species-typical development (West-Eberhard 2003; Badyaev 2008; Badyaev and Uller 2009). Parental effects not only form an entrenched part of offspring development, however; they may also provide a source for expression of novel phenotypic variation (West-Eberhard 2003; Badyaev 2008), influence population dynamics (Inchausti and Ginzburg 2009), be a significant generator of natural and sexual selection (Donohue 2009), affect the rate and direction of phenotypic evolution (Kirkpatrick and Lande 1989; Bonduriansky and Day 2009; Chapter 15), contribute to the persistence of induced phenotypes across generations (Odling-Smee et al. 2003; Jablonka and Raz 2009; Chapter 17), and enable trans-generational transmission of acquired functions or information (Boyd and Richerson 1985; Jablonka and Lamb 1995, 2005; Uller 2008; for simplicity, I use the term 'trans-generational' to include both parental effects that involve only two generations (often referred to as 'intergenerational') and those where the effects of past generations of phenotypes accumulate, interact, or remain stable across more than two generations). These examples emphasize the need for evolutionary theory to take seriously the many ways by which parents reconstruct the developmental niche for their offspring and make parental effects central to the integration of developmental and evolution-

The Evolution of Parental Care. First Edition. Edited by Nick J. Royle, Per T. Smiseth, and Mathias Kölliker.

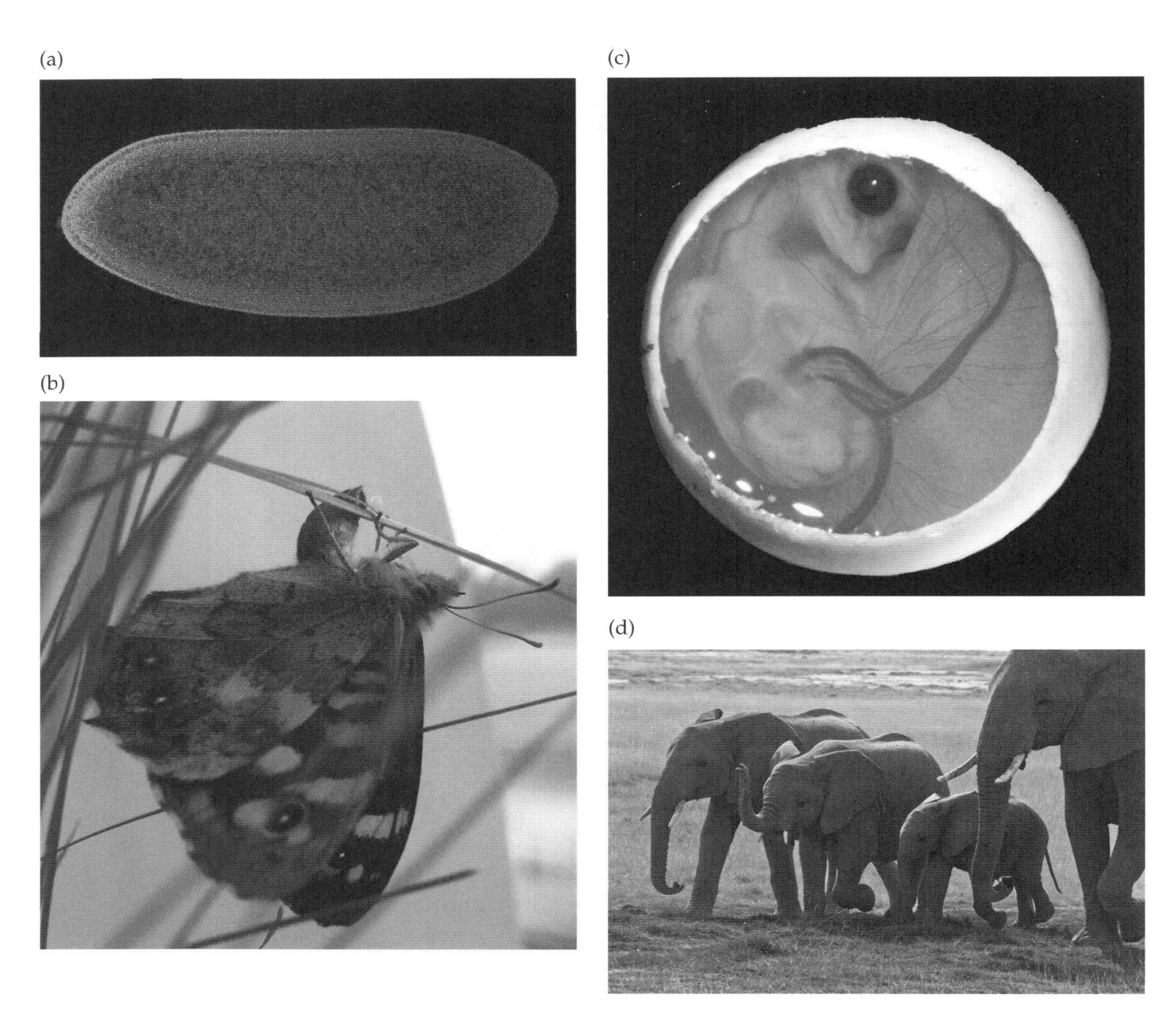

Figure 14.1 Parental effects—causal effects of the parental phenotype on offspring phenotype—are as fundamental to development as DNA. (a) All development starts with a responsive phenotype—an egg—produced by the parental phenotype. Early development occurs without expression of the offspring genome, as in maternal mRNA regulation of the anterior–posterior polarity in Drosophila embryos. (b) The parents also provide offspring with an ecological context for development, for example, by choosing where to lay the eggs. (c). Macro- and micronutrients in the egg of oviparous species, such as turtles, continue to nurse the embryo and contribute to its growth and differentiation. (d) Behavioural interactions between mothers and offspring after birth are necessary for the formation of species-typical behaviours expressed later in life in many social organisms. (Photo credits: (a): Stefan Baumgartner; (b): Casper Breuker; (c): Weiguo Du; (d): Joan Egert.)

ary biology (e.g. Jablonka and Lamb 2005; Badyaev 2008, 2009; Badyaev and Uller 2009; Bonduriansky and Day 2009; Odling-Smee 2010).

This chapter briefly covers three aspects of parental effects that are relevant to understanding their role in the evolution of parental care and offspring development (Fig. 14.2; see also Badyaev 2008, 2009 and Badyaev and Uller 2009). First, I will show how the mechanisms of parental effects can contribute to the generation of evolutionarily significant variation. I will argue that parental effects support the notion that phenotypic change may precede genetic change during adaptive evolution (Baldwin 1902; West-Eberhard 2003), and I suggest that parental effects are particularly well suited to addressing the role of developmental plasticity for

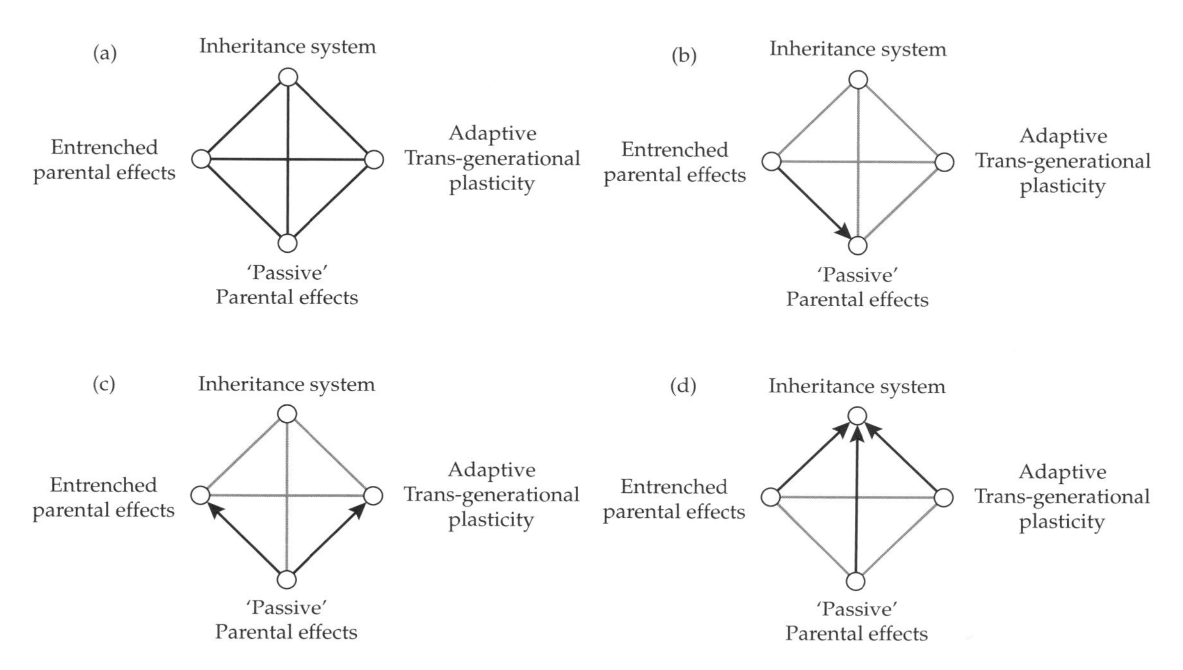

Figure 14.2 The mechanisms of parental effects can take on different 'roles' in development and evolution. (a) The open circles indicate four ways by which parental effects can contribute to the development and evolution of phenotypes. Each state may be connected to other states (black lines). Four of the possible transitions that are discussed in this chapter are shown in panels B–D. (b) The mechanisms of parental effects that contribute to species-typical development provide a source of induction of phenotypic variation via novel genetic or environmental input in the parental generations. The induced variants can be seen as 'passive' consequences of the evolved developmental reliance on parental phenotype. Nevertheless, such parental effects can influence the rate and direction of evolution by affecting the functionality of novel phenotypes and their recurrence (Sections 14.2 and 14.3). (c) Phenotypic variation expressed via parental effects is subject to natural selection, which, if variants are heritable, can result in the evolution of increased or reduced reliance on parental phenotype for normal development under species-typical developmental conditions via genetic assimilation (arrow pointing towards entrenched parental effects; Section 14.3.3), or, if the recurrence of more than one induced variant is sufficiently frequent, become stabilized by natural selection as conditionally expressed variants (arrow pointing towards adaptive trans-generational plasticity; Section 14.4). (d) Trans-generational plasticity can be seen as adaptive transmission of information across generations through a system of inheritance (e.g. chromatin-based systems, behaviours). However, if variants transmitted through non-genetic means become reliably reconstructed and sufficiently stable for natural selection to sort among their associated phenotypic effects, the mechanisms of parental effects that are involved in the replication of life cycles may evolve towards an inheritance system in a more strict sense that shares more features with the DNA-based system of inheritance (Box 14.1).

the origin of adaptive trait variation. Second, I will outline how the parental and offspring phenotypes that comprise parental effects evolve under natural selection and how this, under some circumstances, can lead to precise context-specific effects of the parental phenotype on offspring development in the form of adaptive trans-generational plasticity (Uller 2008). Third, I discuss the relationship between context-dependent parental effects and non-genetic 'systems of inheritance' (Jablonka and Lamb 2005), with the aim to provide an entry into the literature that explores the relationships between development, inheritance, and evolution from an information perspective. Finally, I provide a summary and some suggestions for how the evolutionary dynamics of parental effects can be further explored.

14.2 Parental effects and the origins of variation

Evolutionary change begins with developmental change, providing the phenotypic variation that is necessary for adaptive evolution. Developmental change, in turn, must begin with a phenotype that is responsive to novel genetic or environmental input. A description of the causes of evolution therefore requires an analysis of how existing mechanisms of development give rise to novel phenotypes (e.g. Mivart 1871; Gottlieb 1992; West-Eberhard 2003). The expression of phenotypic variation during development can be described as a two step process, by which novel genetic (via mutation, hybridization etc.) or environmental input is followed by accommodation of this input; that is, mutual adjustment of different parts that produces a functional phenotype (West-Eberhard 2003, 2005). Both genetic and environmental induction can contribute to evolutionarily relevant phenotypic variation since responses to novel environments often vary genetically between individuals, and hence can be heritable. Selection can therefore modify the regulation and form of genetically variable phenotypic accommodations over generations, a process referred to as genetic accommodation (West-Eberhard 2003; Moczek 2007). This perspective emphasizes that a complete understanding of adaptive evolution requires a description of both the developmental origin of adaptive phenotypes and the processes that result in an increase in the frequency of those phenotypes across generations. It also suggests a creative role of developmental plasticity in evolution since plasticity permits phenotypic accommodation and thus facilitates expression of novel, but functional, phenotypes in response to environmental change (West-Eberhard 2003).

Discussions of the role of developmental plasticity for expression of novel phenotypes tend to focus on the direct effect of genetic or environmental input on organisms within a single generation (e.g. Baldwin 1902; Wcislo 1989; Gottlieb 1992; West-Eberhard 2003, 2005; Moczek 2008; Pfennig et al. 2010; but see Badyaev 2009). However, parental effects allow maternal and paternal phenotypic accommodation to have carry-over effects on offspring development, thereby leading to expression of phenotypic variation in the following generation. Thus, genetic and environmental change affecting the parental generation may initiate evolutionary divergence in developmental trajectories (Badyaev 2008, 2009). For example, the reliance on maternally derived mRNA and proteins for the earliest stages in development implies that genetic or environmental modification of the regulation of oogenesis is involved in reorganization of developmental pathways (Sun et al. 2005; Minelli and Fusco 2010). Indeed, experimental studies of marine invertebrates suggest that both egg size and maternally derived factors (e.g. mRNA) that regulate embryonic development have contributed to evolutionary diversification of larval forms (Sinervo and McEdward 1988; Raff and Byrne 2006; Minelli and Fusco 2010).

The potential role of parental effects in the origin of novel variation is not restricted to regulation of early developmental patterning, but encompasses all parental influences on offspring phenotypes, including all forms of parental care (Chapter 1). In mammals, hormones of maternal origin play an important role in regulation of receptor densities, enzymes, growth factors, and other signalling molecules that are necessary for organ differentiation, including the brain (Fowden and Forehad 2009). Prenatal hormone exposure also plays a role in long-term regulation of gene expression via modification of patterns of DNA methylation (Weaver et al. 2004; Harris and Seckl 2011; Chapter 17). Variation in hormone exposure during development resulting from genetic variation between mothers or variation in maternal environment can cause short- and long-term physiological variation in the offspring, with concomitant effects on morphology, behaviour, and life-history. For example, prenatal exposure to high levels of glucocorticoids can reduce birth weight, cause hypertension, and increase the activity of the hypothalamus–pituitary–adrenal (HPA) axis, which is associated with changes in stress-related behaviours such as anxiety (reviewed in Meaney et al. 2007; Harris and Seckl 2011; Chapter 17). Experimental studies on fish, lizards, and birds suggest that many of the maternal effects observed in mammals are phylogenetically conserved and reflect shared developmental mechanisms among vertebrates (e.g. McCormick 1998; Uller and Olsson 2006; Love and Williams 2008).

Species-typical development also relies on maternal uptake and transmission of micronutrients directly from the environment. For example, development of the olfactory system in mammals requires olfactory stimuli obtained from the amniotic fluid (Michel 2011). The composition of amniotic fluid is derived from the maternal plasma and is therefore directly influenced by maternal physiology and diet (Robinson and Méndez-Gallardo 2011). Consequently, maternal diet influences offspring response to potential food items by influencing the development of sensory neurons associated with particular olfactory receptors (Robinson and Méndez-Gallardo 2011; Todrank et al. 2011). Finally, behavioural interactions between parents and offspring are also a source of phenotypic novelties; offspring growing up in unusual family structures or under novel patterns of parental care may show changes in, for example, sociality, mate preferences, and parental behaviour (e.g. Bradshaw and Schore 2007; Hansen et al. 2010).

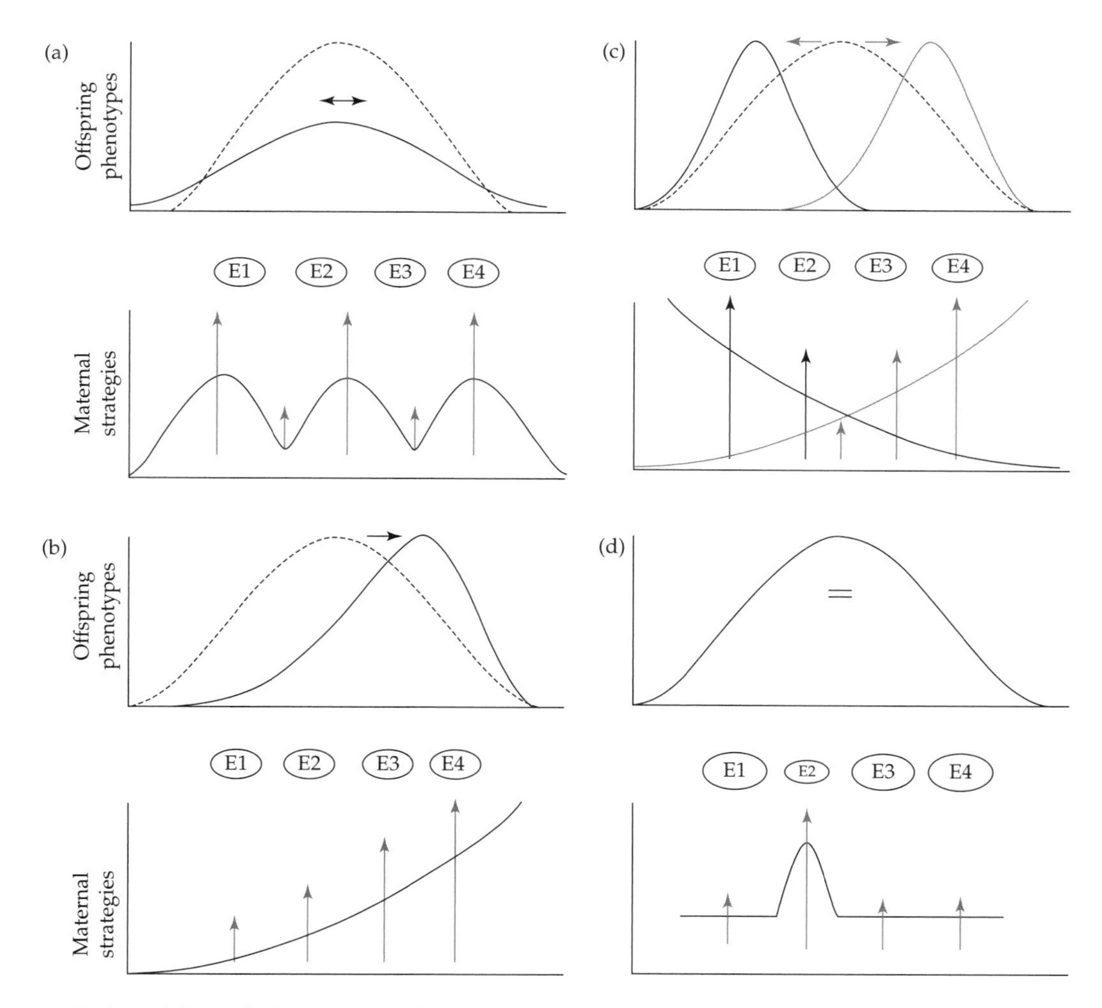

Figure 14.3 Parental effects in development. Each panel shows the distribution of offspring phenotypes (upper graphs) with (solid line) and without (dashed line) the distribution of parental effects (vertical arrows) on neonates (ovals labelled E1–E4). Horizontal grey arrows show the direction of change in the distribution of offspring phenotypes. (a) Variable parental transmission of developmental factors can increase variation in offspring phenotype. (b) Spatial or temporal variation in transmitted factors can cause directional change in offspring phenotype. (c) Morph-specific transmission can contribute to polymorphisms. (d) Context- or offspring-specific transmission can compensate for differences in developmental trajectories. (Redrawn from Fig. 2 in Badyaev 2008; reprinted with permission from the author and John Wiley & Sons.)

14.2.1 Patterns of phenotypic variation

These examples demonstrate how developmentally entrenched parental effects can contribute to modifications in the regulation and form of phenotypes. Badyaev (2008) has suggested that such parental induction can be captured in four ways (Fig. 14.3). First, parental effects may increase the variance in offspring phenotype (Fig. 14.3a). For example, maternal stress can result in increased variation in transfer of developmental resources among offspring with concomitant variation in offspring development (e.g. Badyaev 2005a). Increased variation among offspring may also result from maternal effects on the integration of developmental modules. Disruption of phenotypic integration results in greater variance in the connection of modules and thus greater among-individual variance in developmental outcomes (reviews in Hallgrímsson and Hall 2005).

Second, temporal or spatial variation in maternal transmission of resources can produce directional variation in offspring phenotypes (Fig. 14.3b). Such directional variation is common in birds, where egg size and yolk hormones often vary with laying order (Christians 2002; Groothuis et al. 2005). In several passerines, for example, environmental effects on ovarian activity and circulating levels of maternal hormones influence the accumulation of hormones by developing oocytes and results in within- and between-clutch variation in offspring phenotypes (e.g. Schwabl 1993; Badyaev et al. 2003). Highly divergent allocation of resources may also contribute to discontinuous phenotypic variation (Fig. 14.3c). A large number of polymorphisms depend on resource availability (e.g. Smith and Skúlason 1996), which provides substantial scope for maternal effects on morph expression in such systems. In spade-foot toads (*Spea multiplicata*), changes in maternal body condition under interspecific competition affect maternal egg investment and result in strong maternal effects on the development of resource-use polymorphism in the offspring (Pfennig and Martin 2009; Martin and Pfennig 2010). Furthermore, studies of mammals have shown that spatial or temporal variation in maternal resource allocation between sons and daughters contributes to morphological and behavioural divergence between the sexes (Moore 1995; Chapter 10). Thus, differential maternal allocation to offspring with different genotypes or developmental histories can exaggerate such differences and contribute to the expression of alternative phenotypes (Fig. 14.3c).

Finally, parental effects can constrain the production of novel phenotypes by limiting the effect of novel genetic and environmental input on offspring development (Fig. 14.3d). Female lizards are able to compensate for poor thermal conditions by adjusting their thermoregulatory behaviour or nest site choice. This can reduce the effect of ambient temperature on offspring development (Uller et al. 2011), maintain stasis in offspring reaction norms to temperature across climatic conditions (Doody et al. 2006), and may contribute to the build-up of cryptic genetic variation (which may prove important for a future response to selection; Schlichting 2008). Similarly, parental effects may compensate for genetic variation by genotype-specific allocation to offspring, which can limit the phenotypic effects of genetic variation and contribute to the resolution of constraints imposed by, for example, sexually antagonistic variation.

In summary, because development always relies on a parental phenotype, environmental or genetic changes to parent–offspring relationships can contribute—via phenotypic accommodation in both generations—to evolutionarily relevant variation. Parental effects can both facilitate expression of novel phenotypes and retard it, depending on to what extent parental accommodation of novel input results in differential transmission of resources in ways that influence pre-existing developmental mechanisms (Badyaev 2008).

14.3 Parental effects and adaptive evolution

The spread of a novel variant in a population partly depends on its fitness effects. Processes that positively affect the likelihood that novel genetic or environmental input will produce a functional phenotype should therefore promote adaptive evolution. However, even beneficial variants are likely to be lost by stochastic processes if they are rare. Thus, processes that increase recurrence of heritable variation will also facilitate evolutionary change (West-Eberhard 2003). Parental effects can have a positive impact on both functionality and recurrence of novel phenotypes.

14.3.1 Parental effects can increase functionality of novel variation

Genetic or environmental induction of phenotypic variation via parental effects may initially be accommodated by a functional parental phenotype. Parental phenotypic accommodation of novel input should therefore reduce the risk of severe disruption of offspring development, and may even facilitate expression of functional phenotypes by capitalizing on pre-existing developmental mechanisms (Badyaev 2008). Although it may be tempting to interpret context-dependent parental effects as reflecting adaptations to past fluctuations in the environment (see Section 14.4), incorporation of recurrent environmental and parental components into development is a fundamental feature of evolution and is more likely when environments show low, rather than high, variation (Gottlieb 1992; West-Eberhard 2003; Lickliter and Harshaw 2011). This is because evolution of development capitalizes on resources that are reliably available in the environment. However, this reliance of development on maternally transmitted gene products and environmental components may promote biased responses to conditions not previously encountered and enable phenotypic responses to a broader range of conditions. For example, maternally transmitted immunoglobulins and other immune factors activate and regulate development of the offspring immune system with long-lasting consequences on, for example, B and T cell repertoires (Lemke et al. 2004). Maternal exposure to novel pathogens may therefore contribute to directional, and functional, change in offspring phenotype (e.g. resistance) in environments not previously encountered. Similarly, the integration of maternally derived substances in the neuro-anatomical development of olfactory organ in mammals (see Section 14.2 above) suggests that a change in diet can generate directional food preferences in the offspring, even if the mechanisms that are involved have not been selected as a system for transmission of information about food regimes between generations per se. Thus, despite the fact that directional, apparently adaptive, phenotypic change in novel environments is facilitated by a pre-existing developmental reliance on the maternal phenotype, this reliance need not have been selected for because it increases offspring fitness in *fluctuating* environments (see also Section 14.5 below). It could simply form a part of a developmentally entrenched mechanism that under most conditions contributes to development of a species-typical, invariant, phenotype.

14.3.2 Parental effects can increase recurrence of novel variation

Selection on rare variants is very inefficient as they are likely to be lost due to stochastic processes. Factors that promote recurrence of a novel phenotype should therefore increase the likelihood that the rare phenotype is being selected and, if it is heritable, facilitate adaptive evolution. This argument led West-Eberhard (2003) to conclude that environmentally induced phenotypes have greater evolutionary potential than those induced via mutation. Parental effects contribute to the recurrence of novel phenotypes in at least two ways. First, a rare genetic or environmental input can influence more than one individual if it is accommodated via a parent. Genetic variation, for example due to multiple paternity, allows recurrence in a diversity of genetic backgrounds and thus enhances the opportunity for genetic accommodation. Similarly, within-brood variation in a developmental context, for example due to laying order effects, may increase the probability of a favourable match between phenotype and selection (e.g. Badyaev 2005a), and thereby the likelihood that the rare variants persist and can spread.

Second, parental effects can contribute to persistence of induced phenotypes across several generations (Jablonka and Lamb 1995, 2005; West-Eberhard 2007; Fig. 14.4). For example, persistence of a novel food preference is facilitated by mechanisms that enable offspring to copy their parents' diet, which results in the incorporation of novel food types into development in each generation (Fig. 14.4a). Parental effects may also allow environmentally induced phenotypic variation to be transmitted more or less stably down lineages via behavioural and epigenetic mechanisms even after the initial stimulus has ceased to exist (Fig. 14.4b). This is exemplified by research on the mechanisms of developmental plasticity and maternal care in rats (*Rattus norvegicus*) (Chapter 17). Cross-fostering of pups between mothers that direct high versus low levels of parental care towards their offspring (in the form of licking and grooming) showed that female offspring inherit the parental care behaviour of their foster mother, suggesting that maternal effects contribute to stability of between-lineage differences in maternal care.

The research on maternal care in rats emphasizes that the crucial element of trans-generational persistence of an environmentally induced phenotype is that the phenotype contributes to the reconstruction of the developmental niche, thus favouring its own expression (Fig. 14.4b). In the absence of germ-line transmission of induced variants (which does occur; see reviews by Youngson and Whitelaw 2008; Jablonka and Raz 2009; Fig. 14.4c), a limited period of parent–offspring interactions implies that parental effects must persist into adulthood to be maintained. For example, within- and trans-generational persistence of the effects of maternal care on pups have been linked to effects of maternal licking and grooming on the methylation status of the promoter regions of the estrogen receptor alpha and the glucocorticoid receptor genes, which remain stable throughout ontogeny and influence parental care in adulthood (Chapter 17). Stable inheritance of environmentally induced variants may often involve parental transmission of substances or behavioural interactions that affect epigenetic marks, which enables early environments to have long-lasting consequences via cellular epigenetic inheritance (Weaver et al. 2004; Gluckman et al. 2009; Chapter 17).

14.3.3 Parental effects and genetic accommodation

Environmental induction followed by reconstruction of the developmental niche via parental effects can initiate and maintain population differences without genetic divergence, as exemplified by dietary preferences in mammals (e.g. Avital and Jablonka 2000). However, if there is genetic variation in, for example, uptake, digestion, or circulation of novel odorants, or in offspring sensitivity to compounds circulating in the amniotic fluid (or in the milk), natural selection can fine-tune responses to local conditions and cause genetic divergence between populations. Mutation accumulation due to weak selection on rarely expressed alleles and costs associated with plastic responses can also contribute to further population divergence initiated by developmental plasticity (reviewed in Pfennig et al. 2010; Snell-Rood et al. 2010). This process in which changes in gene frequencies within populations is secondary to the origin of novel variation is referred to as genetic accommodation (West-Eberhard 2003). With respect to parental effects, genetic accommodation may occur in response to selection on the parental phenotype, offspring phenotype, or both, which should result in co-adaptation of parental and offspring phenotypes (Chapters 15 and 16; see also Section 14.4). Furthermore, initially deleterious effects can be eliminated via genetic accommodation (West-Eberhard 2003; Grether 2005). For example, the negative effects of maternal stress on offspring phenotype and fitness in mammals may gradually be reduced via selection on genetic variation in maternal stress response, transfer of corticosterone across the placenta, or offspring sensitivity to prenatal hormone exposure.

The evolution of integration of environmental input and offspring response capitalizes on pre-existing sensory systems in females (e.g. detection of photoperiod, or diet composition in the environment), physiological responses associated with reproduction (e.g. shared hormonal regulation of responses to environmental variation and breeding), and offspring sensitivity to maternal physiology (e.g. maternal hormones triggering expression of hormone receptors in the embryo) (Nijhout 2003; Badyaev 2009; Uller and Badyaev 2009). Changes in the environmental context of breeding, such as photoperiod, temperature, or food availability, will often be associated both with expression of novel variation in the offspring and selection on this variation, which enhances the scope for selection to effectively sort between phenotypes and hence

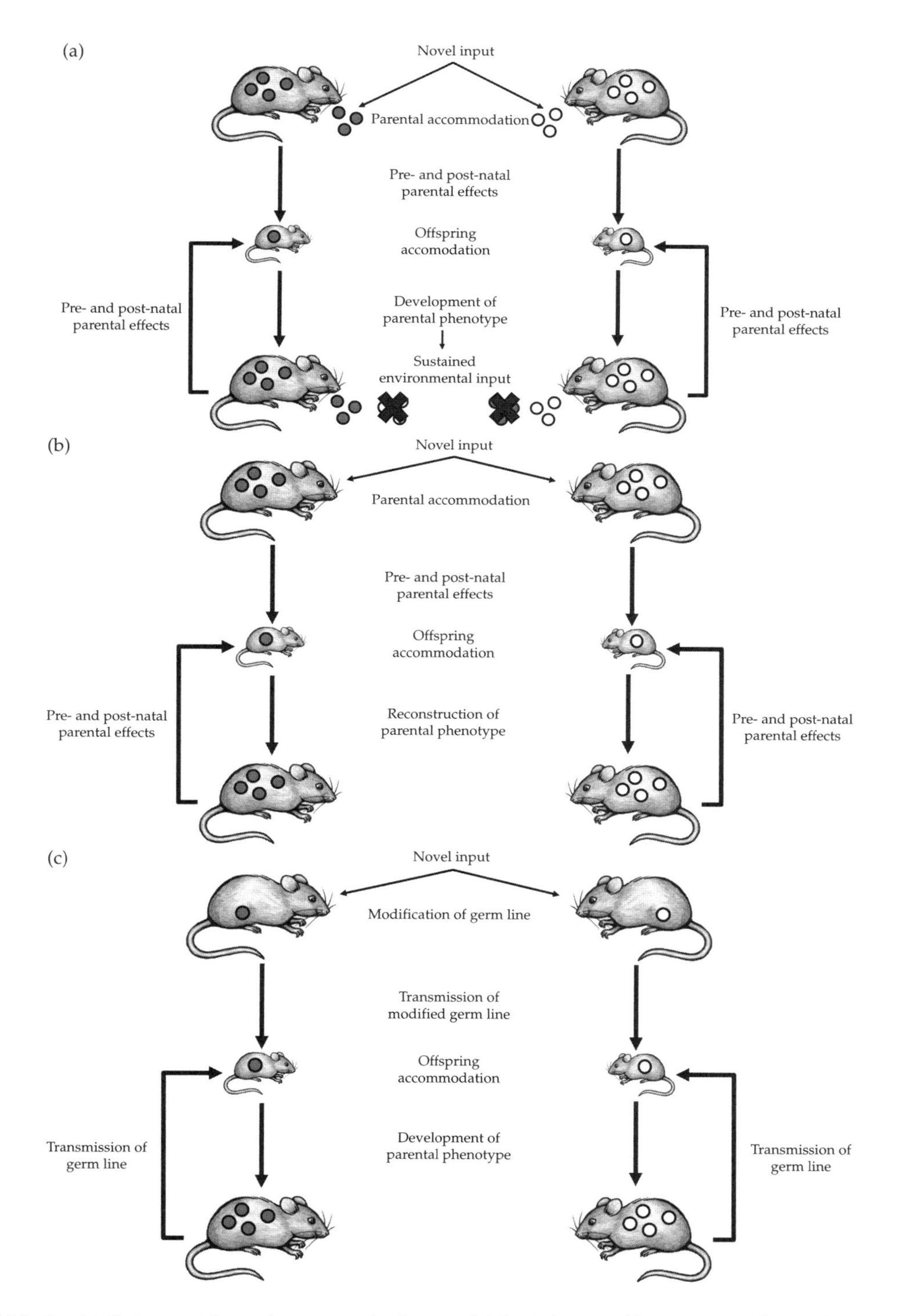

Figure 14.4 Parental effects can contribute to the recurrence of environmentally induced phenotypes. (a) An environmentally induced phenotypic change (a grey or white phenotype) can persist in a population because offspring from 'grey mothers' prefer 'grey environments' as adults and offspring from 'white mothers' prefer 'white environments' as adults. The induced phenotype only persists as long as all inducing environment persists. (b) An environmentally induced phenotypic change (a grey or white phenotype) can persist in a population even if the inducing factor is no longer is present in the population if the offspring develop a phenotype that reconstructs the parental effect that favoured its expression. (c) An environmentally induced phenotypic change can persist if it is transmitted via the germ line, either as a genetic variant or as an epigenetic variant that is not being erased during epigenetic reprogramming in the germ cells or during development.

increases the rate of evolution (West-Eberhard 2003; Badyaev 2005b; Badyaev 2009). The role of parental effects in evolution of local adaptation may be particularly important when the most reliable cue, like photoperiod, does not have a causal effect on offspring development unless it is mediated via parental responses.

In summary, parental effects facilitate development of functional phenotypes in response to novel input and increase the recurrence of those phenotypes by enabling cross-generational stability of environmentally induced variation. Both processes increase the potential for adaptive evolution and suggest substantial scope for environmentally induced variation to have evolutionary consequences via genetic accommodation. Furthermore, parental effects can enable persistence of induced phenotypes even if they are initially deleterious, which may result in genetic accommodation of the regulation of developmental processes to restore fitness.

14.4 Evolution of trans-generational plasticity

The evolution of developmentally entrenched parental effects can be seen as a process in which organisms accommodate and accumulate environmental input to pass on the most recurrent organism–environment configurations (Badyaev 2008, 2009; Badyaev and Uller 2009). Passive context-dependent parental effects may thus represent a transient period of phenotypic accommodation of environmental input that exposes phenotypic (and genetic) variation to selection. However, research in a wide range of disciplines—including behavioural and evolutionary ecology, evolutionary anthropology, and microbiology—emphasize that parental effects may also enable adaptive transfer of information about coming selective regimes across one or several generations (e.g. Feldman and Laland 1996; Mousseau and Fox 1998; Falkner and Falkner 2003; Gluckman et al. 2005; Jablonka and Lamb 2005; Uller 2008). Behavioural transmission of functional solutions to local environments from parents to offspring in humans is a familiar example, but similar processes also occur in organisms without sophisticated cognitive abilities. For example, the timing of germination in the herb *Campanulastrum americanum* depends on maternal light regime during seed production (Galloway 2005). Seeds from plants grown in light gaps tend to germinate in autumn and develop as annuals, whereas seeds from plants in shady conditions germinate in spring and develop as biennials. Because seeds tend to fall close to the maternal plant and light regimes are relatively stable across generations, but variable at the population level, adjustment of the timing of germination in relation to the maternal light environment is favoured over fixed or bet-hedging strategies (Galloway and Etterson 2007).

Maternal effects on seed germination can be seen as adaptive, trans-generational, plasticity (Mousseau and Fox 1998; Marshall and Uller 2007; Uller 2008). Furthermore, its similarity to behavioural transmission of adaptive strategies in animals shows that trans-generational plasticity can involve very different mechanisms and occur at different life-history stages (Fig. 14.1). This raises at least three questions. First, under what conditions can parental effects form an adaptive channel of transmission of information between generations? Second, are the mechanisms involved in adaptive trans-generational plasticity different from those of developmentally entrenched parental effects, and, finally, should they be seen as alternative systems of inheritance?

14.4.1 Adaptive evolution of trans-generational plasticity

In keeping with the standard framework of phenotypic plasticity (e.g. Schlichting and Pigliucci 1998), the evolution of trans-generational plasticity can be visualized as a set of two evolving reaction norms (Fig. 14.5). The parental phenotype may be responsive to some aspect of its environment by changing its morphology, physiology, or behaviour. Variation in the parental phenotype constitutes a fluctuating environment for the offspring, which is associated with a corresponding norm of reaction. Evolution of trans-generational plasticity is captured by the co-evolution of those two reaction norms so that a particular environmental context experienced by parents induces a particular phenotypic response in the offspring (Fig. 14.5).

Whether evolution of parental effects will be primarily driven by changes in the parental or offspring reaction norms is a question that has apparently received scant interest. However, the response is likely to depend on the relative amount of heritable variation, the genetic architecture of the phenotypes involved, and the relative strength of selection. If offspring plasticity is constrained, the offspring phenotype evolves only via evolution of the parental norm of reaction (i.e. via indirect genetic effects; Smiseth et al. 2008). Conversely, if parental plasticity is prevented from evolving, the evolution of parental effects reduces to evolution of within-generation phenotypic plasticity; the offspring reaction norm is expected to evolve to maximize fitness within the set of costs and constraints specified (reviewed in Berrigan and Scheiner 2004). Evolved parental strategies or 'passive' parental effects may impose selection on offspring development, thereby contributing to evolutionary diversification in developmental trajectories (Badyaev 2005a). For example, variation in egg size simultaneously affects offspring phenotype (e.g. size) and exercises selection on this phenotype, which provides substantial scope for evolution of alternative developmental strategies mediated via maternal effects (Badyaev 2005a; Pfennig and Martin 2009; Lancaster et al. 2010).

Making predictions regarding the conditions that promote evolution of trans-generational plasticity is currently hampered by the limited number of theoretical models, but should be subject to similar considerations as within-generation plasticity (Berrigan and Scheiner 2004) but framed within the context of the evolution of parental care (Smiseth et al. 2008). In particular the nature of fluctuating selection, the availability of cues, and benefits and costs of the transmission of information across generations for offspring and parents are expected to affect the evolution of trans-generational plasticity (Uller 2008). It can be useful to conceptualize the evolution of parental effects as evolution of developmental responses to different sources of input that carry information about coming selective regimes (Leimar et al. 2006; Fig. 14.5b, Box 14.1). Under this perspective, the parental phenotype is a source of information for the offspring when the parental phenotype correlates with current or future conditions experienced by the offspring (Shea et al. 2011; Box 14.1). Such correlations can arise in heterogeneous environments (e.g. Galloway 2005) or because of genetic or environmental variation in parental traits (e.g. Love & Williams 2008; Lancaster et al. 2010), both of which should favour evolution of trans-generational plasticity (Revardel et al. 2010; Shea et al. 2011). Assuming no parent–offspring conflict, we would expect selection to maximize information transfer between generations. Thus, not only should offspring evolve to adjust their phenotype according to the parental phenotype, but selection on parents may also favour transfer of developmental factors that are informative about the conditions that the offspring experience, or will experience. This should result in a tight integration between parental environment, parental phenotype, and offspring phenotype (Fig. 14.5c).

Empirical examples of potentially adaptive trans-generational plasticity mirror how parental phenotypes can modify offspring phenotypes more generally (see Fig. 14.3). Facultative diversifying maternal effects can be favoured in heterogeneous environments with different degrees of predictability (Crean and Marshall 2009). Directional changes in maternal allocation of androgens across the laying sequence in altricial birds might mitigate detrimental effects on hatching asynchrony (which is determined by parental onset of incubation; Groothuis et al. 2005). Discrete polyphenisms in insects also involve maternal effects. For example, the phase shift from solitary to gregarious morphs in locusts is initiated within generations but becomes progressively stronger across generations as a result of maternal transfer to the egg froth of a compound produced by the accessory glands (reviewed in Simpson and Sword 2009). Although the adaptive significance of trans-generational effect in locusts remains to be verified, it is possible that it might maximize the efficiency of phase shift in a gradually changing environment (Simpson and Sword 2009), conditions that potentially could favour parental effects that accumulate or persist across more than two generations (Jablonka et al. 1995).

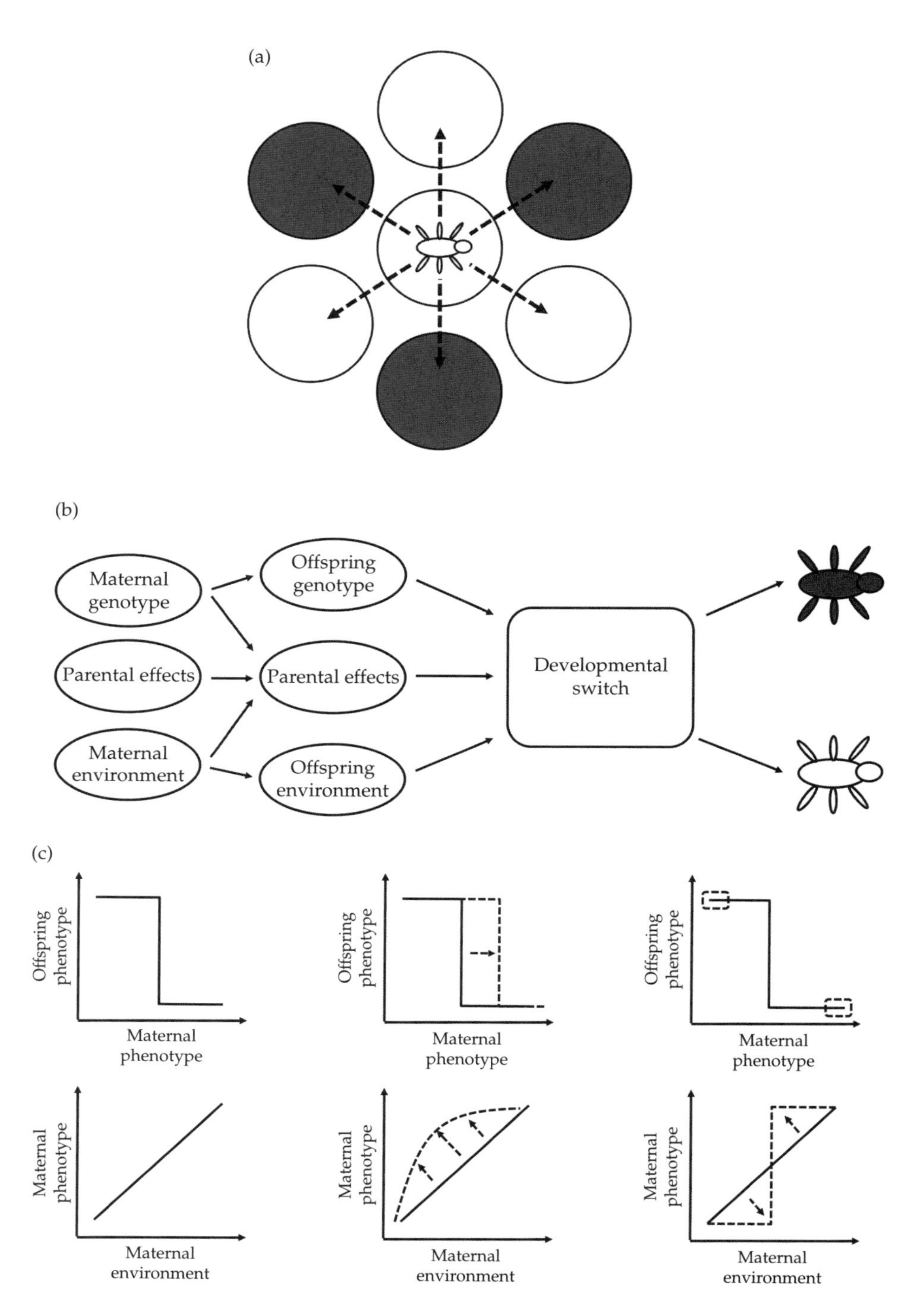

Figure 14.5 Evolution of trans-generational plasticity. (a) Evolution of adaptive plasticity requires temporal and/or spatial heterogeneity, for example, a meta-population structure with dispersal between patches with different environmental conditions. (b) Evolution of development can be conceptualized as evolution of sensitivity of a developmental switch to different sorts of input—genetic, parental, and environmental (modified from Leimar et al. 2006). Here the offspring phenotype is a polymorphism with only two possible states. Environmental conditions can vary at temporal or spatial scales so that the parental environment/phenotype carries information about the environment likely to be experienced by offspring. (c) If the maternal phenotype correlates with selection on the population of offspring phenotypes, this developmental switch can evolve to be responsive to variation in the parental phenotype (left panel). A shift in the parental phenotype (dashed lines) can change selection on the offspring reaction norm, which may evolve to maintain the fit between offspring phenotype and offspring environment (dashed lines, middle panel). Concordant selection across generations may allow evolution of parental reaction norms that maximize the information transfer between parents and offspring and thus minimize the risk for mismatch between offspring phenotype and their environment (right panel).

Box 14.1 Parental effects as systems of inheritance

From a perspective of modern biology, it may come as a surprise that the separation of development and heredity is a fairly recent innovation. Into the twentieth century, biologists saw inheritance (i.e. 'like begets like') as taking place throughout epigenesis via between-generation recurrence of the factors that build bodies (Amundson 2005). Only with the new field of genetics did heredity become the passing of traits between generations and, more specifically, transmission genetics (Amundson 2005). The discovery of a materialistic basis of the 'gene' thus turned DNA into the primary 'system of inheritance' upon which evolution relies. The unique position of DNA is challenged, however; parental effects show that phenotypic stability within lineages and differences between lineages can be maintained via parental transmission of non-genetic factors as well as DNA (Jablonka and Lamb 2005; Jablonka and Raz 2009). But are parental effects alternative systems of inheritance and, if so, how do those inheritance systems compare to genetic inheritance?

Biologists and philosophers alike often think about heredity as the passing of information between generations (e.g. Williams 1992; Maynard-Smith 2000; Jablonka 2002; Shea 2007; Bergstrom and Rosvall 2010). The concept of information in this context is not unproblematic (e.g. Oyama 2000), but one interpretation is that 'a source becomes an informational input when an interpreting receiver can react to the form of the source (and variations in this form) in a functional manner' (Jablonka and Lamb 2006; see also Jablonka 2002). This puts the study of inheritance systems into a broader context of signals and communication (Skyrms 2010; Shea in press), and focuses on the evolution of developmental responses to different sources of input, regardless of their origin (e.g. genetic versus environmental; Jablonka 2002; Shea et al. 2011). Using this approach, Jablonka and Lamb (2005) separated inheritance systems—ways to transmit information—as genetic, epigenetic, behavioural, and symbolic. By definition, parental effects do not fall under genetic inheritance (Chapter 1). However, they may fall under any of the other inheritance systems—epigenetic, behavioural, and symbolic (Helanterä and Uller 2010).

What makes the genetic inheritance system special and different from the mechanisms that contribute to parental effects (Shea 2007; Helanterä and Uller 2010)? One distinction between inheritance systems can be made based on the processes that generate correlational information between what is being transmitted and an adaptively relevant feature of the environment (Shea et al. 2011). The variants transmitted through inheritance systems involved in adaptive trans-generational plasticity carry information because the parental phenotype responds to some aspect of its environment that correlates with a feature that is of adaptive relevance to the offspring. This correlational information can be exploited by developmental processes because of the continuity between parental and offspring phenotypes (Fig. 14.5). In the genetic inheritance system, on the other hand, correlational information requires a process of selection that builds up gene frequency differences between environments (Leimar et al. 2006; Shea et al. 2011). Build-up of information through selection requires both stable transmission of developmental resources (such as genes) and sufficiently long time scales. Also, the DNA-based inheritance system seems to have the adaptive function to enable the transmission of heritable phenotypes down generations (Maynard Smith 2000; Shea 2007; Bergstrom and Rosvall 2010). These features may not be unique to DNA, however, and it is possible that epigenetic or behavioural mechanisms that initially contribute to parental effects (e.g. DNA methylation) can evolve to take on a similar role in heredity as the DNA-based system.

14.4.2 Trans-generational plasticity under parent–offspring conflict

As demonstrated by examples in this book, selection is not always concordant across generations. The implications of such parent–offspring conflict for the evolution of parental care in the form of resource provisioning have been explored in detail elsewhere (Chapter 7). Here I focus on how parent–offspring conflict will affect the evolution of trans-generational plasticity, that is the joint evolution of parental and offspring reaction norms in a heterogeneous environment, without assuming that the investment is costly to parents or beneficial to offspring (i.e. not only parental care).

Parent–offspring conflict is relevant for both continuous and discontinuous phenotypes that are of interest to a wide range of biologists. For example, in both plants and animals, competition between kin can generate parent–offspring conflict over natal dispersal. The parental inclusive fitness is often maximized at a higher dispersal rate than the offspring inclusive fitness because offspring that do not disperse compete with their parents and siblings for access to limited resources, and the cost of dispersal in terms of survival or reproductive success is usually paid for solely by the offspring (e.g. Frank 1986). A similar scenario may apply to diapause, which can be seen as dispersal in time (Tauber et al. 1986). Interestingly, related species often differ in the extent to which dispersal and diapause are genetically, maternally, or environmentally influenced (Tauber et al. 1986; Braendle et al. 2006), which provides opportunities for comparative tests of the adaptive significance of trans-generational plasticity and the role of parent–offspring conflict for the evolution of parental effects. For example, poor maternal nutrition and crowding increase the proportion of winged offspring in some aphid species, whereas in others genetic or direct environmental effects predominate (Braendle et al. 2006). Among vertebrates, research on the common lizard, *Lacerta vivipara*, has emphasized the role of kin competition for dispersal and provided experimental evidence that this conflict is modulated by the maternal environment (e.g. density; Meylan et al. 2007) and the maternal phenotype (e.g. maternal age; Ronce et al. 1998). Similar parent–offspring conflicts may arise over morphological and behavioural phenotypes associated with maternal hormone transfer, including offspring size, begging behaviour, and growth (Groothuis et al. 2005; Müller et al. 2007; Chapter 7).

Although context-dependent maternal effects on offspring behaviour can be interpreted as maternal 'manipulation' of offspring phenotype (e.g. Schwabl et al. 1997; Love and Williams 2008), parental manipulation will often be evolutionarily unstable (Müller et al. 2007; Uller and Pen 2011). The reason for this instability is that offspring can evolve to respond to the maternal phenotype in ways that maximize their own inclusive fitness, not that of the mother. In the absence of constraints on evolutionary counter-responses by the offspring, evolved patterns of trans-generational plasticity therefore often mirror those under offspring 'control' of trait expression (Uller and Pen 2011). This may apply even when offspring are unable to assess their own environment since the parental phenotype provides an additional source of information about local conditions (Müller et al. 2007; Uller and Pen 2011). Thus, even when the parent is the only individual who can directly detect the environment, context-dependent parental effects on traits that do not involve variation in costly resource allocation might most commonly represent the offspring optima. However, the simplifying assumptions of theoretical treatments (Revardel et al. 2010; Uller and Pen 2011)—for example two discrete environments, binary offspring response—question the generality of the conclusions. Indeed, 'deception' can be evolutionarily stable in signalling systems under certain conditions (Searcy and Nowicki 2005; Skyrms 2010) and may play a role also in signalling between parents and offspring (e.g. maternal manipulation of offspring phenotype; Chapter 7). Furthermore, costs associated with expression of particular phenotypes, such as parental investment (Chapter 3), complicate the interpretation of patterns of trans-generational plasticity, as the effect of the (parental) environment on offspring phenotype can be intermediate to that under complete maternal or offspring control (Uller and Pen 2011). Models trying to predict the shape of parental and offspring reaction norms will therefore have to carefully specify the temporal and spatial variation in environmental heterogeneity, limits on detection of environmental cues, costs involved with expression of particular phenotypes, the underlying genetic architecture, and the extent to which selection is concordant between parents and offspring.

14.4.3 Mechanisms of trans-generational plasticity

The often discrete and precise induction of offspring phenotype by parental effects may give the impression that the mechanisms of adaptive trans-generational plasticity, such as maternal effects on seed germination (Galloway and Etterson 2007), are different from developmentally entrenched parental effects and those that promote diversifying or directional phenotypic variation in novel or stressful environments. However, adaptive trans-generational plasticity is more likely to represent a stable evolutionary state arising from emergent parental effects (e.g. initially induced by stress) that become stabilized by natural selection as maintenance of alternative phenotypes of both adults and developing offspring (Badyaev and Uller 2009; Fig. 14.2). For example, there is some evidence from birds and mammals that the same hormonal mechanisms that form an integral part of species-typical development also contribute to stress-induced, non-adaptive, variation and highly precise and adaptive environment-specific maternal effects on offspring phenotype (e.g. reviews by Fowden and Forhead 2009; Badyaev 2009; Uller and Badyaev 2009). Although mechanisms acting late in ontogeny may provide greater scope for information transfer between generations (Jablonka and Lamb 2005; Badyaev 2008), several authors emphasize the adaptive significance of environment-dependent reprogramming of epigenetic processes early in development (e.g. Mousseau and Fox 1998; Gluckman et al. 2005; Galloway and Etterson 2007). Thus, the mechanisms of adaptive trans-generational plasticity are not only similar to those involved in developmentally entrenched parental effects but may span the entire continuum from epigenetic modification of gene expression, maternal transfer of micro- and macro-nutrients to the egg yolk or the developing fetus, to post-natal behavioural interactions between parents and offspring.

14.5 Exploring the evolutionary dynamics of parental effects

A recent perspective (Badyaev 2009; Badyaev and Uller 2009) views parental effects as part of an evolutionary process in which the most recurrent parental resources are retained and eventually may become developmentally entrenched, visible only through genetic or environmental disruption of the species-typical developmental system. When such disruption is sufficiently recurrent for natural selection to stabilize the expression of induced alternative phenotypes, the result is adaptive trans-generational plasticity. Thus, developmentally entrenched and context-dependent parental effects are different outcomes of the same evolutionary process involving the same developmental mechanisms (Fig. 14.2). Badyaev (2009) has suggested that this process is an example of the Baldwin effect. The main tenet of the Baldwin effect is that phenotypic accommodation to environmental input can eventually become 'internalized' without any need for inheritance of acquired characters—all that is required is heritable variation in the initial response or that the initial response allows the population to persist until heritable variation accumulates (Baldwin 1902; see Weber and Depew 2002; West-Eberhard 2003 for discussion). This will result in a pattern of evolutionary diversification that reflects the ontogenetic flexibility of ancestral phenotypes. This chapter has emphasized three aspects of this process: 1) the role of parental effects for the origin of phenotypic variation via phenotypic accommodation of genetic or environmental input; 2) how particular aspects of parental effects (directionality and recurrence of novelties) can increase the likelihood that environmentally induced phenotypes can spread in the population, and 3) how selection on parents and offspring can sometimes maintain alternative phenotypes within a population in the form of adaptive trans-generational plasticity.

Parental effects in general, and parental care in particular, may thus contribute to evolutionary diversification or adaptation to novel environments in several ways. Evidence that maternal effects facilitate persistence in novel environments comes from studies of the seed beetle *Stator limbatus*, where offspring survival on a novel host species is facilitated by maternal plasticity in egg size (Fox and Savalli 2000). Diversifying maternal effects resulting from stress-induced changes in female reproductive physiology have contributed to the rapid and successful colonization of challenging climatic regions by house finches (Badyaev et al. 2003, 2008). Nevertheless, the available evidence that phenotypic accommodation via parental effects allows directional changes in response to novel environments that increase fitness and form the basis for local adaptation is often circumstantial, inferential, or based on laboratory conditions only.

Furthermore, the extent to which parental effects contribute to release of cryptic genetic variation that can enable evolution via genetic accommodation has not yet received much attention. Finally, although evidence for trans-generational persistence of novel phenotypic variation is rapidly accumulating (Jablonka and Raz 2009), whether or not this has played an important role in evolution by facilitating genetic accommodation (including the Baldwin effect), by enabling adaptive transfer of information across generations or as an inheritance system remains poorly understood.

Specific tests of whether parental effects have contributed to the origin and evolution of adaptation require investigation of the relationship between mechanism and regulation of developmentally entrenched and context-dependent parental effects in a historical context of environmental recurrence (Badyaev and Uller 2009). Support may be sought by assessing whether plastic responses in the ancestral state correspond to the developmental regulation of adaptive strategies in populations living under derived conditions. For example, evolution of relatively high developmental rate in colonial birds may have involved direct effects of high density and aggression on circulating levels of androgens in breeding females, which results in high androgen levels in eggs and an associated faster developmental rate (Gil et al. 2007). A similar scenario has been proposed for the evolution of alternative morphs in spadefoot toads (Pfennig and Martin 2009; Martin and Pfennig 2010). However, to show that developmental plasticity and parental effects played a role in the evolution of a particular phenotype, it is necessary to capture the ongoing process to document the transition from stress-induced variation to the evolution of local adaptation via phenotypic and genetic accommodation (Moczek 2007; Uller and Helanterä 2011). This requires an explicit focus on the developmental basis for evolutionary change on short time scales, which calls for innovative research programmes at the interface of developmental biology and evolutionary ecology. One such example is a long-term study of the house finch colonization of North America. Close integration of endocrinological regulation of female reproduction, oogenesis, and offspring growth facilitated evolution of local adaptation in sexual size dimorphism under novel climatic conditions via phenotypic accommodation of stress-induced variation, followed by cross-generational transfer of a subset of locally favoured phenotypes (summarized in Badyaev 2009).

As the house finch example demonstrates, human activities, such as species introductions, pollution, and habitat change can sometimes provide ideal settings for following populations as they encounter novel environments and gradually adapt (or go extinct). Such systems provide opportunities to link environmental induction of context-dependent parental effects and evolutionary change, and enables assessment of how important and general this process may be. For example, Marshall (2008) showed that in the bryozoan, *Bugula nerita*, a brief exposure to high levels of copper (a pollutant) resulted in offspring with reduced survival. This is expected since copper is toxic at high doses. However, offspring from copper-exposed mothers were relatively more tolerant to copper stress per se than offspring from non-exposed mothers. Although this may reflect an evolved adaptive trans-generational plastic response and a form of parental care (the history and heterogeneity of copper exposure in this species is unknown; Marshall 2008), it may also be a passive outcome of phenotypic accommodation and developmentally entrenched maternal effects. For example, an increase in circulation of metallothionein mRNA in response to copper exposure during oogenesis could have concomitant effects on egg composition, with carry-over effects on the development of heavy metal resistance in the offspring (e.g. Lin et al. 2000). Thus, the context-dependent parental effects observed in studies like this may be passive, capitalizing on pre-existing entrenched parental effects and representing a transient stage in the environmental induction of novel phenotypes, which may be followed by genetic accommodation in populations where the novel environmental factor is a recurrent feature (Fig. 14.2). A shift in focus from treating parental effects only as patterns of phenotypic variation, or as adaptive transfer of information across generations, towards viewing them as part of a process that connects environmental induction and adaptation will allow us to gain novel insights into the mechanisms of evolutionary change.

Acknowledgements

My views on the role of parental effects in development and evolution have been strongly influenced by discussions with Alex Badyaev, Heikki Helanterä, Ido Pen, and Nick Shea. I thank them for their time. I am also grateful to Mathias Kölliker, Camilla Hinde, Heikki Helanterä, Tom Ezard, and two anonymous reviewers for comments on this chapter.

References

Amundson, R. (2005). *The Changing Role of the Embryo in Evolutionary Thought*. Cambridge University Press, New York, NY.

Avital, E. and Jablonka, E. (2000). *Animal Traditions. Behavioural Inheritance in Evolution*. Cambridge University Press, New York, NY.

Badyaev, A. V. (2005a). Maternal inheritance and rapid evolution of sexual size dimorphism: Passive effects or active strategies? *American Naturalist* 166, S17–30.

Badyaev, A. V. (2005b). Stress-induced variation in evolution: from behavioural plasticity to genetic assimilation. *Proceedings of the Royal Society of London, Series B* 272, 877–86.

Badyaev, A. V. (2008). Maternal effects as generators of evolutionary change a reassessment. *Year in Evolutionary Biology 2008* 1133, 151–61.

Badyaev, A. V. (2009). Evolutionary significance of phenotypic accommodation in novel environments: an empirical test of the Baldwin effect. *Philosophical Transactions of the Royal Society of London, Series B* 364, 1125–41.

Badyaev, A. V., Beck, M. L., Hill, G. E., and Whittingham, L. A. (2003). The evolution of sexual size dimorphism in the house finch. V. Maternal effects. *Evolution* 57, 384–96.

Badyaev, A. V. and Uller, T. (2009). Parental effects in ecology and evolution: mechanisms, processes and implications. *Philosophical Transactions of the Royal Society of London, Series B* 364, 1169–77.

Badyaev, A. V., Young, R. L., Hill, G. E., and Duckworth, R. A. (2008). Evolution of sex-biased maternal effects in birds. IV. Intra-ovarian growth dynamics can link sex determination and sex-specific acquisition of resources. *Journal of Evolutionary Biology* 21, 449–60.

Baldwin, J. M. (1902). *Development and Evolution*. Macmillan & co. Ltd, New York.

Bergstrom, C. T. and Rosvall, M. (2010). The transmission sense of information. *Biology and Philosophy* **26**, 159–176.

Berrigan, D. and Scheiner, S. M. (2004). Modeling the evolution of phenotypic plasticity. In T. J. DeWitt and S. M. Scheiner, eds. *Phenotypic Plasticity. Functional and Conceptual Approaches*. Oxford University Press, New York, NY.

Bonduriansky, R. and Day, T. (2009). Nongenetic inheritance and its evolutionary implications. *Annual Review of Ecology, Evolution, and Systematics* 40, 103–25.

Boyd, R. and Richerson, P. (1985). *Culture and the Evolutionary Process*. University of Chicago Press, Chicago.

Bradshaw, G. A. and Schore, A. N. (2007). How elephants are opening doors: Developmental neuroethology, attachment and social context. *Ethology* 113, 426–36.

Braendle, C., Davis, G. K., Brisson, J. A., and Stern, D. L. (2006). Wing dimorphism in aphids. *Heredity* 97, 192–9.

Christians, J. K. (2002). Avian egg size: variation within species and inflexibility within individuals. *Biological Reviews* 77, 1–26.

Crean, A. J. and Marshall, D. J. (2009). Coping with environmental uncertainty: dynamic bet hedging as a maternal effect. *Philosophical Transactions of the Royal Society of London, Series B* 364, 1087–96.

Donohue, K. (2009). Completing the cycle: maternal effects as the missing link in plant life histories. *Philosophical Transactions of the Royal Society of London, Series B* 364, 1059–74.

Doody, J. S., Guarino, E., Georges, A., Corey, B., Murray, G., and Ewert, M. (2006). Nest site choice compensates for climate effects on sex ratios in a lizard with environmental sex determination. *Evolutionary Ecology* 20, 307–30.

Falkner, R. and Falkner, G. 2003. Distinct adaptivity during phosphate uptake by the cyanobacterium *Anabaena variabilis* reflects information processing about preceding phosphate supply. *Journal of Trace Microprobe Techniques* 21, 363–75.

Feldman, M. W. and Laland, K. N. (1996). Gene-culture coevolutionary theory. *Trends in Ecology and Evolution* 11, 453–7.

Fox, C. W. and Savalli, U. M. (2000). Maternal effects mediate host expansion in a seed-feeding beetle. *Ecology* 81, 3–7.

Fowden, A. L. and Forhead, A. J. (2009). Hormones as epigenetic signals in developmental programming. *Experimental Physiology* 94, 607–25.

Frank, S. A. (1986). Dispersal polymorphisms in subdivided populations. *Journal of Theoretical Biology* 122, 303–9.

Galloway, L. (2005). Maternal effects provide phenotypic adaptation to local environmental conditions. *New Phytologist* 166, 93–100.

Galloway, L. and Etterson, J. (2007). Transgenerational plasticity is adaptive in the wild. *Science* 318, 1134–6.

Gil, D., Biard, C., Lacroix, A., Spottiswoode, C. N., Saino, N., Puerta, M., and Moller, A. P. (2007). Evolution of yolk androgens in birds: Development, coloniality, and sexual dichromatism. *American Naturalist* 169, 802–19.

Gilbert, S. F. (2010). *Developmental Biology*. 9th ed. Sinauer Associates Inc, Sunderland, MA.

Gluckman, P. D., Hanson, M. A., Buklijas, T., Low, F. M., and Beedle, A. S. (2009). Epigenetic mechanisms that underpin metabolic and cardiovascular diseases. *Nature Reviews Endocrinology* 5, 401–8.

Gluckman, P. D., Hanson, M. A., Spencer, H. G., and Bateson, P. (2005). Environmental influences during development and their later consequences for health and disease: implications for the interpretation of empirical studies. *Proceedings of the Royal Society of London, Series B* 272, 671–7.

Gottlieb, G. (1992). *Individual Development and Evolution. The Genesis of Novel Behavior*. Oxford University Press, NY.

Gottlieb, G. (1997). *Synthesizing Nature-Nurture. Prenatal Roots of Instinctive Behaviour*. Lawrence Erlbaum Associates, Mahwah, NJ.

Grether, G. F. (2005). Environmental change, phenotypic plasticity, and genetic compensation. *American Naturalist*166, E115–23.

Groothuis, T. G. G., Muller, W., Von Engelhardt, N., Carere, C., and Eising, C. (2005). Maternal hormones as a tool to adjust offspring phenotype in avian species. *Neuroscience and Biobehavioral Reviews* 29, 329–52.

Hallgrímsson, B. and Hall, B. K. (2005). *Variation. A Central Concept in Biology*. Elsevier Academic Press, Burlington, MA.

Hansen, B. T., Johannessen, L. E., and Slagsvold, T. (2010). Interspecific cross-fostering of great tits (*Parus major*) by blue tits (*Cyanistes caeruleus*) affects inter- and intraspecific communication. *Behaviour* 147, 413–24.

Harris, A. and Seckl, J. (2011). Glucocorticoids, prenatal stress and the programming of disease. *Hormones and Behavior* 59, 279–89.

Helanterä, H. and Uller, T. 2010. The Price equation and extended inheritance. *Philosophy & Theory in Biology 2, e101.*

Hogarth, P. J. (1976). *Viviparity*. Edward Arnold (Publishers) Ltd, London, UK.

Inchausti, P. and Ginzburg, L. R. (2009). Maternal effects mechanism of population cycling: a formidable competitor to the traditional predator–prey view. *Philosophical Transactions of the Royal Society of London, Series B* 364, 1117–24.

Jablonka, E. (2002). Information: Its interpretation, its inheritance, and its sharing. *Philosophy of Science* 69, 578–605.

Jablonka, E. and Lamb, M. J. (1995). *Epigenetic Inheritance and Evolution. The Lamarckian Dimension*. Oxford University Press, New York, NY.

Jablonka, E. and Lamb, M. J. (2005). *Evolution in Four Dimensions. Genetic, Epigenetic, Behavioral, and Symbolic Variation in the History of Life*. MIT, Cambridge, MA.

Jablonka, E. and Lamb, M. J. (2006). The evolution of information in the major transitions. *Journal of Theoretical Biology* 239, 236–46.

Jablonka, E., Oborny, B., Molnar, I., Kisdi, E., Hofbauer, J., and Czaran, T. (1995). The adaptive advantage of phenotypic memory in changing environments. *Philosophical Transactions of the Royal Society of London, Series B* 350, 133–41.

Jablonka, E. and Raz, G. (2009). Transgenerational epigenetic inheritance: prevalence, mechanisms, and implications for the study of heredity and evolution. *The Quarterly Review of Biology* 84, 131–76.

Kirkpatrick, M. and Lande, R. 1989. The evolution of maternal characters. *Evolution* 43, 485–503.

Lancaster, L. T., Mcadam, A. G., and Sinervo, B. (2010). Maternal adjustment of egg size organizes alternative escape behaviours, promoting adaptive phenotypic integration. *Evolution* 64, 1607–21.

Leimar, O., Hammerstein, P., and Van Dooren, T. J. M. (2006). A new perspective on developmental plasticity and the principles of adaptive morph determination. *American Naturalist* 167, 367–76.

Lemke, H., Coutinho, A., and Lange, H. (2004). Lamarckian inheritance by somatically acquired maternal IgG phenotypes. *Trends in Immunology* 25, 180–6.

Li, L., Zheng, P., and Dean, J. (2010). Maternal control of early mouse development. *Development* 137, 859–70.

Lickliter, R. and Harshaw, C. (2011). Canalization and malleability revisited: the developmental basis of phenotypic stability and variability. In K. E. Hood, C. T. Halpern, G. Greenberg, and R. M. Lerner, eds. *Developmental Science, Behaviour, and Genetics*, pp. 491–526. Wiley-Blackwell, MA.

Lin, H. C., Hsu, S. C., and Hwang, P. P. (2000). Maternal transfer of cadmium tolerance in larval Oreochromis mossambicus. *Journal of Fish Biology* 57, 239–49.

Love, O. P. and Williams, T. D. (2008). The adaptive value of stress-induced phenotypes: effects of maternally derived corticosterone on sex-biased investment, cost of reproduction, and maternal fitness. *American Naturalist* 172, E135–49.

Maestripieri, D. (2009). Maternal influences on growth, reproduction, and behaviour in primates.

In D. Maestripieri and J. M. Mateo, eds. *Maternal Effects in Mammals*, pp. 256–91. Chicago University Press, Chicago.

Marshall, D. and Uller, T. (2007). When is a maternal effect adaptive? *Oikos* 116, 1957–63.

Marshall, D. J. (2008). Transgenerational plasticity in the sea: Context-dependent maternal effects across the life history. *Ecology* 89, 418–27.

Martin, R. A. and Pfennig, D. W. (2010). Maternal Investment Influences Expression of Resource Polymorphism in Amphibians: Implications for the Evolution of Novel Resource-Use Phenotypes. *PLoS ONE* 5, e9117.

Maynard Smith, J. (2000). The concept of information in biology. *Philosophy of Science* 67, 177–94.

Mayr, E. (1982). *The Growth of Biological Thought. Diversity, Evolution and Inheritance*. Belknap Press, Harvard University Press, Cambridge, MA.

Mccormick, M. (1998). Behaviorally induced maternal stress in a fish influences progeny quality by a hormonal mechanism. *Ecology* 79, 1873–83.

Meaney, M. J., Szyf, M., and Seckl, J. R. (2007). Epigenetic mechanisms of perinatal programming of hypothalamic-pituitary-adrenal function and health. *Trends in Molecular Medicine* 13, 269–77.

Meylan, S., Clobert, J., and Sinervo, B. (2007). Adaptive significance of maternal induction of density-dependent phenotypes. *Oikos* 116, 650–61.

Michel, G. F. (2011). The roles of environment, experience, and learning in behavioural development. In K. E. Hood, C. T. Halpern, G. Greenberg, and R. M. Lerner, eds. *Developmental Science, Behaviour, and Genetics*, pp. 123–65. Wiley-Blackwell, MA.

Minelli, A. and Fusco, G. (2010). Developmental plasticity and the evolution of animal complex life cycles. *Philosophical Transactions of the Royal Society of London, Series B* 365, 631–40.

Mivart, St. G. (1871). *On the Genesis of Species*. D. Appleton and co., New York, NY.

Moczek, A. P. (2007). Developmental capacitance, genetic accommodation, and adaptive evolution. *Evolution & Development* 9, 299–305.

Moczek, A. P. (2008). On the origins of novelty in development and evolution. *Bioessays* 30, 432–47.

Moore, C. L. 1995. Maternal Contributions to Mammalian Reproductive Development and the Divergence of Males and Females. *Advances in the Study of Behavior* 24, 47–118.

Mousseau, T. A. and Fox, C. W. (1998). *Maternal Effects as Adaptations*. Oxford University Press, New York, NY.

Müller, W., Lessells, C. M., Korsten, P., and Von Engelhardt, N. (2007). Manipulative signals in family conflict? On the function of maternal yolk hormones in birds. *American Naturalist* 169, E84–96.

Nijhout, H. F. (2003). Development and evolution of adaptive polyphenisms. *Evolution and Development* 5, 9–18.

Odling-Smee, F. J. (2010). Niche inheritance. In M. Pigliucci and G. B. Müller, eds. *Evolution. The Extended Synthesis*, pp. 175–208. MIT Press, Cambridge, MA.

Odling-Smee, F. J., Laland, K. N., and Feldman, M. W. (2003). *Niche Construction. The Neglected Process in Evolution*. Princeton University Press, Princeton, NJ.

Oyama, S. (2000). *The Ontogeny of Information. Developmental Systems and Evolution*. Duke University Press.

Pelegri, F. (2003). Maternal factors in zebrafish development. *Developmental Dynamics* 228, 535–54.

Pfennig, D. W. and Martin, R. A. (2009). A maternal effect mediates rapid population divergence and character displacement in spadefoot toads. *Evolution* 63, 898–909.

Pfennig, D. W., Wund, M. A., Snell-Rood, E. C., Cruickshank, T., Schlichting, C. D., and Moczek, A. P. (2010). Phenotypic plasticity's impacts on diversification and speciation. *Trends in Ecology and Evolution* 25, 459–67.

Raff, R. A. and Byrne, M. (2006). The active evolutionary lives of echinoderm larvae. *Heredity* 97, 244–52.

Revardel, E., Franc, A., and Petit, R. (2010). Sex-biased dispersal promotes adaptive parental effects. *BMC Evolutionary Biology* 10, 217.

Reynolds, J., Goodwin, N., and Freckleton, R. (2002). Evolutionary transitions in parental care and live bearing in vertebrates. *Philosophical Transactions of the Royal of London, Series B* 357, 269–81.

Robinson, S. R. and Méndez-Gallardo, V. (2011). Amniotic fluid as an extended milieu intérieur. In K. E. Hood, C. T. Halpern, G. Greenberg, and R. M. Lerner, eds. *Developmental Science, Behaviour, and Genetics*, pp. 234–84. Wiley-Blackwell, MA.

Ronce, O., Clobert, J., and Massot, M. (1998). Natal dispersal and senescence. *Proceedings of the National Academy of Sciences of the USA* 95, 600–5.

Schlichting, C. D. (2008). Hidden reaction norms, cryptic genetic variation, and evolvability. *Year in Evolutionary Biology 2008, Annals of the New York Academy of Sciences*, 1133, 187–203.

Schlichting, C. D. and Pigliucci, M. (1998). *Phenotypic Evolution. A Reaction Norm Perspective*. Sinauer Associates Inc, Sunderland, MA.

Schwabl, H. (1993). Yolk is a source of maternal testosterone for developing birds. *Proceedings of the National Academy of Sciences of the USA* 90, 11446–50.

Schwabl, H., Mock, D. W., and Gieg, J. A. (1997). A hormonal mechanism for parental favouritism. *Nature* 386, 231–231.

Searcy, W. A. and Nowicki, S. (2005). *The Evolution of Animal Communication. Reliability and Deception in Signalling Systems*. Princeton University Press, Princeton, NJ.

Shea, N. (2007). Representation in the genome and in other inheritance systems. *Biology & Philosophy* 22, 313–31.

Shea, N. (in press). Cue, signal, inherited representation. In B. Calcott, R. Joyce, and K. Sterelny, eds. *Signaling, Commitment, and Emotion*. MIT Press.

Shea, N., Pen, I., and Uller, T. (2011). Three epigenetic information channels and their different roles in evolution. *Journal of Evolutionary Biology*, 24, 1178–87.

Simpson, A. J. and Sword, G. A. (2009). Phase polyphenism in locusts: mechanisms, population consequences, adaptive significance and evolution. In D. W. Whitman and T. N. Ananthakrishnan, eds. *Phenotypic Plasticity of Insects. Mechanisms and Consequences*, pp. 147–90. Science Publishers, Enfield, NH.

Sinervo, B. and McEdward, L. R. (1988). Developmental consequences of an evolutionary change in egg size—an experimental test. *Evolution* 42, 885–99.

Skyrms, B. (2010). *Signals. Evolution, Learning and Information*. Oxford University Press, New York, NY.

Smith, T. B. and Skúlason, S. (1996). Evolutionary significance of resource polymorphism in fishes, amphibians, and birds. *Annual Review of Ecology and Systematics* 27, 111–33.

Smiseth, P. T., Wright, J., and Kölliker, M. (2008). Parent-offspring conflict and co-adaptation: behavioural ecology meets quantitative genetics. *Proceedings of the Royal Society of London B* 275, 1823–30.

Snell-Rood, E. C., Van Dyken, J. D., Cruickshank, T., Wade, M. J., and Moczek, A. P. (2010). Toward a population genetic framework of developmental evolution: the costs, limits, and consequences of phenotypic plasticity. *Bioessays* 32, 71–81.

Sun, Y. H., Chen, S. P., Wang, Y. P., Hu, W., and Zhu, Z. Y. (2005). Cytoplasmic impact on cross-genus cloned fish derived from transgenic common carp (Cyprinus carpio) nuclei and goldfish (Carassius auratus) enucleated eggs. *Biology of Reproduction* 72, 510–15.

Tauber, M. J., Tauber, C. A., and Masaki, S. (1986). *Seasonal Adaptations of Insects*. Oxford University Press, New York, NY.

Todrank, J., Heth, G., and Restrepo, D. (2011). Effects of in utero exposure on neuroanatomical development of the olfactory bulb and odour preferences. Proceedings of the Royal Society of London B, 278, 1949–55.

Uller, T. (2008). Developmental plasticity and the evolution of parental effects. *Trends in Ecology and Evolution* 23, 432–8.

Uller, T. and Badyaev, A. V. (2009). Evolution of 'determinants' in sex determination: A novel hypothesis for the origin of environmental contingencies in avian sex bias. *Seminars in Cell & Developmental Biology* 20, 304–12.

Uller, T. and Helanterä, H. (2011). When are genes 'leaders' or 'followers' in evolution? *Trends in Ecology and Evolution* 26, 435–36.

Uller, T. and Olsson, M. (2006). Direct exposure to corticosterone during embryonic development influences behaviour in an ovoviviparous lizard. *Ethology* 112, 390–7.

Uller, T. and Pen, I. (2011). A theoretical model for the evolution of maternal effects under parent-offspring conflict. *Evolution* 65, 2075–84.

Uller, T., While, G. M., Cadby, C. D., Harts, A., O'Connor, K., Pen, I., and Wapstra, E. (2011). Altitudinal divergence in maternal thermoregulatory behaviour may be driven by differences in selection on offspring survival in a viviparous lizard. *Evolution* 65, 2313–24.

Wcislo, W. T. (1989). Behavioral environments and evolutionary change. *Annual Review of Ecology and Systematics* 20, 137–69.

Weaver, I. C. G., Cervoni, N., Champagne, F. A., D'alessio, A. C., Sharma, S., Seckl, J. R., Dymov, S., Szyf, M., and Meaney, M. J. (2004). Epigenetic programming by maternal behavior. *Nature Neuroscience* 7, 847–54.

Weber, B. H. and Depew, D. J. (2002). *Evolution and Learning. The Baldwin Effect Reconsidered*. MIT Press, Cambridge, MA.

West-Eberhard, M. J. (2003). *Developmental Plasticity and Evolution*. Oxford University Press, NY.

West-Eberhard, M. J. (2005). Phenotypic accommodation: Adaptive innovation due to developmental plasticity. *Journal of Experimental Zoology Part B-Molecular and Developmental Evolution* 304B, 610–18.

West-Eberhard, M. J. (2007). Dancing with DNA and Flirting with the Ghost of Lamarck. *Biology and Philosophy* 22, 439–51.

Williams, G. C. (1992). *Natural Selection. Domains, Levels, and Challenges*. Oxford University Press, New York, NY.

Wolf, J. B. and Wade, M. J. (2009). What are maternal effects (and what are they not)? *Philosophical Transactions of the Royal Society B: Biological Sciences* 364, 1107–15.

Youngson, N. and Whitelaw, E. (2008). Transgenerational epigenetic effects. *Annual Review of Genomics and Human Genetics* 9, 233–57.

CHAPTER 15

The quantitative genetic theory of parental effects

Jarrod Hadfield

15.1 Introduction

There are many theoretical approaches for studying the evolution of parental care and parent–offspring interactions (Chapters 2, 7, 9, and 16; Mock and Parker 1997), but here I focus on theory developed in the field of quantitative genetics. The reasons for this are twofold; first, they allow tractable dynamic models for phenotypes determined by multiple genes and the environment. Second, theory and application are so entwined in quantitative genetics that the development of theory is nearly always followed, or sometimes even preceded, by methods to estimate the relevant parameters from data—a useful resource for empiricists. However, my main aim is not to champion the quantitative genetic approach over others, but to clarify how quantitative genetic models of parent–offspring interaction work, and how the key concepts fit with more familiar ideas from behavioural ecology. Traditionally, the two approaches have often focused on fundamentally different things; with behavioural ecology focusing on parent–offspring conflict and quantitative genetics focusing on parent–offspring co-adaptation (reviewed in Smiseth et al. 2008; Hinde et al. 2010). The goal of this chapter is to dispel the perception that the tension between the interests of the individual and the interests of kin does not have a natural place in the quantitative genetic approach, and to clarify that its omission from much recent theoretical and empirical work is not warranted. Its omission seems to be inadvertent and may have arisen because those applying the quantitative genetic approach have continued to associate concepts from behavioural ecology with concepts of the same name from quantitative genetics, particularly those pertaining to fitness and selection (Chapter 1).

Quantitative genetic models of parental effects are designed to predict evolutionary change in suites of traits that affect traits expressed in offspring and/or are affected by traits expressed in parents. In the first section I give a detailed exposition of the Kirkpatrick–Lande model (hence forth the K–L model; Kirkpatrick and Lande 1989, 1992; Lande and Kirkpatrick 1990), a model that generalized a great deal of previous theory in which the phenotype and fitness of an individual was influenced by the phenotypes of its parents (Dickerson 1947; Willham 1963, 1972; Falconer 1965; Cheverud 1984). The model is difficult to understand and so the intention is to derive and explain it in a way that is both didactic and complementary to the original work, with special emphasis on clarifying what is meant by selection. To facilitate this, I work through a simple biological example in the second section and highlight the relationship between the selection parameters of the K–L model and concepts from behavioural ecology and life-history evolution. By doing this I argue that recent theoretical and empirical work in quantitative genetics has assumed values for these selection parameters that contradict central ideas from behavioural ecology that have wide empirical support. In the third section I describe the Willham model (Willham 1963, 1972), a special case of the K–L model widely used by empiricists, and show that by changing assumptions about the form of selection we come to very different conclusions about what types of genetic architecture act as constraints to evolutionary change. Following

The Evolution of Parental Care. First Edition. Edited by Nick J. Royle, Per T. Smiseth, and Mathias Kölliker.

Cheverud (1984) I place the Willham model in the context of Hamilton's rule in order to further elucidate the meaning of selection, and relatedness, in quantitative genetic models of parent–offspring interaction.

In most quantitative genetic models, values for the genetic parameters, such as genetic correlations, are assumed and the main focus is on evolutionary change in the mean. However, explaining why certain values for the genetic parameters are more likely than others is an interesting avenue of research, particularly in the context of social interactions where they appear in the relatedness term of Hamilton's rule (see also Chapter 16). In the fourth section I discuss why we might expect the genetic parameters of traits involved in parent–offspring interactions to be different from those of other traits, but suggest that current expectations about the genetic architecture of traits involved in parent–offspring interactions (Wolf and Brodie 1998) may be challenged when we entertain more reasonable patterns of selection and mutation.

In order to understand parent–offspring interactions fully the K–L model has one important short-coming: it fails to account for the fact that offspring are not passive vessels receiving parental care, but often express traits such as begging that modify parental behaviour. A general framework for modelling a wide range of interactions is the indirect genetic effect (IGE) approach (Moore et al. 1997; Wolf et al. 1999; McGlothlin et al. 2010) which has recently been used to analyse a model in which parents affect offspring *and* offspring affect parents (Kölliker et al. 2005). Although such an approach may become the quantitative genetic method of choice for modelling these types of interactions, in the fourth section I identify a conceptual difficulty with the IGE approach that arises when it is applied to parent–offspring interactions. Although this contradiction may have few practical consequences it is hoped that more theoretical work identifying any consequences are pursued before the IGE approach is more widely used.

In order to prevent the chapter from becoming too turgid, readers can find various notes concerning the smaller and less relevant technical details in Box 15.1. These are referenced in the body of the text.

15.2 The K–L model

The model of Kirkpatrick and Lande in its most general form follows the evolutionary dynamics of a suite of traits using a quantitative genetic approach (Kirkpatrick and Lande 1989, 1992; Lande and Kirkpatrick 1990). However, unlike the multivariate breeders' equation (Lande 1979) an individual's trait values can be, in part, determined by the trait values expressed in that individual's parents (Box 15.1, note 1). The K–L model generalizes a great deal of previous work and remains the most comprehensive theoretical treatment of the subject. However, it is not easy to understand. In part, this is because theory is never easy, but it is also because some of the notation is ambiguous, the discussion of key concepts and terms is often cursory, and there are some confusing mistakes only some of which are corrected in a little known erratum (Kirkpatrick and Lande 1992). In order to understand how the K–L model works it will be useful to take a concrete example, and so for the majority of the chapter I will consider two traits: weight at independence (henceforth *body-size*) and the amount of food that an individual provisions each of its offspring (henceforth *provisioning*).

For clarity I will use the words *influence* and *affect* and their derivatives in a precise way throughout the chapter: if by provisioning its offspring a parent can make it larger, and by being larger that offspring has higher fitness I say that parental provisioning *affects* offspring body-size, and by this *influences* offspring fitness. This is what a quantitative geneticist means by parental effect—parental provisioning has no parental effect on fitness in this instance, because there is no direct causal link. Another source of confusion is that the word parental can refer to a generation or a role, and sometimes traits when they are specific to a role. In the context of generation I will try and use the words 'previous generation' $(t-1)$, 'current generation' (t), and 'future generation' $(t+1)$.

The trait values of an individual from the future generation follow the model:

$$\mathbf{z}(t+1) = \mathbf{a}(t+1) + \mathbf{e}(t+1) + \mathbf{M}\bar{\mathbf{z}}(t) \qquad (15.1)$$

where z denotes phenotype, a additive genetic value (or breeding value), and e a non-heritable

Box 15.1 Additional notes on quantitative genetic models of parental effects

Note 1: Much of the work that I discuss actually considers maternal effects only, but here I elucidate the theory in the context of parental effects since the extension is straightforward.

Note 2: Although it will be obvious to many it is worth pointing out that $\underset{t}{E}[\mathbf{z}(t)] = \underset{t}{E}[\bar{\mathbf{z}}(\mathbf{t})]$ since each individual has two parents.

Note 3: These are given incorrectly in Kirkpatrick and Lande (1989) but are corrected in Kirkpatrick and Lande (1992).

Note 4: A special case of this equation is given incorrectly as Eq.10 in Kirkpatrick and Lande (1989) and appears corrected as Eq.5 in Kirkpatrick and Lande (1992). Nevertheless, the section dealing directly with the Willham model remains very confusing because they use the term β_m and refer to it as a parental selection gradient before they introduce the concept of parental selection. However, β_m is *not* a parental selection gradient in the Willham section, it is the direct selection gradient on trait m—in my notation $\beta_{I,2}$. Part of the difficulty with Kirkpatrick and Lande (1989) is that the subscripts o and m switch meaning throughout the manuscript: sometimes they refer to the role of the individual in which the trait is expressed (i.e. o indicates body-size, and m indicates provisioning) as in the Willham section, and sometimes they refer to the role an individual is playing (i.e. o indicates a trait in an individual, and m indicates a trait expressed by an individual's mother) as in the distinction between direct and parental selection.

Note 5: When the selection gradients are zero for the traits that parentally effect body-size (i.e. the Willham model), evolutionary change in the total parental effect (parental performance) caused by a correlated response to selection is correctly predicted, as is the change in body-size. However if the selection gradients are non-zero (i.e. Cheverud's extension), then evolutionary change cannot be predicted by measuring a composite selection gradient on parental performance (Kirkpatrick and Lande 1989), except under very specific assumptions. Technically then, Cheverud's extension only applies to cases where a trait is parentally affected by a *single* trait expressed in parents, thus undermining the strength of the Willham model to empiricists studying natural selection. However, from a practical perspective, estimating the selection gradient on the composite parental performance and applying it to the genetic parameters of the Willham model may prove to be more precise and more accurate than attempting the full K–L approach. Exploring the bounds of error on Cheverud's extension to the Willham model when empiricists are challenged by modest sample sizes and the danger of a misspecified K–L model would seem like a worthwhile task (See Discussion).

Note 6: Cheverud (1984) mistakenly uses selection intensities rather than standardized selection gradients through out the paper. In addition, the derivation is not exactly equivalent to the Willham model because what is called parental performance is actually something *proportional* to parental performance such that $m_{1,2}$ is positive but not necessarily equal to one. This may sound like a small point, but the relationship between the Willham model and Hamilton's rule given by Cheverud (1984) is easier to interpret when $m_{1,2} = 1$ (see below).

Note 7: Note that in Cheverud (1984) parental performance is sex-limited (it is maternal performance) and so the LHS ($-\beta_{I,2}$) is divided by 2. If a trait is sex-limited then *direct* selection gradients associated with that trait should be halved, but parental selection gradients should remain untouched. For example, if only females provision then only half the individuals (females) will experience fitness variation caused directly by the trait. However, under parental selection all individuals will experience fitness variation caused by the trait because all individuals have a mother, even males.

Note 8: In fact, rather than setting $m_{1,2} = 1$, Cheverud (1984) has this as a free parameter and defines the relatedness term in Hamilton's rule as $m_{1,2}\left(\frac{g_{1,2}}{g_{2,2}} + 0.5\right)$. However, I think it makes more sense to keep the relatedness term as I have done, and think of the benefit as $m_{1,2}\beta_{I,1}$. The advantage of this is that it puts the cost and benefit in the same units; the benefit is how much does a unit change in parental performance change offspring fitness (via a change in the offspring trait).

Note 9: I use the (matrix) notation $\mathbf{A}$ for the *vector* of total breeding values following McGlothlin et al. (2010).

residual (Falconer and Mackay 1996; Bulmer 1985; Lynch and Walsh 1998). They appear in bold face because they are vectors and include terms for both traits: body-size (z_1) and provisioning (z_2). $\mathbf{M}$ is the parental effect coefficient matrix and $\bar{z}(t)$ is the average trait value expressed by an individual's parents. The product $\mathbf{M}\bar{\mathbf{z}}(t)$ is a vector of parental effects for the traits, with the ith element being the total effect of traits expressed in the individual's parents on the ith trait of the individual: $\sum_j m_{i,j}\bar{z}_j(t)$, where $m_{i,j}$ is the effect that a unit change in trait j expressed in the individual's parents has on trait i.

It is important to state at which point in the future generation's life-cycle *all* quantities appearing in Equation 15.1 are measured. Conventionally, the most natural point is to measure the traits when the future generation are zygotes, because then we are measuring the future generation's traits before they have been exposed to selection. After all, quantitative geneticists are interested in predicting the mean phenotype of the future generation without having to specify the form of selection that may act on them. Of course, zygotes express neither body-size nor provisioning, so the trait values at this point are hypothetical: it would be the expected trait values if selection ceased until expression.

In order to work out the mean phenotype in the future generation we can simply sum the expectations for each term:

$$\underset{t+1}{E}[\mathbf{z}(t+1)] = \underset{t+1}{E}[\mathbf{a}(t+1)] + \underset{t+1}{E}[\mathbf{e}(t+1)] + \mathbf{M}\underset{t+1}{E}[\bar{\mathbf{z}}(t)] \tag{15.2}$$

where I use subscripts to denote the generation over which the expectation is taken.

This notation singles out $\underset{t+1}{E}[\bar{\mathbf{z}}(t)]$ as being something odd—we're taking the expectation of a trait expressed in the current generation, $\bar{\mathbf{z}}(t)$, but over individuals in the future generation, $\underset{t+1}{E}$. Indeed, it has been one source of confusion. Because of selection in the current generation the average trait values of individuals in the current generation when they were zygotes, $\underset{t}{E}[\bar{\mathbf{z}}(t)]$, differ from the mean trait values of the current generation that go on to become parents (Box 15.1, note 2). K–L often use the term 'individuals measured after selection' to denote $\underset{t+1}{E}[\bar{\mathbf{z}}(t)]$, which at first reading may suggest that the mean trait value of the current generation after selection is the mean trait value of those individuals that go on to be parents. This is not what is intended. More correctly, the mean of the current generation after selection is the mean trait value of the current generation that an individual of the future generation experiences. When the trait is independent of fertility then this quantity is the same as measuring the mean trait value of individuals that go on to be parents, but generally 'after selection' includes both viability selection and fertility selection. To take an example, let's imagine four zygotes (two of each sex) from the current generation form two pairs, and that the provisioning values of these pairs are -1 and 1. Consequently, $\underset{t}{E}[\bar{z}_2(t)] = 0$. Let's imagine that all zygotes survive to be parents and that pair 1 contributes 1 zygote to the future generation and pair 2 contributes 2 zygotes. The three individuals of the future generation will then experience provisioning values, -1, 1, and 1 giving $\underset{t+1}{E}[\bar{z}_2(t)] = \frac{1}{3}$. Essentially, selection modifies the distribution of traits in the current generation *experienced* by individuals of the future generation because fit parents interact with more individuals of the next generation than less fit parents.

Equation 15.2 does not appear to be very predictive: terms with $t+1$ are still appearing on the right hand side and it would be nice to write them down using terms with t only. We can do so using the following approximation (i), assumption (ii), and identity (iii):

i) One of the foundations of quantitative genetics developed by Fisher (1918) is the concept of the breeding value, which has the property:

$$\underset{t+1}{E}[\mathbf{a}(t+1)] \approx \underset{t+1}{E}[\mathbf{a}(t)]$$

since breeding values are transmitted without bias from parents to their offspring under weak selection and random mating (Falconer 1985).

ii) We will assume that the residual environment of the future generation at conception is the same as the residual environment that the current generation experienced at conception:

$$\underset{t+1}{E}[\mathbf{e}(t+1)] = \underset{t}{E}[\mathbf{e}(t)]$$

This may seem untenable, but the idea is that if there are differences in the average environment between generations such as temperature, parasites, or food availability, their effects on the traits could be controlled for, at least hypothetically.

iii) One of the most powerful identities in evolutionary biology, first shown by Robertson (1966) is:

$$\underset{t+1}{E}[\mathbf{z}(t)] = \underset{t}{E}[\mathbf{z}(t)] + \underset{t}{COV}(\mathbf{z}(t), w(t))$$
$$= \underset{t}{E}[\mathbf{z}(t)] + S(\mathbf{z}(t))$$

where w is the relative fitness of an individual, and fitness in our case is measured as the number of zygotes. $S()$ is known as a selection differential and is the change in the mean value measured before and 'after selection'. Although I have expressed Robertson's (1966) result in terms of trait value, z could be exchanged with anything measurable (for example breeding value) and the result would still hold. It is also important to realize that this result makes no assumption about the relationship between relative fitness and z—it could be linear, it could be loop-the-loop.

We can substitute these three results into Equation 15.2, to give:

$$\underset{t+1}{E}[\mathbf{z}(t+1)] = \underset{t}{E}[\mathbf{a}(t)] + S(\mathbf{a}(t))$$
$$+ \underset{t}{E}[\mathbf{e}(t)] + \mathbf{M}\left(\underset{t}{E}[\bar{\mathbf{z}}(t)] + S(\bar{\mathbf{z}}(t))\right)$$

To obtain the mean phenotype in the current generation we can also replace t with $t-1$ in Equation 15.2:

$$\underset{t}{E}[\mathbf{z}(t)] = \underset{t}{E}[\mathbf{a}(t)] + \underset{t}{E}[\mathbf{e}(t)] + \mathbf{M}\underset{t}{E}[\bar{\mathbf{z}}(t-1)]$$

and then use Robertson's (1966) identity (iii):

$$\underset{t}{E}[\mathbf{z}(t)] = \underset{t}{E}[\mathbf{a}(t)] + \underset{t}{E}[\mathbf{e}(t)]$$
$$+ \mathbf{M}\left(\underset{t-1}{E}[\bar{\mathbf{z}}(t-1)] + S(\bar{\mathbf{z}}(t-1))\right)$$

The change in mean phenotype is therefore given as:

$$\underset{t}{\Delta}\mathbf{z} = S(\mathbf{a}(t)) + \mathbf{M}\left(\underset{t-1}{\Delta}S(\bar{\mathbf{z}}) + \underset{t-1}{\Delta}\bar{\mathbf{z}}\right) \tag{15.3}$$

where $\underset{t}{\Delta}$ indicates a change in a quantity from generation t to $t+1$.

In words, $S(\mathbf{a}(t))$ represents evolutionary change *sensu stricto*; caused by selection altering gene frequencies (see also Bonduriansky and Day 2009). However, it is apparent that changes in the parental effects can also cause change, not because the effect of the parental traits change (**M** is fixed) but because the actual values of the parental traits experienced by offspring are altered. This alteration can be the result of two processes; it can arise because the trait values of individuals that *may* have gone on to become parents differ between the two generations, $\underset{t-1}{\Delta}\bar{\mathbf{z}}$, but it can also arise because different patterns of selection modify the parental traits that offspring experience even when the different generations are identical as zygotes, $\underset{t-1}{\Delta}S(\bar{\mathbf{z}})$.

$S(\mathbf{a}(t))$ are the genetic covariances between traits and relative fitness, and the most widely known corollary of Robertson's (1966) result is that these are equal to evolutionary change in standard quantitative genetic models (see also Price 1972). Unfortunately, they are not always very useful for understanding the biology that underlies evolutionary processes because they substitute causation for correlation, and conflate inheritance with selection, both of which may be of interest. In many respects evolutionary biologists are not interested in what these models tell them about currently changing gene frequencies (Grafen 1988), but the insight they give into the adaptive significance of parental effects and their genetic basis. For these reasons we can rewrite these covariances in terms of inheritance (**C**) and selection (β):

$$S(\mathbf{a}(t)) = \mathbf{C}(t)\beta(t) \tag{15.4}$$

By doing this it should be understood that we make some very strong assumptions. These assumptions have been discussed many times before, usually in the context of the multivariate breeders' equation (Lande 1979; Mitchell-Olds and Shaw 1987; Grafen 1988; Rausher 1992; Hadfield 2008; Morrissey et al. 2010) but also in the context of kin selection (Queller 1992). In essence, a sufficient condition for Equation 15.4 to be valid is that β represents the *causal* effects of the traits on fitness and that all genetically correlated traits directly affecting fitness have been included in the analysis.

In our example $\mathbf{C}(t)$ is a matrix with two rows (because there are two traits) and an arbitrary number of columns. The element c_{ij} is the covariance between the breeding values for trait i in individuals in the current generation and some characteristic j which selection acts upon. β_j are selection gradients; the causal effect of characteristic j on the relative fitness of an individual in the current generation. Kirkpatrick and Lande (1989) choose two types of characteristics—the trait values of the individuals themselves, but also the trait values of the individual's parents. In this case, **C** would have four columns, two associated with traits in the individuals and two associated with traits in the individual's parents.

For clarity we can rewrite this as $\mathbf{C}(t)\beta(t) = \mathbf{C}_I(t)\beta_I(t) + \mathbf{C}_M(t)\beta_M(t)$, where terms involving the individual's own traits (I) are separated from those involving the traits of the individual's parents (M). I will call $\beta_I(t)$ *direct* selection gradients, and following K–L I will call $\beta_M(t)$ *parental* selection gradients, which are conceptually equivalent to a trait's parental effect on fitness.

The meaning of these selection gradients appears to be the single biggest source of confusion. We can think of these selection gradients as regression coefficients from a multiple regression with the response variable being the number of zygotes an individual from the current generation produces, and the predictor variables being, in this example, the four characteristics; the individual's body-size, the individual's provisioning, the individual's parent's body size and the individual's parent's provisioning. Since the regression coefficients represent the *effect* of a particular characteristic on fitness we might question whether all four gradients are likely to be non-zero, a question we will return to. For now, we will assume they could be:

$$\underset{t}{\Delta}\mathbf{z} = \mathbf{C}_I(t)\beta_I(t) + \mathbf{C}_M(t)\beta_M(t) + \mathbf{M}\left(\underset{t-1}{\Delta}S(\mathbf{z}) + \underset{t-1}{\Delta}\mathbf{z}\right) \tag{15.5}$$

The model results in complicated dynamical behaviour because change at time t depends on what happened at time $t-1$, which in turn depends on what happened at time $t-2$. To simplify matters we can pretend that a constant pattern of selection ($\beta(t) = \beta$) has been operating on the suite of traits such that $\underset{t}{\Delta}S(\mathbf{z}) = 0$ and consequently the rate of evolutionary change has become constant $\underset{t}{\Delta}\mathbf{z} = \Delta\mathbf{z}$. If we are also willing to assume that this selection is weak then the covariances between breeding values and phenotype values will also, over time, become constant ($\mathbf{C}(t) = \mathbf{C}$), and we can derive a simplified version of Equation 15.5:

$$\begin{aligned}\Delta\mathbf{z} &= \mathbf{C}_I\beta_I + \mathbf{C}_M\beta_M + \mathbf{M}\Delta\mathbf{z}\\ &= (\mathbf{I} - \mathbf{M})^{-1}(\mathbf{C}_I\beta_I + \mathbf{C}_M\beta_M)\end{aligned} \tag{15.6}$$

We can use this result to explore other models and the conclusions drawn from them. However, before we do so it will be useful to dig a bit deeper into what is meant by inheritance, and to do this we will work through the example of $\mathbf{C}_I$. Each element of $\mathbf{C}_I$ is the covariance between the breeding value of a trait and the phenotypic value of a trait. In the absence of parental effects $\mathbf{C}_I = \mathbf{G}$, the familiar **G** matrix that appears in the multivariate breeders equation (Lande 1979). However in the presence of parental effects complications arise. Imagine a mutation arising in a zygote that increases provisioning. This mutation will increase both the individual's breeding value and trait value for provisioning, and will contribute positively to the covariance between breeding value and phenotype. If we imagine that this mutation has no pleiotropic effect on body-size then the mutation cannot contribute to the covariance between breeding value for provisioning and phenotypic body-size, or at least not immediately. In the following generation, half the descendants of this individual are expected to carry the mutation and will therefore have a greater breeding value for provisioning. However, they will also have larger body sizes because the mutation present in their parent has had a parental effect on their body size. This will contribute positively to the covariance between breeding value for provisioning and phenotypic body-size in spite of there being no direct causal effect of the mutation on both traits (i.e $g_{1,2} = 0$ but $c_{1,2} > 0$). $\mathbf{C}_I$ must therefore capture the direct effects of genes, but also the indirect effects of genes expressed in parents. As this example shows, these covariances also take time to equilibrate, but K–L show that given certain assumptions regarding stationarity $\mathbf{C}_I$ can be expressed in terms of the genetic covariances (**G**) and a geometric series of the parental effect matrix (**M**) that modifies these (co)variances:

$$\mathbf{C}_I = \mathbf{G}(\mathbf{I} - \frac{1}{2}\mathbf{M}^\top)^{-1} \tag{15.7}$$

and

$$\mathbf{C}_M = \frac{1}{2}\mathbf{C}_I \tag{15.8}$$

When quantitative geneticists discuss genetic variances and genetic correlations they are referring to quantities that can be derived from **G** rather than **C**.

The relationship between $\mathbf{C}_I$ and $\mathbf{C}_M$ (Equation 15.8) also allows us to substitute $\mathbf{C}_M$ in Equation 15.6 to give:

$$\Delta \mathbf{z} = (\mathbf{I} - \mathbf{M})^{-1}\,\mathbf{C}_I\left(\beta_I + \frac{1}{2}\beta_M\right)$$
$$= (\mathbf{I} - \mathbf{M})^{-1}\,\mathbf{C}_I\beta_N \qquad (15.9)$$

where K–L call $\beta_N = \beta_I + \frac{1}{2}\beta_M$ the *net* selection gradients (Box 15.1, note 3).

15.3 An example and its relation to behavioural ecology

The derivation given above is very general, with no restrictions placed on the form of **M** or the sign and magnitude of the selection gradients. However, the example of body-size and provisioning suggests that restrictions could be made and that certain patterns of selection and parental effect are biologically more reasonable than others. Here, I argue that in this example, and examples like it, we can often use previous experimental and theoretical work in behavioural ecology and life-history evolution in order to make *a priori* predictions about the form of selection. However, in doing so we need to be careful how ideas from these fields are translated into related concepts in quantitative genetics because mistranslations are easy.

First we will assume that the amount of provisioning by a parent positively affects offspring body-size ($m_{1,2} > 0$), but the extent to which an individual is provisioned does not have a causal effect on how much that individual goes on to provision its own offspring. Likewise we assume that parental body size has no causal effect on traits expressed in offspring, giving:

$$\mathbf{M} = \begin{bmatrix} 0 & m_{1,2} \\ 0 & 0 \end{bmatrix} \qquad (15.10)$$

In the more general derivation we allowed direct selection and parental selection on both traits. However, in many cases it would seem reasonable to set the parental selection gradients to zero. This is not equivalent to saying that parents do not influence the fitness of their offspring—they do—only that they do so by affecting (**M**) some aspect of their offspring's phenotype which then has a causal effect (β_I) on their offspring's fitness (see also Chapter 16). For many aspects of parent–offspring interaction it would seem possible to posit some traits expressed in the individuals themselves that mediate the influence that parental traits have on offspring fitness (Price 1998). There are of course examples, such as brood defence or infanticidal behaviour, where it is natural to think of parental selection operating, but personally I think it is possible to distil the central features of the K–L model without it. Bearing this in mind, let us assume that being larger confers a fitness advantage ($\beta_{I,1} > 0$; Kingsolver and Pfennig 2004) but any additional fitness benefits of being provisioned, or having large parents, are absent ($\beta_{M,2} = 0$ and $\beta_{M,1} = 0$). Let's also assume that the direct selection gradient for provisioning is *negative* ($\beta_{I,2} < 0$) reflecting the fact that an individual pays a cost (in terms of current or future zygote production) by provisioning its offspring more. Although the opposing signs of these direct selection gradients are assumed, it should be emphasized that they are based on broad empirical support demonstrating the trade-off between offspring size and number (Smith and Fretwell 1974), one of the best supported of all life-history trade-offs (Stearns 1992).

Much of the recent work in quantitative genetics has been derived or interpreted under the assumption that there is no net directional component to selection on provisioning ($\beta_{N,2} = 0$). This assumption may be tenable in models in which the parents' influence on offspring fitness is solely captured through the parental selection gradients (e.g. Kölliker et al. 2010), but is unlikely to hold in models where parents are able to influence their offspring's fitness via their effect on their offspring's traits, as shown above. Changing the assumptions of such models to something more reasonable may alter the conclusions and insights drawn from such models considerably. Moreover, the assumption that there is no net directional selection on provisioning (or related traits such as litter size) also has putative empirical support (e.g. McAdam and Boutin 2004, using results from Réale et al. 2003) although close inspection of these types of study often reveal two problems: 1) offspring survival has been included in the fitness measure and 2) an individual's traits have been dropped in favour of parental traits when calculating selection gradients such that direct selection gradients are effectively set to zero.

Although it is common in behavioural ecology to define individual fitness so that it includes the survival, and sometimes even the fecundity, of the individual's offspring (Clutton-Brock 1988), this definition of fitness—which I will call *weighted* fitness (Grafen 1982)—is not compatible with the quantitative genetic approach (Chapter 1; Cheverud and Moore 1994; Wolf and Wade 2001): individual fitness in these discrete-generation quantitative genetic models is how many zygotes an individual produces over its lifetime. The notion that at equilibrium, or conflict resolution (Godfray 1995), selection on provisioning should be stabilizing rather than directional may have arisen because these different definitions of fitness have been used out of context. Indeed, a paraphrase of Lack (1954) and related work (Charnov and Krebs 1974; Smith and Fretwell 1974): 'there is an optimal amount of provisioning that parents should engage in, and it is that which *maximises individual fitness*' certainly suggests stabilizing selection should be the norm. However, the idea that current levels of provisioning may be optimal are derived from the fact that individual fitness in this statement is the number of *surviving* offspring—weighted fitness—not fitness as a quantitative geneticist should define it.

These opposing direct selection gradients represent antagonistic selection across life-stages and at face value suggest that individuals are selected to behave like cuckoos; to take as much parental care as possible ($\beta_{I,1} > 0$) but at the same time minimize their own parental investment in order to maximize egg production ($\beta_{I,2} < 0$). Of course, in many species this selfishness is limited by kin selection, and in the behavioural ecological models described above this is dealt with, in part, by using weighted fitness. Kin selection enters into quantitative genetic models through other routes, which can most easily be understood by putting them in the context of Hamilton's (1964) rule. Before doing this however, it will be instructive to work through the Willham model (a special case of the K–L model similar in form to this example), not only because it is an empirically tractable and well used model, but also because it was in the context of this model that Cheverud (1984) developed the key insight that direct selection on parental care would be negative.

15.4 The Willham model

Incorporating parental effects, or more specifically maternal effects, into quantitative genetic models has a long history in animal breeding dating at least back to Dickerson (1947). Rather than work chronologically through the developments I choose to describe directly the work of Kirkpatrick and Lande in which much of the previous work can be subsumed. In Table 15.1, using the notation of the K–L model employed above, I represent key developments in the quantitative genetic theory of parental effects. The modelling framework now known as the Willham model (Willham 1963, 1972) deserves special mention, as it has been the focus of much theoretical work and is the basis for a great deal of applied work in both plant and animal breeding (Lynch and Walsh 1998, pp. 687–714, Walsh and Lynch 2012, Chapter 21). The model has been explored from an evolutionary perspective (Cheverud 1984) and several empirical studies on wild species of plant (Platenkamp and Shaw 1993; Thiede 1998; Byers et al. 1997; Galloway et al. 2009) and animal (Wilson et al. 2005a; Kruuk and Hadfield 2007) have employed it.

Imagine a case where body-size (z_1) is parentally affected by a set of other traits ($z_2, z_3 \ldots z_n$) which are not themselves parentally affected by each other or body-size. In this case the parental effect coefficient matrix looks like:

$$\mathbf{M} = \begin{bmatrix} 0 & \mathbf{m}^\top \\ \mathbf{0} & \mathbf{0} \end{bmatrix} \tag{15.11}$$

where the vector of coefficients $\mathbf{m}$ contains the effect that a unit change in traits $z_2, z_3 \ldots z_n$ expressed in an individual's parents have on that individual's body-size. In the context of the K–L model the parental effect on body-size is obtained by identifying those traits that have a parental effect (i.e $z_2, z_3 \ldots z_n$) and measuring the strength of those effects ($\mathbf{m}$) to obtain the weighted sum: $\mathrm{m}^\top \bar{\mathrm{z}}_{2:n}$.

Table 15.1 A short history of maternal effects models in matrices. The K–L model places no restrictions on the number of traits, how they maternally affect each other, or whether they are under directional selection or not. However, the earlier models are special cases of the K–L model as can be seen by the size of the vector/matrices and by which maternal effect coefficients or selection gradients are zeroed out. Representing the model of Cheverud (1984) as a two-trait model may seem surprising given the text, but see Box 15.1, note 5

	Willham (1963, 1972)	Cheverud (1984)	Falconer (1965)	Kirkpatrick and Lande (1989)
M	$\begin{bmatrix} 0 & m_{1,2} & \dots & m_{1,n} \\ 0 & 0 & \dots & 0 \\ \vdots & \vdots & \ddots & 0 \\ 0 & 0 & \dots & 0 \end{bmatrix}$	$\begin{bmatrix} 0 & m_{1,2} \\ 0 & 0 \end{bmatrix}$	$\begin{bmatrix} m_{1,1} \end{bmatrix}$	$\begin{bmatrix} m_{1,1} & m_{1,2} & \dots & m_{1,n} \\ m_{2,1} & m_{2,2} & \dots & m_{2,n} \\ \vdots & \vdots & \ddots & \vdots \\ m_{n,1} & m_{n,2} & \dots & m_{n,n} \end{bmatrix}$
β_I	$\begin{bmatrix} \beta_1 \\ 0 \\ \vdots \\ 0 \end{bmatrix}$	$\begin{bmatrix} \beta_1 \\ \beta_2 \end{bmatrix}$	$\begin{bmatrix} \beta_1 \end{bmatrix}$	$\begin{bmatrix} \beta_1 \\ \beta_2 \\ \vdots \\ \beta_n \end{bmatrix}$

In reality, there are likely to be many traits that have parental effects on traits such as body-size (n is large), and identifying them all and measuring their effect would be a daunting task. The Willham model sidesteps this problem elegantly, although with certain limitations. Imagine a pair of individuals with multiple offspring. These offspring all have the same parental effect for body-size since **m** is a constant and they all share the same parental phenotypes: $\bar{z}_{2:n}$. Consequently, we could imagine obtaining an estimate of the parental effect by seeing how much more similar these offspring are to each other than they are to another set of offspring from different parents. Although the offspring will also resemble each other because their breeding values are positively correlated (if there is genetic variation for body-size) it is possible to control for this source of variation if a multi-generational pedigree is available and/or manipulative reciprocal cross-fostering techniques are used (Rutledge et al. 1972). Animal breeders often use the term 'parental performance' to denote the deviation of these parental effects from the population mean, and what we can do, with some abuse of notation, is to define the second trait (i.e. z_2) as parental performance. This reduces the problem to a two trait model, similar in form to our assumed model for body-size and provisioning, although the parental effect coefficient is set to one:

$$\mathbf{M} = \begin{bmatrix} 0 & 1 \\ 0 & 0 \end{bmatrix} \tag{15.12}$$

Not only does the Willham model allow a tractable empirical framework for estimating the combined effects of many parental traits, the pattern of zeros in **M** makes mathematical analysis easier. For example, evolutionary change in body size (from Equation 15.6) simplifies to (Box 15.1, note 4):

$$\Delta z_1 = (g_{1,1} + \frac{3}{2}g_{1,2} + \frac{1}{2}g_{2,2})\beta_{I,1} + (g_{1,2} + g_{2,2})\beta_{I,2} \tag{15.13}$$

where $g_{1,1}$ is the additive genetic variance for body-size, $g_{2,2}$ the additive genetic variance in parental performance and $g_{1,2}$ the additive genetic covariance between the two traits. Although parental selection gradients can be non-zero with certain forms of artificial selection (e.g. if a calf is allowed to breed because her mother had high milk yield) I have omitted them, and retain the direct selection gradients only. The objective of animal breeders was to select on traits such as body-size, and so naturally they set $\beta_{I,2} = 0$. Under this assumption, genetic variance in parental performance amplifies the response of body-size to selection, but a negative genetic correlation between parental performance and body-size ($g_{1,2} < 0$) constrains the response when individuals are selected to be larger ($\beta_{I,1} > 0$) (Dickerson 1947; Willham 1972). This notion has been widely taken up in the evolutionary literature (e.g. McAdam and Boutin 2004; Wilson et al. 2005a).

In an important paper, Cheverud (1984) considered the case where selection on parental performance exists (Box 15.1, notes 5–6), and emphasized the situation where it is negative ($\beta_{I,2} < 0$) and opposite in sign to selection on body-size; a situation which, as I argue above, is much more likely to be the case than selection on body-size alone. He refers to this pattern of selection as *altruistic selection*. However, parental performance is under negative direct selection, implying selection for greater selfishness or lower provisioning, and hence, *less* altruistic values. Cheverud's (1984) choice of words appears to stem from the idea that this negative direct selection is maintained because of kin-benefits, and so the adjective *altruistic* refers to the conditions that maintain negative selection on parental provisioning rather than referring to selection itself. Rearrangement of Equation 15.13 to give:

$$\Delta z_1 = g_{1,1}\beta_{I,1} + g_{1,2}\left(\frac{3}{2}\beta_{I,1} + \beta_{I,2}\right) + g_{2,2}\left(\frac{1}{2}\beta_{I,1} + \beta_{I,2}\right) \tag{15.14}$$

shows that under this pattern of selection a negative genetic covariance is only a constraint on the evolution of larger body-size when the strength of selection on body-size is more than two thirds that of selection on parental performance $\frac{|\beta_1|}{|\beta_2|} > \frac{2}{3}$. Likewise, the parental genetic variance only facilitates an evolutionary response when selection on body size is more than twice that of selection on parental performance $\frac{|\beta_1|}{|\beta_2|} > 2$. It seems likely that the strength of selection on traits such as body-size will be stronger than selection on traits linked to parental performance, given that the latter often involve traits that are expressed at later life stages (Medawar 1952; Charlesworth 1994). Consequently, the notion that a negative genetic correlation acts as a constraint may have some generality, although it is unclear whether genetic variance in parental performance will always facilitate a response to selection. This may be particularly so for species with 'slow' life-histories where the relative magnitude of the two selection gradients are likely to be more equal, and this warrants more attention.

15.5 Hamilton's rule

Given that the direct selection gradients appear to represent the selfish benefits afforded by a trait, it is natural to ask how kin-selection enters into the quantitative genetic framework. In the most general setting kin-selection enters through two distinct routes—directly through the effect of parental traits on offspring fitness (the parental selection gradients) and indirectly through the influence that parental traits have on offspring fitness via their effect on offspring traits. In keeping with the sections above we will initially show the relationship between the Willham model and Hamilton's rule under the second process in isolation.

Equation 15.13 describes evolutionary change in body-size in the Willham model. The equivalent equation for parental performance is:

$$\Delta z_2 = (g_{1,2} + \frac{1}{2}g_{2,2})\beta_{I,1} + g_{2,2}\beta_{I,2} \tag{15.15}$$

from which Cheverud (1984), giving the conditions under which parental performance increases, derives a version of Hamilton's rule (Box 15.1, note 7):

$$\begin{aligned} 0 &< (g_{1,2} + 0.5g_{2,2})\beta_{I,1} + g_{2,2}\beta_{I,2} \\ -g_{2,2}\beta_{I,2} &< (g_{1,2} + 0.5g_{2,2})\beta_{I,1} \\ -\beta_{I,2} &< \left(\frac{g_{1,2}}{g_{2,2}} + 0.5\right)\beta_{I,1} \end{aligned} \tag{15.16}$$

where $-\beta_{I,2}$ is the cost, $\beta_{I,1}$ the benefit, and $\frac{g_{1,2}}{g_{2,2}} + 0.5$ a form of relatedness. The benefit represents the fitness advantage of increasing body-size by one unit, and the cost represents the decrease in fitness caused by increasing parental performance by one unit. Because body-size is in the same units as parental performance (and the parental coefficient is 1), the benefit can also be interpreted as the increase in fitness caused by receiving an additional unit of parental performance, thereby placing the cost and benefit on the same scale (Box 15.1, note 8). Although we may expect the relatedness term to be simply 0.5, rather than involving genetic (co)variances, this definition of relatedness is consistent with the concept in Hamilton's rule (Michod and Hamilton 1980). When the genetic covariance between body-size and parental performance is zero Equation 15.16 reduces to a more familiar version of Hamilton's rule, $-\beta_{I,2} < 0.5\beta_{I,1}$, thus highlighting an assumption that underpins many arguments based on a simple version of Hamilton's rule (Chapter 16; Cheverud 1984).

15.6 The evolution of *G*

As we have seen, not only do the genetic variances and covariances play a key role in determining the rate and direction of evolutionary change, they also enter directly into the relatedness term of Hamilton's rule. However, in the models we have used above these parameters (**G**) are fixed quantities and our focus has been on the predictions regarding the direction and rate of evolutionary change ($\Delta\mathbf{z}$). In an interesting paper, Wolf and Brodie (1998) asked a slightly different question: given a certain pattern of selection how should patterns of inheritance change and what should they be at equilibrium (i.e. when evolutionary change ceases; $\Delta\mathbf{z} = 0$)? To understand why this may differ when parental effects exist, imagine two bi-allelic loci, the first affecting body-size and the second affecting provisioning. To make the verbal argument simpler we will assume complete dominance so the effects at each locus are either + or −, and we will also assume that the parental effects only depend on the maternal genotype (i.e maternal effects only). There are four possible combinations of genotypic effects: −−, −+, +−, and ++ where the first sign is the effect of the genotype on body-size and the second sign is the effect of the genotype on provisioning. Since the genotype at the second locus affects provisioning, it is the genotype *carried* by the mother that affects offspring body-size, rather than the genotype *transmitted* by the parents. We will denote the genotype carried by an individuals mother with an M, giving offspring of types $-M^-$, $-M^+$, $+M^-$, and $+M^+$ with body-sizes of -2, 0, 0, and 2 respectively. Imagine then, that there is an absence of direct selection on provisioning but direct stabilizing selection operates on offspring body-size such that a body-size of zero is optimal. In this case offspring of types $-M^+$ and $+M^-$ do better, and because mothers pass their genes to their offspring this implies that on average −+ and +− offspring do better. This form of selection should generate a negative genetic correlation between the traits through linkage-disequilibrium.

To get a more quantitative idea of how selection generates evolutionary change in genetic (co)variances, we can obtain the difference between the genetic (co)variances 'after selection' and before selection by modifying Equations 9b and 12a from Lande (1980) and Equation 15a from Lande and Arnold (1983):

$$\underset{t+1}{COV}(\mathbf{a}(t)) - \underset{t}{COV}(\mathbf{a}(t)) = \mathbf{C}(\gamma - \beta\beta^\top)\mathbf{C}^\top \quad (15.17)$$

where γ is a matrix of quadratic selection gradients. $\underset{t}{COV}(\mathbf{a}(t)) = \mathbf{G}(t)$ and so the left-hand side of this equation is often denoted $\underset{t}{\Delta}\mathbf{G}$ (Phillips and Arnold for example, Equation 2 from 1989), but this can be misleading. Unfortunately there is no robust result which allows $\mathbf{G}(t+1) = \underset{t+1}{COV}(\mathbf{a}(t+1))$ to be obtained from $\underset{t+1}{COV}(\mathbf{a}(t))$ as there was for the mean breeding value (i.e. approximation i) (Walsh and Lynch see chapter 31 of 2012, for an excellent review).

Wolf and Brodie (1998) assume that pleiotropic mutations do not exist, and that changes in **G** are due to patterns of linkage-disequilibrium generated by selection. Using a multivariate extension of the Bulmer (1971) Equation derived by Tallis (Tallis and Leppard 1987, 1988; Tallis 1989) they find:

$$\mathbf{G}(t+1) = \frac{1}{2}\left(\underset{t+1}{COV}(\mathbf{a}(t)) + \mathbf{G}(0)\right) \quad (15.18)$$

where $\mathbf{G}(0)$ is a diagonal matrix representing the fact that initially, in the absence of selection, the traits cannot be genetically correlated if mutations do not have pleiotropic effects. Equation 15.18 can be solved iteratively to find **G** at equilibrium. In accordance with the example above Wolf and Brodie (1998) assume no parental selection and direct stabilizing selection on offspring body size only:

$$(\gamma_I - \beta_I\beta_I^\top) = \begin{bmatrix} \gamma_{I,1} & 0 \\ 0 & 0 \end{bmatrix} = \begin{bmatrix} - & 0 \\ 0 & 0 \end{bmatrix} \quad (15.19)$$

where the right hand matrix indicates the sign of the elements. The motivation behind this choice is in part justified by the fact that *'data from humans demonstrate clear stabilizing selection for birth weight'* (Karn and Penrose 1951: see also Schluter and

Nychka 1994) (Wolf and Brodie 1998). However, Schluter and Nychka (1994) state *'Survival in human infants rises steeply with increasing birth mass ... to a broad flat dome'* thus echoing the earlier statement that *'most selection appeared to be directional'* (Schluter 1988). As discussed above, we do not expect stabilizing selection to be a dominant feature of the system, but rather positive direct selection on body-size ($\beta_{I,1} > 0$) and negative direct selection on provisioning ($\beta_{I,2} < 0$) which gives a very different pattern:

$$(\gamma_I - \beta_I \beta_I^\top) = \begin{bmatrix} -\beta_{I,1}^2 & -\beta_{I,1}\beta_{I,2} \\ -\beta_{I,2}\beta_{I,1} & -\beta_{I,2}^2 \end{bmatrix} = \begin{bmatrix} - & + \\ + & - \end{bmatrix} \tag{15.20}$$

Figure 15.1 shows the expected genetic correlation between body-size and provisioning under these two regimes. When offspring body-size is under direct stabilizing selection then a negative genetic correlation is expected when provisioning positively affects offspring body-size ($m_{2,1} > 0$) (Wolf and Brodie 1998). However, when directional selection is prevalent, we can see from Figure 15.1 that the sign of the genetic correlation depends critically on the relative magnitudes of selection on body-size and provisioning, with a negative genetic correlation only expected when selection on body-size is twice the magnitude of selection on provisioning. The evolution of the genetic correlation in this instance should not be seen as a result of co-adaptation, but rather the result of selection eroding 'useful' genetic variation leaving segregating variation that defines a genetic trade-off (Hazel 1943; Lande 1982; Roff 1992; Stearns 1992; Blows and Walsh 2009). This trade-off differs from the traditional concept of a trade-off in quantitative genetics because the trade-off is cross-generational, occurring between traits such as offspring number (generation t) and offspring size (generation $t + 1$).

The absolute magnitudes of the genetic correlation in Figure 15.1 (and Figure 1 of Kölliker et al. (2005)) are small, and indeed the effect of linkage-disequilibrium is likely to be weak and transitory in out-bred populations compared to the effect of pleiotropic mutations (Bulmer 1971; Lande 1980, 1984). Although it has been suggested that the effect of segregating pleiotropic mutations on the genetic correlation will follow the same pattern as that found for linkage-disequilibrium (Chapter 16; Wolf and Brodie 1998; Kölliker et al. 2005) this fact is far from clear (chapter 31 in Walsh and Lynch 2012) and under certain models directional selection has no impact on genetic correlations in the absence of linkage disequilibrium (Hill 1982). Consequently,

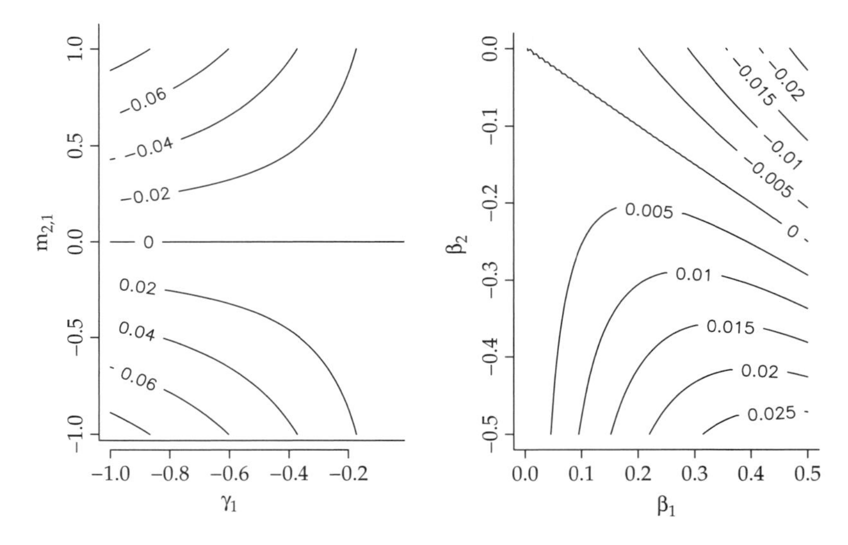

Figure 15.1 Left: Equilibrium genetic correlation when the body-size is under stabilizing selection ($\gamma_1 < 0$), after Wolf and Brodie (1998). Right: Equilibrium genetic correlation when body-size is under positive directional selection ($\beta_1 > 0$) and provisioning under negative directional selection ($\beta_2 < 0$). $m_{2,1}$ is the effect that provisioning has on body-size, which is fixed at one in the right plot in accordance with the Willham model.

explaining any interspecific variation in the sign and magnitude of the genetic correlation needs to be done in the context of interspecific variation in patterns of selection, and even then it should be borne in mind that the genetic correlation may not be shaped by selection but may simply be the outcome of functional constraints on the types of (pleiotropic) mutations that can exist.

15.7 General indirect genetic effect models and parental effect models

In many taxa, offspring also affect the phenotypes of their parents, and one of the most obvious examples of this is the modulation of parental provisioning caused by offspring begging (Chapter 7). A major shortcoming of the K–L model is that it does not allow traits in offspring to affect traits in parents. However, parental effects and offspring effects can be viewed as part of a wider class of indirect effects in which an individual's phenotype may be affected by traits expressed by many different parties such as partners, competitors, predators, and parasites. A general quantitative genetic framework—the IGE (indirect genetic effect) approach—for analysing evolutionary dynamics in the presence of these interactions has been proposed (Moore et al. 1997; Wolf et al. 1999; McGlothlin et al. 2010) and was used by Kölliker et al. (2005) to explore a quantitative genetic model of offspring begging and parental provisioning. Following Wolf and Brodie (1998) they analysed a model showing that the sign of the genetic correlation between provisioning and begging would change depending on a) the parental effect that provisioning has on begging, b) the offspring effect that begging has on provisioning, and c) the relative strengths of selection on the two traits. Like Wolf and Brodie (1998) they assumed that selection on both traits was stabilizing. Rather than restate why stabilizing selection should not be the default assumption in such models and provide alternative analyses under different assumptions (e.g that begging is costly), I would like to show why the IGE approach in its current form is not obviously applicable as a general framework for modelling parent–offspring interactions. During this section, I hope that readers do not lose sight of the fact that the IGE approach has successfully extended quantitative genetic models into new and profitable areas of research, and that richer and more realistic models such as those proposed by Kölliker et al. (2005) should be pursued. Hopefully, this short section will be a first step in delineating the types of models that can be analysed using the IGE approach and identifying the assumptions under which the interpretation given to model parameters remains valid and the models give correct predictions.

Although parent–offspring interactions are not analysed in McGlothlin et al. (2010) I will use it as the reference text for the IGE approach for two reasons. First, it is the third paper in the IGE series synthesizing results from the two earlier papers (Moore et al. 1997; Wolf et al. 1999). Second, a form of Hamilton's rule is put forward in McGlothlin et al. (2010) which is at odds with that formulated above, and is likely to cause confusion without greater clarification.

Equation 4b of McGlothlin et al. (2010) (in my notation) gives evolutionary change as:

$$\Delta \mathbf{z} = \mathbf{C}_{A_I}\beta_I + \mathbf{C}_{A_M}\beta_M \qquad (15.21)$$

where direct selection, β_I, is called non-social selection and parental selection, β_M, is called social selection. $\mathbf{C}_A$ is the covariance between *total* breeding values ($\mathbf{A}$) (Box 15.1, note 9) and trait values, where $\mathbf{A} = (\mathbf{I} - \mathbf{M})^{-1}\mathbf{a}$ (see also Bijma et al. 2007).

Given that $\mathbf{C}_{A_I} = (\mathbf{I} - \mathbf{M})^{-1}\mathbf{C}_I$, it is apparent that Equation 15.21 cannot be a general solution to problems of this sort, because it is equivalent to K–L's result (Equation 15.6) which required quite restrictive assumptions regarding stationarity. In itself this does not seem to be a major problem, but on closer investigation the correspondence between the K–L model under stationarity and Equation 15.21 seems to be surprising given the definition of a central concept in the IGE approach: the total breeding value. Let's assume that our model conforms to the Willham model, in which case the total breeding value of an individual for body-size is simply $A_1 = a_1 + a_2$, and represents the contribution of an individual's breeding value to the average phenotype of *its* offspring. However, an individual's total breeding value should be the contribution of an individual's breeding value to the average phenotype of the offspring generation. Under certain types of interaction these two statements are equivalent, but when the interactions are between parent and offspring they cannot be equivalent when selection occurs. Because individuals with high fitness interact with more individuals of the offspring generation (their offspring) they contribute more to the population mean than the average of their offspring value. Consequently the change in total breeding value (as currently defined) only predicts the change in mean in the absence of selection unless one is willing to entertain hypothetical cross-fostering experiments: we would have to reallocate zygotes at random to individuals of the current generation, such that individuals that failed to breed and those that had the highest number of zygotes both raised on average equal numbers of offspring. Although the indirect genetic effect approach can accommodate group sizes greater than two (Bijma et al. 2007; McGlothlin et al. 2010), having group size as a random variable that depends on the phenotypes of the interacting individuals would require more work. I should emphasize that McGlothlin et al. (2010) do *not* consider cross-generational indirect genetic effects in their paper and that for within-generation effects the issue raised above may be a moot point given the number of interactants is not equivalent to fitness.

Not only does the IGE approach offer a general way of dealing with the quantitative genetics of social interactions, it has been suggested that the approach also extends and generalizes Hamilton's rule (McGlothlin et al. 2010, but see Gardner et al. 2011). McGlothlin et al. (2010) derive their version of Hamilton's rule (see also Queller 1992) in the univariate case which I merely state again using my own notation (Equation 22b from McGlothlin et al. 2010):

$$-\beta_I < \frac{C_{A_M}}{C_{A_I}} \beta_M \qquad (15.22)$$

where terms involving direct (non-social) selection are identified with the cost of Hamilton's rule and terms involving parental (social) selection are identified with the benefit. As McGlothlin et al. (2010) only derive their version of Hamilton's rule in a single-trait case it is hard to establish whether they proscribe equating non-social/social selection with Hamilton's cost/benefits in a more general setting. Since McGlothlin et al. (2010) are not explicit on this matter the unwary reader may come away with the understanding that these two concepts are the same. However, earlier in this chapter I have given plausible examples where parental (social) selection gradients are zero and following Cheverud (1984) derived a version of Hamilton's rule in which both the costs and benefits are equated with different aspects of direct (non-social) selection. Consequently, it should be emphasized that social and non-social selection cannot always be identified in a straightforward way with the costs and benefits of Hamilton's rule, and treating them as such is likely to generate a great deal of confusion.

15.8 Discussion

Currently, the quantitative genetic approach provides a rich framework for understanding, predicting, and measuring evolutionary dynamics when parents directly affect the phenotypes of their offspring. In spite of this there is still a lot of work to do, both theoretically and empirically. In particular, the effect of offspring on parents has largely been ignored by theoretical quantitative geneticists with work by Kölliker et al. (2005) being a rare exception. Further development and refinement of these models is sorely needed. However, the assumptions of these models, and quantitative genetic models generally, need to be better aligned with ideas and beliefs held in other fields, particularly behavioural ecology (see also Chapters 7 and 16). Some of the theory outlined in this chapter needs to be re-evaluated under more reasonable assumptions regarding selection, and empiricists can play a leading role in this by estimating appropriate selection gradients from field data. Currently, much of our understanding about selection in the presence of parental effects comes from analyses that use definitions of fitness incompatible with the quantitative genetic approach and that estimate parental selection gradients while omitting direct selection gradients (e.g. McAdam and Boutin 2004; Wilson et al. 2005b). A better understanding of these processes may well give us insight into how persistent directional selection (Kingsolver and Pfennig

2004; Morrissey and Hadfield 2011) can be observed in the presence of genetic variation despite evolutionary change being small or absent (Garant et al. 2004).

The Willham model liberates empirical quantitative geneticists from having to identify and measure all the parental traits that effect the focal trait in an individual. However, when these traits are under selection the Willham model, as formulated by Cheverud (1984), fails except under the most restrictive of circumstances (Box 15.1, note 5). However, misspecification of the K–L model is a real danger in empirical studies where it may be hard or even impossible to identify and measure all traits that cause parental effects. Moreover, even if all n traits that cause parental effects could be identified, estimating n selection gradients, n^2 parental effect coefficients and $n(n+1)/2$ genetic parameters would be a daunting task if n was even moderately large. Comparing the bias and power between an (incomplete) K–L model and Cheverud's extension to the Willham model would seem like a worthwhile task, as estimating selection on parental performance would be relatively straightforward in a mixed model framework by estimating the covariance between parental performance effects and fitness (Hadfield 2008). Of particular interest is the introduction of simultaneous-recursive mixed models into quantitative genetics from econometrics (Gianola and Sorensen 2004). Not only do these methods generalize other approaches for fitting the trait-based K–L model (e.g. Lande and Price 1989; McGlothlin and Brodie 2009) they would also allow hybrid models to be fitted where parental performance can be separated into a part explained by traits measured in parents and a part explained by a Willham-like residual parental performance. All methods are data hungry and assumption-laden and empiricists need to rise to the difficult challenges of obtaining sufficient sample sizes to estimate quantitative genetic parameters with precision, having the necessary pedigree structure to estimate those parameters with minimal bias, being able to identifying and measure the relevant traits, and designing experiments to obtain estimates of causal relationships that are meaningful outside of the context in which the experiment was performed.

I have not reviewed the existing empirical literature on the quantitative genetics of parental care, as the subject has been reviewed explicitly multiple times before in both plants (Roach and Wulff 1987; Shaw and Byers 1998) and animals (Cheverud 1984; Räsänen and Kruuk 2007; Wilson and Réale 2006; Kruuk et al. 2008). Although sampling errors on genetic correlations are large it does seem like there is genuine variation across taxa in the magnitude and sign of the genetic correlation between offspring traits and traits linked to parental performance (Chapter 16). In this respect, the work of Wolf and Brodie (1998) and Kölliker et al. (2005) are welcome attempts at trying to explain this diversity, although it would seem prudent to push this work beyond the assumptions of the infinitesimal model into more realistic assumptions about the genetic basis of traits. Given that the genetic parameters also appear in the relatedness term of Hamilton's rule it would also be intriguing to see how the evolution of **G**, in the context of parent-offspring interactions, connects with ideas about the evolution of relatedness in social theory (Gardner et al. 2007).

15.9 Acknowledgements

I am indebted to Per Smiseth & Mathias Kölliker for the many suggestions and discussions on how to make the chapter clearer. I also thank Bill Hill, Joel McGlothlin, Laura Ross, Craig Walling, Ian White, Peter Korsten, and two anonymous reviewers for either reading earlier drafts of this manuscript or for useful discussions.

References

Bijma, P., Muir, W. A., and Van Arendonk, J. A. M. (2007). Multilevel selection 1: Quantitative genetics of inheritance and response to selection. *Genetics* 175, 277–88.

Blows, M. W. and Walsh, B. (2009). Spherical cows grazing in flatland: Constraints to selection and adaptation. In J. van der Werf, H. Graser, F. Frankham, and C. Gondro, eds. *Adaptation and Fitness in Animal Populations*, pp. 83–101. Springer.

Bonduriansky, R. and Day, T. (2009). Nongenetic inheritance and its evolutionary implications. *Annual Review of Ecology Evolution and Systematics* 40, 103–25.

Bulmer, M. G. (1971). Effect of selection on genetic variability. *American Naturalist* 105, 201–11.

Bulmer, M. G. (1985). *The mathematical theory of quantitative genetics*. Oxford University Press, New York.

Byers, D. L., Platenkamp, G. A. J., and Shaw, R. G. (1997). Variation in seed characters in *Nemophila menziesii*: Evidence of a genetic basis for maternal effect. *Evolution* 51, 1445–56.

Charlesworth, B. (1994). *Evolution in Age-Structured Populations*. 2nd edition. Cambridge University Press, Cambridge, UK.

Charnov, E. L. and Krebs, J. R. (1974). Clutch-size and fitness. *Ibis* 116, 217–19.

Cheverud, J. M. (1984). Evolution by kin selection - a quantitative genetic model illustrated by maternal performance in mice. *Evolution* 38, 766–77.

Cheverud, J. M. and Moore, A. J. (1994). Quantitative genetics and the role of the environment provided by relatives in behavioural evolution. In C. R. B. Boake, ed., *Quantitative Genetic Studies of Behavioural Evolution*, pp. 67–100. The University of Chicago Press.

Clutton-Brock, T. H., ed. (1988). *Reproductive success*. Univeristy of Chicago Press, Chicago.

Dickerson, G. E. (1947). Composition of hog carcasses as influeneced by heritable differences in rate and economy of gain. Iowa Agricultural Experiment Station Research Bulletin 354, 492–524.

Falconer, D. S. (1965). Maternal effects and selection response. In S. J. Geerts, ed., *Genetics Today, Proceedings of the XI International Congress on Genetics, volume 3*, pp. 763–74. Pergamon, Oxford.

Falconer, D. S. (1985). A note on Fisher's average effect and average excess. *Genetical Research* 46, 337–47.

Falconer, D. S. and Mackay, T. F. C. (1996). *Introduction to Quantitative genetics*. 4th edition. Longman, Harlow, UK.

Fisher, R. A. (1918). The correlation between relatives on the supposition of Mendelian inheritance. *Transactions of the Royal Society of Edinburgh* 52, 399–433.

Galloway, L. F., Etterson, J. R., and McGlothlin, J. W. (2009). Contribution of direct and maternal genetic effects to life-history evolution. *New Phytologist* 183, 826–38.

Garant, D., Kruuk, L. E. B., McCleery, R. H., and Sheldon, B. C. (2004). Evolution in a changing environment: A case study with great tit fledging mass. *American Naturalist* 164, E115–E129.

Gardner, A., West, S. A., and Barton, N. H. (2007). The relation between multilocus population genetics and social evolution theory. *American Naturalist* 169, 207–26.

Gardner, A., West, S. A., and Wild, G. (2011). The genetical theory of kin selection. *Journal of Evolutionary Biology* 24, 1020–43.

Gianola, D. and Sorensen, D. (2004). Quantitative genetic models for describing simultaneous and recursive relationships between phenotypes. *Genetics* 167, 1407–24.

Godfray, H. C. J. (1995). Evolutionary theory of parent-offspring conflict. *Nature* 376, 133–8.

Grafen, A. (1982). How not to measure inclusive fitness. *Nature* 298, 425–6.

Grafen, A. (1988). On the uses of data on lifetime reproductive success. In T. H. Clutton-Brock, ed., *Reproductive success*, pp. 454–71. Univeristy of Chicago Press, Chicago.

Hadfield, J. D. (2008). Estimating evolutionary parameters when viability selection is operating. *Proceedings of the Royal Society B-Biological Sciences* 275, 723–34.

Hamilton, W. D. (1964). Genetical evolution of social behaviour: I. *Journal of Theoretical Biology* 7, 1–16.

Hazel, L. N. (1943). The genetic basis for constructing selection indexes. *Genetics* 28, 476–90.

Hill, W. G. (1982). Rates of change in quantitative traits from fixation of new mutations. *Proceedings of the National Academy of Sciences of the United States of America-Biological Sciences* 79, 142–45.

Hinde, C. A., Johnstone, R. A., and Kilner, R. M. (2010). Parent-offspring conflict and coadaptation. *Science* 327, 1373–6.

Karn, M. N. and Penrose, L. S. (1951). Birth weight and gestation time in relation to maternal age, parity and infant survival. *Annals of Eugenics* 16, 147–64.

Kingsolver, J. G. and Pfennig, D. W. (2004). Individual-level selection as a cause of Cope's rule of phyletic size increase. *Evolution* 58, 1608–12.

Kirkpatrick, M. and Lande, R. (1989). The evolution of maternal characters. *Evolution* 43, 485–503.

Kirkpatrick, M. and Lande, R. (1992). The evolution of maternal characters: Errata. *Evolution* 46, 284.

Kölliker, M., Brodie, E. D., and Moore, A. J. (2005). The coadaptation of parental supply and offspring demand. *American Naturalist* 166, 506–16.

Kölliker, M., Ridenhour, B. J., and Gaba, S. (2010). Antagonistic parent-offspring co-adaptation. *PloS ONE* 5, e8606.

Kruuk, L. E. B. and Hadfield, J. D. (2007). How to separate genetic and environmental causes of similarity between relatives. *Journal of Evolutionary Biology* 20, 1890–1903.

Kruuk, L. E. B., Slate, J., and Wilson, A. J. (2008). New answers for old questions: The evolutionary quantitative genetics of wild animal populations.

Annual Review of Ecology Evolution and Systematics 39, 525–48.

Lack, D. (1954). *The Natural Regulation of Animal Numbers*. Oxford University Press, Oxford.

Lande, R. (1979). Quantitative genetic analysis of multivariate evolution, applied to the brain:body size allometry. *Evolution* 33, 402–416.

Lande, R. (1980). The genetic covariance between characters maintained by pleiotropic mutations. *Genetics* 204, 203–15.

Lande, R. (1982). A quantitative genetic theory of life-history evolution. *Ecology* 63, 607–15.

Lande, R. (1984). The genetic correlation between characters maintained by selection, linkage and inbreeding. *Genetical Research* 44, 309–20.

Lande, R. and Arnold, S. J. (1983). The measurement of selection on correlated characters. *Evolution* 37, 1210–26.

Lande, R. and Kirkpatrick, M. (1990). Selection response in traits with maternal inheritance. *Genetical Research* 55, 189–97.

Lande, R. and Price, T. (1989). Genetic correlations and maternal effect coefficients obtained from offspring-parent regression. *Genetics* 122, 915–22.

Lynch, M. and Walsh, B. (1998). *Genetics and analysis of quantitative traits*. Sinauer, Sunderland, MA.

McAdam, A. G. and Boutin, S. (2004). Maternal effects and the response to selection in red squirrels. *Proceedings of the Royal Society of London, Series B* 271, 75–9.

McGlothlin, J. W. and Brodie, E. D. (2009). How to measure indirect genetic effects: The congruence of trait-based and variance-partitioning approaches. *Evolution* 63, 1785–95.

McGlothlin, J. W., Moore, A. J., Wolf, J. B., and Brodie, E. D. (2010). Interacting phenotypes and the evolutionary process: III. social evolution. *Evolution* 64, 2558–74.

Medawar, P. B. (1952). *An Unsolved Problem of Biology*. H. K. Lewis, London.

Michod, R. E. and Hamilton, W. D. (1980). Coefficients of relatedness in sociobiology. *Nature* 288, 694–7.

Mitchell-Olds, T. and Shaw, R. G. (1987). Regression analysis of natural selection: Statistical inference and biological interpretation. *Evolution* 41, 1149–61.

Mock, D. W. and Parker, G. A. (1997). *The Evolution of Sibling Rivalry*. Oxford Univeristy Press, Oxford.

Moore, A. J., Brodie, E. D., and Wolf, J. B. (1997). Interacting phenotypes and the evolutionary process: I. Direct and indirect genetic effects of social interactions. *Evolution* 51, 1352–62.

Morrissey, M. B. and Hadfield, J. (2011). Directional selection in temporally replicated studies is remarkably consistent. *Evolution* 66, 435–42.

Morrissey, M. B., Kruuk, L. E. B., and Wilson, A. J. (2010). The danger of applying the breeder's equation in observational studies of natural populations. *Journal of Evolutionary Biology* 23, 2277–88.

Phillips, P. C. and Arnold, S. J. (1989). Visualizing multivariate selection. *Evolution* 43, 1209–22.

Platenkamp, G. A. J. and Shaw, R. G. (1993). Environmental and genetic maternal effects on seed characters in *Nemophila menziesii*. *Evolution* 47, 540–55.

Price, G. R. (1972). Extension of covariance selection mathematics. *Annals of Human Genetics* 35, 485–90.

Price, T. (1998). Maternal and paternal effects in birds: Effects on offspring fitness. In T. Mousseau, and C. W. Fox, eds. *Maternal Effects as Adaptations*. Oxford University Press, New York.

Queller, D. C. (1992). A general model for kin selection. *Evolution* 46, 376–80.

Räsänen, K. and Kruuk, L. E. B. (2007). Maternal effects and evolution at ecological time-scales. *Functional Ecology* 21, 408–21.

Rausher, M. D. (1992). The measurement of selection on quantitative traits—biases due to environmental covariances between traits and fitness. *Evolution* 46, 616–26.

Réale, D., Berteaux, D., McAdam, A. G., and Boutin, S. (2003). Lifetime selection on heritable life-history traits in a natural population of red squirrels. *Evolution* 57, 2416–23.

Roach, D. A. and Wulff, R. D. (1987). Maternal effects in plants. *Annual Review of Ecology and Systematics* 18, 209–35.

Robertson, A. (1966). A mathematical model of culling process in dairy cattle. *Animal Production* 8, 95–108.

Roff, D. (1992). *Life History Evolution*. Sinauer, Sunderland, MA.

Rutledge, J. J., Eisen, E. J., Robison, O. W., and Legates, J. E. (1972). Dynamics of genetic and maternal effects in mice. *Journal of Animal Science* 35, 911–18.

Schluter, D. (1988). Estimating the form of natural selection on a quantitative trait. *Evolution* 42, 849–61.

Schluter, D. and Nychka, D. (1994). Exploring fitness surfaces. *American Naturalist* 143, 597–616.

Shaw, R. G. and Byers, D. L. (1998). Genetics of maternal and paternal effects. In T. Mousseau, and C. W. Fox, eds. *Maternal Effects as Adaptations*. Oxford University Press, New York.

Smiseth, P. T., Wright, J., and Kolliker, M. (2008). Parent-offspring conflict and co-adaptation: behavioural ecology meets quantitative genetics. *Proceedings of the Royal Society of London, Series B* 275, 1823–30.

Smith, C. C. and Fretwell, S. D. (1974). Optimal balance between size and number of offspring. *American Naturalist* 108, 499–506.

Stearns, S. (1992). *The evolution of life-histories*. Oxford University Press, Oxford.

Tallis, G. M. (1989). The effects of selection and assortative mating on genetic-parameters. *Journal of Animal Breeding and Genetics-Zeitschrift Fur Tierzuchtung Und Zuchtungsbiologie* 106, 163–79.

Tallis, G. M. and Leppard, P. (1987). The joint effects of selection and assortative mating on a single polygenic character. *Theoretical and Applied Genetics* 75, 41–5.

Tallis, G. M. and Leppard, P. (1988). The joint effects of selection and assortative mating on multiple polygenic characters. *Theoretical and Applied Genetics* 75, 278–81.

Thiede, D. A. (1998). Maternal inheritance and its effect on adaptive evolution: A quantitative genetic analysis of maternal effects in a natural plant population. *Evolution* 52, 998–1015.

Walsh, B. and Lynch, M. (2012). *Evolution and Selection of Quantitative Traits: I. Foundations*. Sinauer, Sunderland, MA.

Willham, R. L. (1963). The covariance between relatives for characters composed of components contributed by related individuals. *Biometrics* 19, 18–27.

Willham, R. L. (1972). The role of maternal effects in animal breeding: III. Biometrical aspects of maternal effects in animals. *Journal of Animal Science* 35, 1288–93.

Wilson, A. J. and Réale, D. (2006). Ontogeny of additive and maternal genetic effects: Lessons from domestic mammals. *American Naturalist* 167, E23–E38.

Wilson, A. J., Coltman, D. W., Pemberton, J. M., Overall, A. D. J., Byrne, K. A., and Kruuk, L. E. B. (2005a). Maternal genetic effects set the potential for evolution in a free-living vertebrate population. *Journal of Evolutionary Biology* 18, 405–14.

Wilson, A. J., Pilkington, J. G., Pemberton, J. M., Coltman, D. W., Overall, A. D. J., Byrne, K. A., and Kruuk, L. E. B. (2005b). Selection on mothers and offspring: Whose phenotype is it and does it matter? *Evolution* 59, 451–63.

Wolf, J. B. and Brodie, E. D. (1998). The coadaptation of parental and offspring characters. *Evolution* 52, 299–308.

Wolf, J. B. and Wade, M. J. (2001). On the assignment of fitness to parents and offspring: whose fitness is it and when does it matter? *Journal of Evolutionary Biology* 14, 347–56.

Wolf, J. B., Brodie, E. D., and Moore, A. J. (1999). Interacting phenotypes and the evolutionary process: II. Selection resulting from social interactions. *American Naturalist* 153, 254–66.

CHAPTER 16

Parent–offspring co-adaptation

Mathias Kölliker, Nick J. Royle, and Per T. Smiseth

16.1 Introduction

Many traits expressed in animal families can be described as interacting phenotypes (Moore et al. 1997) that emerge from the social interactions among individuals in the family. Such traits depend not only on the individual's own genotype (termed direct genetic effects), but also on the genotypes that make up the family environment (termed indirect genetic effects or IGEs; Moore et al. 1997; Wolf et al. 1998; Wolf et al. 1999). IGEs occur when the genotype of one individual affects the expression of genes in another individual, with the effect mediated by the social interaction. They thereby add to the phenotypic variation between individuals upon which selection acts. Interacting phenotypes cannot be attributed unambiguously to a genotype or individual, so the phenomenon of IGEs complicates the common perspective held in behavioural ecology that the evolution of (behavioural) phenotypes can be studied 'as if there were a haploid locus at which each distinct strategy was represented by a distinct allele' (Grafen 1984; often referred to as the phenotypic gambit assumption).

There are several reasons why IGEs may be particularly important in the evolution of parental care and family interactions. First, the often highly repeated transfers of parental resources combined with offspring traits to solicit care means that indirect genetic effects should be particularly effective (Cheverud and Moore 1994). Second, parental care is associated with a highly stable and structured group composition comprising one or both parents and (normally) their genetic offspring. In contrast, many other social groups may show less stability and greater variation in group composition in terms of genotypes, in which case the effect of IGEs are expected to be more ephemeral in nature. Third, interactions within families can take place in three different social dimensions, offering ample scope for IGE's: between parents and offspring, between competing/cooperating offspring, and between male and female parents (Parker et al. 2002). These interactions can involve a wide array of traits, ranging from resources and signals to aggression (Mock and Parker 1997). Many of these traits have evolved as adaptations to family life, and they are likely to have fitness consequences for other family members because they are adapted to influence each other. For instance, parental care evolved because it enhances offspring fitness, and offspring subsequently evolved means for influencing the provision of that parental care affecting parental fitness in turn (Godfray 1995; Mock and Parker 1997; Godfray and Johnstone 2000; Parker et al. 2002; Royle et al. 2002; Chapters 7 and 8). Finally, what distinguishes families from other kin-groups is the fact that parents and offspring are not different categories of related individuals, but two life-stages of individuals that overlap across generations. Individuals compete with kin during one life-stage (as offspring), and then act altruistically towards kin in another (as parent). Parental care as the altruistic trait is therefore only expressed in individuals that were successful competitors in the offspring life-stage (Alexander 1974; Lynch 1987; Lundberg and Smith 1994; Kölliker et al. 2010). This aspect raises the question of how alleles at loci that affect offspring and parental traits co-segregate within genomes and are co-inherited from one generation to the next, and how such linkage matters in the evolution of parental care and family interactions (Cheverud and Moore 1994).

The Evolution of Parental Care. First Edition. Edited by Nick J. Royle, Per T. Smiseth, and Mathias Kölliker.

16.2 Offspring performance traits that convert parental care into offspring fitness

The definition of parental care as 'any parental trait that increases the fitness of a parent's offspring, [...]' (Chapter 1) does not specify the form of care provided by parents or how the fitness benefit of care is mediated. In reality, parental care typically affects offspring fitness indirectly by influencing an offspring trait that converts care into offspring fitness (see also Chapter 15). Such offspring traits are an integral part of the functional context for a particular form of parental care that specifies how exactly offspring benefit from care. Examples for such offspring performance traits (Arnold 1983) include growth and developmental rate in the context of food provisioning (Cheverud 1984) and aggregation with parents in the context of antipredator defence (Cocroft 1999).

Early evolutionary genetic models of parental effects incorporated offspring performance traits (Cheverud 1984; Lynch 1987; Kirkpatrick and Lande 1989; Queller 1992; Cheverud and Moore 1994; Wade 1998), and assumed that they were heritable (i.e. through direct genetic effects) and affected by a heritable parental effect (i.e. through indirect genetic effects) (Chapter 15). The genetic covariance between the parental trait and the corresponding offspring performance trait (the direct-maternal genetic covariance: Cheverud and Moore 1994) was theoretically shown to have a profound impact on the evolution of parental care and the co-evolution of the parent and offspring traits. For instance, under certain assumptions for sign and magnitude of this covariance, the models predicted that the offspring trait should evolve in the opposite direction of that predicted by selection (Cheverud and Moore 1994).

16.3 Trait-based Hamilton's rule and parent–offspring covariances

In its most general form, parental care is predicted to evolve according to Hamilton's rule; that is, if $r^*B > C$ (Hamilton 1964; Chapter 1), where r is the genetic relatedness between parent and offspring, B the fitness benefit of care to offspring, and C the cost of care to the parent. When focusing on a particular form of care and the corresponding offspring performance trait converting care into offspring fitness, a trait-based version of Hamilton's rule can be defined as

$$\left(r + \frac{G_{op}}{G_{pp}}\right) B > C \qquad (16.1)$$

(modified from Cheverud 1984; Wolf 2003). In this equation, G_{op} is the genetic covariance between a parental care trait and the corresponding offspring performance trait, and G_{pp} the additive genetic variance in the parental care trait. A positive covariance (e.g. when individuals that grow faster as offspring also provide more food as parents) facilitates the evolution of care. This is because offspring expressing a high fitness phenotype (e.g. high growth rate) receive the added benefits from more care, and become parents that provide high levels of care themselves. Conversely, a negative value for this covariance (e.g. when individuals that grow less as offspring provide more food as parents) impedes the evolution of parental care (Cheverud 1984; Lynch 1987; Cheverud and Moore 1994). The added benefit of more care is reduced by the offspring's own lower fitness phenotype, which reduces the chance that individuals with high-care genotypes survive into adulthood.

In summary, the genetic covariance between parental care and the corresponding offspring performance trait converting care into offspring fitness is key to understanding the evolution of traits that reflect specific forms of parental care (Eqn 16.1; Wolf 2003). Traditionally genetic relatedness is the core parameter for many predictions about the evolution of parental care, family conflict, and other social interactions (e.g. Gardner et al. 2011). Compared to genetic relatedness, the genetic covariance between parental care and offspring traits has largely been ignored, both theoretically and experimentally.

16.4 Evolution of the parent–offspring covariance

One likely reason why the covariance between parental care and offspring traits has not played a more prominent role is the lack of explicit explanations for why a genetic covariance between parental care and offspring traits should be expected. One justification for genetic dependencies between parental and offspring traits was that they may occur by pleiotropy and act as an evolutionary constraint in the evolution of parent and offspring traits (e.g. Lynch 1987). This argument is unsatisfactory because the genetic architecture of trait expression, including their genetic variances and covariances or epigenetic signatures, are also expected to evolve in response to selection (e.g. Grafen 1984; Hansen 2006). If so, genetic covariances between parental care and offspring traits may at least in part reflect the outcome of past patterns of selection. As a consequence, and contrary to genetic relatedness that can be inferred from pedigrees or molecular markers, it is not possible *a priori* to predict the sign and magnitude of the genetic covariance, and their inclusion in quantitative genetic models has therefore been based on rather arbitrary reasoning.

Co-adaptation models address this problem by exploring how the genetic variances and covariances of interacting parent and offspring traits are predicted to evolve, and how they reflect an adaptive evolutionary outcome as a result of selection acting on family interactions. It is in principle possible to predict the covariances required to parametrize the trait-based Hamilton's rule to study the evolution of parental care and family interactions using such models.

In the following sections, we explore the theoretical framework of co-adaptation models in more detail and discuss the ultimate reasons for selection on variances and covariances in the evolution of family interactions. We further discuss critical tests and review the empirical evidence for parent–offspring co-adaptation. Finally, we discuss the relationship between co-adaptation and conflict resolution models, how they are related, how they differ, and how they may be integrated to provide a more general theory of parent–offspring co-evolution.

16.5 Co-adaptation models

The term parent–offspring co-adaptation portrays the reciprocal nature of adaptation in animal families: as offspring, individuals are under selection to adapt to the component of the family environment defined by their parents' care, and as parents they are under selection to adapt to the component of the family environment defined by their offspring's traits.

16.5.1 Levels of analysis

For any population there is selection both for an adaptive offspring genotype given the average environment created by parents, and for an adaptive parental genotype given the average environment created by offspring. It is generally expected that ecology, life-histor,y and mating system should affect the optimal parent and offspring phenotype as determined by the benefits of care to offspring and the cost of care to parents. This includes selection for well-integrated parent and offspring phenotypes through, for example, effective communication of resource requirements by offspring following birth and/or hatching (Horn and Leonard 2002), hormonal interactions in the placenta (Wolf and Hager 2006), and selection arising from parent–offspring conflict. Resolution models of parent–offspring conflict make predictions about mean phenotypes involved in offspring begging and the provisioning of parental care, and these mean phenotypes are assumed to also reflect the underlying mean genotypes (Chapter 7). Hence, the state of an evolutionarily resolved conflict also reflects parent and offspring genotypes co-adapted within the genome (Kilner and Hinde 2008; Hinde et al. 2010; Kölliker et al. 2010).

At this level of analysis, co-adaptation may have occurred when variation between populations, with respect to mean trait values of parental care and offspring traits, are due to genetic differences (Linksvayer 2007; Qvarnström et al. 2007; Schrader and Travis 2009). However, this level does not tackle the main goal of co-adaptation theory which is to understand how selection acts on the genetic variances and covariances of parent and offspring traits within populations. To predict the selective forces on the covariances, the essential evolutionary component is correlational selection (Lande and Arnold 1983; Phillips and Arnold 1989), which occurs when selection favours particular combinations of traits rather than individual traits in isolation from selection on other traits.

16.5.2 Sources of correlational selection in families

There are two main sources of correlational selection in the case of parent–offspring interactions (see Box 16.1 for details). The first is due to ongoing directional selection on both parent and offspring traits, which results in selection for or against particular trait combinations during the evolutionary process (Brodie and McGlothlin 2006). Given that this pattern of selection is determined by the combined effects of directional selection acting on different traits, it is expected to occur in populations that are not in evolutionary equilibrium.

A second source of correlational selection is expected to be effective irrespective of whether a population is at equilibrium or not. It occurs

Box 16.1 Effects of selection on genetic variances and covariances

How directional and non-linear selection affects the evolution of genetic variances and covariances among multiple phenotypic traits is defined by the equation

$$\Delta\mathbf{G} = \mathbf{G}\left(\boldsymbol{\gamma} - \boldsymbol{\beta}\boldsymbol{\beta}^{\mathrm{T}}\right)\mathbf{G}. \quad (16.1.1)$$

G is the matrix of genetic variances and covariances (**G**-matrix), $\Delta\mathbf{G}$ the change in the **G**-matrix due to selection but before recombination (i.e. before the next generation of individuals in the population is produced), $\boldsymbol{\gamma}$ the matrix of non-linear selection gradients, and $\boldsymbol{\beta}$ the vector of directional selection gradients (Lande 1980; Tallis and Leppart 1988; Philipps and McGuigan 2006; see also Chapter 15). The superscript 'T' denotes vector transposition. This equation isolates the effect of selection on the variances and covariances. However, recombination breaks down genetic covariances, and the quantitative evolution of the G-matrix across generations towards a state of stable mutation-selection balance can be solved iteratively (Wolf and Brodie 1998; Kölliker et al. 2005), or by approximation (Lande 1980; Tallis and Leppart 1988). More extended versions include the explicit effect of mutation including pleiotropic mutations, assortative mating, migration, and recombination and are presented and explained in (Philipps and McGuigan 2006).

Assuming the simplest possible multivariate case of two traits (an offspring trait indicated by the subscript 'o', and a parental trait indicated by the subscript 'p'), and expanding the inner bracket of the equation above (i.e. the selection part) we obtain

$$\left(\boldsymbol{\gamma} - \boldsymbol{\beta}\boldsymbol{\beta}^{\mathrm{T}}\right) = \left[\begin{pmatrix} \gamma_{oo} & \gamma_{op} \\ \gamma_{po} & \gamma_{pp} \end{pmatrix} - \begin{pmatrix} \beta_o \\ \beta_p \end{pmatrix} \begin{pmatrix} \beta_o & \beta_p \end{pmatrix}\right]$$
$$= \begin{pmatrix} \gamma_{oo} - \beta_o^2 & \gamma_{op} - \beta_o\beta_p \\ \gamma_{po} - \beta_p\beta_o & \gamma_{pp} - \beta_p^2 \end{pmatrix}$$

The gradients on the diagonal of the matrices specify the total selection on genetic variances of the two traits, the gradients on the off-diagonal selection on the genetic covariance between the two traits. γ_{oo} and γ_{pp} are the non-linear selection gradients on an offspring and a parental trait (stabilizing for negative values or disruptive for positive values), and β_o and β_p the directional selection gradients. γ_{op} (equivalent to γ_{po}) is the selection on trait combinations ('correlational selection') (Lande and Arnold 1983). This equation shows several things. First, and not surprisingly, non-linear stabilizing and directional selection reduces the genetic variance of traits (RHS: diagonal elements). Second, net correlational selection affecting genetic covariances depends on if and how the trait interaction affects fitness (γ_{op}), and on the patterns of directional selection on the two traits (Brodie and McGlothlin 2006) (see also Chapter 15). For example, in the absence of correlational selection ($\gamma_{op} = 0$), antagonistic directional selection on the two traits favours a positive genetic correlation, while directional selection in the same direction favours a negative genetic covariance (note that this prediction only strictly applies to non-interacting phenotypic traits. In the case of interacting phenotypes, the sign of the predicted covariance also depends on the nature of parent–offspring interaction, for example the strength of a parental effect; Chapter 15). Conversely, at evolutionary equilibrium no directional selection on the two traits is expected ($\beta_o = \beta_p = 0$), and the direction and extent of non-linear selection determines selection on the covariance (Fig. 16.1). While the population mean for traits may be evolutionarily stable, the **G**-matrix can continue to evolve in a population at apparent equilibrium, provided heritable variation and a source of correlational selection is maintained. Co-adaptation models show that social epistasis induces non-zero gradients for γ_{op} in the absence of extrinsic correlational selection (Fig. 16.1).

whenever the fitness-related traits of parents and offspring are heritable and interact in the sense that they reciprocally influence one another, and when the outcome of this interaction is under selection. From a genetic perspective, offspring traits affected by parental care are determined by an interaction between direct influences from genes expressed in offspring and indirect influences from genes expressed in parents. As a result, different combinations of parental and offspring genotypes may yield similar phenotypes or fitness (Wolf and Brodie III 1998). A number of terms have been used to name this socially mediated form of genetic interaction between loci, such as 'parent–offspring intermixing ability' (Wade 1998), 'intergenomic epistasis' (Linksvayer 2007), 'among-genotype epistasis' (Wolf 2000), 'social epistasis' (Linksvayer 2007), or, in the specific context of animal communication, 'signaller-receiver gene epistasis' (Sinervo et al. 2008). Here we will use the term 'social epistasis' because this term is not restricted to communicative interactions, specifies that the interaction is analysed at the genetic level, and that the traits parents and offspring use in interactions, including their sensitivities to each other, have evolved due to a social benefit (i.e. providing and obtaining care). Co-adaptation models study the consequences of social epistasis as a source for correlational selection on the genetic variances and covariances of parent and offspring traits.

16.5.3 Co-adaptation models

Feldman and Eshel (1982) were the first to propose the possibility of co-adaptive evolution in parent–offspring interactions, although the first formal model of parent–offspring co-adaptation was developed by Wolf and Brodie (1998) more than a decade later. Wolf and Brodie's model considers the evolution of the genetic covariance between an offspring trait under selection and a parental effect that influences the expression of the offspring trait. Because the parental effect was also assumed to be heritable, the parent had an indirect genetic effect on the offspring phenotype (denoted by the maternal effect coefficient m). The core feature of the model is that the genetic variance-covariance matrix (G-matrix) is allowed to evolve under selection on the offspring trait. In multivariate quantitative genetic theory, selection on the variances and covariances is defined by the matrix of non-linear selection gradients (Lande and Arnold 1983), with the elements on the diagonal representing selection on the trait variances, and the elements on the off-diagonal representing selection on the trait covariances (Box 16.1). Wolf and Brodie (1998) showed that a parental effect on this offspring trait generated correlational selection through social epistasis due to an interaction between a direct genetic effect on the offspring trait and an indirect genetic effect of the maternal effect on the offspring trait (compare off-diagonal elements of Fig. 16.1a and 16.1b). The sign and magnitude of selection on the covariance depended on the strength of selection on the parentally affected offspring trait, genetic relatedness r, and the sign and magnitude of the parental effect coefficient m.

Wolf and Brodie (1998) assumed that relatedness between the parent and its own offspring is 0.5. However, extra-pair paternity or intra-specific brood parasitism can lead to variation in average relatedness between parent and offspring below this value, limiting the scope for effective co-adaptation (Fig. 16.1). Relatedness has a trait-based meaning in co-adaptation models and determines the predictability of the different components of the family environment (i.e. their indirect genetic effects). This is different from inclusive fitness models, in which relatedness is defined as the likelihood of identity by descent for a particular allel (Michod 1982). In co-adaptation models, higher genetic relatedness implies a more predictable value for the heritable component of the parental environment for offspring (or of the offspring environment for parents) and, hence, stronger correlational selection for particular combinations of parent and offspring traits.

The initial co-adaptation model by Wolf and Brodie (1998) assumes that selection acted on the offspring trait only, and that the offspring are passively influenced by the parental effect. Kölliker et al. (2005) extended the model by including selection on the parent and reciprocal feedback between the offspring trait and parental care. In this model, parental care affects an offspring trait converting care into offspring fitness, and the offspring trait

(a) Non-interacting parent and offspring traits

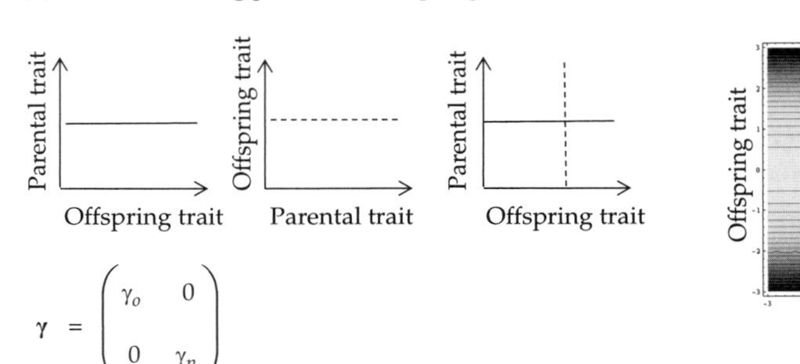

$$\gamma = \begin{pmatrix} \gamma_o & 0 \\ 0 & \gamma_p \end{pmatrix}$$

(b) Parental effect on offspring trait

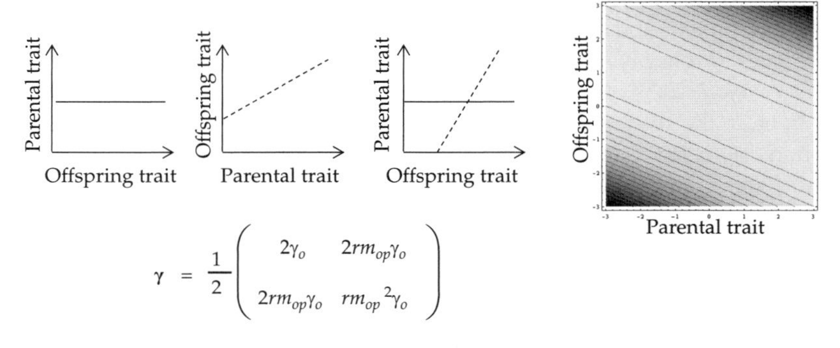

$$\gamma = \frac{1}{2}\begin{pmatrix} 2\gamma_o & 2rm_{op}\gamma_o \\ 2rm_{op}\gamma_o & rm_{op}{}^2\gamma_o \end{pmatrix}$$

(c) Interacting parent and offspring traits

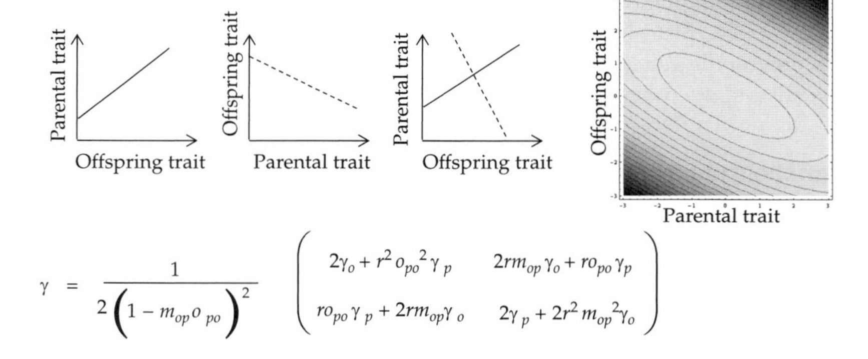

$$\gamma = \frac{1}{2\left(1 - m_{op}o_{po}\right)^2}\begin{pmatrix} 2\gamma_o + r^2 o_{po}{}^2\gamma_p & 2rm_{op}\gamma_o + ro_{po}\gamma_p \\ ro_{po}\gamma_p + 2rm_{op}\gamma_o & 2\gamma_p + 2r^2 m_{op}{}^2\gamma_o \end{pmatrix}$$

Figure 16.1 Social epistasis and correlational selection. (a), (b), and (c) are composed similarly, starting from left with reaction norm plots for how the parental trait depends on genetic variation in the offspring trait (solid line), how the offspring trait depends on genetic variation in the parent trait (dashed line), and finally what the expected trait expression is when the parent and offspring trait interact (corresponding to the intersection point). The equation below the reaction norm plots illustrates a simple formulation of stabilizing selection on the parent and offspring trait, respectively, and the contour plots graphically display this equation in terms of a fitness surface (light colour—high fitness, dark colour—low fitness). (a) Non-interacting parent and offspring traits. For the contour-plot we assumed selection on the offspring trait only (γ_o = -0.7). (b) Parental effect on offspring trait (following Wolf and Brodie 1998). Note the off-diagonal elements in the matrix of non-linear selection gradients which are solely due to social epistasis. The contour-plot shows how the orientation of the ridge changed due to the parental effect (compared to a). (c) Reciprocally interacting parent and offspring traits. Note how the off-diagonal element in the selection matrix becomes more complex and now also incorporates a term due to selection on the parent trait. Social epistasis operates now not only through the offspring trait, but also through the parental trait (following Kölliker et al. 2005). m_{op} denotes how the parental trait affects the offspring trait, and o_{po} how the offspring trait affects the parental trait.

in turn also affects parental care (a parental trait converting the offspring effect into parental fitness) resulting in reciprocal feedback (Kölliker 2003; Kölliker et al. 2005). The model was framed in terms of the interaction between provisioning parents and begging offspring, although it applies more generally to parent and offspring traits that influence each other. The correlational selection on parental care and the offspring trait now depends on the relative magnitude of selection on parent versus offspring and the signs and magnitudes of the coefficients of the parent and offspring effects (Fig. 16.1c).

How social epistasis generates correlational selection can be visualized by first plotting fitness surfaces and then exploring their orientation (Fig. 16.1; Phillips and Arnold 1989). Selection favours the alignment of the genetic covariance between parental care and offspring traits to the ridge of the fitness surface (Blows 2006) because the trait value combinations along this ridge yield highest and similar fitness. How well this alignment is achieved depends not only on selection, but also on the mode of reproduction and the genetic mechanisms underlying the covariance. For example, the orientation favoured by selection is opposed by the constraint of genetic recombination. A genetic covariance evolves most readily in the absence of genetic recombination, as under asexual reproduction. Under sexual reproduction and free recombination (i.e. unlinked loci), it can only evolve by linkage disequilibrium maintained by selection, which is disrupted to a substantial degree each generation (Sinervo and Svensson 2002; Conner and Hartl 2004; p. 159). Co-adaptation models show that selection from social epistasis can be strong enough to select for covariances of relevant magnitude despite free recombination (Wolf and Brodie 1998; Kölliker et al. 2005), but relatively strong selection and interaction effects are required. Clearly, from the perspective of parent–offspring co-adaptation, factors limiting recombination between the interacting parental and offspring traits should have a selective advantage by improving the possible alignment of the covariance to the orientation of the fitness surface. For instance, co-adaptation may select for genomic or transcriptomic reorganization of parent and offspring genes by favouring physical linkage of the loci or pleiotropy, which would effectively limit recombination and allow maximal alignment to the fitness surface. Alternatively, assortative mating due to limited dispersal and/or mating preferences may enhance linkage disequilibrium with respect to the loci expressing the parent and/or offspring trait independent of selection, and can thereby substantially enhance their co-adaptation (Kölliker et al. 2005). Finally, the evolution of epigenetic inheritance mechanisms of parent and offspring traits may have evolved partly to limit the consequences of genetic recombination weakening parent–offspring co-adaptation (see Section 16.4.4).

At the present time, our knowledge of the molecular genetic bases of parent–offspring interactions is still limited, except for a few model species. In rodents, evidence for pleiotropic effects of genes influencing maternal and offspring behaviour is increasing (Chapter 17). For example, in mice, the same gene affecting female milk let-down also affects the suckling efficiency of pups (*Peg3*; Curley et al. 2004), and in rhesus macaques the same gene affects both the degree of female maintenance of physical contact with its infant and the infant's demand for physical contact (mu-opioid receptor—*OPRM1*; Barr et al. 2008; Higham et al. 2011; Chapter 17). Co-adaptation models predict pleiotropic or linked genes (or the linked expression of genes) to be common, and further research on the genomic and transcriptomic organization of parent–offspring interaction will be of great interest and importance as key tests of co-adaptation models.

Co-adaptation models have usually treated the sensitivities of offspring traits to parental care and of parental care to offspring traits as evolutionarily fixed coefficients (Smiseth et al. 2008; but see Kölliker et al. 2010). This assumption is limiting since we expect parent and offspring effect coefficients to evolve as they can be thought of as adaptive strategies for how to respond to, and be affected by, other family members (Kölliker 2005; Bleakley et al. 2010). Future models should incorporate the coefficients of the parent and offspring effects as evolving traits. Using an approach based on interacting (behavioural) reaction norms where both the elevation and slope of the reaction norm can evolve (Smiseth et al. 2008; Dingemanse et al. 2010) would allow for analysis of correlational selection among these reaction norm components. Furthermore, current co-adaptation models have remained implicit about the ultimate reasons for particular assumed patterns of selection because the functional context of why parents and offspring respond to each other has not been explicitly incorporated (Smiseth et al. 2008; Chapter 15). The fusion of co-adaptation with conflict resolution models will be a major step ahead in the development of a more complete evolutionary theory for the evolution of parental care and family interactions.

16.5.4 Co-adaptation and epigenetic inheritance

Co-adaptation can generate not only evolutionary changes in genetic covariances between parental and offspring traits, but also in the epigenetic signature of gene expression of these traits. A recent model showed that selection via co-adaptation can favour the evolution of genomic imprinting; that is, the parent-of-origin specific expression of genes in offspring (Wolf and Hager 2006). Co-adaptation favours genomic imprinting because it provides a way of eliminating the disruptive effect of genetic recombination on the covariance. Thus, if imprinting evolved due to co-adaptation, the allele inherited from the caring parent is predicted to be expressed in offspring (Wolf and Hager 2006). This prediction is the opposite to the pattern of imprinting predicted by the kinship theory of genomic imprinting where the allele inherited from the non-caring parent (usually the male) should be expressed in offspring (provided there is multiple mating; Wilkins and Haig 2003). The mammalian placenta is characterized by a particularly high frequency of imprinted genes and both maternal and paternal imprinting are common. Wolf and Hager (2006) speculated that this variation may reflect epigenetic signatures of selection on particular loci from co-adaptation and on other loci from genetic conflict.

16.5.5 Other implications of parent–offspring co-adaptation

Parent–offspring co-adaptation is potentially important for additional reasons than those discussed above. First, in most species parental care comprises multiple forms of care. Each of them may affect a different offspring performance trait converting parental care into offspring fitness, and/or they are expressed at different stages of offspring development. As a consequence, co-adaptation is to be expected not only for each combination of parent–offspring traits, but also across different parental behaviours and offspring traits and across different stages of care. For example, in the burying beetle *Nicrophorus vespilloides*, the relative contribution of pre- and post-hatching parental care to offspring growth and survival was found to be co-adapted with a corresponding offspring stage-specific capacity for converting the particular form of care into growth and survival (Lock et al. 2007).

Second, as already mentioned, the typical parent–offspring interaction consists of a series of repeated interactions of resource provisioning by parents and resource demand by offspring. Parental supply and offspring demand should not be thought of as fixed traits, but rather as the outcome of interacting behavioural reaction norms (i.e. a supply reaction norm expressed in parents and a demand reaction norm expressed in offspring; Hussell 1988; Mock and Parker 1997; Smiseth et al. 2008). These behavioural reaction norms merely reflect the genetic basis of the rules that determine how parents and offspring react to variation in each other's traits. If these rules determine behaviours (e.g. parental food provisioning and offspring begging), they play a role in the behavioural dynamics over time (Dobler and Kölliker 2009). In reality, it appears unlikely that the repeated interactions precisely follow these rules, and deviations may occur regularly in the dynamics due to environmental noise or perceptual error. From a co-adaptation perspective, we should expect selection on behavioural reaction norms that allow parents and offspring to behaviourally stabilize interactions rapidly after such perturbations, thereby avoiding fitness penalties due to deviations from the behavioural equilibrium (Dobler and Kölliker 2009). Some support for this prediction comes from humans where unstable family environments (measured as insecure parent–offspring attachment) are correlated with traits of low reproductive expectations such as early menarche and first birth (Chisholm et al. 2005). The prediction that parent and offspring fitness are associated not only with mean levels of behaviour, but also with the variance in behaviour that is generated in the interaction over time, is a novel prediction based on co-adaptation logic (Dobler and Kölliker 2009), and requires further experimental work. Behavioural stability (low variance over time) may be one explanation for why co-adapted (or matched; see Section 16.3) parent–offspring interactions often yield higher parent and/or offspring fitness.

Third, co-adaptation theory has been framed in terms of genetic values for parental care and offspring traits. This choice is made primarily for reasons of tractability and to allow for specific focus on how social epistasis through parental care and parent–offspring interactions generates correlational selection. The level of complexity could of course be increased by substituting the fixed genetic value for parental care for a reaction norm determining how the expression of parental care is modified by extrinsic environmental conditions, such as food availability (Chapter 14). Hinde et al. (2010) present the first model for the evolution of trans-generational phenotypic plasticity through co-adaptation by including parental quality as an environmental term. They show that variation in this term leads to a pattern of co-adaptation between parental food provisioning and offspring begging that depends on parental quality. Under their assumption that parental quality is environmentally determined, this prediction reflects trans-generational plasticity, that is, an environmentally induced covariance between parent and offspring traits. However, if one assumes a heritable component to parental quality, the predicted covariance in their model could also be partly genetic. Thus, we need further theoretical development to better understand the conditions under which we expect parent—offspring co-adaptation to select for covariances that are due to trans-generational plasticity or genetic linkage.

Finally, co-adaptation models illustrate how parental care can be viewed as a 'socially complex phenotype', emerging as an outcome of family interactions and the heritable components of variation underlying them. From the perspective of direct and indirect genetic effects, the family can be viewed as a socially complex gene-expression network that may show some of the developmental and evolutionary properties that are well understood for traditional complex phenotypes that are expressed based on within-genome gene-expression networks. These properties may include canalization, robustness, and evolutionary stasis (Wagner 2005; Kölliker et al. 2010). Because interactions within families are usually partitioned into dyadic relationships (parent–offspring, sib–sib, parent–parent) there has been little explicit consideration of families as socially complex phenotypes to date, despite its potential importance to explain some of the diversity in family interactions.

16.6 Experimental evidence for parent–offspring co-adaptation

Experimental tests of co-adaptation theory are essentially quantitative genetic experiments (Bleakley et al. 2010) that aim to disentangle how combinations of parent and offspring traits create social epistasis and correlational selection, and that test for covariances between parent and offspring traits and epigenetic influences. Such experiments can be carried out without prior knowledge of the molecular basis underlying trait expression, and/or without prior knowledge of the heritability of parent and offspring traits (although evidence for co-adaptation hinges on the assumption that the underlying traits have a heritable basis). A very useful distinction in such experiments centres on whether the genetic origins of parents and offspring, and the family environment in which they are expressed, are matched or mis-matched. Matching and mis-matching are trait-based definitions based on similarity between genetic origin and the actual family environment, directly relating to the meaning of r (coefficient of relatedness) in co-adaptation models (Table 16.1). Matched parent–offspring combinations are similar by origin in trait value, while mis-matched combinations are dissimilar. Variation in matching can be best experimentally manipulated by cross-fostering.

A critical distinction has to be made with regard to the level at which variation in the genotypic origin is tested (see Section 16.4.1 above). For example, experimental designs using crosses between populations/inbred lines or among species test for co-adaptation at the level of differences in mean traits. Thus, they test for co-adaptation reflecting consequences of local adaptation, drift, or speciation. Although such crosses provide important insights into co-adaptation among parental and offspring traits, they are not directly addressing the key question as to whether correlational selection on parent and offspring traits within a population shapes the genetic architecture of parent–offspring interactions. To this end, it is important to focus on heritable variation in family environments within populations, and correlational selection is expected if there is an interaction between offspring genotype and the family environment (or social epistasis if the family environment is heritable).

16.6.1 Genotype x family environment interactions, social epistasis, and co-adaptation

Genotype x family environment interactions: We will first consider the broader field of study on genotype by family environment (G x FE) interactions. G x FE may reflect social epistasis if variation in the family environment is heritable. In other words, social epistasis is a sub-category of GxFE interactions. For example, van Noordwijk (1988) suggested that sibling competition could be a source of G x FE interactions if certain offspring genotypes perform better in competition with siblings in small broods, while other genotypes perform better in larger broods. If brood size is determined by the environment in which females reproduce (e.g. food availability), then there would be an offspring genotype x parental environment interaction on offspring competitive performance. However, if there is heritable variation in brood size, then social epistasis would generate selection for a genetic covariance between the offspring's competitive ability and brood size.

The study of G x FE interactions as a factor promoting variation between individuals in human and primate behavioural development has a relatively long tradition (Rutter 2006). Examples where the molecular basis of the interaction has been identified come from humans (*Homo sapiens*) and rhesus macaques (*Macaca mulatta*). In humans, breastfeeding has multiple environmental effects on child development. A recent study found that the effects of breastfeeding, as compared to the effects of bottle-feeding, on a child's intellectual development, measured as IQ scores of children at ages 7–13 years, depended on the genotype at the *FADS2* locus, a gene that is involved in fatty acid metabolism (Caspi et al. 2007). Breastfeeding was only correlated with IQ scores if the child was heterozygous or homozygous for the C allele, while there was no such correlation if the child was homozygous for the G allele. If variation between women in fatty acid content of breast milk is heritable, this form of a G x FE interaction could reflect social epistasis between the mother's fat metabolism, breastmilk production, and the offspring's cognitive development.

Another example involves polymorphism at the monoamine oxidase A (*MAOA*) gene. In both humans and rhesus macaques, offspring aggressiveness is partly determined by the environment provided by parents (parental maltreatment in humans, and maternal absence in macaques). The specific effect of the parental environment on the development of offspring aggressiveness depends upon allelic variation at the regulatory promoter region of the *MAOA* locus (Caspi et al. 2002; Newman et al. 2005). In humans, offspring aggression is more strongly influenced by parental maltreatment in children with a genotype for low *MAOA* expression activity than in children with a genotype for high *MAOA* expression activity. Similarly, in rhesus macaques, parental absence affects offspring aggression only in offspring with a low *MAOA* expression activity genotype (Newman et al. 2005). From a co-adaptation perspective, it would be very interesting to know if this same locus is associated with parental care styles in both humans and rhesus macaques.

Social epistasis: Crosses between inbred lines provide a straightforward experimental design for manipulating the genetic background of the family environment. In an experiment on mice specifically designed to disentangle the genetics of parental care (lactation) and mother–pup interactions, females and males from two inbred strains were first cross-bred and then provided with cross-fostered offspring from their own or a different strains (CBA and C57/B6; Hager and Johnstone 2003). The study found that, when female–pup interactions were mis-matched with respect to the maternal line-background (CBA nursing female and B6 maternal origin of pups, and *vice versa*), offspring received smaller quantities of milk than when female–pup interactions were matched (Fig. 16.2a). This result clearly shows social epistasis between mothers and their pups in terms of strategies for providing and demanding milk, respectively. The use of inbred crosses is a powerful experimental design for testing social epistasis between strains, but interpretation of the observed patterns in terms of co-adaptation requires some caution. Potential problems associated with long-term artificial inbreeding and inbreeding depression, or major life-history differences between lines which arose for unknown reasons during inbreeding (e.g. different litter sizes and birth weights; Hager and Johnstone 2003), may at least partly lead to strain differences in parent and offspring behaviour, making conclusions about the ultimate reason for apparent patterns of co-adaptation uncertain.

A recent study in the viviparous poeciliid fish species *Heterandria formosa* tested for parent–offspring co-adaptation between populations based on crosses within and between progeny of females caught from two distinct populations (Schrader and Travis 2009). In this study, one of the two original populations shows a markedly higher level of matrotrophy (maternal provisioning through a placenta-like structure) than the other. Assuming that this variation represented local adaptation to the respective environments in which the populations live, Schrader and Travis (2009) predicted co-adaptation between fish mothers and their offspring in terms of the amount of transferred nutrients (measured as birth weight of the fry). The study found a significant interaction between paternal and maternal population origin on the weight of newborn fry in the direction predicted by co-adaptation models: fry were lighter when paternal and maternal origins were mis-matched, indicating lower resource transfer through matrotrophy (Fig. 16.2b). However, because no cross-fostering could be done in this study, the effects of the environment provided by the mother (level of matrotrophy) were potentially confounded by the effects of genes passed from the female to the offspring. Hence, despite being consistent with the hypoth-

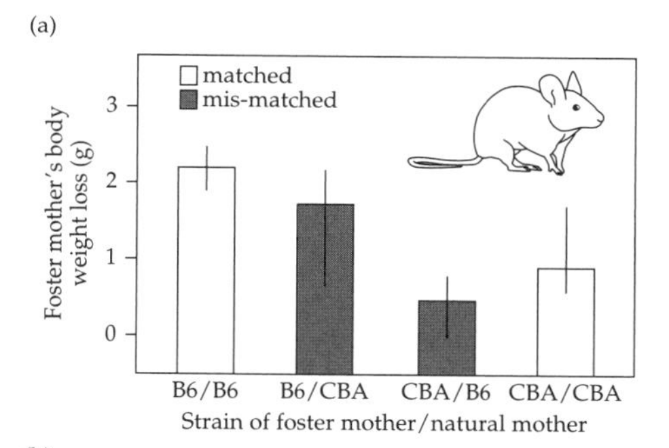

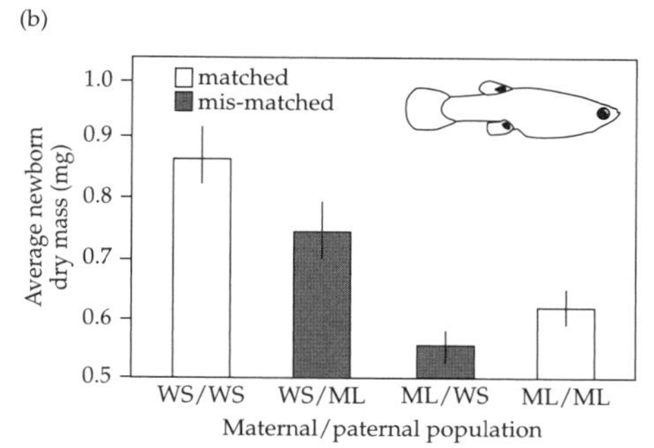

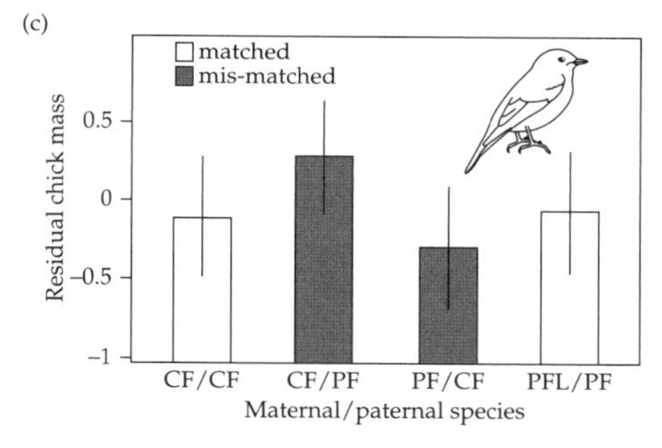

Figure 16.2 Experimental evidence for social epistasis in parent–offspring interactions. (a) Results from a cross-breeding and cross-fostering experiment among two inbred lines in mice (redrawn from Fig. 3 in Hager and Johnstone 2003; used with permission from the authors and Nature Publishing Group). When the offspring genotypes matched the genotype of their foster mother, the pups obtained more milk than when they were mis-mached (as indicated by a corresponding larger weight loss by the females). Shown are medians and interquartile ranges. (b) Results from a cross-breeding experiment in a matrotrophic poeciliid fish (redrawn from Fig. 4 in Schrader and Travis 2009; used with permission from the authors and John Wiley & Sons Ltd). Males and females from two populations differing in the amount of matrotrophy were cross-bred, and larvae born from matched breeding pairs showed higher dry mass (used as a proxy of maternal provisioning) than larvae born from mis-matched breeding pairs. Shown are means and standard errors. (c) Results from a cross-fostering experiment between collared and pied flycatchers (redrawn from Fig. 1a in Qvarnström et al. 2007; reprinted with permission from the authors and The Royal Society). Matching by species had no significant effect on chick growth. Shown are means and standard errors.

esis of social epistasis and co-adaptation, these results cannot be fully separated from the alternative explanation that offspring of mis-matched matings suffer from outbreeding depression due to crosses between individuals from two distinctly locally adapted populations (Lynch 1991).

A cross-fostering study among two closely related altricial bird species, the pied flycatcher (*Ficedula hypoleuca*) and collared flycatcher (*F. albicollis*), tested for between-species variation in parent–offspring co-adaptation (Qvarnström et al. 2007). One half of the chicks from each brood were cross-fostered between the two species shortly after hatching, while the other half was left in the natal nest. Contrary to the predictions from co-adaptation models, the results showed that the species in which nests the chicks had been placed was the only factor influencing offspring growth (Fig. 16.2b), and the species had an effect on chick begging that was independent of the effect of the species in which nest the chick had been placed. Thus, pied flycatchers appear to be the better parents, while collared flycatchers appear to have the more competitive chicks. There are several potential reasons for the lack of evidence for social epistasis in this study. Pied and collared flycatcher parents had very similar food provisioning rates (Qvarnström et al. 2007), limiting the scope for detecting social epistasis in this species comparison. Furthermore, designs based on complete cross-fostering (Hinde et al. 2010) are preferable in tests of social epistasis to the partial cross-fostering design used by Qvarnström et al. (2007). The reason is that complete cross-fostering designs generate experimental variation in the family environment for both offspring and parents, while partial cross-fostering designs generate experimental variation in the family environment for offspring only. Furthermore, the competitive interactions between chicks of the two species may also have changed the family environment in an unpredictable way, potentially blurring patterns of social epistasis and co-adaptation between parents and offspring.

Finally, a recent study used an elegant experimental design to test for co-adaptation based on within-population variation in offspring begging and parental provisioning in canaries (*Serinus canaria*) (Hinde et al. 2010). In this study, each pair was allowed to raise two broods per year, one of which was the parents' own offspring and the other of which was cross-fostered. For the cross-fostered broods, whole clutches were exchanged among nests. Offspring begging and growth was measured in both broods, which allowed any mis-match to be estimated as the deviation between the begging behaviour of cross-fostered broods and the begging behaviour in the parents' own brood. The study found a reduction in chick growth as a function of an increase in the degree of mis-match, and that this effect occurred regardless of the absolute levels of provisioning and begging and the direction of the mis-match (Hinde et al. 2010). Taking offspring growth as a proxy for offspring fitness, this result provides direct evidence of correlational selection favouring particular combinations of offspring begging behaviour and an unknown parental control behaviour that in turn results in matching between parent and offspring traits (Hinde et al. 2009). Further work is needed to establish whether the correlational selection favours covariation between parental provisioning and offspring begging is a result of an environmentally induced maternal effect (the scenario favoured by Hinde et al. 2009; Hinde et al. 2010) or genetic covariance.

Parent–offspring covariances: Estimation of covariances between maternal and offspring traits is commonly used in animal breeding and quantitative genetic studies on parental care (Cheverud 1984). The focus is usually on an offspring trait that is thought to be under selection (most frequently offspring growth or body mass), and that is affected by both direct genetic effects of genes expressed in offspring and indirect genetic effects due to heritable variation in maternal effects. Positive covariances have been reported between offspring growth and parental effects on offspring growth, for example in mice (*Mus musculus*) (Lynch 1987) and pigeons (*Columba livia*) (Aggrey and Cheng 1995). Conversely, paternal half-sib breeding experiments in burying beetles (*N. vespilloides*) (Rauter and Moore 2002) and dung beetles (*Onthophagus taurus*) (Hunt and Simmons 2002) showed small and statistically non-significant covariances, at least with regard to the sire (which estimates the additive genetic component of the covariance). The stronger estimated covariances with dams suggests that non-additive effects such as maternal effects may be important in these cases (Hunt and Simmons 2002).

The evidence for genetic correlations between parent and offspring traits that are involved in the reciprocal transactional interactions between parents and offspring (e.g. provisioning and begging) have been reviewed in some detail previously (Kölliker 2005; Kölliker et al. 2005; Smiseth et al. 2008), and we refer the reader to these reviews for further details. Positive covariances were found between maternal sensitivity to begging calls and offspring begging call intensity in great tits (*Parus major*) (Kölliker et al. 2000), between food provisioning and begging rate in the burying beetle *Nicrophorus vespilloides* (Lock et al. 2004), between maternal milk letdown and pup suckling efficiency in mice (Curley et al. 2004), and between food provisioning and begging intensity in canaries (Hinde et al. 2009). Conversely, negative covariances were found between maternal provisioning and offspring elicitation in the burrower bugs *Sehirus cinctus* (Agrawal et al. 2001), and between maternal and offspring contact behaviour in rhesus macaques (*Macacca mulatta*) (Maestripieri 2004). Variation in the sign of the covariance is partly predicted by patterns of selection on the parent–offspring interactions, that is, if the offspring is predominately under selection to adapt to the parental environment or vice versa. Patterns of selection may partly relate to a species' life-history (family size and longevity; Kölliker et al. 2005) and to whether parents or offspring behaviourally control the parent–offspring interaction (Hinde et al. 2010).

Not all studies testing for a genetic covariance between parental and offspring behaviour have found evidence in support of the prediction of covariance between parent and offspring. A recent artificial selection experiment in house sparrows (*Passer domesticus*) found no evidence for heritable variation in begging call intensity or genetic correlation with parental provisioning, either estimated from the response to artificial selection over four generations, or based on parent–offspring or sibling analysis (Dor and Lotem 2009).

In most of these studies the covariances were estimated based on cross-fostering experiments, which cannot distinguish between covariances induced by environmental maternal effects and those reflecting genetic covariances. Future experimental studies will have to dig deeper to hopefully elucidate the biological bases of the reported patterns of co-adaptation and covariation between parents and their offspring. This will require combined breeding and cross-fostering experiments (like the one by Hager and Johnstone 2003) and molecular studies that explore the genomic and transcriptomic organization of genes affecting parent and offspring traits (Chapter 17). To test for phenotypically plastic parent–offspring co-adaptation, experiments will be required that independently manipulate the environment to which parents are exposed prior to offspring production, and the environmental conditions under which they raise their offspring (Hinde et al. 2009).

16.7 Co-adaptation and conflict resolution

Why is it important to consider parent–offspring co-adaptation in future research on the evolution of family interactions? The question of how parent–offspring interactions should evolve has traditionally been addressed using game-theoretic conflict resolution models (Chapter 7). These models built on Trivers' (1974) seminal insight that evolutionary conflict between parents and offspring over the amount and duration of parental investment is expected due to asymmetries in relatedness. The zone of conflict (Godfray 1995) arises because offspring inclusive fitness is maximized at a higher level of parental investment than parental inclusive fitness (Trivers 1974). More recent models for the resolution of conflict are typically built on the usual form of Hamilton's rule (or a marginal value version of it; Godfray 1991, 1995). These models will have to be modified in order to focus on particular forms of parental care and their corresponding offspring performance traits that convert care into offspring fitness, because the genetic covariance between a specific form of parental care and the corresponding offspring trait can strongly change the evolutionary dynamics and possibly even the optimal endpoints of the co-evolutionary process (Feldman and Eshel 1982; West-Eberhard 1983; Cheverud 1984; Cheverud and Moore 1994; Kölliker et al. 2000). Currently, conflict resolution and co-adaptation models largely complement each other by focusing on different aspects of family interactions (Smiseth et al. 2008; Table 16.1). However, co-adaptation and conflict resolution are conceptually tightly linked (Chapters 7 and 15). Because models of parent–offspring co-adaptation have so far not been incorporated into conflict resolution theory, novel models are needed that explore how conflict resolution is altered by such fusion. As stated by Mock and Parker (1997) '[. . .], the model would be affected very radically if the parental and offspring strategies were prescribed by the same genes at the same locus. In reality, it would be rather unlikely that the very genes causing offspring to beg more or less vigorously would be equivalent to genes causing parents to provide more or less food [. . .]' (p. 33). Note that 'The model' here refers broadly to game-theoretic

Table 16.1 Comparison of assumptions and predictions of current parent–offspring co-adaptation and conflict resolution models

	Type of model	
	Co-adaptation	Conflict resolution
Assumptions		
Function of parental trait	Parental effect, evolutionary function not specified	Parental care traits, enhance offspring fitness
Fitness counting	Individual fitness; benefit component to offspring life-stage, cost (fecundity) component to parental life-stage	Inclusive fitness; separate for parents and offspring
Genetic relatedness *r*	Trait-based relatedness, predictability of trait values from parents to offspring and vice versa, important determinant of social epistasis	Identity-by-descent based relatedness, weighing of indirect fitness; determines the strength of kin selection and, hence, conflict
Effect of interaction	Combined effect of parent and offspring genotype on phenotypes, and in turn fitness	Combined effect of parent and offspring phenotypes on inclusive fitness
Interaction mechanisms	Fixed parental and/or offspring effect coefficients, evolving parental BRNs	Fixed response rules, evolving parental response
Predictions		
Genetic architecture	1) Genetic linkage of interacting parental and offspring traits 2) Evolutionary robustness of parental care	1) Trans-generational phenotypic plasticity of interacting parental and offspring traits 2) n/a
Function of offspring trait	Fitness conversion (e.g. growth/development), begging: sensory exploitation	Honest signal/competitive trait
Effect of sibling rivalry	Enhanced (under escalation) or reduced (under negotiation) selection on genetic linkage*	Escalation of begging, begging restraint under sibling negotiation or cooperation
Effects of reduced relatedness	1) Weakened social epistasis → reduced correlational selection 2) Reduced offspring and/or parent fitness due to poorer match between parental and offspring genotypes	1) Lower indirect fitness → broader zone of conflict, and enhanced selection for exaggerated begging 2) Reduced offspring and/or parent fitness due to kin recognition, higher levels of begging, and/or reduced parental care
'Weaning'	Higher behavioural instability (variance in time dynamics of interactions)	Escalated begging, reduced provisioning
Epigenetics	Maternally imprinted growth enhancers expressed in offspring	Paternally imprinted growth enhancers expressed in offspring

*M.K. unpublished results

conflict resolution models. As argued earlier, not only pleiotropic genes change how selection on parents and offspring can lead to evolutionary change in family interactions, but also any genetic covariance between parent and offspring traits, no matter its molecular basis, at least in the short-term. Mock and Parker (1997) discounted the importance of genetic covariances by arguing that pleiotropy would be unlikely. If we think of genetic covariances as the result of pleiotropic *de novo* mutations that subsequently act as a constraint on adaptive evolution, Mock and Parker's (1997) argument may be warranted. Selection should eliminate such mutations quickly from the gene pool of a population.

However, as argued throughout this chapter, genetic covariances may actually be common in the presence of correlational selection and co-adaptation because they are adaptive and, hence, maintained by selection. If so, a complete theory for the evolution of parental care and family interactions cannot ignore them. Theoretical predictions and accumulating empirical evidence indicate that correlated parent and offspring traits may actually be the rule rather than the exception, so more explicit consideration of genetic trait architecture needs to be given in future work on the resolution of parent–offspring conflict.

16.7.1 Genetic covariance and parent–offspring conflict

A brief sketch of how integration of insights from co-adaptation could affect our understanding of how conflicts are resolved is presented here (without claiming a general solution). If we include the covariance between a particular form of parental care and the corresponding offspring performance trait in models of parent–offspring conflict according to Eqn 16.1, there is a change in fitness functions due to the scaling of the benefits of care B by the parent–offspring covariance (more precisely by G_{op}/G_{pp}) and, consequently, the optima for parents and offspring. This shift occurs because the same amount of change in a particular form of parental care (e.g. protection against predation) leads to higher fitness benefits if the offspring also inherit the corresponding offspring trait that effectively converts this care into offspring fitness (e.g. a strong tendency to aggregate with the parent). This scaling effect on offspring fitness implies that a genetic covariance that evolves due to co-adaptation can facilitate or hinder the evolution of this form of care and change the width of the zone of conflict, depending on its sign and magnitude potentially destabilize the usual fitness optima. These effects are due to an indirect response to selection in the genetically correlated offspring trait, and could be understood as a form of runaway selection between parents and offspring—at least for positive covariances (West-Eberhard 1983; Wade 1998; Kölliker et al. 2000; Kölliker and Richner 2001). However, selection on the covariance is expected to change as the population approaches the fitness optimum from congruent directional selection on parent and offspring traits to antagonistic (within the zone of conflict) and non-linear (stronger curvature) selection. This change in selection should alter the pattern of correlational selection on parent and offspring traits dynamically (Box 16.1) and along with it, the sign and/or magnitude of the genetic covariance. How quickly this can occur will depend on the degree of recombination among genes affecting parent and offspring traits. As a consequence, the functions depicting parent and offspring inclusive fitness cannot be assumed to remain static during evolution of the genetic covariance between parents and offspring. To solve for an evolutionarily stable state under such dynamic feedback between selection, genetic architecture, and the responses to selection will require further theoretical development.

16.8 Conclusions and future directions

In this chapter we have shown that an important component in the evolution of family interactions is correlational selection on parent and offspring traits which favours their co-adaptation; that is, their genetic or phenotypic linkage. In the case of parent–offspring interactions, such covariances quantify associations within genomes between how individuals compete for parental resources as offspring and how they subsequently provide care as parents. At present there is no comprehensive theory for the evolution of family interactions where such effects are accounted for. But by using a trait-based form of Hamilton's rule to study the co-evolution of a particular form of parental care and its corresponding offspring performance trait that converts this form of care into offspring fitness, it can be shown that genetic covariances between parents and offspring can affect the evolutionary origin of care, critically change evolutionary trajectories, the optima for parental care, and the zone of conflict. From a theoretical perspective, further progress in combining parent–offspring co-adaptation and conflict resolution models is important in order to obtain a better understanding of the evolution of parental care and family interactions.

From an experimental perspective, our understanding of parent–offspring co-adaptation would be furthered by research within the four following areas: First, there is a need for further work on the role of social epistasis in the expression of parental care and offspring traits, particularly with respect to parent and offspring fitness-related traits in relation to variation between individuals within populations, as opposed to between populations or species. This is the basis for correlational selection and the ultimate reason why we expect parent and offspring traits to become genetically linked. Furthermore, by measuring the outcome of matched versus mis-matched parent–offspring interactions at the level of both parent and offspring fitness, we gain an insight into whether parents or offspring control the co-evolutionary process (Hinde et al. 2010), a distinction of major importance in understanding how family conflicts are actually resolved (Royle et al. 2002, 2004). Second, more work is also needed on the strength and direction of correlational selection combined with estimates of parent–offspring covariance. Such studies can be based on cross-breeding and cross-fostering experiments where matched and mis-matched parent–offspring interactions are experimentally generated, and would provide direct tests of parent–offspring co-adaptation if the direction of correlational selection and the sign of the covariance are congruent. Third, the mechanism underlying parent–offspring covariance needs to be elucidated more broadly. Except for a few examples of genes in laboratory rodents, little is known about whether parent–offspring covariances are due to pleiotropy, physical gene linkage, linkage disequilibrium, or trans-generational phenotypic plasticity. This will require the full range of quantitative and molecular genetic methods to narrow down patterns of covariance to genetic loci, environmental factors, and hormonal mechanisms involved in the expression of the covariance (see also Chapters 3, 7, and 14). Finally, more studies of epigenetic effects in parent–offspring interactions in addition to those on the mammalian placenta are necessary. Such studies are essential to establish what the role of co-adaptation and conflict are in shaping selection on epigenetic signatures of gene expression.

Acknowledgements

We thank Allen Moore, James Curley, Rufus Johnstone, and Tobias Uller for their helpful comments on earlier drafts of the chapter, and Dita Vizoso for making the animal drawings in Fig. 16.2.

References

Aggrey, S. E. and Cheng, K. M. (1995). Genetic correlation between genetic and parental effects on growth in pigeon squabs. *Journal of Heredity* 86, 70–2.

Agrawal, A. F., Brodie III, E. D., and Brown, J. (2001). Parent-offspring coadaptation and the dual genetic control of maternal care. *Science* 292, 1710–12.

Alexander, R. D. (1974). The evolution of social behavior. *Annual Review of Ecology and Systematics* 5, 325–83.

Arnold, S. J. (1983). Morphology, performance and fitness. *American Zoologist* 23, 347–61.

Barr, C. S., Schwandt, M. L., Lindell, S. G., et al. (2008). Variation at the mu-opioid receptor gene (OPRM1) influences attachment behavior in infant primates. *Proceedings of the National Academy of Sciences USA* 105, 5277–81.

Bleakley, B. H., Wolf, J. B., and Moore, A. J. (2010). The quantitative genetics of social behaviour. In T. Székely, A. J. Moore, and J. Komdeur, eds. *Social Behaviour: Genes, Ecology and Evolution*, pp. 29–54. Cambridge University Press, Cambridge.

Blows, M. W. (2006). A tale of two matrices: multivariate approaches in evolutionary biology. *Journal of Evolutionary Biology* 20, 1–8.

Brodie III, E. D. and McGlothlin, J. W. (2006). A cautionary tale of two matrices: the duality of multivariate abstraction. *Journal of Evolutionary Biology* 20, 9–14.

Caspi, A., McClay, J., Moffitt, T. E., et al. (2002). Role of genotype in the cycle of violence in matreated children. *Science* 297, 851–4.

Caspi, A., Williams, B., Kim-Cohen, J., et al. (2007). Moderation of breastfeeding affects on the IQ by genetic variation in fatty acid metabolism. *Proceedings of the National Academy of Sciences USA* 104, 18860–5.

Cheverud, J. M. (1984). Evolution by kin selection: a quantitative genetic model illustrated by maternal performance in mice. *Evolution* 38, 766–77.

Cheverud, J. M. and Moore, A. J. (1994). Quantitative genetics and the role of the environment provided by relatives in behavioral evolution. In C. R. B. Boake ed. *Quantitative Genetic Studies of Behavioral Evolution*, pp. 67–100. The University of Chicago Press, Chicago.

Chisholm, J. S., Quinlivan, J. A., Petersen, R. W., et al. (2005). Early stress predicts age at menarche and first birthj, adult attachment, and expected lifespan. *Human Nature* 16, 233–65.

Cocroft, R. B. (1999). Offspring-parent communication in a subsocial treehopper (Hemiptera: Membracidae: *Umbonia crassicornis*). *Behaviour* 136, 1–21.

Conner, J. K. and Hartl, D. L. (2004). *A Primer of Ecological Genetics*. Sinauer Associates, Inc., Sunderland, MA.

Curley, J. P., Barton, S., Surani, A., et al. (2004). Coadaptation in mother and infant regulated by a paternally expressed imprinted gene. *Proceedings of the Royal Society of London, Series B* 271, 1303–9.

Dingemanse, N. J., Kazem, A. J. N., Réale, D., et al. (2010). Behavioural reaction norms: animal personality meets individual plasticity. *Trends in Ecology and Evolution* 25, 81–9.

Dobler, R. and Kölliker, M. (2009). Behavioural attainability of evolutionarily stable strategies in repeated interactions. *Animal Behaviour* 77, 1427–34.

Dor, R. and Lotem, A. (2009). Heritability of nestling begging intensity in the house sparrow (*Passer domesticus*). *Evolution* 63, 738–48.

Feldman, M. W. and Eshel, I. (1982). On the thoery of parent-offspring conflict: a two-locus genetic model. *The American Naturalist* 119, 285–92.

Gardner, A., West, S. A., and Wild, G. (2011). The genetical theory of kin selection. *Journal of Evolutionary Biology* 24, 1020–4.

Godfray, H. C. J. (1991). Signalling of need by offspring to their parents. *Nature* 352, 328–30.

Godfray, H. C. J. (1995). Evolutionary theory of parent-offspring conflict. *Nature* 376, 133–8.

Godfray, H. C. J. and Johnstone, R. A. (2000). Begging and bleating: the evolution of parent-offspring signalling. *Philosophical Transactions of the Royal Society of London, Series B* 355, 1581–91.

Grafen, A. (1984). Natural selection, kin selection and group selection. In J. R. Krebs, and N. B. Davies, eds. *Behavioural Ecology*. 2nd ed., pp. 62–84. Blackwell Scientific Publishers, Oxford,

Hager, R. and Johnstone, R. A. (2003). The genetic basis of family conflict resolution in mice. *Nature* 421, 533–5.

Hamilton, W. D. (1964). The genetical evolution of social behaviour. *Journal of Theoretical Biology* 7, 1–52.

Hansen, T. F. (2006). The evolution of genetic architecture. *Annual Review in Ecology, Evolution and Systematics* 37, 123–57.

Higham, J. P., Barr, C. S., Hoffman, C. L., et al. (2011). Mu-opioid receptor (OPRM1) variation, oxytocin levels and maternal attachment in free-ranging rhesus macaques Macaca mulatta. *Behavioral Neuroscience* 125, 131–6.

Hinde, C. A., Buchanan, K. L., and Kilner, R. M. (2009). Prenatal environmental effects match offspring begging to parental provisioning. *Proceedings of the Royal Society of London, Series B* 276, 2787–94.

Hinde, C. A., Johnstone, R. A., and Kilner, R. M. (2010). Parent-offspring conflict and co-adaptation. *Science* 327, 1373–6.

Horn, A. G. and Leonard, M. L. (2002). Efficacy and the design of begging signals. In J. Wright and M. L. Leonard, eds. *The Evolution of Begging: Competition, Cooperation and Communication*, pp. 127–41. Kluwer Academic Publishers, Dordrecht.

Hunt, J. and Simmons, L. W. (2002). The genetics of maternal care: direct and indirect genetic effects on phenotype in the dung beetle *Onthophagus taurus*. *Proceedings of the National Academy of Sciences USA* 99, 6828–32.

Hussell, D. J. T. (1988). Supply and demand in tree swallow broods: a model of parent-offspring food-provisioning interactions in birds. *American Naturalist* 131, 175–202.

Kilner, R. M. and Hinde, C. A. (2008). Information warfare and parent-offspring conflict. *Advances in the Study of Behavior* 38, 283–336.

Kirkpatrick, M. and Lande, R. (1989). The evolution of maternal characters. *Evolution* 43, 485–503.

Kölliker, M. (2003). Estimating mechanisms and equilibria for offspring begging and parental provisioning. *Proceedings of the Royal Society of London, Series B (Suppl.)* 270, S110–13.

Kölliker, M. (2005). Ontogeny in the family. *Behavior Genetics* 35, 7–18.

Kölliker, M., Brinkhof, M. W. G., Heeb, P., et al. (2000). The quantitative genetic basis of offspring solicitation and parental response in a passerine bird with biparental care. *Proceedings of the Royal Society of London, Series B* 267, 2127–32.

Kölliker, M., Brodie Iii, E. D., and Moore, A. J. (2005). The coadaptation of parental supply and offspring demand. *American Naturalist* 166, 506–16.

Kölliker, M. and Richner, H. (2001). Parent-offspring conflict and the genetics of offspring soliciation and parental response. *Animal Behaviour* 62, 395–407.

Kölliker, M., Ridenhour, B. J., and Gaba, S. (2010). Antagonistic parent-offspring co-adaptation. *PLoS ONE* 5, e8606.

Lande, R. (1980). The genetic covariance between characters maintained by pleiotropic mutations. *Genetics* 94, 203–15.

Lande, R. and Arnold, S. J. (1983). The measurement of selection on correlated characters. *Evolution* 37, 1210–26.

Linksvayer, T. A. (2007). Ant species differences determined by epistais between brood and worker genomes. *PLoS ONE* 2, e994.

Lock, J. E., Smiseth, P. T., and Moore, A. J. (2004). Selection, inheritance and the evolution of parent-offspring interactions. *The American Naturalist* 164, 13–24.

Lock, J. E., Smiseth, P. T., Moore, P. J., et al. (2007). Coadaptation of prenatal and postnatal maternal effects. *American Naturalist* 170, 709–18.

Lundberg, S. and Smith, H. G. (1994). Parent-offspring conflicts over reproductive efforts: variations upon a theme by Charnov. *Journal of Theoretical Biology* 170, 215–18.

Lynch, M. (1987). Evolution of intrafamilial interactions. *Proceedings of the National Academy of Sciences USA* 84, 8507–11.

Lynch, M. (1991). The genetic interpretation of inbreeding depression and outbreeding depression. *Evolution* 45, 622–9.

Maestripieri, D. (2004). Genetic aspects of mother-offspring conflict in rhesus macaques. *Behavioral Ecology and Sociobiology* 55, 381–7.

Michod, R. E. (1982). The theory of kin selection. *Annual Review of Ecology and Systematics* 13, 23–55.

Mock, D. W. and Parker, G. A. (1997). *The Evolution of Sibling Rivalry*. Oxford University Press, Oxford.

Moore, A. J., Brodie Iii, E. D., and Wolf, J. B. (1997). Interacting phenotypes and the evolutionary process: I. Direct and indirect genetic effects of social interactions. *Evolution* 51, 1352–62.

Newman, T. K., Syagailo, Y. V., Barr, C. S., et al. (2005). Monoamine oxidase A gene promoter variation and rearing experience influences aggressive behavior in rhesus monkeys. *Biological Psychiatry* 57, 167–72.

Parker, G. A., Royle, N. J., and Hartley, I. R. (2002). Intrafamilial conflict and parental investment: a synthesis. *Philosophical Transactions of the Royal Society of London, Series B* 357, 295–307.

Philipps, P. C. and McGuigan, K. L. (2006). Evolution of genetic variance-covariance structure. In C. W. Fox and J. B. Wolf, eds. *Evolutionary Genetics*, pp. 310–25. Oxford University Press, New York.

Phillips, P. C. and Arnold, S. J. (1989). Visualizing multivariate selection. *Evolution* 43, 1209–22.

Queller, D. C. (1992). A general model for kin selection. *Evolution* 46, 376–80.

Qvarnström, A., Vogel Kehlenbeck, J., Wiley, C., et al. (2007). Species divergence in offspring begging intensity: difference in need or manipulation of parents? *Proceedings of the Royal Society of London, Series B* 274, 1003–8.

Rauter, C. M. and Moore, A. J. (2002). Evolutionary importance of parental care performance, food resources, and direct and indirect genetic effects in a burying beetle. *Journal of Evolutionary Biology* 15, 407–17.

Royle, N. J., Hartley, I. R., and Parker, G. A. (2002). Begging for control: when are offspring solicitation behaviours honest? *Trends in Ecology and Evolution* 17, 434–40.

Royle, N. J., Hartley, I. R., and Parker, G. A. (2004). Parental investment and family dynamics: interactions between theory and empirical tests. *Population Ecology* 46, 231–41.

Rutter, M. (2006). *Genes and Behavior*. Blackwell Publishing, Malden, MA.

Schrader, M. and Travis, J. (2009). Do embryos influence maternal investment? Evaluating maternal-fetal coadaptation and the potential for parent-offspring conflict in a placental fish. *Evolution* 63, 2805–15.

Sinervo, B., Clobert, J., Miles, D. B., et al. (2008). The role of pleiotropy vs signaller-receiver gene epistasis in life history trade-offs: dissecting the genomic architecture of organismal design in social systems. *Heredity* 101, 197–211.

Sinervo, B. and Svensson, E. (2002). Correlational selection and the evolution of genomic architecture. *Heredity* 89, 329–38.

Smiseth, P. T., Wright, J., and Kölliker, M. (2008). Parent-offspring conflict and co-adaptation: behavioural ecology meets quantitative genetics. *Proceedings of the Royal Society of London, Series B* 275, 1823–30.

Tallis, G. M. and Leppart, P. (1988). The joint effects of selection and assortative mating on multiple polygenic characters. *Theoretical and Applied Genetics* 75, 278–81.

Trivers, R. L. (1974). Parent-offspring conflict. *American Zoologist* 14, 249–64.

Van Noordwijk, A. J. (1988). Sib competition as an element of genotype-environment interaction for body size in the great tit. In G. De Jong ed. *Population Genetics and Evolution*, pp. 124–37. Springer Verlag, Berlin.

Wade, M. J. (1998). The evolutionary genetics of maternal effects. In T. A. Mousseau and C. W. Fox, eds. *Maternal*

Effects as Adaptations, pp. 5–21. Oxford University Press, New York.

Wagner, A. (2005). *Robustness and Evolvability in Living Systems.* Princeton University Press, Princeton.

West-Eberhard, M. J. (1983). Sexual selection, social competition, and speciation. *Quarterly Revue of Biology* 58, 155–83.

Wilkins, J. F. and Haig, D. (2003). What good is genomic imprinting: the function of parent-specific gene expression. *Nature Genetics* 4, 1–10.

Wolf, J. B. (2000). Gene interactions from maternal effects. *Evolution* 54, 1882–98.

Wolf, J. B. (2003). Genetic architecture and evolutionary constraint when the environment contains genes. *Proceedings of the National Academy of Sciences USA* 100, 4655–60.

Wolf, J. B. and Brodie Iii, E. D. (1998). The coadaptation of parental and offspring characters. *Evolution* 52, 299–308.

Wolf, J. B., Brodie Iii, E. D., Cheverud, J. M., et al. (1998). Evolutionary consequences of indirect genetic effects. *Trends in Ecology and Evolution* 13, 64–9.

Wolf, J. B., Brodie Iii, E. D., and Moore, A. J. (1999). Interacting phenotypes and the evolutionary process. II. Selection resulting from social interactions. *American Naturalist* 153, 254–66.

Wolf, J. B. and Hager, R. (2006). A maternal-offspring coadaptation theory for the evolution of genomic imprinting. *PLoS Biology* **4**, e380.

CHAPTER 17

Genetics and epigenetics of parental care

Frances A. Champagne and James P. Curley

17.1 Introduction

Studies across a wide variety of taxa have established that variation in parental care is heritable and partly mediated by genetic mechanisms. For example, nesting behaviour in painted turtles (*Chrysemys picta*) (McGaugh et al. 2010), ovipositioning in butterflies (*Euphydryas editha*) (Singer et al. 1988), egg laying in seed beetles (*Callosobruchus maculatus*) (Messina 1993), maternal yolk hormone transfer in collared flycatchers (*Ficedula albicollis*) (Tschirren et al. 2009), parental feeding rates in long-tailed tits (*Aegithalos caudatus*) (MacColl and Hatchwell 2003), and even parental care in humans (Perusse et al. 1994) all show a significant heritable component. Though these studies indicate a significant role for genetic variation in explaining phenotypic variation in parental care, there are only a few model species, such as honeybees, rodents, primates, and humans, in which molecular techniques have been implemented to determine how genes and the environment interact in the development and evolution of parental care.

In this chapter, we will provide an overview of genetic studies utilizing diverse experimental strategies, such as strain variants, selective breeding designs, quantitative trait loci (QTL), and target gene-knockout/manipulation to identify candidate genes/loci which account for variation in parental care and the complex interactions between parents and infants that promote growth and survival. A summary of the genes/loci discussed in this chapter and the behavioural correlates of these genetic influences is provided in Table 17.1. The well characterized genomes of honeybees and laboratory rodents have contributed to more thorough investigations of gene–behaviour relationships in these model species. In this chapter we will focus on how analyses of these species have contributed to our understanding of the genetics of care-giving (in the case of worker bees) and parenting, and highlight the study of gene polymorphisms identified in these species which are associated with variation in parental and offspring behaviour. A unique class of genes that we will discuss specifically are imprinted genes; these genes have been explored in the context of offspring development, growth, survival, and behaviour and exhibit parent-of-origin specific expression due to epigenetic silencing of a parental allele (also see Chapters 7 and 14). There is also increasing support for the role of imprinted genes in the expression of parental care. Finally, the established role of environmental experiences in neurobiological, physiological, and behavioural developmental pathways suggests that consideration of the interplay between genes and the environment will be critical to our understanding of parental care. Recent studies of epigenetic mechanisms—molecular changes that regulate gene activity—suggest that parenting may exert long-term effects on offspring through these non-genomic inheritance pathways. We will provide an overview of this emerging field of study and discuss how epigenetics affects the stability and heritability of variation in behaviour. Overall, it is apparent that the study of genetics and epigenetics in the context of the reciprocal relationship between parents and offspring provides insights into complex and dynamic processes involved in the evolution of parental care, particularly parent–offspring coadaptation (Chapter 16) and the resolution of parent–offspring conflict (Chapter 7).

The Evolution of Parental Care. First Edition. Edited by Nick J. Royle, Per T. Smiseth, and Mathias Kölliker.

Table 17.1 Summary of candidate genes/loci identified across species to be involved in parental care

Species	Candidate genes/loci	Method of discovery	Parental behaviour
Honeybee (*Apis mellifera*)	Hyg1, Hyg2, Hyg3, Uncap1, Uncap2, Rem1	QTL	hygenic behaviour
	Pln1, Pln2, Pln3, Pln4	QTL	termination of nursing and
	VG (Vitellogenin)	RNA-interference	transition to foraging
Prairie vole (*Microtus ochrogaster*)	V1ar (Vasopressin Receptor)	comparative genomics, RNA-interference, transgenic mice	social bonding with female mating partner and biparental care
Domestic Pig (*Sus scrofa*)	fosB (AP1 transcription factor)		maternal infanticide
	Peg3 (Paternally expressed gene 3)		
	OXTR (oxytocin receptor)		
	5-HTR2c (Serotonin receptor 2c)	QTL & polymorphism-phenotype	
	GR (Glucocorticoid Receptor)	association	
	ADRA2a (Alpha-2A adrenergic receptor)		
	ERβ (Estrogen-receptor beta)		
	DRD1 (Dopamine receptor 1)		
House Mouse (*Mus musculus*)	NOS1 (Nitric oxide synthase 1) CRHbp (CRH binding protein)	microarray analysis of gene expression	maternal aggression towards an intruder
	Various [see note]		
	OXTR, OXT (Oxytocin), PRL (Prolactin), PRLR (Prolactin receptor) ERα (Estrogen receptor alpha), fosB, Fox1b (fork-head box gene 1b), CRH (Corticotropin-releasing hormone), CRHr (Crh receptors), 5-HTR5b (Serotonin receptor 5b)	QTL & gene knockout	maternal infanticide, pup retrieval, placentophagia, milk let-down, nursing, licking/grooming, nest building, maternal aggression
	Imprinted Genes – Peg3, Mest (Mesoderm-specific transcript homolog protein), Magel2 (MAGE-like 2)	gene knockout	infanticide, pup retrieval, placentophagia, milk let-down, nursing, pup-licking, nest building, maternal aggression
	ERα	gene knockout	paternal care and infanticide
	PR (Progesterone receptor)		
	PRLR	gene knockout	Paternal–offspring recognition mediated by olfactory neurogenesis
	Genes contained within nonpseudoautosomal region of the Y-chromosome (e.g. Sry—testes determining factor)	transgenic mice with full and partial chromosome duplications and deletions	inhibition of pup retrieval and promotion of infanticide
Rat (*Rattus norvegicus*)	AVP (Vasopressin)	polymorphism-phenotype association between selected strains	nursing
Rhesus macaques (*Macaca mulatta*)	5-HTT (Serotonin transporter)	polymorphism-phenotype association	infant abuse, rejection and contact
	OPMR1 (Mu-opioid receptor)	polymorphism-phenotype association	infant contact and restraining
Humans (*Homo sapiens*)	5-HTT	polymorphism-phenotype association	maternal sensitivity, orientation towards infants
	COMT (Catechol-O-methyltransferase), DRD4 (Dopamine D4 receptor)	polymorphism-phenotype association	maternal sensitivity and infant attachment
	OXTR	polymorphism-phenotype association	maternal sensitivity

*Note—See Kuroda et al. (2011) for a table listing the phenotypes of over 35 different transgenic and knockout mice.

17.2 Genetics of care-giving in honeybees (*Apis mellifera*)

There is a long and distinguished history of the use of honeybees in behavioural genetic research. Recent use of molecular genetic techniques including QTL analysis, gene expression microarrays, and epigenetic analysis (see Section 17.6.1) as well as the sequencing of the honeybee genome in 2006 has significantly advanced our understanding of how phenotypic variation in honeybee social behaviours can be attributable to both inherited genetic sequence variation as well as environmentally mediated changes in the expression of these genes.

Honeybees are eusocial insects, and their societies are structured around one reproductive female (the queen). Care to developing larvae is provided by thousands of other females populating the colony, which are sterile relatives known as 'workers' upon which fitness of the offspring is completely dependent. Workers engage in various forms of care across their lifespan. Within the first three weeks of adult life they participate in offspring attendance inside the hive (i.e. brood care or nursing) after which they switch to food provisioning. They start foraging for nectar and pollen to provide food for developing larvae and other adults. Middle-aged adult workers also perform an additional care-giving behaviour, nest hygiene, which involves the removal of diseased pupae and larvae from the nest.

17.2.1 Genetic factors regulating hygienic behaviour

The genetic basis of the two components of hygienic behaviour, the uncapping of the diseased brood cell followed by the removal of the contents, has been established through classic experiments involving the inter-crossing of inbred honeybee lines (Rothenbuhler 1964). Using this approach, the resistant (Brown) line has been observed to engage in very high levels of hygienic behaviour whereas females from the susceptible (Van Scoy) line exhibit almost no hygienic behaviour. When these two lines were crossed, the F1 population resembled the phenotype of the Van Scoy line. When male drones of this F1 population were backcrossed to the Brown resistant line, approximately 25% of the population demonstrated hygienic behaviour, indicating the involvement of approximately two loci. Based on the observation that 50% of backcrossed colonies did not show uncapping, it was assumed that one locus was responsible for the uncapping behaviour and one locus for removal (Rothenbuhler 1964).

Since this pioneering study, similar crossing experiments have suggested that up to seven loci may be involved in the regulation of hygienic behaviour, and some loci may be involved in both behavioural components (Lapidge et al. 2002). A recent study has sought to discover candidate genes at these loci that may explain some of the phenotypic variation in hygienic behaviour using QTL analysis (Oxley et al. 2010). QTL analysis is a molecular tool that is commonly used to dissect the genetic regulation of complex traits. In these studies, genetically heterogeneous individuals are screened for phenotypic variation, genotyped for many molecular markers, and the association of phenotypic variation and variation at particular marker-sites statistically tested. Depending upon the sensitivity of the test, such an analysis can reveal genomic regions containing candidate genes that may account for a proportion of the phenotypic variation. QTL analysis has identified three loci (Hyg1-3) that significantly predicted whether honeybee workers would engage in hygienic behaviour. These loci accounted for 30% of the observed phenotypic variation in this trait. Two loci (Uncap 1 & 2) were found to predict whether workers performed the uncapping behaviour whereas one locus (Rem1) was found to predict removal behaviour.

Hygienic behaviour is not solely genetically determined but also depends on environmental factors, such as the odour produced by dead larvae. Thus, loci detected by the QTL analysis are most likely to influence variation in the sensitivity of workers to these environmental cues, which is consistent with the observation that four candidate genes associated with Hyg1 in this study are involved in the regulation of olfactory processing and, therefore, sensitivity to the environment (Sambandan et al. 2006).

17.2.2 Transition from nursing to foraging

In terms of food provisioning, workers naturally show a bias in their foraging for either nectar or pollen, which is in part regulated by genetic differences among individuals. Honeybee lines have been selectively bred to show preferences for either nectar or pollen, and cross-fostering has provided further support that these phenotypic differences have a genetic basis (Calderone and Page 1992). Strains that selectively forage for pollen also show an earlier onset of foraging which is associated with larger ovaries and higher early developmental levels of the protein vitellogenin which then rapidly decline prior to becoming foragers (Amdam and Page 2010). Vitellogenin is encoded by the VG gene, and by using RNA-interference to knock down the levels of this protein, it has been experimentally demonstrated that this protein does indeed influence the timing of foraging onset (e.g. workers become foragers sooner when vitellogenin levels are reduced; Marco Antonio et al. 2008). Interestingly, using a population of honeybees produced by backcrossing between high and low pollen hoarding strains, four major QTLs (pln1-4) have been identified that influence the age of onset of foraging as well as the nectar-pollen foraging preference bias (Hunt et al. 2007). Among the candidate genes at these loci, there is a significant over-representation of genes involved in ovarian development and insulin signalling; genes that could potentially govern resource allocation to reproduction and life-history progression.

It has recently been hypothesized that ancestral molecular mechanisms that link ovarian development and foraging have been co-opted in the evolution of honeybee food provisioning and foraging behaviour (Amdam and Page 2010). Across many insects, the foraging preference of females varies according to reproductive state, and although honeybee worker females never mate, they can produce eggs should their ovaries develop sufficiently. In honeybees, the size of ovaries is controlled by nutrient-dependent signalling cascades that initiate the release of steroid hormones, including juvenile hormones which stimulate vitellogenin expression (Hunt et al. 2007). It was further hypothesized that the large ovaries of pollen preferring workers leads to higher VG expression and vitellogenin secretion, which then leads to a negative feedback effect on the ovaries, reduced levels of vitellogenin, and an earlier onset of foraging. Supporting evidence for this hypothesis has come from a follow-up study that investigated the expression of candidate genes detected by QTL analysis. Two genes, HR46 (pln2) and PDK1 (pln3), were found to differ in expression between pollen and nectar preferring workers and were correlated with ovary size, foraging bias, and the age of transition from nursing to foraging (Wang et al. 2009).

The age at which worker bees switch from nursing to foraging is related to the particular needs of the colony. For example, if the colony loses a substantial proportion of its workforce then there are fewer older foragers present and maturation is accelerated amongst young workers. This change in maturation rate and transition in behaviour is related to pheromones produced by adult foragers. The presence or absence of these pheromones is able to induce changes in the expression of hundreds of genes (particularly those encoding transcription factors and metabolic proteins) in young workers advancing the age at which they become foragers. Microarray analysis has revealed that the expression of up to 40% of genes differs between workers who are 'nurses' or 'foragers' independent of their age (Whitfield et al. 2003). Congruent with these findings, treatment of young bees with brood pheromone (a pheromone produced by larvae) up-regulates the expression of genes known to be highly expressed in nurses and down-regulates genes that are highly expressed in foragers (Alaux et al. 2009). However, the gene-expression inducing effects of brood pheromones are reversed in older bees who are already foragers, suggesting a complex interplay between the physiological state of workers and the influence of social interactions on gene expression. To summarize, work on honeybees demonstrates that variation in care-giving can be attributed to both inherited genetic variation as well as environmentally induced variations in gene expression. This is a theme that is also evident for perhaps the most well-studied animal model in behavioural genetics—laboratory rodents.

17.3 Laboratory studies of the genetics of parental care in rodents

Laboratory-based studies of the genetics of parental care have increasingly focused on model species such as rats and mice (typically *Rattus norvegicus* and *Mus musculus*), which have well characterized genomes and can be observed under controlled environmental conditions. Though somewhat limited by the 'artificial' nature of the rearing environment, these approaches have resulted in the development of experimental strategies for defining candidate genes or loci associated with specific aspects of the parent–offspring interaction. Across species, it is evident that olfactory circuits, hormones, and neuroendocrine mechanisms play a significant role in the expression of parental care. This homology suggests that genes identified in studies of the parental care of laboratory rodents may allow for hypothesis driven studies of the genetic bases of parental care across mammalian and non-mammalian species as well as invertebrates. In the following sections, we highlight findings from these laboratory based studies, with a particular focus on the study of differences among strains, selective breeding, QTL analysis, transgenic mice, and manipulations of sex chromosomes.

17.3.1 Strain and species differences in parental care

Amongst laboratory strains of mice and rats there is significant between-strain variation in parental care behaviours such as latency to retrieve pups, nest-building/nest quality, nursing, pup-licking, and overall post-partum contact with pups (Champagne et al. 2007; McIver and Jeffrey 1967). Though shifts in parental care can be achieved experimentally through shifts in the quality of the environment or through the reproductive experience of parents, the stability over generations of strain-specific patterns of parent–offspring interactions suggests a genetic basis to these behaviours. For example, comparisons between the spontaneously hypertensive (SHR) and the non-hypertensive Wistar Kyoto rat (WKY) indicate a higher frequency of post-partum nursing and pup licking in SHR rats and this between-strain variation in care accounts for over 43% of the variation in offspring blood-pressure (Myers et al. 1989). However, reciprocal cross-fostering between SHR and WKY results in shifts in the frequency of maternal care, such that females provide care more similar to the strain of the pups they are rearing (Cierpial et al. 1990). Offspring genotype-induced shifts in maternal care (i.e. indirect genetic effects; Chapters 15 and 16) have also been observed in cross-fostering studies using inbred mice (Hager and Johnstone 2003; Curley et al. 2010; see Fig. 17.1). Thus, though there are characteristic patterns of maternal care amongst specific rodent strains, offspring genotype plays a critical role in maintaining that pattern or in shifting the quality or quantity of care. However, despite observations of stable and heritable maternal strain and pup strain effects on parenting, the specific genes or loci that contribute to this variation have yet to be identified.

Although there are few genetic studies of paternal behaviour in these laboratory based models, evidence exists for strain differences in the responsiveness of males to pups. Comparison of standard laboratory mice strains (AJ, DBA, CF-1, and C57) and wild mice suggest significant variation in the frequency with which males ignore/avoid pups, are infanticidal or paternal. For example, while DBA and AJ mice exhibit low levels of infanticide and are observed to engage in nurturing responses to neonatal pups on approximately 40–60% of experimental trials, C57 and wild mice are infanticidal during 60–80% of trials (Perrigo and Vom Saal 1994; see Fig. 17.2). These between-strain differences are stable, yet experience, such as whether or not mating/copulation occurred (Perrigo et al. 1991), can exert a profound effect on the frequency of parental behaviour when examining within-strain variation. Though it is not known which gene variants lead to strain differences in the overall frequency of infanticide vs. paternal behaviour, the emerging literature on the olfactory and neuroendocrine pathways which are altered by sexual experience in males may provide insights into the genetic mechanisms of strain variation in paternal care.

A rodent model that has been explored in great depth in the context of laboratory and field studies of the genetic contributions to social and parental

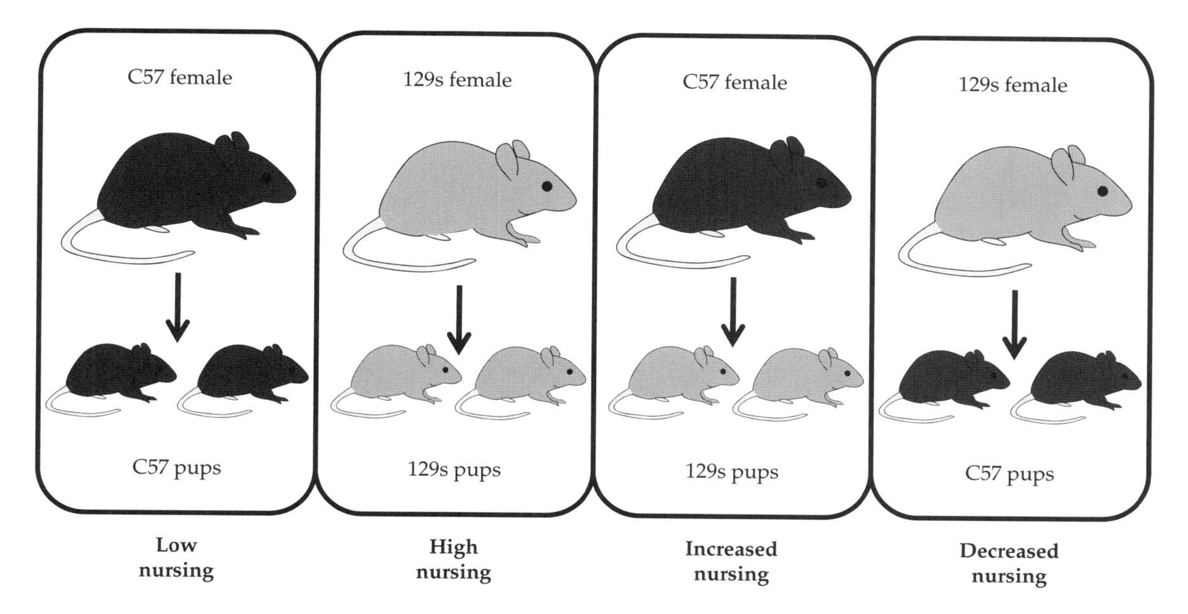

Figure 17.1 The interaction of pup and maternal genotype determines care-giving. Characteristic strain differences (between 129s and C57 mice) in nursing can be altered if mothers rear pups of a different genotype (Curley et al. 2010).

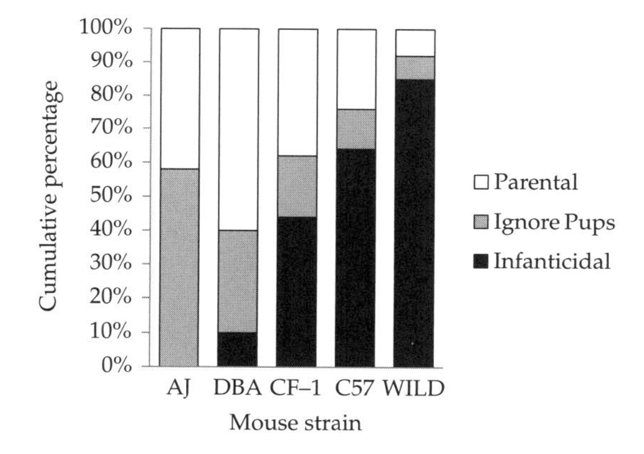

Figure 17.2 Percentage of virgin male mice that are spontaneously parental, infanticidal, or ignore pups in a retrieval test in various inbred strains and wild mice (Perrigo and vom Saal 1994).

behaviour involves comparisons between different species of voles. The prairie vole (*Microtus ochrogaster*) establishes socially monogamous bonds and displays biparental care, whereas the closely related montane vole (*M. montanus*) engages in a polygamous social organization. In addition, juvenile female offspring of prairie voles are more likely to exhibit spontaneous maternal responsiveness when presented with neonatal pups whereas amongst polygamous voles, such a spontaneous maternal response to pups is absent (Olazabal and Young 2006). Comparisons of neuropeptide receptors in the brains of monogamous vs. polygamous voles suggest that variation in the regional distribution of oxytocin (OXTR), vasopressin (V1aR), dopamine, and corticotrophin releasing hormone (CRH) receptors may account for this behavioural variation (McGraw and Young 2010). In particular, female nurturing responses are found to be highly correlated with OXTR density and distribution in the striatum both within- and between-species. However, as yet no polymorphic differences in the OXTR gene have been identified that explain this variation (Olazabal and Young 2006; Young et al. 1996).

The genetic basis of the species differences in male behaviour (polygamous vs. monogamous social bonds and the parental care of offspring) has been linked to variation in the V1aR gene. The coding region of V1aR shares 99% homology between prairie and montane voles, however, variation in the 5′ regulatory region of this gene indicates species differences (Donaldson and Young 2008). There is an expanded microsatellite sequence in the 5′ flanking region of the prairie vole V1aR gene that is absent in the montane vole sequence. This genetic variation has been demonstrated to induce brain-region specific patterns of V1aR gene expression and species differences in affiliative behaviour. However, though brain region specific elevations in V1aR gene expression are observed in other monogamous species, such as the California mouse (*Peromyscus californicus*), and common marmoset (*Callithrix jacchus*) (Insel and Young 2000), across species comparisons suggest that the microsatellite variation observed in the V1aR gene does not necessarily distinguish all monogamous vs. polygamous species (Fink et al. 2006). Thus, although vasopressin, oxytocin, and other neuropeptides are likely to be involved in variation in parenting, there may be diverse pathways that have evolved to allow for the heritability of this variation, reflecting the convergent evolution of particular patterns of behaviour such as monogamy and biparental care.

17.3.2 Selective breeding

One approach used to investigate the genetic contribution to phenotypes, as demonstrated in the Section 17.2.1 discussion of *Apis mellifera*, has been the selective breeding of individuals who demonstrate extremely high or low levels of the trait of interest. Typically, the founder populations for these selective breeding studies are wild-caught animals or outbred stocks (genetically heterogeneous) or are the product of inter-crosses between two or more inbred strains that diverge on several phenotypic measures. While this approach has been used extensively in the investigation of the genetic bases of behaviours such as addiction and aggression, there are only a few examples of this experimental approach for parental care in rodents. In one study, female laboratory mice of the outbred ICR strain were selectively bred for the level of maternal aggression (e.g. number of attacks made towards an intruder male). Following eight generations of selection for high aggression, the realized heritability of maternal aggression was found to be 0.40 (Gammie et al. 2006). This high aggression line was also found to be slower to retrieve pups and to have reduced litter weights/sizes at birth, indicating that genes selected for their role in enhancing aggression likely also contribute to other maternal traits. In a microarray analysis of gene expression of hypothalamic brain tissue from highly aggressive and control females, 200 genes showed significant changes, including neuronal nitric oxide synthase 1 (NOS1) and CRH binding protein (CRHbp); genes that had previously been implicated in the neurobiology of maternal behaviour (Gammie et al. 2006).

In female prairie voles, selective breeding has also been used to examine maternal vs. infanticidal behaviour. Amongst adult virgin prairie voles, approximately 40% of females are spontaneously infanticidal towards newborn pups whereas the remaining females are maternal or ignore pups. Selective breeding for these traits indicated that by the third generation 90% of females in the infanticidal line were non-maternal in their response to pups, whereas only 40% of females from the control or maternal line were non-maternal (Olazabal 2010). Moreover, cross-fostering did not alter the emergence of this behaviour indicating genetic inheritance (though pre-natal maternal effects cannot be excluded). These studies demonstrate that selective breeding is a useful strategy for determining if variation in parental care has a genetic basis are as well facilitating investigation into the target genes which contribute to this variation. However, thus far, this approach has not been broadly applied in the study of the genetics of parental care and parent–offspring interactions.

Selective breeding for behavioural traits such as activity often leads to a correlated response to selection in terms of parental behaviours, providing evidence for genetic correlations between these behaviours. For instance, artificial selection for high levels of anxiety-like or avoidance behaviour has typically resulted in females that exhibit elevated levels of nursing behaviour (Bosch 2011; Holland 1965). However, the direction of these effects is not always consistent for all parental behaviours. For example, rats selected for high levels of depressive-like behaviour show lower levels of nursing and higher levels of licking/grooming (Lavi-Avnon et al. 2005). Furthermore, a mouse line selected for high voluntary wheel running was found to exhibit high levels of maternal aggression (Gammie et al. 2003, Girard et al. 2002). For the following seventeen generations, the more aggressive line was then selected for high maternal aggression only and it was found that in each generation, 17% of females would fail to nurture their offspring. Within this line, the neglect phenotype was associated with altered central dopaminergic functioning (Gammie et al. 2008). In a line of rats selectively bred for high anxiety-like behaviour (HAB), there is a polymorphism in the promoter region of the vasopressin (AVP) gene that results in higher levels of AVP mRNA expression in the paraventricular nucleus (PVN) of the hypothalamus (Bosch 2011). This may explain both the higher anxiety-like behaviour and the elevated nursing in this line, as AVP in the PVN is involved in the regulation of both behaviours.

Overall, these studies suggest an overlap in the genetic basis of parental behaviours and behavioural traits such as activity and anxiety. This is informative with respect to non-laboratory based studies across a variety of taxa that have demonstrated associations between parental care and other adult traits that are also under selection. For example, female great tits (*Parus major*) with bold personalities (more exploratory and aggressive) show higher nest defence, lay eggs earlier in the breeding season and increase the levels of androgens given to each egg with successive egg laying order compared to great tits with shy personalities (Groothuis et al. 2008; Hollander et al. 2008). If, as has been suggested, there are co-adaptive selection pressures for particular suites of adult (personality), parental, and offspring traits (Roulin et al. 2010), then these data from laboratory rodents suggest that genes that modulate common developmental pathways are likely candidates for genes evolving under co-adaptation. Furthermore, artificial selection for offspring traits in rodents can likewise lead to genetic lines that diverge in parental behaviour. For example, amongst outbred rats selected for the production of high levels of ultrasonic vocalizations (USV) during infancy there are line-specific differences in the frequency of post-partum maternal care. In the high USV line, females have been found to nurse and lick/groom their own pups less when compared to females from the low USV line (Brunelli et al. 2010). USVs are a critical feature of a pup's ability to solicit maternal care and it is noteworthy that reduced maternal care is a characteristic of this line, suggesting that this co-occurrence may be adaptive; as is expected under parent–offspring coadaptation (Chapter 16). Future analysis of genetic variation that accounts for these co-adapted phenotypes may provide significant insights into the dynamics of selection within the context of parent–offspring interactions.

17.3.3 Quantitative trait loci (QTL) analysis

QTL analysis has been utilized to associate phenotypic variation in mammalian parental traits. Offspring survival, as an indication of the quality of maternal performance (which correlates with other aspects of maternal care such as nest-building, placentophagia, nursing, licking/grooming of pups, and aggression towards intruders) of female mice produced through inter-crossing the LG/J and SM/J inbred lines, has been assessed using QTL (Peripato et al. 2002). Two loci strongly associated with maternal performance impairments were identified. One locus on chromosome 2 showed over-dominance indicating that females heterozygous for the parental strains had better offspring survival, while the other loci on chromosome 7 showed under-dominance indicating that homozygous females for either parental strain performed better than heterozygous females. Candidate genes were determined for these main effects based upon previously established roles of particular genes in regulating maternal care. Those identified were the oxytocin gene (chromosome 2), fosB and Peg3 (chromosome 7). Interestingly, this study also indentified 23 genomic regions that were found to interact epistatically, meaning that variation in maternal performance could be predicted by the combinatorial effects of particular homozygous or heterozygous genomic regions at different loci. Candidate genes from this analysis included CRH, CRH receptor, Mest/Peg1, prolactin, prolactin receptor, and the serotonin receptor (5-HTR5b). In total, the observed significant direct and epistatic genetic effects accounted for around 35% of the phenotypic variation observed in maternal performance.

Agricultural applications of QTL have been used to determine genetic loci associated with maternal infanticide in sows. Amongst pigs generated from an F2 inter-cross of White Duroc and Erhualian parental breeds, 13% of females kill their offspring after birth (Chen et al. 2009). Seven genomic regions, including three on the X-chromosome, were found to have a direct effect upon maternal infanticide. These regions included the loci homologous to the region of mouse chromosome 7 containing fosB and Peg3 (Peripato et al. 2002). Other candidate genes identified included those known to play a role in the regulation of behaviour including the glucocorticoid receptor (GR), the oxytocin receptor (OXTR), the Alpha-2A adrenergic receptor (ADRA2A), and the serotonin receptor 2C (5-HTR2C) (Chen et al. 2009). In a follow-up study, individual single nucleotide polymorphisms (SNPs) were tested for several of these positional candidate genes with SNPs in some genes (e.g. 5-HTR2C) showing differences between infanticidal and maternal sows whereas SNPs in other genes (e.g. OXTR) did not show any relationship to maternal behaviour (Chen et al. 2011). This study also identified SNPs outside the QTL loci which were related to infanticidal behaviour (in particular, the estrogen receptor beta (ERβ) and the dopamine receptor 1 (DRD1) genes), suggesting that a combination of candidate loci may be most appropriate in establishing how genetic variation contributes to phenotypic variation in maternal infanticidal behaviour.

17.3.4 Gene knockout (KO) studies

Within the field of behavioural genetics, there has been increasing use of transgenic mouse models as a strategy for determining the role of target genes in behavioural variation. In the case of parental care (primarily maternal care), there have been several genes identified that induce significant changes in survival and growth of offspring, timing of onset of maternal care during the post-partum period, maternal motivation to retrieve pups to the nest, and frequency of mother–infant interactions (for a recent review see Kuroda et al. 2011; see Table 17.1). Knockout of the immediate early gene fosB (located on chromosome 7) reduces pup survival associated with impairments in pup retrieval and crouching over pups (Brown et al. 1996). Oxytocin (OXT) knockout dams show impaired milk-ejection (Nishimori et al. 1996) and deletion of the alpha isoform of the estrogen receptor (ERα) is associated with increased infanticide and reduced pup retrieval (Ogawa et al. 1998). Finally, mutation of the prolactin receptor (PRLR) induces impairments in ability to lactate and reduces motivation to retrieve pups (Lucas et al. 1998). In one of the few studies on paternal care, deletion of the progesterone receptor (PR) gene led to reduced aggression toward pups and increased parental behaviour (retrieving, crouching, licking pups) (Schneider et al. 2003). Overall, studies using knockout mice have identified many gene targets that play a critical role in parental care. This list of candidate genes expands considerably when survival, growth, and placental functioning are included as indices of parental care and investment.

In addition to genes that alter parental care, transgenic models illustrate the influence of genetic modifications on the ability of offspring to solicit parental care. For example, mutation of the tyrosine kinase Fyn results in suckling deficits leading to 100% mortality amongst Fyn-KO homozygotes being reared by homozygous-KO females (Yagi et al. 1993). This genetic modification is associated with disruptions to olfactory development in homozygous mutant mice, which may inhibit the ability of offspring to orient and form nipple attachments. Similar suckling deficits have been observed in Brn-3a (a transcription factor located on chromosome 14) KO mice, likely associated with deficits in ability to generate the motor rhythms necessary for suckling (Xiang et al. 1996). The genetic 'match' between mothers and offspring may also be critical to offspring development. For example, deletion of the serotonin 5-HT1a receptor induces decreases in USVs in homozygous mutant offspring being reared by homozygous or heterozygous mutant dams (Weller et al. 2003). However, amongst 5-HT1a KO heterozygous offspring, rearing by a WT dam leads to decreased USVs in infancy and increased anxiety-like behaviour in adulthood compared to 5-HT1a heterozygous offspring being reared by heterozygous dams. This is molecular evidence for social epistasis mediated by the reciprocal nature of parent–offspring interactions and leading to a phenotypic match between parents and offspring (Chapter 16).

17.3.5 Sex chromosomes

In mammals, all individuals inherit autosomes and X chromosomes from both parents, but only sons inherit the non-pseudoautosomal region of the Y chromosome (YNPAR) from fathers (the very small pseudoautosomal region (PAR) of the Y chromosome recombines during meiosis with the X chromosome). It is the presence of the Sry (testes determining factor) gene on the YNPAR that triggers male-specific sexual development, though recently it has been established that other genes carried by this YNPAR may also regulate sex specific behaviours independently of the hormonal organizing effects of Sry (Arnold and Chen 2009). Transgenic mice have been produced whereby the Sry gene on the Y chromosome is removed and reinserted onto an autosome resulting in gonadal sex determination that is independent of sex chromosome complement. Thus, four types of mice are created: XX females (XXF), XY females (XYF), XY males (XYM), and XX males (XXM). In one study, virgin XY females were found to be less parental in a pup retrieval test than XX females, suggesting that genes on the Y chromosome may inhibit parental care. These females also had a more male-like pattern of brain development such as an increased density of AVP neurons in the lateral septum (Gatewood et al. 2006). In a separate mouse model, males have been generated that carry two X chromosomes plus a duplicated short stretch of the Y chromosome containing seven genes including Sry (Reisert et al. 2002). These males have been shown to be significantly better at retrieving pups and less likely to commit infanticide than males that have the full complement of Y chromosome genes. This is further support that genes on the Y chromosome inhibit parental care in mice, though it remains to be determined which genes are responsible for this effect and whether these genes may also play a role in the regulation of parental care in other species.

17.4 Gene polymorphisms in primates and humans

The experimental studies in rodents described in the previous section have demonstrated multiple loci, target genes, and neurobiological pathways that are associated with parental care and these insights have provided a tool with which to explore the genetic basis of parental behaviour in humans and non-human primates. Several gene polymorphisms have been identified which predict variation in multiple behaviours contributing to our understanding of the genetic basis of individual differences in parental care. Here, we will focus on genetic variants of the serotonin transporter, mu-opioid receptor, dopamine D4 receptor, and oxytocin receptor.

17.4.1 Serotonin transporter (5-HTT)

The serotonin transporter is a pre-synaptic transmembrane protein that is responsible for serotonin uptake from the synaptic junction. In both rhesus macaques (*Macaca mulatta*) and humans there exists a insertion/deletion polymorphism (21bp in rhesus, 43bp in humans) which gives rise to short and long forms of this gene. The short form of this gene is associated with reduced transcriptional activity and lowered serotonin uptake (Lesch et al. 1996). Significantly, female rhesus macaques carrying the short allele are more likely to be abusive towards their infants (McCormack et al. 2009). Abusive parenting in these monkeys is characterized by increased rates of rejecting infants, spending less time in ventral contact, and breaking off contact with the infant within the first three months of life. Interestingly, particularly from the perspective of parent–offspring co-adaptation and the selection for pleiotropic parent and offspring influences, infants carrying the short-allele are also more resistant to handling, more difficult to console, and display more tantrums than infants homozygous for the long allele (McCormack et al. 2009).

In humans, two functional variants of the long allele (LG and LA) have been identified, with the LA/LA homozygous genotype being found to lead to higher mRNA expression *in vitro*. Mothers who are LA/LA homozygous have lower scores of maternal sensitivity at six months post-partum when observed at home compared to mothers who carry the S or LG allele (Mileva-Seitz et al. 2011). Conversely, in another study, S-carrying mothers were found to be less sensitive to their 1–3 year old infants when tested in a laboratory setting (Bakermans-Kranenburg and van Ijzendoorn 2008). It is possible that S-carrying mothers may be more sensitive to their infants under low stress situations (at home) but less sensitive in stressful situations (observed at the lab), which would suggest genotype x environment interactions underlying variation in parental sensitivity to infants. Furthermore, variation in the human 5-HTT gene has been associated with parent–offspring attachment behaviours, similar to that found for rhesus macaques. For instance, infants carrying the short allele were found to have attachment relationships at 15 months that could be predicted by their mother's responsiveness to them at 7 months of age, whereas the attachment of infants carrying the long allele could not be predicted by earlier maternal responsiveness (Barry et al. 2008). Moreover, children homozygous for the short allele have also been reported to be at heightened risk for developing disorganized attachment if their mothers also exhibit low maternal responsiveness (Spangler et al. 2009). These results indicate a role for the 5-HTT gene in mediating the behavioural dynamics in repeated parent–offspring interactions.

17.4.2 Mu-opioid receptor (OPRM1)

Humans and rhesus macaques possess unique SNPs in the mu-opioid receptor gene, with individuals who are described as G-carriers having increased affinity for the endogenous ligand beta-endorphin. It has recently been shown that female rhesus macaques that possess the G-allele are much less likely to allow their infants to break contact from them and explore the environment around the time of weaning (Higham et al. 2011). Although no other aspect of parenting was found to differ with regard to genotype, G-carriers were also found to have increased circulating plasma levels of oxytocin during lactation which may further explain their altered parenting style. Infant rhesus monkeys that carry the G-allele cling more to their mothers and make more distress vocalizations when maternally separated, again pointing towards a pleiotropic effect on parent and infant behaviour (Barr et al. 2008). These findings suggest that genetic variation at this locus is regulating both maternal and infant behaviours central to successful mother–infant attachment (Curley 2011).

17.4.3 Dopamine D4 receptor (DRD4)

Within the 3rd exon of the dopamine D4 receptor gene (DRD4) there is a 48-bp variable number tandem (VNTR) repeat polymorphism that can have up to 10 repeats, with the most common variants being the 4, 7, and 2 repeat forms. Short repeats (below 5) appear to code for receptors that are more efficient at binding dopamine, and long repeats (over 6) of the VNTR have been identified as a risk factor for psychopathology. In studies of human parental behaviour, mothers who possess a 7-repeat allele are more sensitive to fussy infants and less sensitive to less fussy infants compared to parents without the 7-repeat allele (Kaitz et al. 2010). When considering variation within DRD4 in conjunction with variation in polymorphisms of the catechol-O-methyltransferase (COMT) gene, which is involved in dopamine metabolism, a gene x gene x environment interaction is observed. Females possessing the 7-repeat of the DRD4 gene as well as the val/val or val/met COMT genotype (i.e. those females with the least efficient dopaminergic functioning), are less sensitive to their children if faced with more daily hassles but more sensitive to their children if they experience fewer daily hassles (van IJzendoorn et al. 2008). This same polymorphism has also been implicated in the development of disorganized attachment amongst infants. A Hungarian study found that 67% of disorganized infants carry the 7-repeat allele whereas only 20% of securely attached infants and 50% of insecurely attached infants have the 7-repeat allele (Lakatos et al. 2002).

17.4.4 Oxytocin receptor (OXTR)

A SNP in the third intron of the human oxytocin receptor gene (rs53576) of unknown functionality has previously been identified as potentially explaining variation in maternal behaviour. In a sample of mothers of 2 year old children at risk of externalizing problems (i.e. aggression, defiance), those mothers with the G/G genotype of the OXTR gene were found to be more sensitive than mothers who carry at least one A allele at the OXTR gene (van IJzendoorn et al. 2008). This result was found to be significant even when controlling for factors such as mother's emotional state, education level, or marital relationship quality. This study also highlights an important point regarding polymorphism association studies. While statistically significant associations can often be found, they often account for a small percentage of the variation in the parental trait being measured. For instance, in one study (Bakermans-Kranenburg and van Ijzendoorn 2008), the OXTR gene and 5-HTT polymorphisms significantly accounted each for only 3% of the variation in maternal sensitivity whereas maternal education itself accounted for 15% of the variation.

17.5 Epigenetic influences on parental care and offspring development: genomic imprinting

Classic Mendelian genetics posits that both mothers and fathers pass on functionally active copies of autosomal nuclear genes to their offspring. However, there is a small subset of mammalian genes where there is monoallelic expression of either the paternally or maternally inherited allele. This parent-of-origin specific expression of genes (mediated by the silencing of genes) is referred to as genomic imprinting. A maternally expressed gene is silenced when passed from a father to both his sons and daughters (the copy being inherited from the mother is active) and will only become active again in the grandchildren of his daughters and not of his sons. Genomic imprinting is achieved mechanistically through chromatin remodelling and epigenetic modifications (described further in Section 17.6) during gametogenesis, which are then maintained following fertilization and throughout somatic development (Ideraabdullah et al. 2008). Approximately 80–100 genes have been identified in mammals that are subject to this form of epigenetic silencing through complete loss of expression, dependent upon their parent-of-origin. Interestingly, recent research has revealed that perhaps hundreds more genes may be subject to subtler parental biases in gene expression, though the mechanism for this process remains unknown (Gregg et al. 2010).

Gene expression and gene knockout studies have contributed to a greater understanding of the functional role of imprinted genes. Strikingly, the vast majority of imprinted genes appear to be expressed in the developing foetus and placenta and are therefore well situated to coordinate resource transfer during gestation between mother and offspring (Constancia et al. 2004). For example, insulin-like growth factor 2 (Igf2) is paternally expressed and encodes a growth factor gene that is expressed in the placenta, promoting placental growth, the transport of nutrients, and embryonic growth. However, the Igf2 receptor (Igf2r), a mannose-6-phosphate receptor uncoupled from growth promoting properties, is maternally expressed and appears to act as a sink for Igf2 (Constancia et al. 2004). In the brain, expression of the gene Rasgrf1 occurs only from the paternal allele up until weaning, after which expression becomes biallelic. Mice lacking Rasgrf1 expression show growth deficits whereas transgenic mice that are made to over-express this gene during the post-natal period show accelerated growth (Drake et al. 2009). Similarly, infant mice lacking the paternally expressed transcript Gnasxl of the Gnas gene (which encodes for the stimulatory G-protein subunit Gsα) suffer from reduced post-natal growth and are also impaired in suckling (Plagge et al. 2004). Other imprinted genes appear to have roles in both the pre-natal and post-natal periods. One such gene is the paternally expressed Peg3 gene, which encodes a large zinc finger protein that is an important regulator of apoptosis (cell death) during development (Broad et al. 2009). Peg3 demonstrates highly conserved imprinting across mammalian species, developmental time -points and tissues. Infant mice lacking a functional copy of Peg3 are growth retarded during embryogenesis, suffer higher embryonic and neonatal mortality, exhibit deficits in suckling, and have impaired post-natal growth (Curley et al. 2004).

Parent-of-origin patterns of gene expression are broadly consistent with genetic conflicts over maternal investment between paternally and maternally inherited genes expressed in offspring (Wilkins and Haig 2003; Chapter 7). Nevertheless, maternally expressed genes do not always act to restrict resource acquisition, and it has been argued that it is difficult to reconcile the biology of molecular imprinting mechanisms with this theory (Keverne and Curley 2008). Furthermore, conflict theory predicts imprinting in offspring genes, but does not explain why imprinted genes would regulate parental care.

The paternally expressed genes, Peg3, Peg1, and Magel2 have been targeted in transgenic mice affecting multiple behavioural phenotypes, including maternal care. Peg3 mutant mice have increased amounts of apoptosis in areas of the forebrain in which Peg3 is normally expressed such as the amygdala, hypothalamus, bed nucleus of the stria terminalis (BNST), caudate putamen, and nucleus accumbens (Broad et al. 2009). Peg3 mutant mice also have a reduced number of oxytocin neurons in the PVN, supraoptic nucleus, and medial preoptic area (MPOA) (Li et al. 1999) and less oxytocin receptor binding in the BNST, MPOA, and lateral septum (Champagne et al. 2009). Consequently, mice that lack a functional copy of this gene have been found to exhibit disruptions to normal maternal care—including reduced food intake during pregnancy, reduced litter size and pup weights at birth, impaired milk let-down, poorer nest quality, reduced pup retrieval, nursing, and pup-licking (Champagne et al. 2009; Curley et al. 2004; Li et al. 1999). Female mice lacking the paternally expressed Peg1 gene (also called mesoderm-specific transcript or Mest) also show deficits in pup retrieval, nest-building, and placentophagia, in addition to having smaller litters with higher mortality (Lefebvre et al. 1998). Moreover, the paternally expressed gene Magel2 is highly expressed in the hypothalamus and regulates cell cycle, differentiation, and apoptosis (Kozlov et al. 2007). Female mice who lack a functional copy of Magel2 have been found to give birth to smaller litters and to exhibit dramatically increased levels of infanticide (Mercer and Wevrick 2009). These behavioural changes may be associated with the reduced expression of orexin, dopamine, and serotonin in the hypothalamus of Magel2 knockout females (Kozlov et al. 2007; Mercer et al. 2009). As with Peg3 knockout mice, infant mice lacking Peg1/Mest or Magel2 also show growth retardation independent of maternal genotype (Bischof et al. 2007; Lefebvre et al. 1998).

In the case of each of these three paternally expressed genes, it is apparent that the normal function of each gene is to regulate offspring growth and resource acquisition (through expression of the gene in the placenta, embryo, and neonatal brain) as well as adult maternal care (through expression in the female hypothalamus). Overall, the parent-of-origin specific expression of these paternally expressed genes is probably best conceptually envisaged in the context of evolutionary co-adaptation. Female offspring who are able to extract adequate maternal care both pre- and post-natally are the same individuals who are subsequently well adapted for and genetically predisposed toward this mothering style in adulthood (Keverne and Curley 2008), although the reasons for paternal rather than maternal expression remain to be elucidated.

17.6 Epigenetic influences on parental care and offspring development: nature via nurture

As discussed earlier with regard to honeybees (see Section 17.2.2), though genetic factors play a critical role in shaping both between- and within-species variation in parental care, the modulating effects of environmental experiences on multiple aspects of reproductive behaviour are also evident. These environmental effects can be both transient and/or sustained over the lifespan of an individual and in many cases can be observed at the level of gene expression. Recent advances in molecular biology have highlighted a range of mechanisms which regulate gene activity. These 'epigenetic' mechanisms are critical for the process of development and there is emerging data to suggest that parental care can induce epigenetic effects in offspring, which can be maintained even into subsequent generations. Though there are many mechanisms that can increase or decrease gene expression, much of the focus of the emerging field of *behavioural epigenetics* has been the exploration of post-translational histone modifications and DNA methylation (Champagne 2008; Keverne and Curley 2008). The condensed chromatin structure evident in the cell nucleus consists of DNA wrapped around clusters of histone proteins. Chemical modification to the histones, through acetylation, methylation, phosphorylation, and ubiquitination, leads to dynamic changes in the interactions between the N-terminal tails of the histones and the DNA. For example, histone acetylation is typically associated with increased gene expression. In contrast, DNA methylation is a chemical modification to cytosines within the DNA sequence typically associated with gene silencing. DNA methylation is achieved through the enzymatic actions of DNA methyltransferases (DNMTs) and can be a stable and mitotically heritable epigenetic modification that is critical in cellular differentiation. Though epigenetic marks within the DNA were thought to have limited plasticity beyond the very early stages of embryonic development, there is increasing evidence for pre-natal, post-natal, and even adult experience-dependent and quantitative changes in gene activity associated with changes in DNA methylation.

17.6.1 Maternal nutrition effects on DNA methylation

Maternal nutrition during pregnancy can have a significant impact on the growth and development of the foetus, with long-term consequences for brain development and metabolism (Godfrey and Barker 2001; Symonds et al. 2007). Studies of pre-natal nutrition indicate that both overall food intake and the specific nutrient composition of the diet during gestation can induce epigenetic changes. In humans, comparison of blood samples from offspring exposed to famine with that of siblings that were not exposed to famine indicates that there is decreased DNA methylation of the Igf2 gene as a result of poor maternal nutrition (Heijmans et al. 2008). Laboratory studies in rodents have subsequently identified specific nutritional deficits, such as pre-natal protein restriction or folic acid/choline deficiency, as having similar, generalized epigenetic consequences. Offspring of female rats placed on a protein deficient diet throughout gestation were found to have elevated hepatic GR and peroxisomal proliferator-activated receptor (PPAR) gene expression associated with decreased DNA methylation of these genes (Lillycrop et al. 2005; Lillycrop et al. 2008). Dietary effects on levels of the DNA methyltransferase DNMT1 may account for these observed modifications in global and gene-specific methylation, as DNMT1 expression is increased as a function of protein/choline restriction (Lillycrop et al. 2007). The maternal nutritional environment can therefore have a sustained impact on development through alterations in gene expression that are maintained through DNA methylation.

Epigenetic modifications in response to the nutritional environment during the early stages of development may also have implications for the morphological changes associated with caste phenotypes in eusocial insects. Honeybees have functional DNA methyltransferases and the degree of methylation of the genome varies during the course of development (Wang et al. 2006). Amongst female honeybees, social/reproductive caste is determined through early nutritional exposure to royal jelly. Larvae provided with a diet composed primarily of royal jelly grow more rapidly, have well-developed ovaries, and emerge as queen bees. In contrast, larvae provided with low levels of this rich nutritional resource and are smaller, have only rudimentary ovaries, and become workers. These caste differences in development are associated with differential gene expression in queen bees vs. workers (Evans and Wheeler 1999) and manipulation of the activity of the DNA methyltransferase DNMT3 in honeybees provides evidence that DNA methylation mediates this caste determination (Kucharski et al. 2008). It is likely that the experiential effects on nursing, foraging, and hygienic behaviour and related brain gene expression discussed in Section 17.2 are mediated by similar epigenetic mechanisms, though this remains to be elucidated.

17.6.2 Post-natal influence of maternal care

Variation in post-natal maternal behaviour may also induce epigenetic changes in offspring development. In rats, post-natal maternal licking/grooming (LG) behaviour has been found to induce long-term changes in neuroendocrine function and behaviour of offspring, with consequences for stress responsivity and cognition, and cross-fostering studies have confirmed that these effects are mediated by the level of maternal care received during post-natal development (Meaney 2001). Analysis of the GR promoter region suggests that variations in GR expression associated with differential levels of maternal care are maintained though altered DNA methylation (Weaver et al. 2004). Thus, offspring who receive high levels of maternal LG during the early post-natal period have decreased hippocampal GR promoter DNA methylation, increased GR expression, and decreased stress responsivity. In contrast, low levels of LG are associated with increased GR DNA methylation, decreased GR expression, and an increased hypothalamic–pituitary–adrenal (HPA) response to stress. Time course analysis has indicated that these maternally-induced epigenetic profiles emerge during the post-natal period and are sustained into adulthood. Maternal LG also affects γ-aminobutyric acid (GABA) circuits and receptor sub-unit composition, and in a recent study, reduced hippocampal levels of glutamic acid decarboxylase (GAD1), the rate-limiting enzyme in GABA synthesis, were found in the male offspring of low LG dams associated with increased DNA methylation within the GAD1 promoter (Zhang et al. 2010). These findings suggest that variation in the frequency of mother–infant interactions can lead to stable changes in gene expression through the modification of epigenetic marks within the genome. Although the adaptive advantage of these epigenetic effects has not been explored extensively, it is likely that this environmental imprinting plays an important role in the developmental plasticity of offspring (Chapter 14).

17.6.3 Transgenerational impact of maternal care

Across species, there is evidence for the transmission of individual differences in maternal behaviour across multiple generations (for review see Champagne 2008). Moreover, when cross-fostering is used within these studies, it becomes clear that the inheritance of maternal behaviour is not necessarily dependent on the transmission of genetic variation from mothers (F0) to daughters (F1) and grand-daughters (F2). Our increasing understanding of early-life influences on the developing maternal brain and the epigenetic effects of maternal care may provide insights into the mechanism of the transgenerational continuity of maternal behaviour. In rodents, exposure of female pups to low levels of LG leads to reduced expression of ERα in the MPOA (Champagne et al. 2003). These effects on ERα expression are sustained into adulthood with consequences for the estrogen sensitivity of these females, resulting in reduced maternal behaviour amongst offspring who have received low levels of LG (Champagne et al. 2001; Champagne et al. 2003). Analysis of the 1B promoter region of the ERα gene in MPOA tissue implicates DNA methylation as a potential mediator of these maternal effects. At several sites within the ERα promoter there is elevated DNA methylation associated with exposure to low levels of LG (Champagne et al. 2006). Maternal effects on DNA methylation within the ER promoter may induce a transgenerational continuity in maternal behaviour such that amongst mothers, daughters, and grand-daughters, there is a high correlation in the frequency of LG behaviour (Champagne 2008).

Variations in mother–infant interactions can also be induced in laboratory rodents with consequences for the epigenetic regulation of genes and the transmission of maternal care. Disruption to the nesting materials within a female rodent's home cage can lead to abusive care-giving characterized by increased frequency of rough handling, dragging, dropping, and stepping on pups (Roth and Sullivan 2005). Daily exposure to this form of abusive social interaction leads to reduced expression of brain-derived neurotrophic factor (BDNF) in the pre-frontal cortex in adulthood associated with increased DNA methylation within the IV BDNF promotor region (Roth et al. 2009). Female rat pups exposed to abusive care-giving in infancy engage in abusive care-giving toward their own offspring and F2 offspring of these F1 females have elevated levels of DNA methylation within the BDNF promoter in the cortex and hippocampus. Females rearing offspring in a communal nest display elevated levels of nursing and LG, and when rearing their own offspring (under standard conditions) both F1 and F2 communal females exhibit increased maternal behaviour (Curley et al. 2009). Experience-dependent inheritance may also be an important consideration in the transmission of genetically induced variations in maternal behaviour. The deficits in maternal behaviour induced by mutation of Peg3 can be observed in F1 and F2 female offspring (Curley et al. 2008) despite the epigenetic silencing of the mutant Peg3 allele in these females (because Peg3 is paternally expressed, see above). In this case, we hypothesize that although genetic factors are critical in inducing the initial behavioural variation, the persistence of these effects in F1 and F2 offspring, that is the transmission across generations, is mediated by the epigenetic influence of maternal care.

17.7 Conclusions

Although diverse forms of parental care have evolved to promote survival, growth, and development of offspring (Chapter 1), when considering the genetics and epigenetics of parental care, there are several common themes that emerge. First, it is apparent that genes which promote or inhibit parental behaviour often also shape the ability of offspring to solicit resources. The parent–offspring relationship, regardless of species, consists of reciprocal and dynamic interactions. Offspring are not passive recipients of care and parents have evolved strategies that match or compensate for the particular features of their biological offspring. This phenomenon is most evident when there is a mis-match between parent and offspring characteristics, as is evident in cross-fostering studies conducted between strains (see Fig. 17.1) and between gene knockout heterozygous, homozygous, and WT mice. This parent–offspring co-adaptation (Chapter 16) will provide individuals with an advantage in terms of growth, survival, and/or reproduction. When considering within species variation in parental and infant genotype/phenotype, particularly in genetically heterogeneous populations, environmental experiences can moderate the genetic match between parents and offspring as exemplified by the studies of gene polymorphisms in the 5-HTT, OPRM1, and DRD4 alleles in humans discussed in this chapter. Genetic constraints on the range of parental care may therefore have evolved to promote compatibility of parental and offspring genotypes whereas environmentally induced plasticity may allow for adaptations within that range to occur when there are more subtle genetic dissimilarities between the parent–offspring dyad.

A second theme that emerges from research on the genetics and epigenetics of parental care, suggests that although there is a certain degree of specificity regarding the facet of parental care regulated by particular genes or loci, these genes are also typically involved in many other non-parental aspects of physiological and behavioural functioning. Rather than being a unitary trait, parental care consists of multiple metabolic, physiological, and behavioural characteristics, a complexity suggesting that there will be a significant overlap in the function of genes associated with parental care with other fitness-related traits. This overlap is particularly apparent in gene knockout studies in which even a single gene mutation can induce widespread changes in reproduction, activity levels, exploratory behaviour, and metabolism. One explanation for the breadth of these effects is the cascade of changes in gene expression that may occur in the presence of a developmental mutation in a single target gene. Conditional gene knockout and RNA interference techniques may provide useful tools for disentangling these developmental effects from the role of target genes at the time of parental care. However, analysis of the non-parental traits which covary with parental care may provide greater insights into the evolutionary processes through which variation in parental care emerges and is maintained.

Finally, it is apparent that both genes and the environment play a critical role in shaping the dynamics of parental care, and epigenetic effects may provide a mechanism through which these influences can interact and lead to stable and potentially heritable variations in phenotype. The interplay between genes and the environment that can lead to altered gene expression through epigenetic processes such as DNA methylation, may play a critical role driving phenotypic variation of both parental and offspring traits. Consequently it seems likely that selection does not act on genes involved in parental care and offspring solicitation in isolation, but also acts on mechanisms involved in the epigenetic regulation of genes during development. These mechanisms in turn may affect how parent and offspring behaviours can be transmitted across generations.

Acknowledgements

This research was supported by Grant Number DP2OD001674 from the Office of the Director, National Institutes of Health.

References

Alaux, C., Le Conte, Y., Adams, H. A., Rodriguez-Zas, S., Grozinger, C. M., Sinha, S., and Robinson, G. E. (2009). Regulation of brain gene expression in honey bees by brood pheromone. *Genes Brain and Behavior* 8, 309–19.

Amdam, G. V. and Page, R. E. (2010). The developmental genetics and physiology of honeybee societies. *Animal Behaviour* 79, 973–80.

Arnold, A. P. and Chen, X. (2009). What does the 'four core genotypes' mouse model tell us about sex differences in the brain and other tissues? *Frontiers in Neuroendocrinology* 30, 1–9.

Bakermans-Kranenburg, M. J., and Van Ijzendoorn, M. H. (2008). Oxytocin receptor (OXTR) and serotonin transporter (5-HTT) genes associated with observed parenting. *Social Cognitive and Affective Neuroscience* 3, 128–34.

Barr, C. S., Schwandt, M. L., Lindell, S. G., Higley, J. D., Maestripieri, D., Goldman, D., Suomi, S. J., and Heilig, M. (2008). Variation at the mu-opioid receptor gene (OPRM1) influences attachment behavior in infant primates. *Proceedings of the National Academy of Sciences USA* 105, 5277–81.

Barry, R. A., Kochanska, G., and Philibert, R. A. (2008). G x E interaction in the organization of attachment: mothers' responsiveness as a moderator of children's genotypes. *Journal of Child Psychology and Psychiatry* 49, 1313–20.

Bischof, J. M., Stewart, C. L., and Wevrick, R. (2007). Inactivation of the mouse Magel2 gene results in growth abnormalities similar to Prader-Willi syndrome. *Human Molecular Genetics* 16, 2713–19.

Bosch, O. J. (2011). Maternal nurturing is dependent on her innate anxiety: the behavioral roles of brain oxytocin and vasopressin. *Hormones and Behavior* 59, 202–12.

Broad, K. D., Curley, J. P., and Keverne, E. B. (2009). Increased apoptosis during neonatal brain development underlies the adult behavioral deficits seen in mice lacking a functional paternally expressed gene 3 (Peg3). *Developmental Neurobiology* 69, 314–25.

Brown, J. R., Ye, H., Bronson, R. T., Dikkes, P., and Greenberg, M. E. (1996). A defect in nurturing in mice lacking the immediate early gene fosB. *Cell* 86, 297–309.

Brunelli, S. A., Zimmerberg, B., and Hofer, M. A. (2010). Developmental effects of selective breeding for an infant trait. In K. E. Hood, C. Tucker Halpern, G. Greenberg, and R. M. Lerner, eds. *Handbook of Developmental Science, Behavior, and Genetics* John Wiley & Sons Ltd, Chichester, UK.

Calderone, N. W. and Page, R. E. (1992). Effects of interactions among genotypically diverse nestmates on task specialization by foraging honey bees (Apis mellifer). *Behavioral Ecology and Sociobiology* 30, 219–26.

Champagne, F., Diorio, J., Sharma, S., and Meaney, M. J. (2001). Naturally occurring variations in maternal behavior in the rat are associated with differences in estrogen-inducible central oxytocin receptors. *Proceedings of the National Academy of Sciences USA* 98, 12736–41.

Champagne, F. A. (2008). Epigenetic mechanisms and the transgenerational effects of maternal care. *Frontiers in Neuroendocrinology* 29, 386–97.

Champagne, F. A., Curley, J. P., Keverne, E. B., and Bateson, P. P. (2007). Natural variations in postpartum maternal care in inbred and outbred mice. *Physiology & Behavior* 91, 325–34.

Champagne, F. A., Curley, J. P., Swaney, W. T., Hasen, N. S., and Keverne, E. B. (2009). Paternal influence on female behavior: the role of Peg3 in exploration, olfaction, and neuroendocrine regulation of maternal behavior of female mice. *Behavioral Neuroscience* 123, 469–80.

Champagne, F. A., Weaver, I. C., Diorio, J., Dymov, S., Szyf, M., and Meaney, M. J. (2006). Maternal care associated with methylation of the estrogen receptor-alpha1b promoter and estrogen receptor-alpha expression in the medial preoptic area of female offspring. *Endocrinology* 147, 2909–15.

Champagne, F. A., Weaver, I. C., Diorio, J., Sharma, S., and Meaney, M. J. (2003). Natural variations in maternal care are associated with estrogen receptor alpha expression and estrogen sensitivity in the medial preoptic area. *Endocrinology* 144, 4720–4.

Chen, C., Guo, Y., Yang, G., Yang, Z., Zhang, Z., Yang, B., Yan, X., Perez-Enciso, M., Ma, J., Duan, Y., Brenig, B., and Huang, L. (2009). A genome wide detection of quantitative trait loci on pig maternal infanticide behavior in a large scale White Duroc x Erhualian resource population. *Behavior Genetics* 39, 213–19.

Chen, C., Yang, Z., Li, Y., Wei, N., Li, P., Guo, Y., Ren, J., Ding, N., and Huang, L. (2011). Association and haplotype analysis of candidate genes in five genomic regions linked to sow maternal infanticide in a white Duroc x Erhualian resource population. *BMC Genetics* 12, 24.

Cierpial, M. A., Murphy, C. A., and Mccarty, R. (1990). Maternal behavior of spontaneously hypertensive and Wistar-Kyoto normotensive rats: effects of reciprocal cross-fostering of litters. *Behavioral and Neural Biology* 54, 90–6.

Constancia, M., Kelsey, G., and Reik, W. (2004). Resourceful imprinting. *Nature* 432, 53–7.

Curley, J. P. (2011). The mu-opioid receptor and the evolution of mother-infant attachment: theoretical comment on Higham et al. (2011). *Behavioral Neuroscience* 125, 273–8.

Curley, J. P., Barton, S., Surani, A., and Keverne, E. B. (2004). Coadaptation in mother and infant regulated by a paternally expressed imprinted gene. *Proceedings of the Royal Society of London, Series B* 271, 1303–9.

Curley, J. P., Champagne, F. A., Bateson, P., and Keverne, E. B. (2008). Transgenerational effects of impaired maternal care on behaviour of offspring and grandoffspring. *Animal Behaviour* 75, 1551–61.

Curley, J. P., Davidson, S., Bateson, P., and Champagne, F. A. (2009). Social enrichment during postnatal development induces transgenerational effects on emotional and reproductive behavior in mice. *Frontiers in Behavioral Neuroscience* 3, 25.

Curley, J. P., Rock, V., Moynihan, A. M., Bateson, P., Keverne, E. B., and Champagne, F. A. (2010). Developmental shifts in the behavioral phenotypes of inbred mice: the role of postnatal and juvenile social experiences. *Behavior Genetics* 40, 220–32.

Donaldson, Z. R. and Young, L. J. (2008). Oxytocin, vasopressin, and the neurogenetics of sociality. *Science* 322, 900–4.

Drake, N. M., Park, Y. J., Shirali, A. S., Cleland, T. A., and Soloway, P. D. (2009). Imprint switch mutations at Rasgrf1 support conflict hypothesis of imprinting and define a growth control mechanism upstream of IGF1. *Mammalian Genome* 20, 654–63.

Evans, J. D. and Wheeler, D. E. (1999). Differential gene expression between developing queens and workers in the honey bee, Apis mellifera. *Proceedings of the National Academy of Sciences USA* 96, 5575–80.

Fink, S., Excoffier, L., and Heckel, G. (2006). Mammalian monogamy is not controlled by a single gene. *Proceedings of the National Academy of Sciences USA* 103, 10956–60.

Gammie, S. C., Edelmann, M. N., Mandel-Brehm, C., D'Anna, K. L., Auger, A. P., and Stevenson, S. A. (2008). Altered dopamine signaling in naturally occurring maternal neglect. *PLoS ONE* 3, e1974.

Gammie, S. C., Garland, T., Jr., and Stevenson, S. A. (2006). Artificial selection for increased maternal defense behavior in mice. *Behavior Genetics* 36, 713–22.

Gammie, S. C., Hasen, N. S., Rhodes, J. S., Girard, I., and Garland, T., Jr. (2003). Predatory aggression, but not maternal or intermale aggression, is associated with high voluntary wheel-running behavior in mice. *Hormones and Behavior* 44, 209–21.

Gatewood, J. D., Wills, A., Shetty, S., Xu, J., Arnold, A. P., Burgoyne, P. S., and Rissman, E. F. (2006). Sex chromosome complement and gonadal sex influence aggressive and parental behaviors in mice. *Journal of Neuroscience* 26, 2335–42.

Girard, I., Swallow, J. G., Carter, P. A., Koteja, P., Rhodes, J. S., and Garland, T. (2002). Maternal-care behavior and life-history traits in house mice (Mus domesticus) artificially selected for high voluntary wheel-running activity. *Behavioural Processes* 57, 37–50.

Godfrey, K. M. and Barker, D. J. (2001). Fetal programming and adult health. *Public Health Nutrition* 4, 611–24.

Gregg, C., Zhang, J., Weissbourd, B., Luo, S., Schroth, G. P., Haig, D., and Dulac, C. (2010). High-resolution analysis of parent-of-origin allelic expression in the mouse brain. *Science* 329, 643–8.

Groothuis, T. G., Carere, C., Lipar, J., Drent, P. J., and Schwabl, H. (2008). Selection on personality in a songbird affects maternal hormone levels tuned to its effect on timing of reproduction. *Biology Letters* 4, 465–7.

Hager, R. and Johnstone, R. A. (2003). The genetic basis of family conflict resolution in mice. *Nature* 421, 533–5.

Heijmans, B. T., Tobi, E. W., Stein, A. D., Putter, H., Blauw, G. J., Susser, E. S., Slagboom, P. E., and Lumey, L. H. (2008). Persistent epigenetic differences associated with prenatal exposure to famine in humans. *Proceedings of the National Academy of Sciences USA* 105, 17046–9.

Higham, J. P., Barr, C. S., Hoffman, C. L., Mandalaywala, T. M., Parker, K. J., and Maestripieri, D. (2011). Mu-opioid receptor (OPRM1) variation, oxytocin levels and maternal attachment in free-ranging rhesus macaques Macaca mulatta. *Behavioral Neuroscience* 125, 131–6.

Holland, H. C. (1965). An apparatus note on A.M.B.A. (automatic maternal behaviour apparatus). *Animal Behaviour* 13, 201–2.

Hollander, F. A., Van Overveld, T., Tokka, I., and Matthysen, E. (2008). Personality and nest defence in the Great Tit (Parus major). *Ethology* 114, 405–12.

Hunt, G. J., Amdam, G. V., Schlipalius, D., Emore, C., Sardesai, N., Williams, C. E., Rueppell, O., Guzman-Novoa, E., Arechavaleta-Velasco, M., Chandra, S., Fondrk, M. K., Beye, M., and Page, R. E., Jr. (2007). Behavioral genomics of honeybee foraging and nest defense. *Naturwissenschaften*, 94, 247–67.

Ideraabdullah, F. Y., Vigneau, S., and Bartolomei, M. S. (2008). Genomic imprinting mechanisms in mammals. *Mutation Research* 647, 77–85.

Insel, T. R. and Young, L. J. (2000). Neuropeptides and the evolution of social behavior. *Current Opinion in Neurobiology* 10, 784–9.

Kaitz, M., Shalev, I., Sapir, N., Devor, N., Samet, Y., Mankuta, D., and Ebstein, R. P. (2010). Mothers' dopamine receptor polymorphism modulates the relation between infant fussiness and sensitive parenting. *Developmental Psychobiology* 52, 149–57.

Keverne, E. B. and Curley, J. P. (2008). Epigenetics, brain evolution and behaviour. *Frontiers in Neuroendocrinology* 29, 398–412.

Kozlov, S. V., Bogenpohl, J. W., Howell, M. P., Wevrick, R., Panda, S., Hogenesch, J. B., Muglia, L. J., Van Gelder, R. N., Herzog, E. D., and Stewart, C. L. (2007). The imprinted gene Magel2 regulates normal circadian output. *Nature Genetics* 39, 1266–72.

Kucharski, R., Maleszka, J., Foret, S., and Maleszka, R. (2008). Nutritional control of reproductive status in honeybees via DNA methylation. *Science* 319, 1827–30.

Kuroda, K. O., Tachikawa, K., Yoshida, S., Tsuneoka, Y., and Numan, M. (2011). Neuromolecular basis of parental behavior in laboratory mice and rats: with special emphasis on technical issues of using mouse genetics. *Progress in Neuropsychopharmacology and Biological Psychiatry* 35, 1205–31.

Lakatos, K., Nemoda, Z., Toth, I., Ronai, Z., Ney, K., Sasvari-Szekely, M., and Gervai, J. (2002). Further evidence for the role of the dopamine D4 receptor (DRD4) gene in attachment disorganization: interaction of the exon III 48-bp repeat and the -521 C/T promoter polymorphisms. *Molecular Psychiatry* 7, 27–31.

Lapidge, K. L., Oldroyd, B. P., and Spivak, M. (2002). Seven suggestive quantitative trait loci influence hygienic behavior of honey bees. *Naturwissenschaften* 89, 565–8.

Lavi-Avnon, Y., Yadid, G., Overstreet, D. H., and Weller, A. (2005). Abnormal patterns of maternal behavior in a genetic animal model of depression. *Physiology and Behavior* 84, 607–15.

Lefebvre, L., Viville, S., Barton, S. C., Ishino, F., Keverne, E. B., and Surani, M. A. (1998). Abnormal maternal behaviour and growth retardation associated with loss of the imprinted gene Mest. *Nature Genetics* 20, 163–9.

Lesch, K. P., Bengel, D., Heils, A., Sabol, S. Z., Greenberg, B. D., Petri, S., Benjamin, J., Muller, C. R., Hamer, D. H., and Murphy, D. L. (1996). Association of anxiety-related traits with a polymorphism in the serotonin transporter gene regulatory region. *Science* 274, 1527–31.

Li, L., Keverne, E. B., Aparicio, S. A., Ishino, F., Barton, S. C., and Surani, M. A. (1999). Regulation of maternal behavior and offspring growth by paternally expressed Peg3. *Science* 284, 330–3.

Lillycrop, K. A., Phillips, E. S., Jackson, A. A., Hanson, M. A., and Burdge, G. C. (2005). Dietary protein restriction of pregnant rats induces and folic acid supplementation prevents epigenetic modification of hepatic gene expression in the offspring. *Journal of Nutrition* 135, 1382–6.

Lillycrop, K. A., Phillips, E. S., Torrens, C., Hanson, M. A., Jackson, A. A., and Burdge, G. C. (2008). Feeding pregnant rats a protein-restricted diet persistently alters the methylation of specific cytosines in the hepatic PPAR alpha promoter of the offspring. *British Journal of Nutrition* 100, 278–82.

Lillycrop, K. A., Slater-Jefferies, J. L., Hanson, M. A., Godfrey, K. M., Jackson, A. A., and Burdge, G. C. (2007). Induction of altered epigenetic regulation of the hepatic glucocorticoid receptor in the offspring of rats fed a protein-restricted diet during pregnancy suggests that reduced DNA methyltransferase-1 expression is involved in impaired DNA methylation and changes in histone modifications. *British Journal of Nutrition* 97, 1064–73.

Lucas, B. K., Ormandy, C. J., Binart, N., Bridges, R. S., and Kelly, P. A. (1998). Null mutation of the prolactin receptor gene produces a defect in maternal behavior. *Endocrinology* 139, 4102–7.

MacColl, A. D. and Hatchwell, B. J. (2003). Heritability of parental effort in a passerine bird. *Evolution* 57, 2191–5.

Marco Antonio, D. S., Guidugli-Lazzarini, K. R., Do Nascimento, A. M., Simoes, Z. L., and Hartfelder, K. (2008). RNAi-mediated silencing of vitellogenin gene function turns honeybee (Apis mellifera) workers into extremely precocious foragers. *Naturwissenschaften* 95, 953–61.

McCormack, K., Newman, T. K., Higley, J. D., Maestripieri, D., and Sanchez, M. M. (2009). Serotonin transporter gene variation, infant abuse, and responsiveness to stress in rhesus macaque mothers and infants. *Hormones and Behavior* 55, 538–47.

McGaugh, S. E., Schwanz, L. E., Bowden, R. M., Gonzalez, J. E., and Janzen, F. J. (2010). Inheritance of nesting behaviour across natural environmental variation in a turtle with temperature-dependent sex determination. *Proceedings of the Royal Society B: Biological Sciences* 277, 1219–26.

McGraw, L. A. and Young, L. J. (2010). The prairie vole: an emerging model organism for understanding the social brain. *Trends in Neuroscience* 33, 103–9.

McIver, A. H. and Jeffrey, W. E. (1967). Strain differences in maternal behavior in rats. *Behaviour* 28, 210–16.

Meaney, M. J. (2001). Maternal care, gene expression, and the transmission of individual differences in stress reactivity across generations. *Annual Review of Neuroscience* 24, 1161–92.

Mercer, R. E., Kwolek, E. M., Bischof, J. M., Van Eede, M., Henkelman, R. M., and Wevrick, R. (2009). Regionally reduced brain volume, altered serotonin neurochemistry, and abnormal behavior in mice null for the circadian rhythm output gene Magel2. *American Journal of Medical Genetics Part B: Neuropsychiatric Genetics* 150B, 1085–99.

Mercer, R. E. and Wevrick, R. (2009). Loss of magel2, a candidate gene for features of Prader-Willi syndrome, impairs reproductive function in mice. *PLoS ONE* 4, e4291.

Messina, F. J. (1993). Heritability and 'evolvability' of fitness components in Callosobruchus maculatus. *Heredity* 71, 623–9.

Mileva-Seitz, V., Kennedy, J., Atkinson, L., Steiner, M., Levitan, R., Matthews, S. G., Meaney, M. J., Sokolowski, M. B., and Fleming, A. S. (2011). Serotonin transporter allelic variation in mothers predicts maternal sensitivity, behavior and attitudes toward 6-month-old infants. *Genes Brain and Behavior* 10, 325–333.

Myers, M. M., Brunelli, S. A., Shair, H. N., Squire, J. M., and Hofer, M. A. (1989). Relationships between maternal behavior of SHR and WKY dams and adult blood pressures of cross-fostered F1 pups. *Developmental Psychobiology* 22, 55–67.

Nishimori, K., Young, L. J., Guo, Q., Wang, Z., Insel, T. R., and Matzuk, M. M. (1996). Oxytocin is required for nursing but is not essential for parturition or reproductive behavior. *Proceedings of the National Academy of Sciences USA* 93, 11699–704.

Ogawa, S., Eng, V., Taylor, J., Lubahn, D. B., Korach, K. S., and Pfaff, D. W. (1998). Roles of estrogen receptor-alpha gene expression in reproduction-related behaviors in female mice. *Endocrinology* 139, 5070–81.

Olazabal, D. E. (2010). Stability and potential inheritance of infanticidal behavior in prairie voles. *Developmental Psychobiology* 52, 825–32.

Olazabal, D. E. and Young, L. J. (2006). Species and individual differences in juvenile female alloparental care are associated with oxytocin receptor density in the striatum and the lateral septum. *Hormones and Behavior* 49, 681–7.

Oxley, P. R., Spivak, M., and Oldroyd, B. P. (2010). Six quantitative trait loci influence task thresholds for hygienic behaviour in honeybees (Apis mellifera). *Molecular Ecology* 19, 1452–61.

Peripato, A. C., De Brito, R. A., Vaughn, T. T., Pletscher, L. S., Matioli, S. R., and Cheverud, J. M. (2002). Quantitative trait loci for maternal performance for offspring survival in mice. *Genetics* 162, 1341–53.

Perrigo, G. and Vom Saal, F. S. (1994). Behavioral cycles and the neural timing of infanticide and parental behavior in male house mice. In F. S. Vom Saal and S. Parmigiani, eds. *Infanticide and Parental Care*. Harwood Academic Publishers Chur, Switzerland.

Perrigo, G., Belvin, L., and Vom Saal, F. S. (1991). Individual variation in the neural timing of infanticide and parental behavior in male house mice. *Physiology and Behavior* 50, 287–96.

Perusse, D., Neale, M. C., Heath, A. C., and Eaves, L. J. (1994). Human parental behavior: evidence for genetic influence and potential implication for gene-culture transmission. *Behavior Genetics* 24, 327–35.

Plagge, A., Gordon, E., Dean, W., Boiani, R., Cinti, S., Peters, J., and Kelsey, G. (2004). The imprinted signaling protein XL alpha s is required for postnatal adaptation to feeding. *Nature Genetics* 36, 818–26.

Reisert, I., Karolczak, M., Beyer, C., Just, W., Maxson, S. C., and Ehret, G. (2002). Sry does not fully sex-reverse female into male behavior towards pups. *Behavior Genetics* 32, 103–11.

Roth, T. L., Lubin, F. D., Funk, A. J., and Sweatt, J. D. (2009). Lasting epigenetic influence of early-life adversity on the BDNF gene. *Biological Psychiatry* 65, 760–9.

Roth, T. L. and Sullivan, R. M. (2005). Memory of early maltreatment: neonatal behavioral and neural correlates of maternal maltreatment within the context of classical conditioning. *Biological Psychiatry* 57, 823–31.

Rothenbuhler, W. C. (1964). Behavior genetics of nest cleaning in honey bees. iv. responses of f1 and backcross generations to disease-killed blood. *American Zoologist* 4, 111–23.

Roulin, A., Dreiss, A. N., and Kolliker, M. (2010). Evolutionary perspective on the interplay between family life, and parent and offspring personality. *Ethology* 116, 787–96.

Sambandan, D., Yamamoto, A., Fanara, J. J., Mackay, T. F., and Anholt, R. R. (2006). Dynamic genetic interactions determine odor-guided behavior in Drosophila melanogaster. *Genetics* 174, 1349–63.

Schneider, J. S., Stone, M. K., Wynne-Edwards, K. E., Horton, T. H., Lydon, J., O'Malley, B., and Levine, J. E. (2003). Progesterone receptors mediate male aggression toward infants. *Proceedings of the National Academy of Sciences USA* 100, 2951–6.

Singer, M. C., Ng, D., and Thomas, C. D. (1988). Heritability of oviposition preference and its relationship to offspring performance within a single insect population. *Evolution* 42, 977–85.

Spangler, G., Johann, M., Ronai, Z., and Zimmermann, P. (2009). Genetic and environmental influence on attachment disorganization. *Journal of Child Psychology and Psychiatry* 50, 952–61.

Symonds, M. E., Stephenson, T., Gardner, D. S., and Budge, H. (2007). Long-term effects of nutritional programming of the embryo and fetus: mechanisms and critical windows. *Reproduction, Fertility and Development* 19, 53–63.

Tschirren, B., Sendecka, J., Groothuis, T. G., Gustafsson, L., and Doligez, B. (2009). Heritable variation in maternal yolk hormone transfer in a wild bird population. *American Naturalist* 174, 557–64.

Van Ijzendoorn, M. H., Bakermans-Kranenburg, M. J., and Mesman, J. (2008). Dopamine system genes associated with parenting in the context of daily hassles. *Genes Brain and Behavior* 7, 403–10.

Wang, Y., Amdam, G. V., Rueppell, O., Wallrichs, M. A., Fondrk, M. K., Kaftanoglu, O., and Page, R. E., Jr. (2009). PDK1 and HR46 gene homologs tie social behavior to ovary signals. *PLoS ONE* 4, e4899.

Wang, Y., Jorda, M., Jones, P. L., Maleszka, R., Ling, X., Robertson, H. M., Mizzen, C. A., Peinado, M. A., and Robinson, G. E. (2006). Functional CpG methylation system in a social insect. *Science* 314, 645–7.

Weaver, I. C., Cervoni, N., Champagne, F. A., D'Alessio, A. C., Sharma, S., Seckl, J. R., Dymov, S., Szyf, M. and Meaney, M. J. (2004). Epigenetic programming by maternal behavior. *Nature Neuroscience* 7, 847–54.

Weller, A., Leguisamo, A. C., Towns, L., Ramboz, S., Bagiella, E., Hofer, M., Hen, R., and Brunner, D. (2003). Maternal effects in infant and adult phenotypes of 5HT1A and 5HT1B receptor knockout mice. *Developmental Psychobiology* 42, 194–205.

Whitfield, C. W., Cziko, A. M., and Robinson, G. E. (2003). Gene expression profiles in the brain predict behavior in individual honey bees. *Science* 302, 296–9.

Wilkins, J.F. and Haig, D. (2003). What good is genomic imprinting: the function of parent-specific gene expression. *Nature Genetics* 4, 1–10.

Xiang, M., Gan, L., Zhou, L., Klein, W. H., and Nathans, J. (1996). Targeted deletion of the mouse POU domain gene Brn-3a causes selective loss of neurons in the brainstem and trigeminal ganglion, uncoordinated limb movement, and impaired suckling. *Proceedings of the National Academy of Sciences USA* 93, 11950–5.

Yagi, T., Aizawa, S., Tokunaga, T., Shigetani, Y., Takeda, N., and Ikawa, Y. (1993). A role for Fyn tyrosine kinase in the suckling behaviour of neonatal mice. *Nature* 366, 742–5.

Young, L. J., Huot, B., Nilsen, R., Wang, Z., and Insel, T. R. (1996). Species differences in central oxytocin receptor gene expression: comparative analysis of promoter sequences. *Journal of Neuroendocrinology* 8, 777–83.

Zhang, T. Y., Hellstrom, I. C., Bagot, R. C., Wen, X., Diorio, J., and Meaney, M. J. (2010). Maternal care and DNA methylation of a glutamic acid decarboxylase 1 promoter in rat hippocampus. *Journal of Neuroscience* 30, 13130–7.

SECTION IV

Conclusions

CHAPTER 18

The evolution of parental care: summary, conclusions, and implications

Nick J. Royle, Per T. Smiseth, and Mathias Kölliker

18.1 Introduction

As the light begins to fade in a broad-leafed woodland in England, a male *Nicrophorus vespilloides* burying beetle, guided by olfactory cues picked up by his feathery antennae, alights on the carcass of a gently decomposing wood mouse. After a thorough inspection of the carcass, he stands on top of the mouse, lowers his head, raises his abdomen, and squirts a cocktail of pheromones into the air. The response to his calls is rapid. A female *N. vespilloides* arrives within minutes. He mates with her almost as soon as she lands. But the calls have also attracted another male. As the female begins the process of burying the mouse the male defends the prized carcass from his rival. They are evenly matched in size but the fight is brief, with the resource-holding male victorious and his defeated rival sent scuttling through the leaf litter to look for opportunities elsewhere. The victorious male and the female jointly bury the carcass, pausing only to mate, lay eggs in the soil nearby, and defend the carcass against other species of marauding beetles and slugs, before they strip the mouse of fur. By the time the eggs hatch, some 60 hours after laying, the larvae wriggle their way to the carcass where the parents have created a bowl-shaped crater in the body of the mouse containing a soup of partially digested meat. In addition to feeding directly on this soup, the tiny larvae also rear up and touch the parents with their legs, begging to be fed. The female obliges by regurgitating food for them. Meanwhile the male industriously patrols the carcass, keeping it free of microbes and mould. Three days after hatching parental interest in the larvae wanes and the rapidly growing larvae self-feed on the carcass for another couple of days before leaving to wander in search of a suitable place to pupate.

If you are familiar with *Nicrophorus* burying beetles, whose common name in England is the sexton beetle after their 'grave-digging' exploits, then the natural history account sketched above will come as no surprise. To most other people however this level of complex parental care in an insect might raise an eyebrow at the very least. Why do burying beetles provide parental care at all? How did it evolve from a non-caring ancestor? Why do parents go to all the trouble of defending the carcass against all comers, and actively provision the prepared food to begging offspring? Why not just lay some eggs and then leave them to hatch, grow, and develop on their own like the blow flies that also buzz around the carcass? Parental care is often costly to the parents, so how and why does it persist once it has evolved? Why do parents stick around after laying eggs next to a substantial source food for their offspring rather than use it to enhance their own survival and the production of a larger number of offspring? What determines whether it is males or females or both parents that stay to provide care? Why do males and females often have different roles? How much care should be provided, how much should each parent contribute, and how should available resources be allocated among individual offspring?

These sorts of questions are applicable to any species with parental care and are by no means limited to burying beetles. Parental care is taxonom-

The Evolution of Parental Care. First Edition. Edited by Nick J. Royle, Per T. Smiseth, and Mathias Kölliker.

ically widespread (see Chapters 1, 4, and 5) and is central to the maintenance of biodiversity through its close association with other phenomena, such as sexual selection, life-history evolution, sex allocation, sociality, cooperation and conflict, growth and development, genetic architecture, and phenotypic plasticity. This book provides an overview of the current state of the study of the evolution of parental care, and the previous 17 chapters provide in-depth analyses of the major themes. In this chapter, we provide a summary of the key points addressed in the previous chapters, draw some conclusions, and consider future directions in the field.

18.2 What is parental care?

Parental care is a purely descriptive term, which does not take account of any costs to parents of providing care, and can be broadly defined as 'any parental trait that increases the fitness of a parent's offspring, and is likely to have originated and/or is currently maintained for this function' (Chapter 1). This definition includes non-behavioural traits, such as gamete provisioning, gestation, viviparity, and nests, but excludes mating behaviours that increase offspring fitness, such as the defence of breeding territories, unless there is evidence that such behaviours evolved and/or are currently maintained because they enhance offspring fitness (Chapter 1). The latter condition is the key to the definition. For instance, many researchers exclude gamete provisioning as a form of parental care because all animals show some form of gamete provisioning. If gamete provisioning were classified as parental care, then all species with anisogamy would potentially be classified as species with female care. So, should all species with anisogamy be considered as showing a form of female care for offspring? The answer is no. Anisogamy and provisioning of resources to gametes by females evolved due to gamete competition leading to disruptive selection favouring the production of small and competitive gametes (sperm) by some individuals (males) and the production of large and nutrient-rich gametes (eggs) by other individuals (females) (Parker et al. 1972). Gamete provisioning should only be included as a form of parental care if it can be demonstrated that the level of provisioning is beyond the minimum level needed for successful fertilization, and that the increase in provisioning evolved or is currently maintained to enhance offspring fitness.

There is an important difference between a conceptual definition of parental care and operational modifications to it. Conceptually, the definition has to be broad and should be based on the ultimate functional context. Operationally, a definition of parental care can include useful proximate criteria that are specifically suited to the question being addressed. The latter approach may be more appropriate when discussing the origins of particular forms of care from a state of no care and when considering patterns of care across species or between the sexes (Chapter 2). Chapter 1 discusses in detail definitions of other key terms used in the study of parental care, and advocates a more consistent usage of these terms and definitions in future studies of parental care.

18.3 Origin and evolution of parental care

18.3.1 Costs and benefits

At the simplest level, parental care is expected to evolve when the benefits of providing care outweigh the costs (Chapters 2 and 3). Providing care is beneficial to offspring as it neutralizes environmental hazards, but is costly to parents as it utilizes resources that parents otherwise could have allocated to their own survival or future reproduction (Chapters 1 and 3). The benefits and costs are split across the generations, so parental care will be under antagonistic selection in parental and offspring life-stages (Chapters 1 and 16). There is selection for receiving care when individuals are offspring because care increases fitness, but there is selection against providing care when individuals are parents because the costs of care reduces fitness (Chapter 1). However, selection on offspring can generate a correlated response in parents due to the close relatedness of parents and offspring (Chapters 15 and 16), which allows parental care to evolve (Chapter 1). Understanding the evolution of parental care therefore requires separating out parental and offspring fitness to avoid double counting (e.g. erroneously assigning offspring survival to both parental and offspring fitness; Chapter 1).

Parental care can increase offspring fitness by affecting offspring survival directly or indirectly via effects on offspring phenotypes that, in turn, affect offspring survival (Chapter 14). Conversely, parental care can reduce parental survival and future mating and reproductive success (Chapter 2). Benefits to offspring may be short term, as in bromeliad crab where parental manipulation of pH in breeding pools through the addition of snail shells and removal of organic matter improves the conditions for offspring development. Alternatively, parental care may have delayed benefits to offspring, as in some species of bird where increased provision of carotenoids in eggs affects the expression of sexual ornaments in adulthood (Chapter 3). Other benefits such as direct provisioning of post-natal offspring by parents may have immediate (reduced probability of starvation), medium (increased growth and development), or long-term effects (e.g. increase in cognitive ability in adult blue tits *Cyanistes caerulens* due to a spider-rich diet, high in taurine, when a nestling; Arnold et al. 2007).

18.3.2 Evolutionary origins

Early attempts to understand the origin of parental care emphasized the role of environmental pressures, such as the harshness of the environment and the use of rich, but ephemeral, resources (Chapters 1 and 5). Whilst ecological conditions appear to be an important factor driving the evolution and diversification of care in some taxonomic groups (e.g. breeding pool size is associated with the evolution of parental care in frogs; Brown et al. 2010), the stability, structure, and harshness of the environment alone does not usually explain the evolutionary origins of care (Chapter 1). For example, extended parental care is relatively rare in insects, even though exposure to harsh environments is common (Chapter 5). Nevertheless, environmental variation may indirectly affect the likelihood that parental care evolved by altering patterns of mortality in parents or offspring, or developmental time in offspring (Chapter 2). For example, theoretical models show that selection favours parental care when egg or juvenile mortality in the absence of care is high, and the duration of the egg or juvenile stage is relatively long (Chapter 2). However, there has been surprisingly little empirical research on the relationship between ecology and the probability of the evolution of parental care, so this is therefore an important topic for future research.

The evolutionary origin of parental care may also be enhanced by the presence of behavioural precursors or incidental parental effects that can be modified into parental care. For example, guarding of eggs and offspring is likely to have evolved from ancestral defensive or aggressive behaviours in non-caring species (Chapter 1), especially when parents recognize and/or regularly encounter their own genetic offspring (and the benefits of care are not outweighed by the costs associated with increased competition with close kin, as may occur in viscous populations; Chapter 2). In some taxa, attendance of eggs is associated with an increase in egg size (Chapter 2), suggesting that large eggs may favour egg attendance and/or that egg attendance may favour large eggs. Once care has originated such co-evolutionary feedback loops may lead to the rapid evolution and diversification of other parental care traits, especially given that natural selection is expected to favour a tight phenotypic integration of parental traits and offspring development (e.g. altriciality) (Chapter 14). Thus, evolutionary loss of care may be rare once complex care has evolved (Gardner and Smiseth 2011) as the social environment provided by family members becomes an important determinant of development and reproduction and, hence, the persistence and maintenance of parental care.

18.3.3 The role of the social environment

The social environment is important to the evolution of parental care because behavioural interactions and the transfer of resources and information between parents and offspring shape the development of phenotypes. Parents transfer a diversity of non-genomic resources to offspring that are essential for development (e.g. maternally-derived mDNA and RNA and various proteins and hormones; Chapter 14). In fact, it is the transcriptional machinery that females add to eggs that allows the inherited DNA to be transcribed and have effects on offspring phenotype. Such parental effects often continue throughout offspring development, may have long-term effects, and take a variety of forms ranging from the pre-natal maternal transfer of transcriptional factors and macro-and micronutrients, through to provisioning of resources to young after hatching or birth and the behavioural transmission of information through learning or imprinting mechanisms. The origin of variation in parental effects begins with a responsive phenotype (phenotypic accommodation; Chapter 14). Such responsiveness to changes in the environment often has a genetic basis that varies among individuals, and is therefore heritable. Consequently, there is considerable scope for the genetic variation underlying phenotypic accommodation to be shaped by selection (genetic accommodation; Chapter 14). Phenotypic accommodation by parents therefore allows parental effects to be carried over across generations (e.g. pre-natal exposure to maternal hormones can epigenetically regulate gene expression in offspring; Chapters 14, 17, and Section 18.6.2), providing a link between the environment experienced by parents and that experienced by their offspring. Parental effects may therefore provide an important source of information for offspring if the environmental conditions experienced by parents and offspring are correlated (Chapter 14).

Parents often respond to environmental unpredictability by producing more offspring than can normally be reared. Such overproduction may represent a bet-hedging strategy, allowing parents to simultaneously track variable resources in the environment and increase their control over resource allocation within the brood. Mothers can potentially manipulate the social environment that offspring experience through hatching asynchrony, which in birds is largely the consequence of early onset of incubation, creating a structured family where first-hatched 'core' chicks enjoy substantial fitness advantages over last-hatched 'marginal' chicks (Mock and Parker 1997; Forbes 2009). The creation of a structured family through phenotypic handicapping of some offspring can ensure that parents can more easily match the brood size to the current conditions (resource tracking) and/or provide a form of insurance against the failure of core chicks (replacement offspring; Mock and Parker 1997). However, overproduction also means offspring demands often exceed parental supply, creating the social environmental conditions for conflicts of interest over parental investment among family members (Mock and Parker 1997; Chapter 8).

18.4 Conflicts and cooperation in parental care

18.4.1 Why are conflicts expected?

Evolutionary conflicts arise because parental care is an altruistic trait that incurs costs to the parents (Chapter 3) and because the providers of care (parents) and recipients (offspring) are not perfectly related to one another in sexual organisms (i.e. do not share all genes in common). This means that a behaviour that maximizes the fitness of one individual does not necessarily maximize the fitness of others (Chapters 7, 8, and 9). Sexual conflict among parents over parental investment is an almost inevitable consequence of sexual reproduction because the two parents are typically unrelated to one another, even if they do have shared interests in their common offspring and may cooperate to raise young together (i.e. biparental care; Chapter 9). Parent–offspring conflict (POC; Chapter 7) and sibling competition (Chapter 8) over parental care occur even though the individuals involved are close relatives. In these cases asymmetries in relatedness between parents and offspring and among siblings generates different optimal levels of care for parents and each individual offspring.

The extent of conflict is determined by the difference among optima for family members (the 'battleground'), or by the amount that each member's fitness is below its optima (the 'conflict load') (Chapter 9). As a result, there is evolutionary conflict even in situations where individuals are not directly interacting with one another and where there is no overt conflict or aggression among interacting individuals (Chapter 7). This is because the conflict refers to the way in which selection acts on family members, not the expression of behaviours such as fighting (Chapters 7 and 9). The outcome of these conflicts of interest (the 'resolution') is the critical determinant of how parental resources are transferred from parents to offspring (Parker et al. 2002), and affects the evolutionary stable mode of care expressed (i.e. uniparental or biparental care; Chapter 9).

18.4.2 Who should provide care?

Which sex provides care depends on a variety of factors including the mode of fertilization, how selection acts on males and females and the certainty of parentage. In external fertilizing species, trade-offs between growth and reproduction are important determinants of sex roles in parental care. In fish, for example, male-only care is more likely in species with external fertilization, which may be because larger females usually have higher fecundity and indeterminate growth, but reproduction diverts resources away from growth, favouring male-biased parental care (Gross 2005). In species with internal fertilization, whichever sex provides parental tissue is best placed to carry on providing care, which is most often, but not always (e.g. seahorses; Paczolt and Jones 2010), females (Chapter 6). However, the primary determinants of sex roles during parental care are sexual selection and the certainty of parentage (Chapters 2, 6, and 11).

In internally fertilizing species, maternity is typically more certain than paternity because females often store sperm from previous mating with other males, which has the effect of reducing the relatedness between a male and the offspring that he might care for (Chapter 11). Thus, sperm competition reduces the certainty of parentage, which in turn reduces the benefits of providing post-zygotic care to males (Chapters 2, 6, and 11). Sexual selection in males to locate unfertilized eggs increases the benefits of mating effort at the expense of parental effort for males. However, selection favours male parental care when the proportion of individuals available to mate in the population (the operational sex ratio; OSR) is very male biased, making the probability of success in mating very low (Chapter 6). In these circumstances it is better on average for males to invest in offspring that already exist (parental effort) rather than investing in future offspring (mating effort). As a result sex roles in parental care are determined by the complex relationships between OSR, sexual selection, certainty of parentage, and the adult sex ratio (see Fig. 18.1 and Chapter 6).

18.4.3 Sexual conflict over care

Models of offspring desertion predict that biparental care will be favoured when two parents are more than twice as good at raising offspring as a single parent (Chapter 9). Desertion of offspring is more likely when a single parent is nearly as effective as a pair, with the deserting parent most likely to be the one that is least effective at parenting and expressing the most strongly sexually-selected traits (typically, but not always, males; Chapters 6 and 9). However, the probability of desertion also depends on available opportunities elsewhere. If the OSR becomes more male biased and sexual selection intensifies, uniparental female care becomes more likely as the initial difference between the sexes in parental care becomes self-reinforcing and the number of successful males gets smaller and smaller (Lehtonen and Kokko 2012; Fig. 18.1). However, biparental care is favoured if sexual selection is relaxed at a high OSR, and the proportion of successful individuals of the more common sex increases (Chapter 6), due to reduced ability to monopolize matings, for example (Chapter 2).

Biparental care is a form of cooperation between parents to rear young together. However, when the costs of providing care are not aligned with the parents' future reproductive interests (i.e. no 'true' monogamy; Parker et al. 2002), there may be conflicts of interest over how much each parent should contribute to this joint investment. The majority of models of sexual conflict over care predict that the focal parent should respond by partial compensation to a change in investment by its partner; a prediction that is broadly supported by experimental empirical data across a wide range of species (Harrison et al. 2009). New models that incorporate costs of negotiation between parents show that when both parents care together for offspring, the ESS levels of relative investment are low early on in breeding. However, the levels increase throughout the period of parental care because opening bids of investment by one parent provides information

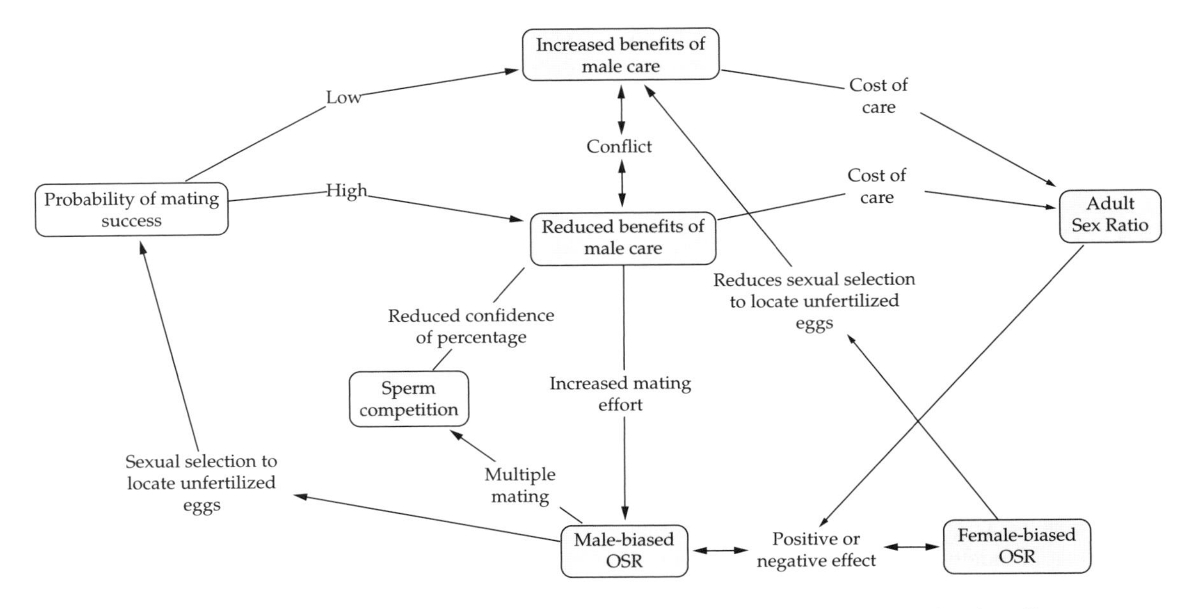

Figure 18.1 Who provides parental care? Anisogamy means that lots of small gametes (male) compete for access to small numbers of large gametes (female), which leads to a male-biased OSR. Sperm from several males seeking out the same egg before fertilization (multiple mating) results in sperm competition, which makes it harder for a male to identify his own young (lowering confidence of parentage), reducing the benefits of male care and increasing the benefits of investing resources in locating unfertilized eggs (mating effort), making the OSR even more male-biased. This leads to stronger sexual selection on males to locate unfertilized eggs, so that a male in a population with a male-biased OSR can prioritize investment either in his offspring that already exist (parental effort) or offspring that may never exist (mating effort). If the OSR becomes male biased this will reduce the probability of success in mating for the average male, which will select for increased male parental effort, but for a subset of males success in mating may increase, selecting for reduced parental effort and greater mating effort. This subset of males is important as selection acts on the parenting decisions of all males that mate. Whether uniparental care or biparental care evolves depends on the strength of the sexual selection on males, and other factors such as the independent effects of the adult sex ratio (ASR) on the OSR. If costs of parental care influence the ASR, through, for example, increased mortality of the caring sex, then this will feedback to the OSR, leading to relaxed sexual selection to locate unfertilized eggs and increased selection for male care (if the caring sex is female) because the caring sex is rarer in the population. Conversely, if the non-caring sex (males) experiences higher mortality then the caring sex becomes more common, strengthening sexual selection on males.

that can be exploited by the other parent, but later on in reproduction this is not possible (Lessells and McNamara 2012). As a result of such sexual conflict, parents are expected to withhold parental investment, thereby lowering the offspring's fitness (e.g. Royle et al. 2002a). Alternatively parents may monitor each other continuously throughout the period of parental care, such that the ESS is for parents to alternate who provides care; in which case the behavioural coordination between parents acts as a form of reciprocity that reduces sexual conflict (Johnstone et al. submitted; Chapter 9). The exact evolutionary outcome of sexual conflict over care depends not just upon the behavioural mechanisms used by parents to negotiate care, but also the shape of the offspring benefit curves and the effects of conflicts between other family members (Chapter 9).

18.4.4 Sibling competition, cooperation, and parental favouritism

The combination of initial overproduction of offspring by parents, relatedness asymmetries, and limited resources means that sibling relationships are often agonistic. Interactions among siblings can involve sublethal competition (e.g. non-aggressive scramble competition for parental resources as in altricial birds competing for access to food provided by parents) and lethal competition (e.g. direct violent attack leading to death as in many birds of prey). However, in some species, siblings may cooperate to secure more resources from parents (e.g. coordinated begging; Black-headed gulls *Larus ridibundus*, Mathevon and Charrier 2004; Banded mongooses *Mungo mungo*, Bell 2007), or reduce the costs of competition through negotiation (e.g. Barn owls *Tyto alba*, Chapter 8). Parental allocation of resources to offspring is rarely equal (Mock and Parker 1997). Parental favouritism can occur whenever there are fitness benefits to parents of differential investment among offspring (e.g. differential allocation of resources to male and female offspring, also termed sex allocation; Chapter 10). Parental favouritism can increase competition among siblings for access to parental resources, leading to conflicts of interest between parents and offspring.

18.4.5 Parent–offspring conflict

Conflict between parents and offspring can occur over both the total amount of parental investment supplied (interbrood conflict) or the distribution of parental investment among offspring (intrabrood conflict). Interbrood conflict leads to selection for offspring to extract more resources from parents at the expense of future siblings (Chapter 7). In contrast, intrabrood conflict leads to selection for mechanisms by which offspring bias the distribution of parental resources at the expense of current siblings (Chapters 7 and 8). Optimal strategies can differ at different stages of life for the same individual (i.e. when a parent compared to when an offspring; Chapters 7 and 16). In order to provide empirical evidence for POC the battleground of conflict must be established (i.e. show that optimal levels of parental investment differ between parents and offspring) and any change to the way in which parental investment is allocated should then lead to an increase in fitness in one party at the expense of the other (Chapter 7). For example, experimental manipulation of gene activation in mice shows that paternally expressed alleles lead to larger offspring via increased demand for maternal resources, because paternally derived alleles are evolutionarily not limited by any costs of maternal care (Haig and Westoby 1989; Chapters 7 and 17).

At the heart of parent–offspring relations is the co-evolution of demand for, and supply of, parental resources (Chapters 7 and 16). Parents may not have full control of resource provisioning because offspring have 'private' information about themselves (e.g. their condition or state) that would be beneficial to the parents to know in order to maximize parental fitness (Chapter 7). Offspring can exploit the parents' lack of information by exaggerating their true needs in order to gain more parental investment than would be optimal for parents to supply (Chapter 7). As a result, the amount of parental resources provided to offspring at conflict resolution depends critically on who controls resource allocation—parents or offspring (Royle et al. 2002b). Parental control is more likely in general when the rate of provisioning is genetically or physiologically constrained or when resource availability is highly predictable (Chapter 7). However, in reality control is likely to be on a continuum, ranging from full parental to full offspring control, and changing dynamically during ontogeny and in response to variation in the (social) environment (Royle et al. 2002b). Co-adaptation models explore how genetic variance and covariance of interacting traits expressed in parents and offspring are expected to co-evolve, with correlational selection essential for this to occur (Chapter 16).

18.5 Co-evolution and correlated responses

Correlational selection occurs when selection favours particular combinations of traits. Co-adaptation models show that correlational selection via social epistasis can be strong enough to select for parent–offspring covariance even if loci are unlinked (Chapter 16). In such cases, offspring traits are affected by an interaction between direct effects of genes expressed in offspring and indirect effects of genes expressed in parents, with selection operating on the outcome of the interaction between parents and offspring. However, selection is required to maintain parent–offspring covariance, otherwise recombination can lead it to break down within a few generations. Hence, factors that limit recombination between correlated parent–offspring traits, such as pleiotropy (e.g. the Peg3 gene that affects maternal milk let-down and offspring suckling behaviour in mice; Curley et al. 2004) or linkage disequilibrium (e.g. via assortative mating), should be favoured by selection (Chapter 16). Empirical evidence for parent–offspring co-adaptation comes mainly from cross-fostering or line-crossing studies, which test whether the genetic origins of parents and offspring, and the social (family) environment in which they are expressed, are matched or mis-matched (Chapter 16). Furthermore, there is also evidence from molecular studies showing an increasing number of genes that influence both maternal care and offspring behaviour in laboratory rodents (Chapter 17). These empirical studies show considerable variation in the sign of the genetic correlation between parental and offspring behaviours, including, in some cases, no correlation between the two despite heritable variation in both parental and offspring behaviours (Chapters 7 and 16). This pattern most likely reflects the dynamic nature of parent–offspring co-adaptation and differences between systems in the extent that parents versus offspring control the interaction (Chapters 7 and 16).

Most co-adaptation models do not allow parent and offspring response rules to evolve (Chapter 16), and therefore provide only limited insights into the stability of parent–offspring interactions. Understanding the dynamics and stability of co-adaptation requires the use of a behavioural reaction norm (BRN) approach, which focuses on the co-evolution of behavioural response rules (Smiseth et al. 2008). Recent models suggest that selection should act on BRNs to stabilize parent–offspring interactions in such a way that parent–offspring co-evolution is associated with variance in behaviours over time, not just mean levels of behaviour (Dobler and Kölliker 2009). These forms of co-evolutionary dynamics are complex because parents exert selection on offspring phenotypes, and, if offspring influence parental care, offspring also exert selection on parental phenotypes (Chapter 15). Further development of indirect genetic effect models is necessary to clarify exactly how selection acts on parents and offspring to determine co-evolution in families (Chapter 15).

Models of co-adaptation illustrate how parental care evolves as a consequence of the outcome of family interactions and the heritable components of variation underlying these relationships (Chapters 15 and 16). Such co-evolutionary effects are not confined to parent–offspring interactions. Despite widespread recognition that behaviour during mating strongly influences patterns of parental care (Chapters 6 and 11), empirical support for predicted relationships between mating behaviour and patterns of parental investment remains equivocal (Alonzo 2010). For example, males are predicted to decrease their parental effort in response to a decrease in paternity if there are opportunities for higher paternity in the future, and females are expected to choose males based on male traits that indicate his genetic quality (Chapters 6 and 11). Neither of these predictions have widespread empirical support (Alonzo 2010; Chapter 11). This may be because research has focused on pairs of traits and ignored social dynamics and co-evolutionary feedbacks among interacting individuals (Kölliker et al. 2005; Alonzo 2010). Traits that protect paternity in particular might be expected to co-evolve with care (e.g. repeated mating rate or nuptial gifts; Chapter 6). Such co-evolution can be examined by applying artificial selection to the traits of interest. For example, selecting for rate of mating is expected to lead to changes in the expression of social traits in males in the context of parental care, which then feeds back to change the expression of behaviour of females in the context of mating, and so on. This social feedback loop is expected to have consequences for the relationship between parents and offspring, with changes in parental care behaviours expected to lead to changes in offspring behaviour, which modify parental behaviour, and so on, forming a feedback loop of co-evolution across traits expressed in different functional contexts (Alonzo 2010).

Feedback loops are likely to be important in the co-evolution of brood parasites and their hosts. Parental care generates a social environment that is highly favourable to the growth, development, and survival of offspring, thereby generating a niche that social parasites can exploit. Because parasitic young are unrelated to their foster siblings, and have no shared interests in the fitness of their hosts, brood parasitism provides a suitable model to examine the limits of selfishness in parental care (Chapter 13). The dynamics of co-evolution between parasites and hosts depends on the virulence of the parasites, as this determines the social environment that the interactions take place in and, by definition, the costs of parasitism to hosts (Chapter 13), thus leading to negative density-dependent selection on the parasitic strategy (the higher the density, the lower the fitness for the parasite due to reduced availability of hosts). Host defences may be constrained by high costs of defence relative to costs of parasitism (Chapter 13). During the offspring-rearing stage more apparently highly virulent parasitic offspring may actually impose lower costs to the host's residual reproductive success than apparently more benign parasites. For example, common cuckoos use vocal and visual signals to procure food from host parents, but the relatively small gape area they present compared to a whole brood of host nestlings limits the rate at which they are fed by their foster parents (Chapter 13). Less obviously virulent parasites, such as brown-headed cowbirds, that are raised alongside host chicks and that use mimicry to procure food, may be more costly to the host parents' future success because they use host chicks to secure more parental investment (Chapter 13). Consequently the relationship between virulence and the costs of parasitism may be complex.

Why do host parents have such difficulty in detecting parasitic offspring? One reason may be that the system is still dynamically co-evolving, and there is an evolutionary lag in the host's defence. Alternatively, the parasite and host may be in evolutionary equilibrium, but the costs of defence may be greater than the costs of accepting brood parasites (Chapter 13). As host susceptibility tends to be the rule, and there are few examples of counter defences by host parents at the chick-rearing stage of development (Chapter 13), it is likely that the latter is true in most systems. Such co-evolutionary relationships are likely to be common features not just of host–parasite dynamics, but also of the network of relationships and interactions among family members (Box 18.1).

Box 18.1 Social networks in the context of parental care

How individuals interact with one another can have important consequences for the expression and evolution of phenotypic traits (e.g. Chapters 7, 8, 9, 15, and 16). The outcomes of such behavioural interactions can affect patterns of organization and structure at a variety of different levels, including colony-level effects (Linksvayer et al. 2009), population-level effects (Plaistow and Benton 2009), and even ecosystem-level effects (Gribben et al. 2009). The genetics and evolutionary dynamics of social effects depend critically on the structure of social interactions (Wolf and Moore 2010; Chapters 15 and 16). However little is known empirically about the structure of social interactions; the processes involved are a 'black box'. The emerging field of social networks analysis provides a potential tool to unlock this box.

The social environment an animal encounters typically consists of a complex network of non-random and highly variable social interactions (Croft et al. 2008; Fig. 18.2). Studies of social behaviour that incorporate a network approach shift the focus from variation in behaviour among individuals per se to how interactions among individuals shapes variation in behaviours and phenotypes (Fewell 2003). This is more realistic as it views the behaviour of individuals as being both the cause and the effect of their social environment (Wolf and Moore 2010; Chapter 15).

A social networks approach has been instrumental in advancing our understanding of numerous areas of behavioural and evolutionary ecology including the evolution of animal societies, cooperation, and the transmission of disease (reviewed in Sih et al. 2009), but has rarely been applied to the study of parental care. In fact there are very few studies that have even examined the relationship between measures of fitness and social network parameters. Two studies have quantified how social position *within* a network is related to the fitness prospects of interacting individuals in the context of mate choice/sexual selection (McDonald 2007; Oh and Badyaev 2010), but we are not aware of any research on the relationship between fitness outcomes and the structure of the network of interactions themselves in any social context. This seems surprising because this information is central to understanding the evolutionary and ecological significance of social interaction networks (Fewell 2003; Sih et al. 2009).

Box 18.1 (Continued)

Understanding the resolution of conflict during interactions among family members in species with parental care is one area that would particularly benefit from integrating information on fitness and network structure. Communication among individuals in animal families involves a network of (often simultaneous) interactions, among siblings, between parents and offspring, and between male and female parents (Horn and Leonard 2005). The resolution of conflicts of interest over the provision of parental investment in families (Chapters 7, 8, and 9) implicitly involves a network of multiple interactions, not just multiple dyadic interactions among individual members (Parker et al. 2002), but is not usually modelled as a network (Godfray and Johnstone 2000; Parker et al. 2002; Kölliker 2005). The reason that family interactions tend to be modelled as multiple dyadic relationships is not only due to mathematical tractability, but also a lack of empirical data. For example, altricial bird nestlings often interact with each other and feeding parents through begging competitions and by jostling for positions close to the feeding parent (Horn and Leonard 2005; Kölliker et al. 1998; Hinde et al. 2010). These begging displays and the dynamics of their movements in relation to other family members within the nest therefore shape brood social structure (Fig. 18.3), but empirical studies do not explicitly account for this (Horn and Leonard 2005).

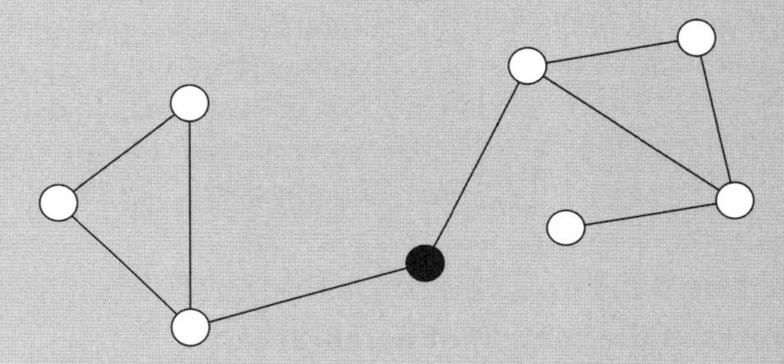

Figure 18.2 A toy social network showing interactions (lines—or 'edges') among individuals (circles—or 'nodes'). One individual (the black node) provides a key link in the network, between the two subgroups, even though it is not directly connected to all individuals in the network. The importance of this link would not be obvious using traditional, but widely-used, techniques that only consider pair-wise (dyadic) interactions between individuals. Such links in the network are likely to be particularly important for the transfer of information and pathogens.

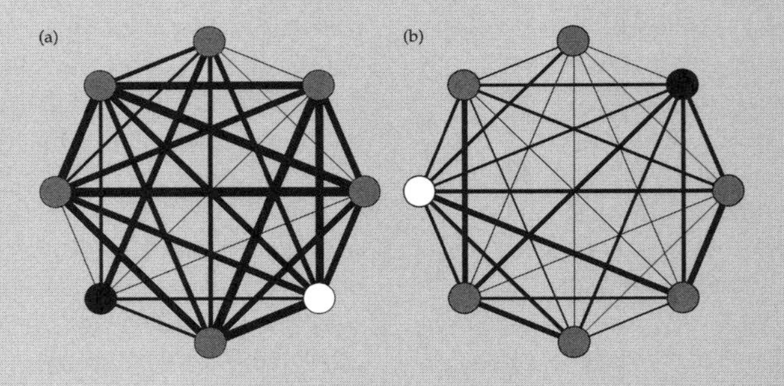

Figure 18.3 Representative networks of begging great tit nestlings associating with one another during feeding by parents: (a) Social network for a strongly-connected brood of eight nestlings, and (b) Social network for a weakly-connected brood of eight nestlings. Experimentally food-deprived nestlings are shown in white, experimentally satiated nestlings in black, and un-manipulated nestlings in grey. The thickness of the lines (edges) indicates the strength of the connection between individuals (nodes). Associations among nestlings are non-random and connectedness within broods is related to hunger (food-deprived nestlings have a greater strength of associations than satiated nestlings) in both networks, even though the networks differ in overall connectedness.

There are several other features of social interactions within families of species that are reared in nursery environments that make them ideal study systems to quantify between-group consequences of variation in social network (family) structure on fitness. In contrast to many social networks analyses, where it can be difficult to assign individuals to groups, there are no 'gambit of the group' issues (Croft et al. 2008). This is because all offspring within each nursery can be clearly and unambiguously assigned to a given group, and there is unlikely to be movement of offspring between nurseries, so each brood represents a discrete network of individuals (or subnetwork if parents interact with other parents during, for example, foraging for food). In addition, because each individual within a brood is likely to interact with multiple individuals in the network on multiple occasions and group membership is clearly defined, it is possible to make use of weighted (as opposed to binary) network metrics. Consequently the quality of information available in constructing the social networks is high, and allows measures of the strength of behavioural interactions to be incorporated (Whitehead 2008). Finally, the use of interacting family members makes it possible to quantify the fitness consequences of variation in social interactions among individuals at levels of selection higher than the individual through the use of replicated networks, which is an important step on the path to establishing how the structure of social interactions co-evolves with traits involved in parental care (e.g. Fehl et al. 2011).

18.6 Mechanisms and constraints in parental care

The proximate mechanisms underlying variation in the costs and benefits of care are important determinants of (co-)evolutionary trajectories that can be broadly classified as being genetic or physiological in origin. Whilst these often provide constraints on evolution the mechanisms themselves can also evolve in response to selection to shape phenotypes.

18.6.1 Physiological mechanisms

Cost–benefit approaches to the study of parental care, where resources are often considered to be limited (i.e. only available to spend once) and the common currency is usually assumed to be energy or time, have been central to the development of theory and underlie most empirical studies (Chapter 3). However, until relatively recently there has been little work on the physiological mechanisms that govern these trade-offs. One of the main problems with the traditional approach is that parents and offspring may not both be drawing from a common pool of resources (known as the 'common currency problem'; Houston and McNamara 1999). The resources required by offspring may differ from those required for somatic maintenance by parents (e.g. parent Palestine sunbirds *Nectarinia osea* feed on flower nectar, but their offspring are fed arthropods; Markman et al. 2002). Although this problem can be addressed in models to some extent by introducing state variables that characterize current physiological state (e.g. hunger, condition etc; Chapter 3), it illustrates the potential issues involved in making simplifying assumptions about resource allocation decisions. Similarly, resource acquisition may involve intrinsic trade-offs between specific components of diet that would not be accounted for if the currency of acquisition was assumed to be just energy (Chapter 3). For example, different ratios of protein to carbohydrates in the diet are optimal for increasing egg production compared to increasing lifespan in *Drosophila* (Lee et al. 2008). Oxidative stress is a potential mechanism underlying these costs that has recently been suggested to be a key proximate constraint on the evolution of life-histories, including parental care (e.g. Metcalfe and Alonso-Alvarez 2010).

Oxidative stress occurs when there is an imbalance between the production of reactive oxygen species, which are byproducts of normal metabolic processes, and the availability of antioxidant defences in favour of the former (Metcalfe and Alonso-Alvarez 2010). There is substantial evidence for oxidative stress as a cost of parental care from a growing number of experimental studies, mostly on birds (Chapter 3). However, there are relatively few studies that demonstrate long-term effects of oxidative stress on correlates of fitness (e.g. Alonso-Alvarez et al. 2006), and most of these are observational rather than experimental studies (Chapter 3). An increase in parental effort can also lead to suppression of the immune system, although this effect is often an indirect consequence of other physiological costs such oxidative stress, depletion of carotenoid reserves, or glucocorticoid stress response (Chapter 3). Hormones in particular provide important mechanistic links between resource availability and parental care (Chapter 3). For example, low levels of glucocorticoid are necessary for normal metabolism, but glucocorticoid secretion is triggered when lipid reserves are exhausted, thereby stimulating glucogenesis and protein breakdown to increase energy production and inducing the cessation of reproduction (i.e. clutch or brood desertion; Wingfield and Sapolsky 2003). The glucocorticoid threshold leading to offspring desertion appears to be modified according to the trade-off between current and future reproduction, at least in birds, and a recent comparative analysis shows that species with current broods of higher reproductive value had weaker acute stress responses and were less likely to desert when exposed to stressors (Bokony et al. 2009).

18.6.2 Genetic mechanisms

Studies of the molecular genetic mechanisms underlying parental care are currently limited to a small number of model systems with well-characterized genomes, such as honeybees and rodents (Chapter 17). In these species, the use of transgenic animals or techniques to alter patterns of gene expression have helped identify candidate genes that are important in parental care. For example, in one study on mice, the Sry gene was removed from the Y chromosome (the testes determining factor) and reinserted on an autosome so that gonadal sex determination was independent of the sex chromosome combination in the transgenic animals (Gatewood et al. 2006). Males with an XX complement of chromosomes were less likely to kill young than XY males, whereas XY females were less attentive parents than XX females, indicating that genes on the Y chromosome are associated with inhibition of parental care (Chapter 17). Studies on sequence variation at particular loci and their relation with parental care often find that, compared to environmental influences, the per-locus phenotypic effects are relatively small (explaining a few per cent of variance in parental behaviour; Chapter 17), but that multiple genes influence parental behaviour. This suggests that parental behaviour conforms well to the assumptions of quantitative genetics (i.e. that traits are affected by many genes, each with small effects on phenotype; Chapters 14 and 16).

Much of the recent excitement in the field of the molecular genetics of parental care involves epigenetics. Epigenetics includes post-translational changes to histone proteins, around which the DNA in the cell nucleus is wrapped, and the DNA itself, for example by methylation (i.e. the addition of methyl groups; Chapter 17). Methylation affects the way in which histones and DNA interact or the extent to which DNA is transcribed, both of which are central to cellular differentiation. Such epigenetic modifications are important because these changes are often stable, and can be mitotically heritable or inherited across generations (Chapter 17). Environmentally induced epigenetic changes to DNA can therefore affect the co-evolutionary dynamics of parents and offspring across generations (Chapter 14). For example, postnatal maternal grooming in rats induces long-term changes in neuro-endocrine function and behaviour of offspring, with the variation in the expression of the genes involved maintained through alterations in the pattern of DNA methylation (Weaver et al. 2004). Thus, epigenetic variation in parent–offspring interactions can be modified and determined by environmental input such as the level of parental care received (Chapter 17). There is also some evidence for transgenerational effects on offspring, as rats that experienced relatively little maternal grooming during development had reduced expression of estrogen receptors, making these females less estrogen sensitive as adults (Champagne et al. 2003). Such epigenetic effects on parent–offspring interactions that are maintained across generations are intriguing because observed heritable variation in parent and offspring strategies are potentially both a cause and an effect of variable outcomes of parent–offspring interactions. Epigenetic signatures may therefore ultimately determine how strategies are passed on from one generation to the next and how heritable variation is maintained. At present the evidence in support of such a conclusion is based on only a handful of model systems and the wider prevalence of epigenetic inheritance remains to be further explored. The rapidly reducing costs of molecular genetics tools means that there is sure to be an explosion of work in this field over the next 5–10 years.

18.7 Evolutionary implications of parental care

Co-evolution and correlated responses mean that selection on parental care traits has important implications for the evolution of other traits and processes. Here we briefly outline some of the consequences of this co-evolution.

18.7.1 Life-history evolution

There is considerable variation in life-histories among species. Some organisms develop very quickly, whilst others are slow developers. Some reproduce just once in a lifetime, producing a large number of offspring in a 'big bang' (semelparity), whilst others produce offspring singly or in batches over multiple reproductive events (iteroparity). Such life-history variation affects the probability of the evolution of parental care (Chapter 2), but co-evolution means that there is also considerable scope for parental care, and associated conflicts of interest, to affect the evolution of life-history traits. For example, comparative analyses indicate that competitive interactions among family members are associated with parentage and the evolution of parent–offspring communication (Briskie et al. 1994), and with variation in pre- and post-natal growth rates in birds (Royle et al. 1999; Lloyd and Martin 2003). In addition, models show that sexual conflict is expected to lead to the co-evolution of biparental care and brood size. This is because the larger the brood size the more valuable the male's care is to the female (Smith and Härdling 2000). Biparental care can therefore lead to an increase in the number, but not necessarily the quality, of offspring raised (Royle et al. 2006).

18.7.2 Social evolution

Parental care is also an important evolutionary transition in the formation of complex social societies. In cooperatively breeding species, care of offspring is provided by other members of the group called 'helpers' in addition to the parents (see Section 18.4.2). The helpers retain the ability to reproduce, either currently or in the future, but are suppressed from breeding by the dominant individuals in the group (i.e. the breeders; Chapter 12). Typically, such groups form when mature offspring delay dispersal and remain on their natal territories to help rear subsequent broods (Chapter 12). This delayed dispersal is associated with intense competition for suitable breeding habitat, which means that cooperation evolves via the effect of ecological constraints on kin structure (Chapter 12). For helpers, staying on the natal territory to assist in rearing their siblings may therefore represent the 'best of a bad job', and there is still considerable scope for conflict among family members over the provisioning of care (Chapter 12).

The most derived form of sociality is eusociality, where suppression of reproduction has evolved from reproductive skew to such an extent that helpers have lost the ability to mate and reproduce and have become specialized care-givers ('workers'). Eusociality is found in some insects, including some species of bees (e.g. the honeybee *Apis mellifera*), wasps, ants, termites, and also—the only known vertebrate systems—the naked mole rat *Heterocephalus glaber* and the Damaraland mole rat *Fukomys damarensis*. Eusociality can evolve from species with parental care via cooperative breeding. Such an evolutionary transition can only have evolved from a form of care that can be provided by non-breeding adults (e.g. progressive provisioning). In addition, care-giving must be able to be uncoupled from mating and reproduction. Despite this uncoupling of mating and care-giving behaviours, relatedness and patterns of parentage still play a key role in explaining the evolution of social complexity (Chapter 11). For example, a comparative analysis by Hughes et al. (2008) shows that low levels of promiscuity predict the origin of eusociality in insects. Similarly, a recent meta-analysis of cooperative breeding in birds indicates that it is associated with low levels of extra-group paternity (Cornwallis et al. 2010). Consequently, relatedness asymmetries and genetic conflicts shape the evolution of complex social systems much as they shape the evolution of parental care.

18.7.3 Sexual selection

Mate choice may be an important driver of the evolution of parental care that has implications for the evolution of signals of parenting ability in particular (Kvarnemo 2010). If offspring fitness is dependent upon male parental investment, for example, and there is variation among males in the amount of care that they provide, then selection is expected to favour female choice for male sexual ornaments that indicate male parental skills (Hoelzer 1989). Males cannot actively demonstrate parenting skills at the time of mating, so there should be sexual selection for indirect, but reliable, indicators of parental ability (Candolin 2000). This may be a visual trait, such as the nuptial colouration of male sticklebacks *Gasteroseus aculeatus*, or a behavioural trait or suite of traits, such as personality.

18.7.4 Evolution of personality

Consistent individual differences in behaviour across contexts (also known as 'personality') are common in non-human animals (Dall et al. 2004). Such consistency may be advantageous if it contributes to individual variation in productivity (Biro and Stamps 2008). However, there is considerable variation in the consistency of behaviour of parents when providing care (Royle et al. 2010). The key to understanding this variation in behaviour, which is an important determinant of how conflicts are resolved (Chapters 7–9), centres on reliable signals of parental ability and the co-evolution of parent–offspring interactions with behaviours expressed in other functional contexts. Personality may provide a means to reliably indicate parental ability during mate choice, and reduce sexual conflict over care, with variation in personalities maintained by frequency-dependent selection on 'trustworthiness' (Royle et al. 2010). Furthermore, because family conflicts impose selection on parental and offspring behaviours via co-adaptation, personality is expected to co-evolve with patterns of communication and conflict resolution within families (Roulin et al. 2010). Variation in personality could therefore emerge and be maintained via a combination of sexual selection and co-adaptation between behavioural development and parental care in families (Roulin et al. 2010).

18.8 Future directions

Hamilton (1964) emphasized that it is association rather than relatedness *per se* that is critical to the evolution of altruistic behaviour. Parental care is an altruistic behaviour as there is a cost to parents of providing care but benefits to offspring of receiving care (Chapter 1). The importance of association for the evolution of parental care can be seen by the fact that most species of animal show no parent-offspring association despite having similarly high relatedness between parents and offspring as species with extensive parent-offspring association. These associations are important, not only in explaining the evolutionary origins of care (Chapters 1 and 2), but also in explaining the maintenance and subsequent evolution of parental care and complex societies. Parent–offspring associations lead to interactions among individuals, which in turn are central to understanding the evolution of parental care. There is increasing realization that the interactions themselves may be as important as the individuals involved in the interactions, but they are closely intertwined. The phenotype (and genotype) of individuals affects how and whom they interact with, and is an important determinant of the outcome of conflicts over care. The outcome of these interactions among individuals in turn may affect the inheritance of phenotypes across generations through epigenetic effects. Parental traits coevolve with traits in offspring (e.g. provisioning and begging; Chapters 7, 15, and 16), with other parental traits (e.g. paternity and parental care; Chapters 6 and 11) and traits expressed in other species (e.g. social parasites; Chapter 13). The recognition of the importance of the social interactions themselves to the evolution of parental traits is possibly the biggest conceptual advance in studies of parental care since the publication of Clutton-Brock's (1991) book, and the recent availability of molecular tools that are opening up new avenues of research into the genetic and physiological mechanisms underlying parental care behaviours provide the likely biggest practical advance (Chapters 3 and 17, Section 18.6.2). We outline some of the key areas for future research below.

The balance between cooperation and conflict: Evolutionary conflicts underlie most, if not all, parental care traits, limiting the scope for cooperation. However, parental care nevertheless involves cooperation within families (e.g. between male and female parents in biparental care or among sibling 'helpers' in cooperatively breeding vertebrates), so cooperative outcomes are clearly possible. Mechanisms facilitating cooperation include behavioural coordination as a form of reciprocity (Johnstone et al. submitted; Chapter 9), negotiation (e.g. sibling negotiation when resources provided by parents are indivisible; Chapter 8) and intergroup conflict (facilitating intragroup cooperation; e.g. Bowles 2008). These mechanisms work by reducing the differential between the fitness optima of family members and most research has focused on the study of these mechanisms. However, we still have a very limited understanding of the conflict battleground and it is also possible that individual fitness optima remain divergent but conflict is resolved at little or no cost (Cant 2012; Chapter 12). Unlike most models of evolutionary conflict, which assume that an individual's success depends on how much it invests into competition, this model allows individuals to either invest in competition or not (Cant 2012). This mirrors the situation in many cooperative societies, where one class of individuals (e.g. dominants) engage in competitive behaviour whilst others do not (i.e. subordinates; Chapter 12). However, mechanisms for reducing the costs of conflict whereby some classes of individuals choose not to compete are not likely to be restricted to cooperatively breeding species. More empirical work examining cooperative outcomes in the context of high potential for conflict (e.g. coordination of begging behaviour towards parents by siblings) and conflict in the context of cooperative breeding (e.g. suppression of reproduction by dominants to subordinates) would be insightful in this regard.

Genetics underlying parental care: More widespread availability of sequenced genomes or transcriptomes across species would facilitate comparative studies examining the dynamics of evolutionary conflicts over time (Chapter 7), and thereby significantly increase our understanding of the evolutionary origins of parental care traits (Chapter 4). For example, a recent study proposes that transposons ('jumping genes'; Lynch et al. 2011) were important in the evolutionary origin of viviparity and placentation, which raises intriguing new questions about the evolutionary origin of this form of parental care because transposons show very different evolutionary dynamics to regular loci. Furthermore, quantitative genetic approaches continue to be important in order to understand how genetic covariance can change evolutionary trajectories and zones of conflict (Chapter 16) or to test hypotheses about patterns of inheritance of parental care and offspring development and behaviour across generations. Finally, compared to other evolutionary approaches, selection lines have been little used in the study of the evolution of parental care, but in order to examine the co-evolution of traits such as paternity and parental care (Chapters 6, 9, and 11) they are likely to prove invaluable (Alonzo 2010). In addition we currently know very little about the quantitative genetics of physiological mechanisms associated with parental care (Chapter 3).

Proximate mechanisms underlying parental care: In species with parental care a full understanding of the evolutionary development of an organism requires knowledge of how parental care interacts with offspring development. Technological advances in molecular genetics are likely to contribute particularly to this understanding, especially with regard to work on the control of mechanisms (Chapter 3), but there are numerous other unresolved issues. These include the role of hormesis (or tipping points) in the regulation of parental care behaviours (Chapters 3 and 6), the physiological factors underlying sib–sib interactions (Chapter 8), the hormonal mechanisms regulating behaviour in cooperatively breeding species (Chapter 12), and the mechanisms involved in adjustments of sex ratios (Chapter 10). In addition, although models make explicit assumptions about the shape of fitness functions there is still very little empirical work to support these assumptions, despite the shape of fitness functions having profound effects on predicted levels of conflict (Mock and Parker 1997;

Chapter 9) and resource allocation decisions such as sex ratio adjustment (Chapter 10). The costs and benefits of parental care are often difficult to establish due to the common currency problem (Chapter 3), but taking a more mechanistic approach may help illuminate this issue. Understanding the proximate basis of costs and benefits of care is also likely to be important in understanding downstream effects.

Downstream effects: Parental care can be considered an early life-history effect in that the (social) environmental conditions experienced during growth and development when cared for by parents can have substantial, long-term, effects on the phenotype of offspring when adults (Chapters 3, 14–16). Despite an upsurge of recent work on parental effects (Chapter 14), including epigenetics (Chapters 14 and 17), there are still key areas we know very little about. These include downstream effects of competition early in life (Chapter 8), of receiving care predominantly from helpers or workers rather than parents (Chapter 12) and the general importance of indirect genetic effects (Chapter 15), and social epistasis (Chapter 16) early in life on the expression of traits in adulthood. Understanding variation in downstream effects involves an understanding of early life conditions, and, in particular, the plasticity of behaviour expressed by both parents and offspring. Thus, developmental pathways in species with parental care cannot be fully understood outside the context of parental care, where phenotypic and/or genotypic accommodation of parental care and offspring development and their co-adaptation are expected to shape parental care trait expression and co-evolution.

Plasticity of behaviour: There is considerable variation among individuals in plasticity of behaviour in the context of parental care. The behavioural stability of parent–offspring interactions may depend on the responsiveness of each of the parties to each other (Chapters 7 and 16), which can be a source of phenotypic novelties upon which selection can act (Chapter 14). However, at the other end of the spectrum, behavioural consistency may be important in the context of sexual selection and the resolution of sexual conflicts (Royle et al. 2010). More studies on the plasticity of parental strategies that account for individual variation in behaviour are now necessary to bridge the gap between quantitative genetic and behavioural ecology approaches to parental care (e.g. BRNs; Chapters 7 and 16). Specifically more studies are required that assess the nature of behavioural mechanisms involved in the negotiation of resources among family members, and the factors that select for task specialization and its plasticity (Chapter 9). This will help increase our understanding of the behavioural mechanisms that stabilize biparental care and cooperative breeding systems, and more generally, the family as a multidimensional social network (Box 18.1). Behavioural plasticity may also be an important determinant of success in brood parasites relating to their virulence—an area of research that is currently under-explored (Chapter 13).

Theory: Closer connections between theory and empirical work are undoubtedly needed to develop models that predict the complex inter-relations among interacting individuals and traits during the provision of parental care (Chapter 11). In addition, it is important for models of the evolution of sex roles and parental care to explicitly incorporate feedback between ecology and the evolution of patterns of parental care (Chapters 2, 6, 9, and 11), and for models to clearly distinguish between origins and maintenance of parental care (Chapter 2). But perhaps the greatest insights will come from the incorporation of parent–offspring trait covariation to models of conflict resolution and more explicit consideration of how family members interact with one another during the provision of parental care. Conflict resolution models address evolutionarily stable outcomes for parent–offspring interactions, but are currently limited in two different ways. They assume that there is no covariation between parent and offspring traits and focus on the stable co-evolutionary endpoint (Chapter 16). We therefore have poor understanding of the evolutionary origin of offspring influences on parental care and the co-evolutionary dynamics of parent–offspring interactions (Chapters 15 and 16). Furthermore, evolutionary models of parent–offspring interactions in general assume that interactions between individuals involve multiple dyadic encounters rather than considering all interactions among individuals simultaneously (i.e. dyads, not networks, of interactions; Box 18.1). A complete theory of family interactions and the evolution of parental care needs to take account of this complexity as it will almost certainly have important implications for how conflicts are resolved and the evolutionary trajectories of parental care traits.

18.9 Conclusions

The dynamic nature of family interactions and the potential for co-evolution among family members has contributed to the huge diversity of parental care behaviours expressed both across and within taxa. The life-history and ecology of a species are also important determinants of the evolution of parental care, which feeds back to alter life-history and the environments species inhabit, and has evolutionary consequences for the evolution of other traits, such as sociality. Co-evolutionary feedback between life-history and parental care traits, mediated by genetic conflicts and (social) environmental variation, generates and maintains diversity in these traits, making parental care a key engine of biodiversity. Its importance in affecting the reproductive output of species means that, although rarely acknowledged as such, understanding parental care may also be central to the conservation of species and evolutionary processes (e.g. Ferrière et al. 2004). Future work on the evolution of parental care is likely to be more holistic and integrative in approach, using a suite of techniques, including comparative analyses, phenotypic and molecular approaches, and theory in addition to experiments, and considering multiple components simultaneously, in order to understand how parental care traits and processes evolve. At present the study of parental care stands on the cusp of an exciting new era of integrated 'social systems biology' that is likely to advance our understanding of the field even more rapidly over the next 20 years than it has advanced in the last two decades.

Acknowledgements

Thanks to Megan Head and Tobias Uller for constructive comments on the chapter and to Tom Pike for drawing Figs 18.2 and 18.3 in addition to providing constructive comments.

References

Alonso-Alvarez, C., Bertrand, S., Devevey, G., et al. (2006). An experimental manipulation of life-history trajectories and resistance to oxidative stress. *Evolution* 60, 1913–24.

Alonzo, S. H. (2010). Social and coevolutionary feedbacks between mating and parental investment. *Trends in Ecology and Evolution* 25, 99–108.

Arnold, K. E., Ramsay, S L., Donaldson, C., and Adam, A. (2007). Parental prey selection affects risk-taking behaviour and spatial learning in avian offspring. *Proceedings of the Royal Society of London, Series B* 274, 2563–9.

Bell, M. B. V. (2007). Cooperative begging in banded mongoose pups. *Current Biology* 17, 717–21.

Biro, P. A. and Stamps, J. A. (2008). Are animal personality traits linked to life-history productivity? *Trends in Ecology and Evolution* 23, 361–8.

Bókony, V., Lendvai, A. Z., Liker, A., Angelier, F., Wingfield, J. C., and Chastel, O. (2009). Stress response and the value of reproduction: are birds prudent parents? *American Naturalist* 173, 589–98.

Bowles, S. (2008). Conflict: Altruism's midwife. *Nature* 456, 326–7.

Briskie, J. V., Naugler, C. T., and Leech, S. M. (1994). Begging intensity of nestling birds varies with sibling relatedness. *Proceedings of the Royal Society of London, Series B* 258, 73–8.

Brown, J. L, Morales V., and Summers, K. (2010). A key ecological trait drove the evolution of biparental care and monogamy in an amphibian. *The American Naturalist* 175, 436–46.

Candolin, U. (2000). Changes in expression and honesty of sexual signalling over the reproductive lifetime of sticklebacks. *Proceedings of the Royal Society of London, Series B* 267, 2425–30.

Cant, M. A. (2012). Suppression of social conflict and evolutionary transitions to cooperation. *American Naturalist* 179, 293–301.

Champagne, F. A., Weaver, I. C., Diorio, J., Sharma, S., and Meaney, M. J. (2003). Natural variations in maternal care are associated with estrogen receptor alpha expression and estrogen sensitivity in the medial preoptic area. *Endocrinology* 144, 4720–40.

Clutton-Brock, T. H. (1991). *The Evolution of Parental Care*. Princeton University Press, Princeton, New Jersey.

Cornwallis, C. K., West, S. A., Davis, K. E., and Griffin, A. S. (2010). Promiscuity and the evolutionary transition to complex societies. *Nature* 466, 969–74.

Croft, D. P., James, R., and Krause, J. (2008). *Exploring Social Networks*. Princeton University Press, Princeton, New Jersey.

Curley, J. P., Barton, S., Surani, A., and Keverne, E. B. (2004). Coadaptation in mother and infant regulated by a paternally expressed imprinted gene. *Proceedings of the Royal Society of London, Series B* 271, 1303–9.

Dall, S. R. X., Houston, A. I., and McNamara, J. M. (2004). The behavioural ecology of personality: consistent individual differences from an adaptive perspective. *Ecology Letters* 7, 734–9.

Dobler, R. and Kölliker, M. (2009). Behavioural attainability of evolutionarily stable strategies in repeated interactions. *Animal Behaviour* 77, 1427–34.

Fehl, K., van der Post, D. J., and Semman, D. (2011). Co-evolution of behaviour and social network structure promotes human cooperation. *Ecology Letters* 14, 546–51.

Ferrierè, R., Dieckmann, U., and Couvet, D. (2004). *Evolutionary Conservation Biology*. Cambridge University Press, Cambridge.

Fewell, J. H. (2003). Social insect networks. *Science* 301, 1867–70.

Forbes, S. (2009). Portfolio theory and how parents manage investment risk. *Oikos* 118, 1561–9.

Gardner, A. and Smiseth, P. T. (2011). Evolution of parental care driven by mutual reinforcement of parental food provisioning and sibling competition. *Proceedings of the Royal Society of London, Series B* 278, 196–203.

Gatewood, J. D., Wills, A., Shetty, S., Xu, J., Arnold, A. P., Burgoyne, P. S., and Rissman, E. F. (2006). Sex chromosome complement and gonadal sex influence aggressive and parental behaviors in mice. *Journal of Neuroscience* 26, 2335–42.

Godfray, H. C. J. and Johnstone, R. A. (2000). Begging and bleating: The evolution of parent-offspring signalling. *Philosophical Transactions of the Royal Society of London, Series B* 355, 1581–91.

Gribben, P. E., Byers, J. E., Clements, M. et al. (2009). Behavioural interactions between ecosystem engineers control community species richness. *Ecology Letters* 12, 1127–36.

Gross, M. R. (2005). The evolution of parental care. *Quarterly Review of Biology* 80, 37–45.

Haig, D. and Westoby, M. (1989). Parent-specific gene-expression and the triploid endosperm. *American Naturalist* 134, 147–55.

Hamilton, W. D. (1964). The genetical evolution of social behaviour. *Journal of Theoretical Biology* 7, 1–52.

Harrison, F., Barta, Z., Cuthill, I., and Székely T. (2009). How is sexual conflict over parental care resolved? A meta-analysis. *Journal of Evolutionary Biology* 22, 1800–12.

Hinde, C. A., Johnstone, R. A., and Kilner, R. M. (2010). Parent-offspring conflict and coadaptation. *Science* 327, 1373–6.

Hoelzer, G. A. (1989). The good parent process of sexual selection. *Animal Behaviour* 38, 1067–78.

Horn, A. G. and Leonard, M. L. (2005). Nestling begging as a communication network. In P. K. McGregor ed. *Animal Communication Networks*, pp. 170–90. Cambridge University Press, Cambridge.

Houston, A. I. and McNamara, J. M. (1999). *Models of Adaptive Behaviour.* Cambridge University Press, Cambridge.

Hughes, W. O. H., Oldroyd, B. P., Beekman, M., and Ratnieks, F. L. W. (2008). Ancestral monogamy shows kin selection is key to the evolution of eusociality. *Science* 320, 1213–16.

Kölliker, M. (2005). Ontogeny in the family. *Behav. Genet.* 35, 7–18.

Kölliker, M., Richner, H., Werner, I., and Heeb, P. (1998). Begging signals and biparental care: nestling choice between parental feeding locations. *Anim. Behav.* 55, 215–22.

Kölliker, M., Brodie, E. D., and Moore, A. J. (2005). The coadaptation of parental supply and offspring demand. *American Naturalist* 166, 506–16.

Kvarnemo, C. (2010). Parental care. In D. F. Westneat and C. W. Fox, eds. *Evolutionary Behavioural Ecology*, pp. 451–67. Oxford University Press, Oxford.

Lee, K. P., Simpson, S. J., Clissold, F. J. et al. (2008). Lifespan and reproduction in *Drosophila*: new insights from nutritional geometry. *Proceedings of the National Academy of Science USA* 105, 2498–503.

Lehtonen, J. and Kokko, H. (2012). Positive feedback and alternative stable states in inbreeding, cooperation, sex roles and other evolutionary processes. *Philosophical Transactions of the Royal Society of London, Series B* 367, 211–21.

Lessells, C. M. and McNamara, J. M. (2012). Sexual conflict over parental investment in repeated bouts: negotiation reduces overall care. *Proceedings of the Royal Society of London, Series B* 279, 1506–14.

Linksvayer, T. A., Fondrk, M. K., and Page, R. E. (2009). Honeybee social regulatory networks are shaped by colony-level selection. *American Naturalist* 173, E99–107.

Lloyd, J. D. and Martin, T. E. (2003). Sibling competition and the evolution of prenatal development rates. *Proceedings of the Royal Society of London, Series B* 270, 735–40.

Lynch, V. J., Leclerc, R. D., May, G., and Wagner, G. P. (2011). Transposon-mediated rewiring of gene regulatory networks contributed to the evolution of pregnancy in mammals. *Nature Genetics* 43, 1154–61.

Markman, S., Pinshow, B., and Wright, J. (2002). The manipulation of food resources reveals sex-specific trade-offs between parental self-feeding and offspring care. *Proceedings of the Royal Society of London, Series B* 269, 1931–8.

Mathevon, N. and Charrier, I. (2004). Parent-offspring conflict and the coordination of siblings in gulls. *Proceedings of the Royal Society of London, Series B* 271, 145–7.

McDonald, D. B. (2007). Predicting fate from early connectivity in a social network. *Proceedings of the Natural Academy of Science USA* 104, 10910–14.

Metcalfe, N. B. and Alonso-Alvarez, C. (2010). Oxidative stress as a life-history constraint: the role of reactive oxygen species in shaping phenotypes from conception to death. *Functional Ecology* 24, 984–96.

Mock, D. W. and Parker, G. A. (1997). *The Evolution of Sibling Rivalry*. Oxford University Press, Oxford.

Oh, K. P. and Badyaev, A. V. (2010). Structure of social networks in a passerine bird: consequences for sexual selection and the evolution of mating strategies. *American Naturalist* 176, E80–90.

Paczolt, K. A. and Jones, A. G. (2010). Post-copulatory sexual selection and sexual conflict in the evolution of male pregnancy. *Nature* 464, 401–04.

Parker, G. A., Baker, R. R., and Smith, V. G. F. (1972). The origin and evolution of gamete dimorphism and the male-female phenomenon. *Journal of Theoretical Biology* 36, 529–53.

Parker, G. A., Royle, N. J., and Hartley, I. R. (2002). Begging scrambles with unequal chicks: interactions between need and competitive ability. *Ecology Letters* 5, 206–15.

Plaistow, S. J. and Benton, J. G. (2009). The influence of context-dependent maternal effects on population dynamics: and experimental test. *Philosophical Transactions of the Royal Society of London, Series B* 364, 1049–58.

Roulin, A., Dreiss, A. N., and Kölliker, M. (2010). Evolutionary perspective on the interplay between family life, and parent and offspring personality. *Ethology* 116, 787–96.

Royle, N. J., Hartley, I. R., Owens, I. P. F., and Parker, G. A. (1999). Sibling competition and the evolution of growth rates in birds. *Proceedings of the Royal Society of London, Series B* 266, 923–32.

Royle, N. J., Hartley, I. R., and Parker, G. A. (2002a). Sexual conflict reduces offspring fitness in zebra finches. *Nature*, 416, 733–6.

Royle, N. J., Hartley, I. R., and Parker, G. A. (2002b). Begging for control: When are offspring solicitation behaviours honest? *Trends in Ecology and Evolution* 17, 434–40.

Royle, N. J., Hartley, I. R., and Parker, G. A. (2006). Consequences of biparental care for begging and growth in zebra finches *Taeniopygia guttata*. *Animal Behaviour* 72, 123–30.

Royle, N. J., Schuett, W., and Dall, S. R. X. (2010). Behavioral consistency and the resolution of sexual conflict over parental investment. *Behavioral Ecology* 21, 1125–30.

Sih, A., Hanser, S. F., and McHugh, K. A. (2009). Social network theory: new insights and issues for behavioural ecologists. *Behavioural Ecology and Sociobiology* 63, 975–88.

Smiseth, P. T., Wright, J., and Kölliker, M. (2008). Parent-offspring conflict and co-adaptation: behavioural ecology meets quantitative genetics. *Proceedings of the Royal Society of London, Series B* 275, 1823–30.

Smith, H. G. and Härdling, R. (2000). Clutch size evolution under sexual conflict enhances the stability of mating systems. *Proceedings of the Royal Society of London, Series B* 267, 2163–70.

Weaver, I. C., Cervoni, N., Champagne, F. A., D'Alessio, A. C., Sharma, S., Seckl, J. R., Dymov, S., Szyf, M., and Meaney, M. J. (2004). Epigenetic programming by maternal behavior. *Nature Neuroscience* 7, 847–54.

Whitehead, H. (2008). *Analyzing Animal Societies*. Chicago University Press, Chicago.

Wingfield, J. C. and Sapolsky, R. M. (2003). Reproduction and resistance to stress: when and how. *Journal of Neuroendocrinology* 15, 711–24.

Wolf, J. B. and Moore, A. J. (2010). Interacting phenotypes and indirect genetic effects. In D. F. Westneat and C. W. Fox, eds. *Evolutionary Behavioural Ecology*, pp. 225–45. Oxford University Press, Oxford.

Index

Page numbers in *italics* refer to figures or tables